Role of IoT in Green Energy Systems

Vasaki Ponnusamy
Universiti Tunku Abdul Rahman, Malaysia

Noor Zaman
Taylor's University, Malaysia

Low Tang Jung
Universiti Teknologi PETRONAS, Malaysia

Anang Hudaya Muhamad Amin
Higher Colleges of Technology, UAE

IGI Global
PUBLISHER of TIMELY KNOWLEDGE

A volume in the Advances in Environmental Engineering and Green Technologies (AEEGT) Book Series

Published in the United States of America by
IGI Global
Engineering Science Reference (an imprint of IGI Global)
701 E. Chocolate Avenue
Hershey PA, USA 17033
Tel: 717-533-8845
Fax: 717-533-8661
E-mail: cust@igi-global.com
Web site: http://www.igi-global.com

Library of Congress Cataloging-in-Publication Data

Names: Ponnusamy, Vasaki, 1974- editor. | Zaman, Noor, 1972- editor. | Low, Tang Jung, 1956- editor. | Muhamad Amin, Anang Hudaya, editor.
Title: Role of IoT in green energy systems / Vasaki Ponnusamy, Noor Zaman, Tang Jung Low and Anang Hudaya Muhamad Amin.
Description: Hershey, PA : Engineering Science Reference, 2021. | Includes bibliographical references and index. | Summary: "This book presents updated research trends in energy harvesting, energy management and green technology as well as highlighting new product and application developments towards a wider acceptance of smart and green energy and what additions of ICT such as WSN and smart applications have contributed toward green technology"-- Provided by publisher.
Identifiers: LCCN 2020026771 (print) | LCCN 2020026772 (ebook) | ISBN 9781799867098 (h/c) | ISBN 9781799867104 (s/c) | ISBN 9781799867111 (eISBN)
Subjects: LCSH: Renewable energy sources--Data processing. | Clean energy--Data processing. | Internet of things.
Classification: LCC TJ808 .R64 2021 (print) | LCC TJ808 (ebook) | DDC 621.0420285/4678--dc23
LC record available at https://lccn.loc.gov/2020026771
LC ebook record available at https://lccn.loc.gov/2020026772

This book is published in the IGI Global book series Advances in Environmental Engineering and Green Technologies (AEEGT) (ISSN: 2326-9162; eISSN: 2326-9170)

British Cataloguing in Publication Data
A Cataloguing in Publication record for this book is available from the British Library.

Advances in Environmental Engineering and Green Technologies (AEEGT) Book Series

Sang-Bing Tsai
Zhongshan Institute, University of Electronic Science and Technology of China, China & Wuyi University, China
Ming-Lang Tseng
Lunghwa University of Science and Technology, Taiwan
Yuchi Wang
University of Electronic Science and Technology of China Zhongshan Institute, China

ISSN:2326-9162
EISSN:2326-9170

Mission

Growing awareness and an increased focus on environmental issues such as climate change, energy use, and loss of non-renewable resources have brought about a greater need for research that provides potential solutions to these problems. Research in environmental science and engineering continues to play a vital role in uncovering new opportunities for a "green" future.

The **Advances in Environmental Engineering and Green Technologies (AEEGT)** book series is a mouthpiece for research in all aspects of environmental science, earth science, and green initiatives. This series supports the ongoing research in this field through publishing books that discuss topics within environmental engineering or that deal with the interdisciplinary field of green technologies.

Coverage

- Biofilters and Biofiltration
- Renewable Energy
- Electric Vehicles
- Waste Management
- Radioactive Waste Treatment
- Sustainable Communities
- Pollution Management
- Green Transportation
- Policies Involving Green Technologies and Environmental Engineering
- Green Technology

IGI Global is currently accepting manuscripts for publication within this series. To submit a proposal for a volume in this series, please contact our Acquisition Editors at Acquisitions@igi-global.com or visit: http://www.igi-global.com/publish/.

The Advances in Environmental Engineering and Green Technologies (AEEGT) Book Series (ISSN 2326-9162) is published by IGI Global, 701 E. Chocolate Avenue, Hershey, PA 17033-1240, USA, www.igi-global.com. This series is composed of titles available for purchase individually; each title is edited to be contextually exclusive from any other title within the series. For pricing and ordering information please visit http://www.igi-global.com/book-series/advances-environmental-engineering-green-technologies/73679. Postmaster: Send all address changes to above address.

Titles in this Series

For a list of additional titles in this series, please visit:
http://www.igi-global.com/book-series/advances-environmental-engineering-green-technologies/73679

Artificial Intelligence and IoT-Based Technologies for Sustainable Farming and Smart Agriculture
Pradeep Tomar (Gautam Buddha University, India) and Gurjit Kaur (Delhi Technological University, India)
Engineering Science Reference • © 2021 • 400pp • H/C (ISBN: 9781799817222) • US $215.00

Handbook of Research on Waste Diversion and Minimization Technologies for the Industrial Sector
Ashok K. Rathoure (M/s Akone Services, Lucknow, India)
Engineering Science Reference • © 2021 • 455pp • H/C (ISBN: 9781799849216) • US $295.00

Smart Farming Techniques and Knowledge-Based Agriculture Emerging Research and Opportunities
Petraq Papajorgji (European University of Tirana, Albania) and Francois Pinet (Irstea Centre de Clermont-Ferrand, France)
Engineering Science Reference • © 2021 • 230pp • H/C (ISBN: 9781799845409) • US $195.00

Solar Concentrating Modules With Louvered Heliostats Emerging Research and Opportunities
Dmitry Strebkov (Federal State Budget Scientific Institution, Russia & Federal Scientific Agroengineering Center VIM, Russia) Natalya Filippchenkova (Federal State Budget Scientific Institution, Russia & Federal Scientific Agroengineering Center VIM, Russia) and Anatoly Irodionov (Federal State Budget Scientific Institution, Russia & Federal Scientific Agroengineering Center VIM, Russia)
Engineering Science Reference • © 2021 • 267pp • H/C (ISBN: 9781799842767) • US $190.00

Smart Agricultural Services Using Deep Learning, Big Data, and IoT
Amit Kumar Gupta (Amity University Jaipur, India) Dinesh Goyal (Poornima Institute of Engineering and Technology, India) Vijander Singh (Manipal University Jaipur, India) and Harish Sharma (Rajasthan Technical University, India)
Engineering Science Reference • © 2021 • 280pp • H/C (ISBN: 9781799850038) • US $195.00

Eco-Friendly Energy Processes and Technologies for Achieving Sustainable Development
Mir Sayed Shah Danish (University of the Ryukyus, Japan) and Tomonobu Shah Senjyu (University of the Ryukyus, Japan)
Engineering Science Reference • © 2021 • 340pp • H/C (ISBN: 9781799849155) • US $195.00

Precision Agriculture Technologies for Food Security and Sustainability
Sherine M. Abd El-Kader (Electronics Research Institute, Cairo, Egypt) and Basma M. Mohammad El-Basioni (Electronics Research Institute, Cairo, Egypt)
Engineering Science Reference • © 2021 • 437pp • H/C (ISBN: 9781799850007) • US $195.00

IGI Global
PUBLISHER of TIMELY KNOWLEDGE

701 East Chocolate Avenue, Hershey, PA 17033, USA
Tel: 717-533-8845 x100 • Fax: 717-533-8661
E-Mail: cust@igi-global.com • www.igi-global.com

Editors dedicated this book to their respective families.

Editorial Advisory Board

Table of Contents

Detailed Table of Contents

of river, lake, reservoir, and marine environment. The sensed data from IoT sensors are communicated wirelessly via acoustic channels to a data collection center for further processing and interpretation. It is foreseeing that judicious deployment of IoT-based UWSN is a promising solution for long-term water quality surveillance.

Internet of things (IoT) is in the forefront of many existing smart applications, including autonomous systems and green technology. IoT devices have been commonly used in the monitoring of energy efficiency and process automation. As the application spreads across different kinds of applications and technology, a large number of IoT devices need to be managed and configured, as they are capable of generating massive amount of sensory data. Looking from this perspective, there is a need for a proper mechanism to identify each IoT devices within the system and their respective applications. Participation of these IoT devices in complex systems requires a tamper-proof identity to be generated and stored for the purpose of device identification and verification. This chapter presents a comprehensive approach on identity management of IoT devices using a composite identity of things (CIDoT) with permissioned blockchain implementation. The proposed approach described in this chapter takes into account both physical and logical domains in generating the composite identity.

This chapter reviews present usage of vibration-based energy harvesting (VEH) devices and their applications. The evolution of energy resources and advancement in electronic technologies has resulted in the need for a self-sustainable wireless/portable electronic device in current modern society. Batteries are non-beneficial in the miniaturization process of electronic designing, and alternative power supplies are desperately needed to drive the advance of the wireless/portable development further. VEH has emerged as one of the most promising alternatives to replace conventional batteries and as the solution for the bottleneck. Consideration of creating an optimal vibration energy harvester is suggested through an analytical model of a mechanical transducer, including a relatively new method defined as triboelectricity. Useful applications and usages of VEH are presented, and some suggestions for improvement are also given. Lastly, the trend of energy harvesting is annotated and commented in-line with the demand of electronic sensors market.

Wireless sensor networks (WSNs) have been popular due to their wide range of applications in almost all walks of life including industry controls, environmental monitoring, health, transportation, military. Usually conventional sensor networks are dedicated, and private networks have little or no communication with the outside world; hence, when connected to the external world by using internet protocol (IP), they form a network of connected devices, sensors, and systems and form an internet of things (IoT). Due to rapid development of IoT infrastructures worldwide, energy demand has been increased significantly to meet the power requirements of billions of connected devices. Since WSN is a foundation of IoT; hence, IoT also inherited with the challenge of providing consistent energy and to maintain hazard-free environments. This chapter is the extension of "Energy Harvesting Models and Techniques: A Review," which focuses on possible energy sources available in ambient environment and the technological mechanism to harvest energy for WSN and IoT that promotes green energy concept.

Different types of energy which generally fulfill the requirements of computing are mostly from thermal, mechanical, solar, wind, acoustic, and wave. Typically, IoT devices are powered by batteries that have limited lifetime, and thus these IoT devices need to be self-powered or require supportive energy sources that uninterruptedly power IoT devices. Energy harvesting is one of the techniques that can be applied to power these devices, which is a procedure of apprehending energy from lone or more energy from renewable sources in the proximate atmosphere known as environmental energy which can be renovated into usable electrical energy. Numerous researches are being carried out to harvest energy. This chapter is the extended version of the previous work carried out and analyses the present works on the application of IoT in energy harvesting systems and extant different research works carried out by the investigators to classify them.

The expansion and development of advanced technology by mankind have a great substantial impact on the natural environment. The buildings consume a large number of natural resources during their construction, design, and operation process. Due to the manipulation of natural resources by the buildings on a large scale, there is a need of better-designed buildings for effective and efficient use of resources. The concept of "green buildings" is an innovative solution for the aforementioned issues and promotes eco-friendly activities. Internet of things (IoT) transforms the idea of green building into real. Considering a large number of IoT applications, the objective of this chapter is to survey the major IoT concepts for green building management. The survey is conducted by summarizing a large number of scientific contributions to the field (i.e., internet of things for green buildings) and presented in this study. This chapter also describes the open research challenges for future work in this emergent area.

The internet of things (IoT) is a system consisting of computing, mechanical, and electronic devices, which are having ability to transfer data in network without human interaction. The sensors used in IoT collect and transfer the data to the cloud, which is further processed using software to perform an action. The IoT is one of the fastest growing industries, and in recent years, it is most widely used in HVAC systems in residential and commercial applications to reduce the energy consumption as building consumes by approximately 40% of total energy. The IoT reduces the energy consumption of the building by optimizing the process variables of HVAC system components, increases life of system components, enhances the comfort of the occupants, and provides remote control of the system. However, there are challenges in data security and privacy, and also there is a lack of IoT platforms specifically oriented towards the proper processing, management, and analysis of such large and diverse data.

In the present era of technology drift, the Internet of Things (IoT) is a promising technology that shows the considerable transition from enterprises to businesses. In this context, renewable energy has played critical roles in developing the agricultural sector with its functionality to generate power over distant areas and different terrain. There has been a transformation in the way we consume electricity and the way it is operated by the utilities. The smart monitoring features are an efficient way to utilize energy and minimize losses. These losses might be the transmission and distribution losses or losses by the inefficient household wiring or appliances. These losses can be minimized if they are monitored well with relevant data and analytics. This book chapter presents an application of the Internet of Energy (IoE) technology in the renewable sector. The study was carried applying Maximum Power Point Tracking (MPPT) functions and corresponding Remote Monitoring System (RMS) functions. The undertaken site for Solar Water Pump (SWP) installation for irrigation is Bihar Sharif (Nalanda, Bihar). The proposed

IoE based technology served maximum power output with high efficiency during operating hours. Here, the farmers may do automatic start/shut, remote start/shut from SMS/web with a health check, and self-diagnostic mechanisms for easy maintenance. Thus, the SWP system emerged as a cost-effective and environment-friendly irrigation solution for remote farmers throughout the year. The solar potential assessment of the site and circuital development were done using PVsyst 7.0 and KiCad software tools, respectively. The successful working of the developed product has been observed at the site.

Green energy infrastructure with the internet technologies relies on five important domains: green machine to machine (M2M), green cloud computing (CC), green data center (DC), green ICT, and green cellular. The ever-increasing demand for cloud computing and heavy dependence on cloud for storage, processing, and applications results in the need for more data centers with high capacity. Power management using wireless sensor networks (WSN) can be a potential solution as there has been a lot of works suing WSN for power management for green buildings, green home, and green farming. The same design can be applied to data centers with modifications to cater for data centers. Since WSN is part of IoT, various IoT-related solutions can be proposed for green data center solutions. A hybrid model that consists of virtualization, cooling systems, and IoT shows energy efficient data center designs. There have been various efforts as such, and this research will present green energy designs and mainly IoT-related initiatives for green-aware data centers.

The ability to sustain communication between nodes as per Industry 4.0 standard within the waterways is still not mature enough. It is prudent to make use of natural resources that are abundantly found in the form of strong wind (wind energy) and underwater tidal waves generation (tidal energy). Climate change and environmental sustainability are now becoming more important for a sustainable smart world. This is discussed in detail along with the decentralized mode of crop management. In this chapter, it is shown that energy efficiency can lead to the selection of different system models for the internet of underwater things (IoUT) where systems are sustainable and can tap previously unexplored resources. This resource is useful as a reference to further explore IoUT and green energy harvesting techniques.

In the computing grid environment, jobs scheduling is fundamentally the process of allocating computing jobs with choices relevant to the available resources. As the scale of grid computing system grows in size

Foreword

Due to the popular demand in mobile and sensing technology, the deployment of smart objects are increasing drastically which flourishes the advancement of Internet of Things (IoT). IoT connects various objects such as sensor, actuators, mobile devices and integrates with communication protocols. With the advent of IoT, Cyber Physical Systems (CPS) are calling for research directions although the two terms might have the same core functionalities. But CPS calls for more integrated solutions with the physical devices and physical world. With this, CPS can combined with IoT to communicate with the physical world to provide reliable services such as healthcare and vehicular systems. Since IoT and CPS are going towards the direction of mobility and mobile world, energy consumption, energy storage and energy management should be investigated. Energy optimization mobile devices in mobile CPS has been investigated in order to extend the nodes lifetime. But future direction should be focusing towards utilizing the flexibility and connectivity IoT to be integrated into mobile CPS to move towards valuable green IoT for energy harvesting, energy management and smart energy solutions.

This book encourages the integration of new trends in ICT into the green technology in the era of Industry 4.0 through smart concepts and ideas. I believe that this book holds the primary objective in providing academic and research community an insight of focusing on the role of ICT applications to sustain ICT supported green environment. Ultimately, this volume is to provide the academic, industrial, and local government communities with recent advances in development, application, and impact of the role of ICT with the mobile elements for the initiative in supporting sustainable green technology. The contents of this book would assist the academics and industrial communities with up-to-date information related to the recent technologies for improved ICT digital ecosystem, mainly focusing on several niche areas of research to encourage more advanced research in this domain.

Rehan Akbar
Universiti Tunku Abdul Rahman, Malaysia

Preface

We are in the era of Industry 4.0 with the world going towards everyTHING, everywhere and anytime with "things" being enabler of technology. Our world is becoming smarter with everything (mobile phones, cars, TVs) connected to each having unique addresses and communication mechanisms. It is foreseen that by 2025 every little thing like pen, paper, food packages etc. would be operated over the Internet by Internet of Things (IoT) towards smarter world. However, in order to enable the smart world to be sustainable, IoT should be embarking into energy efficient (green) paradigms. Green IoT would be a moving factor contributing towards energy efficiency by reducing energy utilization either by using software or hardware procedures

This book is an attempt to collect and publish innovative ideas, emerging trends, implementation experience and use-cases pertaining to different enabling technologies for IoT in general and their application to green energy. Overall, this book is a collection of ideas and experience mainly focusing on the green energy usage towards the realization of IoT based systems and Industry 4.0, which is destined to a sustainable green energy world.

THE ORGANIZATION OF THE BOOK

Chapter 1: Green IoT (G-IoT) Ecosystem for Smart Cities

We are in the era of Industry 4.0 with the world going towards everything, everywhere and anytime with "things" being enabler of technology. Our world is becoming smarter with everything (mobile phones, cars, TVs) connected to each having unique addresses and communication mechanisms. It is foreseen that by 2025 every little thing like pen, paper, food packages etc. would be operated over the Internet by Internet of Things (IoT) towards smarter world. However, in order to enable the smart world to be sustainable, IoT should be embarking into energy efficient (green) paradigms. Since IoT is also key enabler for smart cities, a green aware design of smart cities could see a potential to create a Green IoT Ecosystem for Smart Cities. This ecosystem would comprise the green IoT lifecycle, the six pillars of smart cities, and focusing on the green infrastructure, green applications and green energy. This conceptual study would motivate researchers embarking into smart city projects and keeping green concepts intact in their design and implementations.

Chapter 2: IoT Underwater Wireless Sensor Network Monitoring

Water is a crucial resource for all life on earth and it is fast becoming one of the limited natural resources to humankind especially the clean drinking water and water for agricultural uses. Sensor technologies and wireless communications (both terrestrial and underwater) have been seriously investigated by researchers to find ways to integrate these technologies for a novel data-sensing and data-collecting network for long-term water pollution monitoring purposes. This chapter describes an IoT-based underwater wireless sensor network (UWSN) which is believed to have a huge potential for monitoring the health of river, lake, reservoir, and marine environment. The sensed data from IoT sensors are communicated wirelessly via acoustic channels to a data collection center for further processing and interpretation. It is foreseeing that judicious deployment of IoT-based UWSN is a promising solution for long-term water quality surveillance.

Chapter 3: Composite Identity of Things (CIDoT) on Permissioned Blockchain Network for Identity Management of IoT Devices in Smart Building Application

Internet-of-Things (IoT) is in the forefront of many existing smart applications, including autonomous systems and green technology. IoT devices have been commonly used in the monitoring of energy efficiency and process automation. As the application spreads across different kinds of applications and technology, large number of IoT devices need to be managed and configured, as they are capable of generating massive amount of sensory data. Looking from this perspective, there is a need for a proper mechanism to identify each IoT devices within the system and their respective applications. Participation of these IoT devices in complex-systems requires a tamper-proof identity to be generated and stored for the purpose of device identification and verification. This chapter presents a comprehensive approach on identity management of IoT devices using a composite Identity-of-Things (CIDoT) with permissioned blockchain implementation. The proposed approach described in this chapter takes into account both physical and logical domains in generating the composite identity

Chapter 4: Applications of Vibration-Based Energy Harvesting (VEH) Devices

This chapter reviews present usage of vibration-based energy harvesting (VEH) devices and its applications. The evolution of energy resources and advancement in electronic technologies has resulted in the need for a self-sustainable wireless/portable electronic device in current modern society. Batteries are non-beneficial in the miniaturization process of electronic designing and alternative power supplies are desperately needed to drive the advance of the wireless/portable development further. VEH has emerged as one of the most promising alternatives to replace conventional batteries and as the solution for the bottleneck. Various consideration of creating an optimal vibration energy harvester are suggested through an analytical model of a mechanical transducer, including a relatively new method defined as triboelectricity. Useful applications and usages of VEH are presented and some suggestion for improvement are also given. Lastly, the trend of energy harvesting is annotated and commented in-line with the demand of electronic sensors market.

Chapter 5: Energy Harvesting Models and Techniques for Green IoT – A Review

Wireless sensor networks (WSNs) has been popular due to their wide range of applications in almost all walks of life including industry controls, environmental monitoring, health, transportation, military. Usually conventional sensor networks are dedicated and private networks and have little or no communication with outside world hence when connected to the external world by using internet protocol (IP) they form a network of connected devices, sensors and systems and form an internet of things (IoT). Due to rapid development of IoT infrastructures worldwide, energy demand has been increased significantly to meet the power requirements of billions of connected devices. Since WSN is a foundation of IoT, hence IoT also inherited with the challenge of providing consistent energy and to maintain hazardous free environment. This chapter is the extension of Energy Harvesting Models and Techniques: A Review which focuses on possible energy sources available in ambient environment and the technological mechanism to harvest energy for WSN and IoT that promotes green energy concept.

Chapter 6: Realizing the Benefits of Energy Harvesting for IoT

Different types of energy which generally fulfill the requirements of computing are mostly from thermal, mechanical, solar, wind, acoustic, and wave. Typically, IoT devices are powered by batteries that have limited lifetime and thus these IoT devices need to be self-powered or require supportive energy sources that uninterruptedly power IoT devices. Energy harvesting is one of the techniques that can be applied to power these devices, which is a procedure of apprehending energy from lone or more energy from renewable sources in the proximate atmosphere known as Environmental energy which can be renovated into usable electrical energy. Numerous researches are being carried out to harvest energy. This paper is the extended version of the previous work carried out and analyses the present works on the application of IoT in energy harvesting systems and extant different research works carried out by the investigators to classify them

Chapter 7: Internet of Things for Green Building Management – A Survey

The expansion and development of advanced technology by mankind have a great substantial impact on the natural environment. The buildings consume a large number of natural resources during their construction, design, and operation process. Due to the manipulation of natural resources by the buildings on a large scale, there is a need for better-designed buildings for effective and efficient use of resources. The concept of "Green Buildings" is an innovative solution for the aforementioned issues and promotes eco-friendly activities. Internet of Things (IoT) transforms the idea of green building into real. Considering a large number of IoT applications, the objective of this chapter is to survey the major IoT concepts for green building management. The survey is conducted by summarizing a large number of scientific contributions, to the field (i.e., Internet of Things for Green Buildings) and presented in this study. This chapter also describes the open research challenges for future work in this emergent area.

Chapter 8: Challenges and Issues of IoT Application in Heating Ventilating Air Conditioning Systems – Energy Conservation Using IoT

The internet of things (IoT) is a system consists of computing, mechanical and electronic devices which are having ability to transfer data in network without human interaction. The sensors used in IoT collect and transfers the data to the cloud which is further processed using software to perform an action. The IoT is one of the fastest growing industries and in recent years it is most widely used in HVAC Systems used in residential and commercial applications to reduce the energy consumption as building consumes approximately 40% of total energy consumed. The IoT reduces the energy consumption of the building by optimizing the process variables of HVAC system components, increase life of system components, enhances the comfort of the occupants and provides remote control of the system. However there are challenges in data security and privacy and also there is a lack of IoT platforms specifically oriented towards the proper processing, management and analysis of such large and diverse data.

Chapter 9: MPPT-Based Solar Powered Water Pumping System With RMS – Augmentation of IoE Technology for Irrigation: Importance of Irrigation Water Management

This chapter presents an application of the internet of energy (IoE) technology in the renewable sector. The study was carried out applying maximum power point tracking (MPPT) functions and corresponding remote monitoring system (RMS) functions. The undertaken site for solar water pump (SWP) installation for irrigation is Bihar Sharif (Nalanda, Bihar). The proposed IoE based technology served maximum power output with high efficiency during operating hours. By this solution, the farmers may do automatic start/shut, remote start/shut from SMS/web with health check and self-diagnostic mechanisms for easy maintenance. Thus, the SWP system emerged as a cost-effective and environment-friendly irrigation solution for remote farmers throughout the year. The solar potential assessment of the site and circuital development were done using PVsyst 7.0 and KiCad software tools, respectively.

In the present era of technology drift, the Internet of Things (IoT) is a promising technology that shows the considerable transition from enterprises to businesses. In this context, renewable energy has played critical roles in developing the agricultural sector with its functionality to generate power over distant areas and different terrain. There has been transformation in the way we consume electricity and the way it is operated by the utilities. The smart monitoring features are an efficient way to utilize energy and minimize losses. These losses might be the transmission and distribution losses or losses by the inefficient household wiring or appliances. These losses can be minimized if they are monitored well with relevant data and analytics.

Chapter 10: Green Energy in Data Centers Using Internet of Things

Green energy infrastructure with the Internet technologies relies on five important domains as Green Machine to Machine (M2M), Green Cloud Computing (CC), Green Data Center (DC), Green ICT and Green Cellular. With the ever-increasing demand for cloud computing and heavy dependence on cloud for storage, processing and applications results in the need for more data centers with high capacity. Power management using wireless sensor networks (WSN) can be a potential solution as there has been a lot of works suing WSN for power management for green buildings, green home and green farming.

The same design can be applied to data centers with modifications to cater for data centers. Since WSN is part of IoT, various IoT related solutions can be proposed for green data center solutions. A hybrid model which consists of virtualization, cooling systems and IoT let designs can be foreseen for energy efficient data center designs. There have been various efforts as such and this research will present green energy designs and mainly IoT related initiatives for green aware data centers.

Chapter 11: Green Energy Harvesting and Energy Efficient Routing Protocol in Internet of Underwater Things

The ability to sustain communication between nodes as per Industry 4.0 standard within the waterways is still not mature enough. It is prudent to make use of natural resources that are abundantly found in the form of strong wind (wind energy) and underwater tidal waves generation (tidal energy). Climate change and environmental sustainability are now becoming more and more important for a sustainable smart world. This is discussed in detail along with the decentralized mode of crop management. In this chapter, it is shown that energy efficiency can lead to the selection of different system models for the Internet of Underwater Things (IoUT) where systems are sustainable and can tap previously unexplored resources. This resource is useful as a reference to further explore IoUT and Green Energy Harvesting techniques.

Chapter 12: Incentive-Based Scheduling for Green Computational Grid

In computing grid environment, jobs scheduling is fundamentally the process of allocating computing jobs with choices relevant to the available resources. As scale of grid computing system grows in size over time, exponential increase in energy consumption is foreseen. As such, large data centers (DC) are embarking on green computing initiatives to address the IT operations impact on the environment. The main component within a computing system consuming the most electricity and generates the most heat is the microprocessor. The heat generated by these high-performance microprocessors is emitting high CO2 footprint. Therefore, jobs scheduling with thermal considerations (thermal-aware) to the microprocessors is important in DCs grid operations. An approach for jobs scheduling is proposed in this chapter for reducing electricity usage (green computing) in DC grid. This approach is the outcome of the R&D works based on the DC grid environment in Universiti Teknologi PETRONAS, Malaysia

Chapter 13: Self-Healing and Green Energy in Wireless Sensor Network – A Survey

Green energy can be classified into many domains such as green aware applications, green aware protocols, green energy harvesting and many more. The main focus of this paper is to analyze green energy in terms of energy aware routing protocols and mainly focusing on wireless sensor networks. So, in this domain greening the routing protocols with energy efficient algorithms are investigated. Energy efficient routing algorithms can be deisgned by having self-healing concepts whereby routing algorithms can heal themselves when a routing hole is found or nodes can heal themselves when energy depletion occurs. Having that concept in mind, this paper investigates both the green energy protocols and the self-healing protocols in the literature.

Chapter 14: Globally Renewable Energy Consumption Sources – Globally Renewable Energy Consumption Sources

Electricity consumption will encompass a large converse about connected with international electricity demand while in the next two decades. Renewable Energy Sources and Climate Change Modify Minimization offers a good estimation on the chapter for the technological, scientific, environmentally friendly, financial and also societal aspects of this factor connected with six renewable energy (RE) options for the minimization connected with weather adjust. This research is the extended version of the research presented in IGI global book "Role of IoT in Green Energy Systems", 2015. A full assessment related to any profile minimization options will likely involve an evaluation of respective potential alongside minimization with his bargain with sustainable development as well as all associated risks, and costs

Vasaki Ponnusamy
Universiti Tunku Abdul Rahman, Malaysia

Noor Zaman
Taylor's University, Malaysia

Low Tang Jung
Universiti Teknologi PETRONAS, Malaysia

Anang Hudaya Muhamad Amin
Higher Colleges of Technology, UAE

Acknowledgment

We would like to express our great appreciation to all of those with whom we have had the pleasure to work during this project. The completion of this project could not have been accomplished without the support of them.

First, the editors would like to express deep and sincere gratitude to all the authors who shared their ideas, expertise and experience by submitting chapters to this book and adhered to its timeline.

Second, the editors wish to acknowledge the extra ordinary contributions of the reviewers for their valuable and constructive suggestions and recommendations to improve quality, coherence and content presentation of chapters. Most of the authors also served as referees. Their willingness to give time so generously is highly appreciated.

Finally, our heartfelt gratitude go to our family members and friends for their love, prayers, caring and sacrifices in completing this project well in time.

Vasaki Ponnusamy
Universiti Tunku Abdul Rahman, Malaysia

Noor Zaman
Taylor's University, Malaysia

Low Tang Jung
Universiti Teknologi PETRONAS, Malaysia

Anang Hudaya Muhamad Amin
Higher Colleges of Technology, UAE

Chapter 1
Green IoT (G-IoT) Ecosystem for Smart Cities

Vasaki Ponnusamy
Universiti Tunku Abdul Rahman, Malaysia

Bobby Sharma
Assam Don Bosco University, India

Waqas Nadeem
Lahore Garrison University, Pakistan

Goh Hock Guan
Universiti Tunku Abdul Rahman, Malaysia

N. Z. Jhanjhi
https://orcid.org/0000-0001-8116-4733
Taylor's University, Malaysia

ABSTRACT

We are in the era of Industry 4.0 with the world going towards everything, everywhere, and anytime with "things" being enablers of technology. Our world is becoming smarter with everything (mobile phones, cars, TVs) connected to each, having unique addresses and communication mechanisms. It is foreseen that by 2025 every little thing, like pen, paper, food packages, etc., would be operated over the internet by internet of things (IoT) towards a smarter world. However, in order to enable the smart world to be sustainable, IoT should be embarking into energy efficient (green) paradigms. Since IoT is also a key enabler for smart cities, a green-aware design of smart cities could see a potential to create a green IoT ecosystem for smart cities. This ecosystem would comprise the green IoT lifecycle, the six pillars of smart cities, and focusing on the green infrastructure, green applications, and green energy. This conceptual study would motivate researchers embarking into smart city projects and keeping green concepts intact in their design and implementations.

DOI: 10.4018/978-1-7998-6709-8.ch001

INTRODUCTION

Green IoT would be a moving factor contributing towards energy efficiency by reducing energy utilization either by using software or hardware procedures. When the term Green IoT is stated, there could be two possible ways one could interpret. Are we going to utilize IoT in order to reduce the greenhouse effect or are we looking at the impact of IoT contributing to the greenhouse effect? In the earlier term we should investigate sustainable green applications or infrastructures by incorporating IoT into the design whereas the later focuses more on making the IoT devices more energy efficient. In either case, the entire green IoT lifecycle should focus on green design, green communication, green production, green utilization and green disposal. The design and production of the entire green architecture (device, communication, networking) should be power aware and the utilization and disposal of green devices and technology in order to reduce the greenhouse effects. With this, the global carbon emission reduction and energy consumption savings can be achieved as anticipated by the European Union. A global smart Green IoT Ecosystem (Figure 1) can be foreseen, focusing on the three main IoT domains in the area of application, infrastructure and energy management.

Figure 1. Green IoT Ecosystem

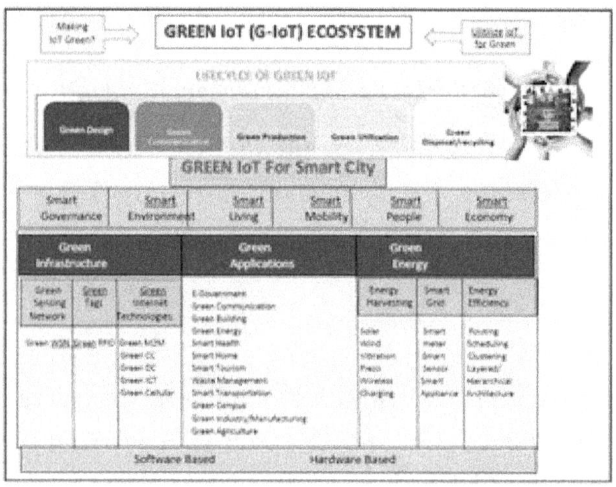

This green IoT ecosystem (Figure 1) is going to be the main contributor in achieving a smarter city with the key focus on sustainability. Smart city framework (Figure 2) consists of smart governance, smart environment, smart living, smart people, smart mobility and smart economy. These six pillars of smart city should be incorporated into the IoT domain (Figure 3)) to create a synergy between Green IoT and Smart City for a generic green ecosystem. While making "THINGS" smart, the green ecosystem should be incorporated to make "THINGS" green (energy aware) as well. To anticipate such vision, research focus should be on the challenges faced by energy hungry IoT devices like Radio Frequency Identification (RFID). RFID devices are very much power hungry and these inter-connected tiny devices are sharing huge amounts of data to the cloud and data centers leading to huge power consumption as well. So innovative ideas are required on how to make these tiny RFID devices green while also making the cloud

(G-CC) and data centers green (G-DC). Reducing the sizes of RFID tags, investing into biodegradable, printable or paper-based RFID tags could be some solutions. Virtualization of data centers and cloud can reduce the amount of data transmission whilst making them more energy efficient. Green Wireless Sensor Network (G-WSN) is another main domain in IoT that needs attention although there are a lot of researches going on in this domain. More energy focused realms need attention like energy efficient/aware/hierarchical routing algorithms, scheduling algorithms, radio optimization techniques, data reduction/aggregation, clustering and efficient communication like 5G (G-cellular). The sensor designs themselves should be more power focused, utilizing green sources of energy/harvesting, wireless energy charging and reducing transmission power. Smart grid systems can be deployed for this purpose by automatically monitoring and controlling energy flows between the energy network and smart alliances or sensor/actuator nodes (G-M2M). The vision of Smart City cannot be materialized without exploring into the application domain. Smart city is transforming with emerging applications like e-government, waste management, smart water management, green building, smart building, smart parking, traffic management and environment monitoring.

Figure 2. Smartcity Framework

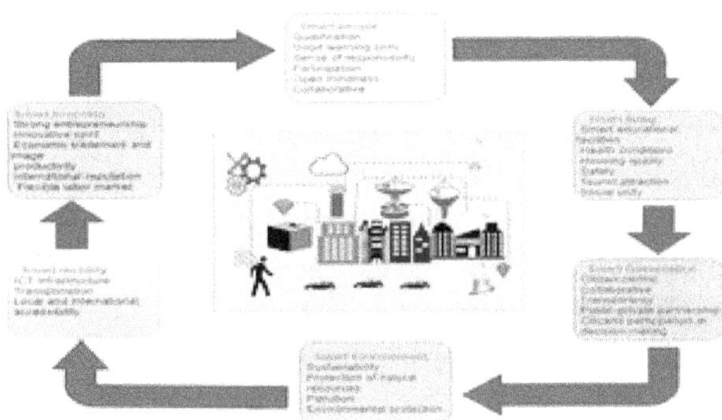

As far as the green IoT is concerned, the entire lifecycle (Figure 3) should be focusing on green design, green communications, green production, green utilization and green disposal (Zhu et al., 2015) The design of IoT devices should be looking into a greener paradigm like greener smartphones, greener RFID designs, greener chips and employing eco-friendly designs. Green communication should be focusing on shorter communication path, energy efficient route selection etc. In the manufacturing context, the use of bio-products (Maksimovic et al.,2017; Leung, 2015; Pazowski, 2015; Zhu et al., 2015) and biodegradable products could be adopted. The application of IoT should be moving towards green utilization of technology, products and energy with efficient and green aware trends. Lastly the disposable itself should be in the 3R principles of reduce, reuse and recycle like smartphone recycling, IoT devices recycling etc.

The chapter is organized as follows: Section 2 surveys literature relevant to various energy harvesting and transmission techniques. A complete discussion about energy harvesting application of IoT is given in Section 3. Section 4 outlines several design challenges and guidelines for future research in energy harvesting for IoT and section 5 concludes the chapter.

Figure 3. Lifecycle of Green IoT

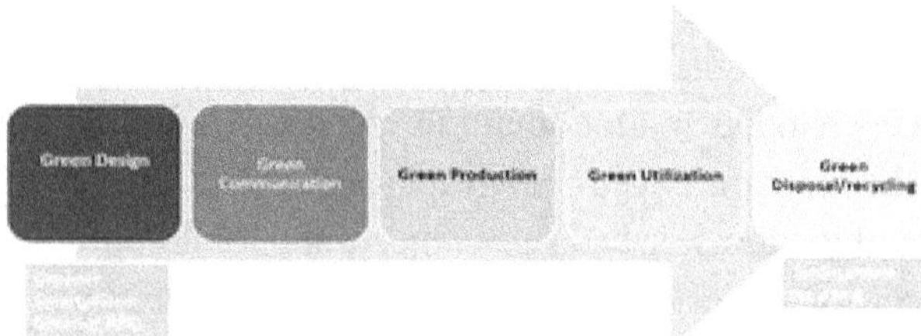

Motivating Factors To Green Computing

Growth rate of the Internet – The internet is growing at a tremendous rate which requires huge computing, communication, networks, servers and data centers. This requires a lot of energy requirements and green computing could be an energy efficient solution.

The demand for cooling facilities – As the number of databases and data centers are increasing, the demand for cooling facilities increases as well. Data centers stores and processes huge amount data which generates a lot of heat. This requires good cooling facilities and moving forward into greener solutions could help to tackle energy hungry data center issues.

Equipment require huge power – With the sophistication of equipment manufacturing, the demand for power increases due to huge processing. This power issue can be tackled efficiently by coming out with greener energy solutions like energy harvesting, wireless charging and proper energy management.

The growth rate of energy costs – The extreme demand for energy leads to increase in costs, that calls for cost effective energy solutions like environmental energy harvesting and replenishments techniques.

Growing awareness on ICT impacts towards the environment – As more and more awareness is created towards the impacts on ICT solutions on environment, there are more efforts in place to reduce the greenhouse effects. The solutions comes as in software as well as hardware dimensions.

NETWORK OF THINGS FOR IoT

Internet of Things is gradually increasing the unity of people and things around us bringing a new trend towards "ambient intelligence". Ambient Intelligence takes into account of sensitive and adaptive electronic environment by responding to the needs of the things around (Aarts, & Wichert, 2009). According to Marcelo (2014), there are six fundamental building blocks that constitute the components of the Internet of Things:

- sensors – for tracking and measuring activity
- connectivity – internet or cloud infrastructure
- processors – contain some computing power
- energy efficiency – difficult for access to charge and replace battery
- cost-effectiveness – easily available
- quality and reliability – need to operate in harsh environments for longer period
- security – need to relay sensitive information

Providing Internet connectivity to everything is the key feature of IoT and this leads to three main categories as network of things as shown in Figure. 4. Application domain identifies the areas of IoT applications that could range from industry, home, healthcare, commerce and more. Energy management for IoT becomes crucial as engineers are trying to figure out how these devices could be powered to provide uninterrupted service and that leads to energy harvesting, energy transmission and storage. Providing suitable infrastructure such as cloud computing for IoT is the next focus area. This chapter examines the energy management aspects and the application domain is restricted to energy harvesting applications only.

Figure 4. IoT Domains

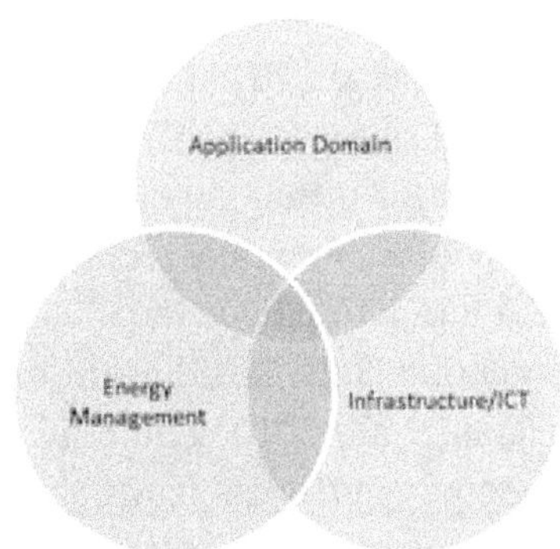

GREEN IoT DOMAIN

Energy efficiency in IoT domain has becoming one of the key challenges for researchers as such devices are expensive to maintain especially being deployed in inaccessible locations where replenishment of energy is almost impossible. Since these devices should be designed to operate unattended for very long duration, green technology concepts should be adopted into the main IoT domain presented in Figure 4. Greening the infrastructure, innovative way of green energy management and looking into green applications should be the way forward towards achieving greener IoT systems. Figure 5 shows the main IoT domain has been redesigned into a green IoT domain. In the following section, each of these three domain (Infrastructure, Applications and Energy Management) will be presented separately with some key implementation and recommendations.

Green Infrastructure

Green Infrastructure (Figure 6) in IoT is categorized into Green Cloud Computing (G-CC), Green IoT in Communication, Green Cellular, Green WSN (G-WSN), Green Data Center (G-DC) and Green Big Data and Cloud (G-CC)

Figure 5. Green IoT (G-IoT) Domain

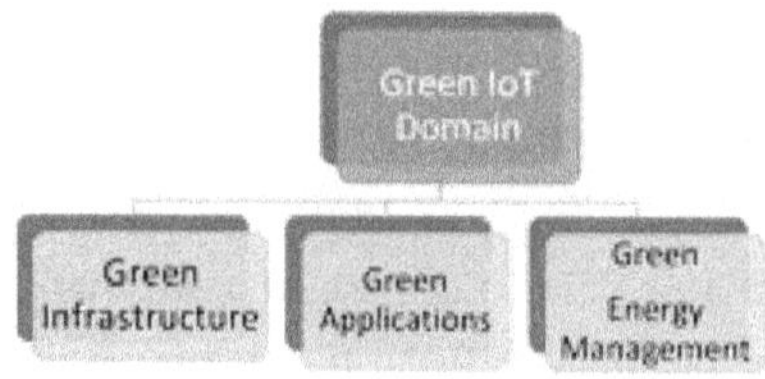

Green Cloud Computing (G-CC)

Could computing typically works as a service, platform as a service (PaaS), infrastructure as a service (IaaS) and software as a service (SaaS) (Zhu et al., 2015). All the resources are shared in the cloud with huge computing and processing power and storage capabilities. With the growth in cloud applications and services, power requirement increases for managing cloud resources. A greener cloud computing will be the way forward, and even as for mobile computing is concerned, mobile cloud computing (MCC) (Chean et al., 2018) could be a better solution to reduce the load of mobile devices for processing and storage. In order to reduce energy computing at the cloud environment many possibilities are available follows (Shaikh et al., 2015; Farahnakian et al., 2014; Zhu et al., 2015; Song et al., 2013; Bruneo et al., 2014; Xiong et al., 2012) such as i) energy efficient hardware and software designs, ii) adoption of virtual machine (VM) techniques, iii) energy efficient resource allocations (scheduling, auction-based, gossip based), iv) energy saving models and policies, v)green network and communication techniques. FOG computing is becoming another better solutions for tackling energy issues in IoT domains, Since IoT nodes and networks are very local centric and processes huge amounts of data, a local centric storage and processing would be a good solution. Since FOG network can be deployed in a home-grown network, all the huge computing, storage and communication can be done locally and only relevant data to be sent to the cloud. IoT network would consist a lot of tiny powered devices which can work better when the communication is done locally instead of transmitting to the cloud. Sarka and Misra (2016) have proposed a mathematical formulation-based fog computing platform (Figure 7) in collaboration

Figure 6. Green Infrastructure

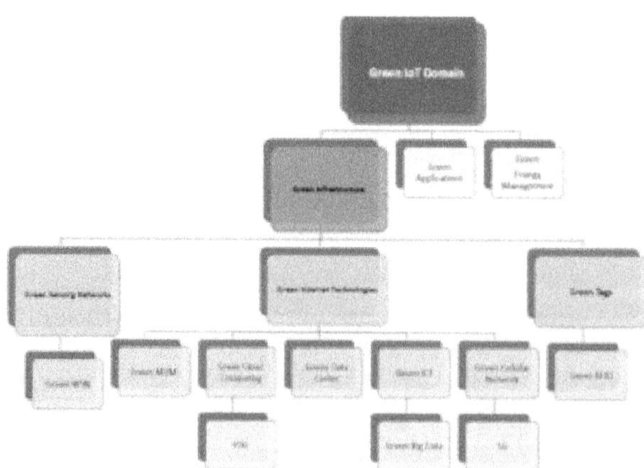

with traditional cloud computing paradigm which yields better service latency and energy consumption. This fog computing based model establishes an efficient green computing platform to support energy and resource hungry IoT devices. The results produce low latency services by IoT and the mean energy consumption was at 40.48% less than the traditional cloud computing model.

Figure 7. FOG based Green IoT Cloud Platform (Sarka and Misra, 2016)

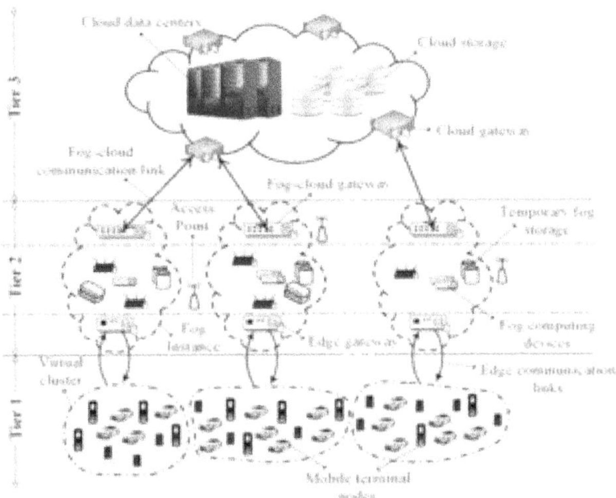

Green IoT in Communication

According to Albreem et al., (2017) green communication will materialise in 2020 and it is predicted that it would be able to handle 1000 more mobile data and todays cellular architecture (Albreem, 20151). Green communication should be a sustainable, energy-aware and environment aware design of communication platforms (Alsamhi et al., 2018). The entire idea should be a low carbon emission mechanism with energy efficient communications. Various efforts and researches haven been conducting reducing carbon emission, efficient path planning, low exposure to radiation, maximizing data rate and reducing Koutitas, 2010; Chan et al., 2013; Feng et al., 2010; Chih-Lin et al., 2014; Mao, 2017; Abrol et al., 2016; Naeem et al., 2013). 5G is expected to contribute vastly towards green IoT (Figure 8) for increasing reliability and QoS for communication between human and machines (robots, drones, sensor).

Wu et al., (2017) proposes an energy efficient solutions for 5G networks with the focus on energy harvesting as well. Among the solutions proposed are: i) by reducing the transmission and receiving distance, ii) by enhancing spectrum availability, iii) by enhancing special degrees of freedom vis M-MIMO (Wu et al., 2017) and by promoting energy harvesting. Figure 9 presented among some of the proposed ideas and the author proposes that with the growth of communication, user traffic, channel, power consumption, and content popularity models need to be investigated.

Figure 8. 5G for Green IoT (Alsamhi et al., 2018).

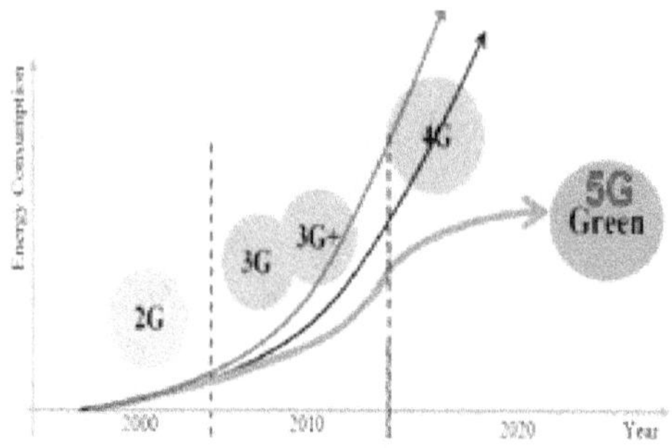

Green Wireless Sensor Network (G-WSN)

As far as wireless sensor network in concerned, some of the following suggestions can be adopted (Zhu et al., 2015; Shaikh et al., 2015; Lin et al., 2015; Anastasi et al., 2009; Rault et al., 2014). i. managing the sensor node active and sleep patterns in which nodes would go into sleep mode while inactive and go into active mode while performing sensing and transmission, ii. Energy replenishment strategy of sensor nodes by harvesting energy from the environment and RF sources, iii. Enhancing radio transmission like transmission power, iv. Reducing the size of data transmission by improving compression, data aggregation algorithms and designing an efficient shortest paths, v. energy efficient routing algorithms by introducing scheduling, hierarchical routing, multipath routing and clustering techniques. A review of various class of routing algorithms such as flat based, cluster based, hierarchical based and mobile agent based can be found in Ponnusamy et al, 2012; Ponnusamy et al., 2014. Arshad et al., (2017) suggests that wireless smart sensor network could be a path to implementing IoT and energy efficiency in sensor network design is a way forward to green IoT. Medium Access Control (MAC) is employed to reduce the communication between nodes and thus yields better lifetime and decreased energy usage (Sun and Ryoo, 2015). Alsamhi et al. (2018) proposed that sensor nodes should be sleep mode (Figure 10) for most of the time and be in idle mode only when needed and this work has been supported by other researchers as well (Ye et al.,2002; Anastasi et al., 2006). Rekha and Sekar, (2016) has done a comprehensive work green WSN by investigating on the energy efficiency, network lifetime, and im-

Figure 9. Green 5G Recommendations (Wu et al., 2017)

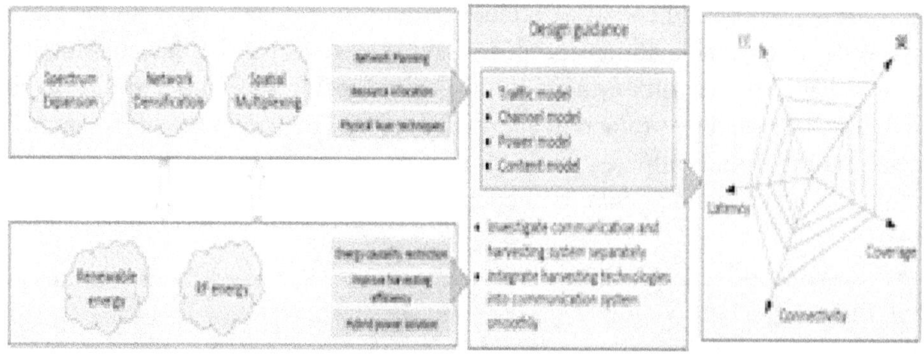

proving communication and routing. The author created a hierarchical framework with efficient deployment of sensor and actuator nodes with clustering mechanisms. The proposed work was a good model of energy efficient and cost effective green IoT realization and it was well suited. Energy harvesting in WSN could be another dimension towards green WSN. A review of energy harvesting applications in

Figure 10. Sensor Modes for Green IoT

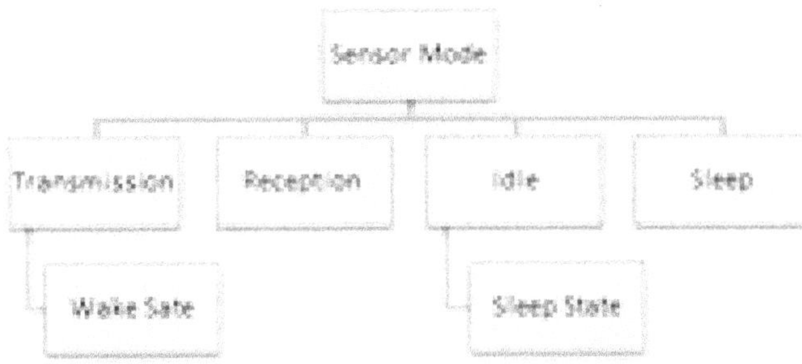

presented in later sections and a lot of work on energy harvesting by Azevedo and Santos, 2012; Eu et al., 2011; Shaikh and Zeadally, 2016 in WSN and biologically inspired ideas for energy harvesting can be found in Ponnusamy, (2016) and a review of energy analysis can be found in Ponnusamy, (2014). Energy harvesting in IoT is presented in Ponnusamy et al., (2020).

As far as WSN is concerned, there are a lot of research directions towards green WSN designs as shown in Figure 11. A much energy efficient routing algorithms could be deployed by employing hierarchical designs and mobility-based relay nodes. In terms of reducing energy consumption by reducing data packet sizes, a better aggregation-based algorithms and compression techniques could be deployed. Energy harvesting should become the main element in ant WSN ecosystems so that sensor nodes can continuously harness energy from the environment, and this would solve most of the energy deficiency issues in WSN. At the data link level, better MAC protocols could be introduced and scheduling algorithms should be introduced for MAC layer communication. Lastly the sensor design itself should be green aware with less carbon emission materials and less energy consuming chips.

Figure 11. Green WSN Research Directions

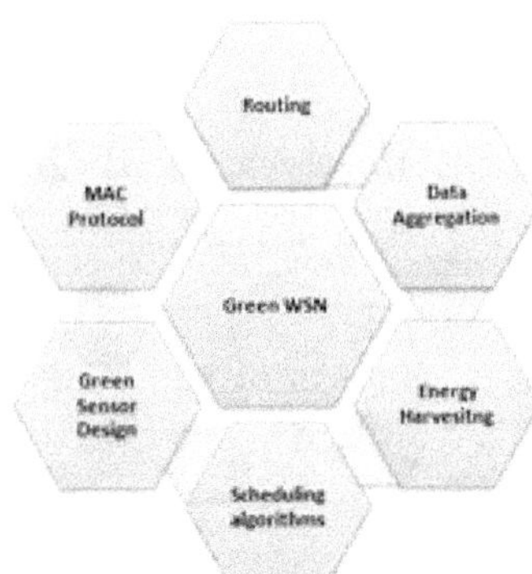

Green Data Center (G-DC)

As for the data centers, the greener DC (Shuja et al., 2014; Dayarathna et al., 2015; Wang et al., 2014; Chong et al., 2013) vision can be accomplished by using renewable energy sources with dynamic power management techniques. The design of hardware itself should be energy aware together with an energy efficient data center model with virtualization concepts. According to Mardamutu et al.,(2018), the following are some of the green data center techniques that can adapted to green IoT in data centers.

i. Power-aware Database
ii. Data Management Through Software And Hardware
iii. Data Management Through Disk Power Management
iv. Virtual Machine -The Problem May Arise In The Angle Of Automatic Scheduling Decision On Dynamic Migration Or Consolidating The VMS Among Physical Servers To Meet The Workload Requirements Meanwhile Saving Energy, Especially For Performance Sensitive
v. Renewable Energy
vi. Datacenter Cooling System
vii. Power Management Using Wireless Sensor Network
viii. Hybrid Model as shown in Figure 12

Figure 12. Hybrid Model Mardamutu et al.,(2018)

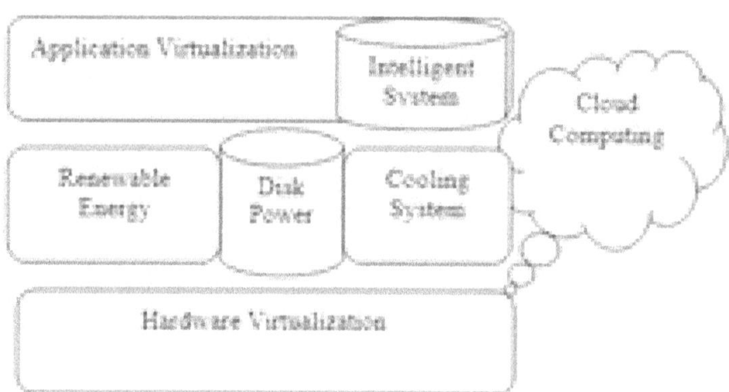

Figure 13. Green WSN Research Direction Mardamutu et al.,(2018)

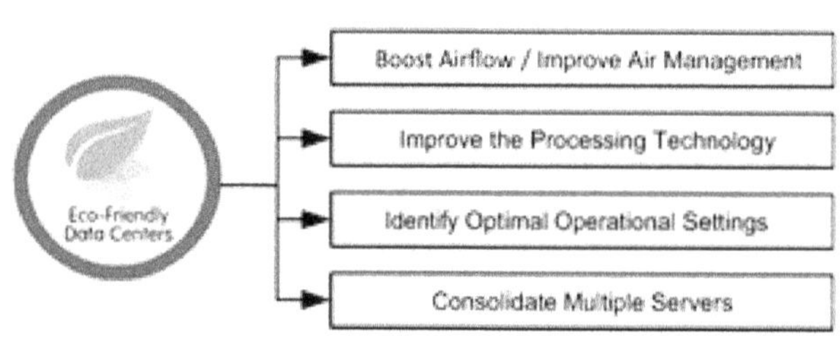

According to Alsamhi et al., 2019, Roy et al., (2016) came out with effective power consumption system in data centers without compromising on the cooling effectiveness. The authors state that effective routing and searching transactions are the better mechanisms for energy efficiency in data servers. Peoples et al., (2013) proposed a Context-Aware Broker (e-CAB) framework for managing next generation DCs. In Liu et al., (2016) a cloud-based air conditioning system was proposed that consists of environmental monitoring, temperature control, communication, air conditioning and ventilation with the cloud platform for data analytics. An Ant Colony System (ACS) was proposed by Farahnakian et al., and Ashraf and Porres (2018) to reduce power consumption at the data center and the system was effective in obtaining near optimal solutions. Several other initiatives towards green data center can be found in Shuja et al., (2014); Dayarathna et al., (2015); Zhu et al., (2015). Green WSN research direction is shown in Figure 13.

Green M2M

Green machine to machine basically the coordination between devices, for example between two sensor node. It is mainly applicable in manufacturing environment for communication between machines. The call for green M2M inspired by some of the reasons as stated by Pacis et al., (2017); i. there no proper sustainability initiatives in the manufacturing sector, ii.monitoring systems (waste and quality)are outdated, iii. No proposal waste disposable strategy and iv. Too much depending on non-renewable energy sources. Therefore Alsamhi et al., (2019) has highlighted that according to Popa et al., (2017); Gapchup et al., (2017); Liu and Ansari, (2019) the following methods could help towards achieving Green M2M.:

i. The transmission power to be adjusted
ii. Efficient routing and communication protocols
iii. Scheduling operation by putting some nodes into sleep mode when not active
iv. Energy saving by efficient resource allocation
v. Deploy energy harvesting mechanisms

Figure 14. Fraemework of IoT For Eco Efficiency in Manufacturing (Miragliotta and Shrouf, 2012)

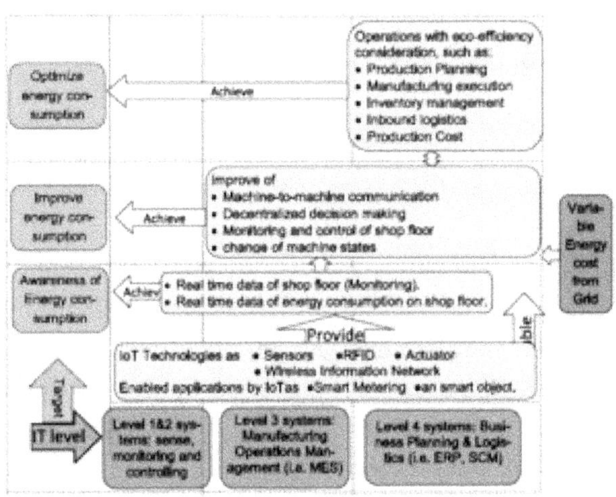

Figure 15. Green Smart City Framework

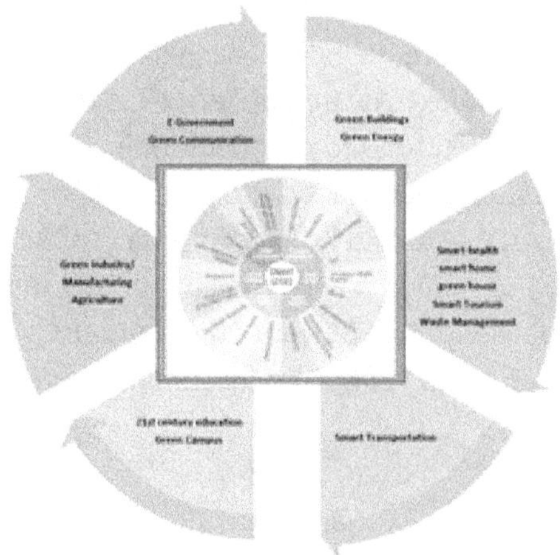

(Giovanni) has proposed an eco-efficient IoT model (Figure 14) for manufacturing with the following outcomes to be expected:

i. Improving energy-aware production scheduling,
ii. Reducing energy bill,
iii. Efficient maintenance management,
iv. Improving environmental reputation, and
v. Supporting decentralization in decision-making at production level to increase energy efficiency

Green RFID

As far as RFID is concerned, the following are some of ways how green RFID can be materialised:

- Towards RFID biodegradable systems
- RFID and wireless sensor networks
- Localization
- Chipless tags
- Sensor tags
- Material aspects

Green Applications

Green applications extend from smart people, smart mobility, smart environment, smart economy, smart living and smart governance which are the main six pillars of smart cities (Figure 15). For the smart governance, E-government applications reducing paper-based transactions and green communication mechanism using electronic mails and electronic applications could be smarter solutions towards greener governance. As for the smart living, smart home concepts, smart health, green house, smart tourism and

Figure 16. Green Applications

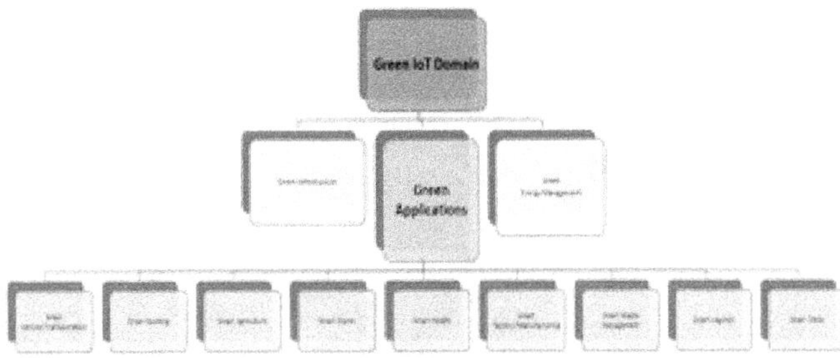

proper waste management could be some of the solutions looking for greener concepts and achieving smartness as well. In the smart mobility domain, smart transportations using ICT solutions for smart parking, ticket management, traffic management etc could be some of the solutions. The smart people would embark into green campus designs and 21st century education towards ICT4E solutions. In the environment sector, effective green designs in buildings like smart meter, energy harvesting, rain water harvesting, proper heat insulators and eco friendly designs. In the smart economy sectors, green industry and manufacturing can be materialized by using green aware materials and designs. Agriculture should be moving towards utilizing IoT devices for water management, pest control and soil moisture control.

With the smart city framework in the context, several IoT based solutions haven been invented in the green concepts in mind. As shown in Figure 16, smart vehicle/transportation, smart building, smart agriculture, smart home, smart health, smart factory, smart water management, smart logistics and smart cities are some of the examples moving towards greener solutions.

G-IoT Supported Agriculture

Green IoT Agriculture and Healthcare Application (GAHA) introduced by Nandyala et al., (2016) is a new domain whereby sensor-cloud computing (CC) is introduced in the agriculture and health applications (Figure 17). This is a new paradigm whereby physical sensors would collect data and communicate the data to CC infrastructure. GAHA shows good performance in terms of energy efficiency and data communication. Maksimović and Omanović-Mikličanin,(2017) evaluated G-IoT and Green nanotechnology and is role towards smart and sustainable agriculture (Figure 18). The IoT devices in this project are the various types of sensors for sensing and the use of RFID tags and global positioning systems for

Figure 17. GAHA Architecture (Nandyala et al., 2016)

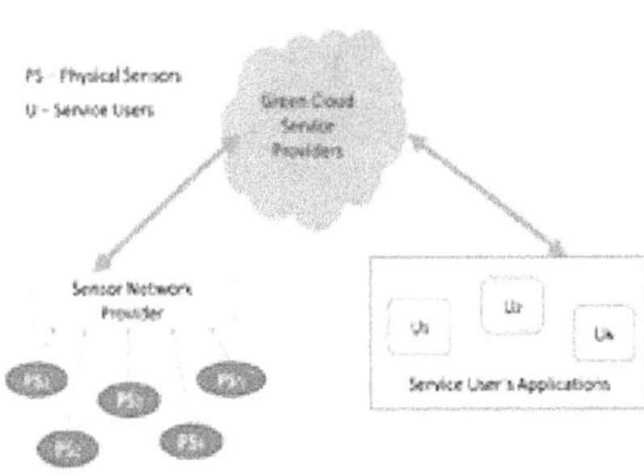

Figure 18. Sustainable Agriculture Platform (Maksimović and Omanović-Mikličanin,(2017)

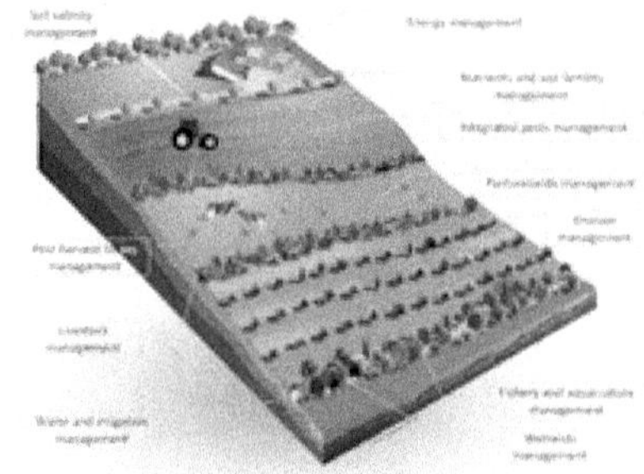

Figure 19. IoT and Machine Learning Based Green Building(Tushar et al., 2018)

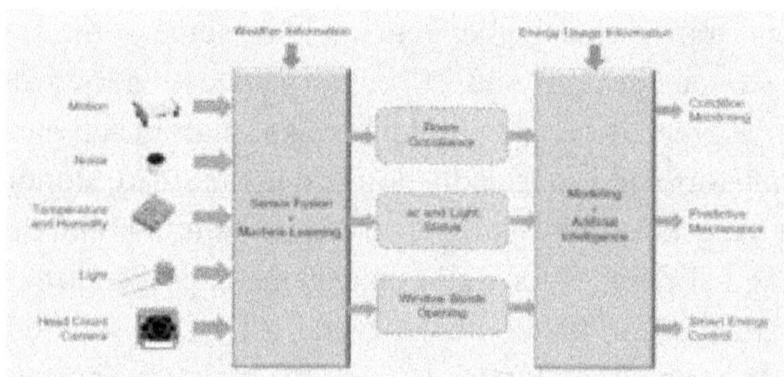

effective communication. The author summarizes that utilizing IoT in agriculture brings multitude of benefits such as real time monitoring (soil, water, pests), increased productivity, food safety, reduced cost and environment protection. With the deployment of IoT based platforms like WSN, RFID, GPS, GIS (Geographical Information System) and data analytics can significantly increase the efficiency of farming and contributes to environment and social impacts (Sarni et al., 2016).

Figure 20. IoT Based Green Campus (Wang, 2014)

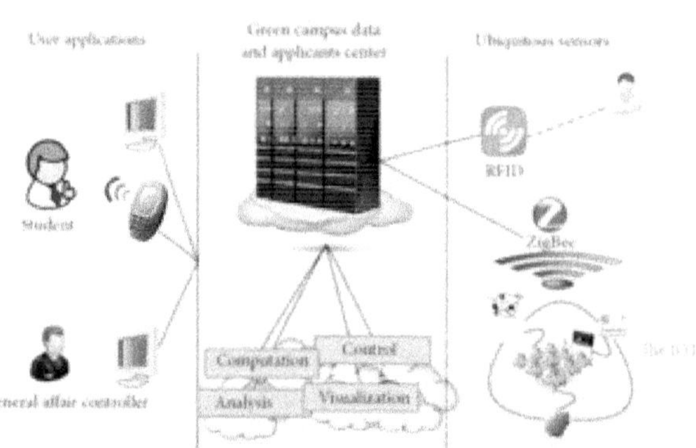

Figure 21. Smart IoT Education Environment (Maksimovic, 2017)

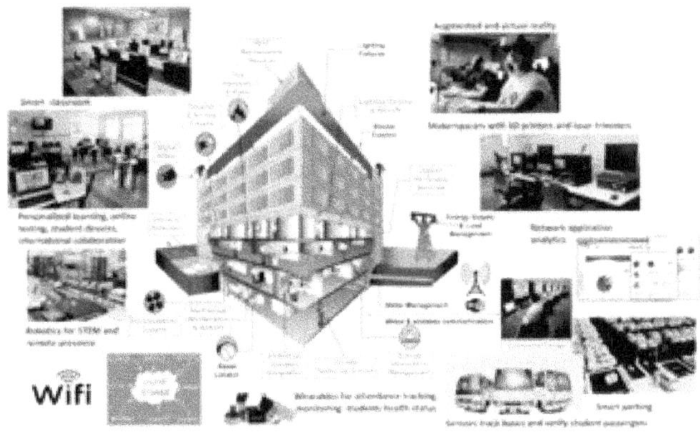

G-IoT Supported Green Building

Tushar et al., (2018) came out with a initiative of Iot based green guilding using machine learning techniques. As seen in Figure 16, the system in integrated with various types of sensors to sense the weather information and to be analyzed by machine learning algorithm. The decision is made based on the room occupancy rate, ac and the light status as well as whether window blinds is open. Based on this energy usage information is made using modelling and AI techniques. The final system is used for condition monitoring, predictive maintenance and smart energy control. There are other initiatives on green campus management as shown in Figure 19, Figure 20 and Figure 21.

Figure 22. Green Energy Management

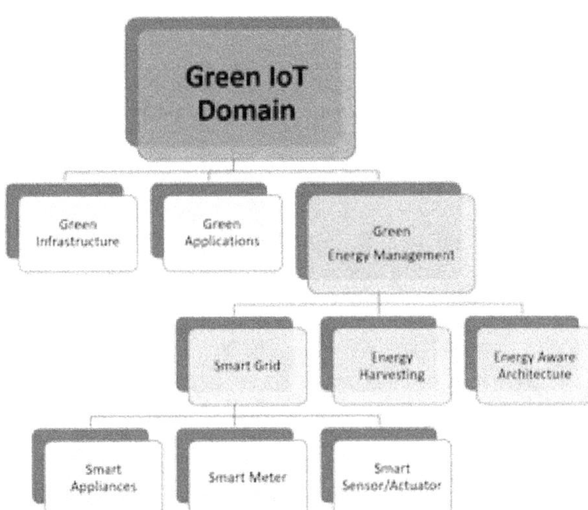

Figure 23. Energy Harvesting Domains

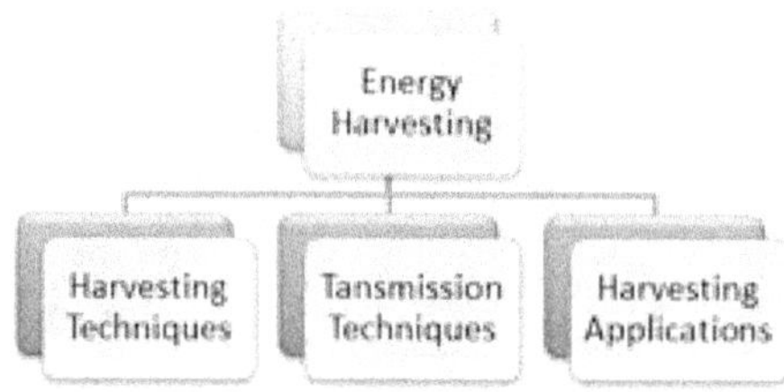

Green Energy Management

Green energy management generally falls into smart grid solutions, energy harvesting and energy aware architecture (Figure 22).

Energy harvesting is a process of scavenging or hunting energy from the environment such as solar, wind or other sources like body heat, foot strike or vibration and converting it into electricity. The harvesting rate determines the performance of wireless network and may typically increase or extend the lifetime indefinitely. This chapter aims at presenting several mechanisms and methods involved in energy harvesting in wireless and sensor network as they are becoming the prominent features for IoT. Various techniques of energy harvesting, including assessing the variety sources of energy, methods of wireless energy transfer and possible application scenarios will be discussed as shown in Figure 23.

To the best of our knowledge, there is little survey conducted on the application areas of energy harvesting in IoT domain. Although there have been a number of excellent literature reviews surrounding energy harvesting for wireless sensor network (Hande et al., 2007; Alippi & Galperti, 2008; Eu et al., 2009; Tan et al., 2009; Simjee & Chou, 2008; Liu et al., 2006; Kumar et al., 2008; Mainwaring et al,. 2002) this chapter breaks new ground by examining potential energy harvesting application for IoT which is beyond wireless sensor network (WSN) alone as IoT can be any wireless communication enabled nodes. In doing so, it is hoped that the review provided in this chapter will lead future researchers to venture on energy efficiency for IoT rather than considering energy aware routing, better MAC protocols, battery replacement, higher battery capacity or low transmission range mechanisms.

Figure 24. Energy Harvesting Techniques

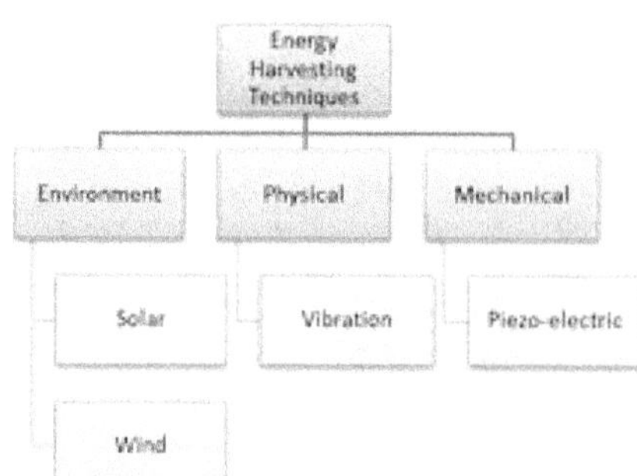

Figure 25. Adaptive system for optimal solar energy harvesting in wireless sensor network nodes (Alippi & Galperti, 2008)

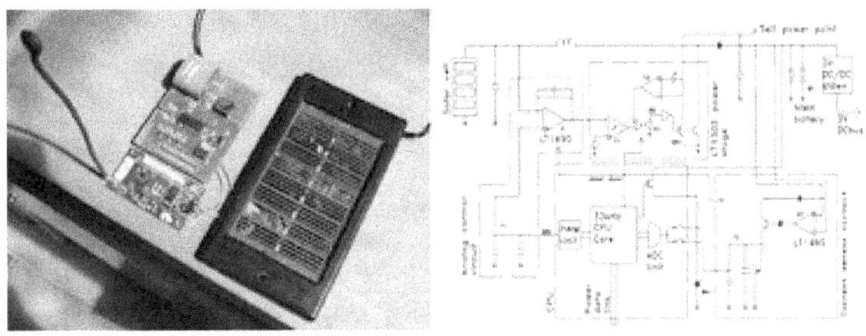

ENERGY HARVESTING TECHNIQUES

There are various methods and sources of energy harvesting among which includes solar harvesting, piezo-electric harvesting, wind harvesting, motion energy harvesting, vibration harvesting and radio frequency harvesting (Figure 24). These techniques differ in the sense that the source of each harvesting mode is different as some are derived from the environment and some of which are from mechanical and physical approach. The following section highlights some of the possible energy harvesting mechanisms with potential applications associated with it.

Solar Energy Harvesting

Solar energy harvesting is an approach of converting solar energy into electricity which consists of monitoring, storing and transferring energy (Figure 26). The harvesting capacity differs in terms of the type of solar panels, battery type and capacity, and complexity of recharge circuit. There are five different modes of solar energy harvesting namely, (i) Heliomote, (ii) Prometheus, (iii) Everlast, (iv) Sunflower and (v) Ambimax. Of these, Heliomote (Kansal et al. 2007) is a single-storage solar energy harvesting scheme using a Mica2 platform. It has an energy monitoring component to track the amount and variance of energy extracted. Figure 25 shows an example of solar power wireless sensor network and the detailed design.

Figure 26. Solar Energy Harvesting Components

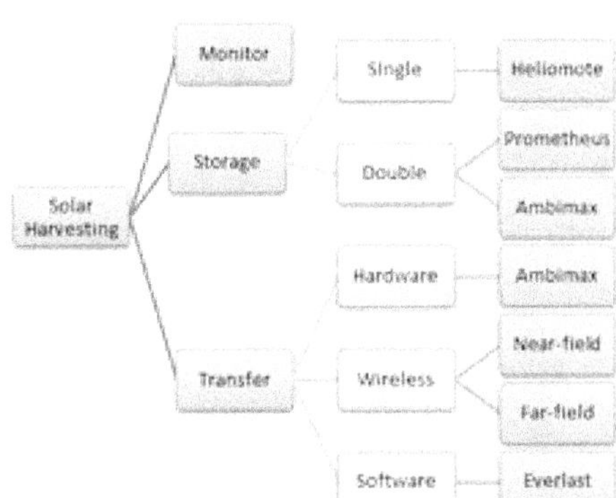

Table 1. Summary of energy harvesting by Piezoelectric, Electrostatic and Electromagnetic

Piezoelectric	Electrostatic	Electromagnetic
Advantages No separate voltage source, no mechanical stops, highest energy density, voltages of 2 to 10 volts	Easier to integrate with electronics and Microsystems, Voltages of 2 to 10 volts.	No separate voltage source, no mechanical stops.
Disadvantages Micro fabrication processes are not compatible with standard CMOS processes and piezo thin films have poor coupling.	Separate voltage source needed, mechanical stops needed.	Max. voltage of 0.1 volts, difficult to integrate with electronics and microsystems.

Prometheus (Jiang et al. 2005) is a double-storage harvesting scheme using the Telos B platform and it has both primary and secondary buffers. The primary storage charges the secondary storage when access energy is available and sensor nodes extract energy from the primary storage after which it falls back to the secondary buffer when energy at the primary buffer falls below a threshold value. Everlast (Sinje & Chou, 2006) is an integrated wireless sensor node equipped with a radio, micro-controller and energy harvesting sub-system, which does not use batteries. AmbiMax (Park et al. 2006) is a double-storage energy harvesting system using the Econode (Park et al. 2005) that can be used to harvest solar and wind energy. The design characteristic of this system is that it is capable of holding other sources like water flow and vibration. Similar to Everlast, AmbiMax has both the primary and secondary buffers to retain energy but the energy is transferred through hardware as opposed to Everlast whereby the energy is transferred through software.

Vibration Energy Harvesting

Roundy (2003) stated that electricity generation through vibration is possible using several techniques such as electromagnetic, electrostatic, or piezoelectric vibrations. These vibration methods are used for inertial sensors as well as for actuators. The power output is usually determined by the magnitude and frequency of input vibrations and the size of the devices used. The type of vibrations in this research mainly focuses on vibrations in office buildings, manufacturing and plant environments and homes. Typical examples are door frames just after the door closes, washing machines, wooden decks with people walking etc. The researcher conducted a comparison of the three conversion techniques mentioned above to narrow down the range of design possibilities. Each technique has its own advantages and disadvantages, and each of these is suitable for different applications. Based on the research, electromagnetic converters are more suitable for larger systems or systems that require wider acceleration magnitudes.

Figure 27. Piezoelectric Effect

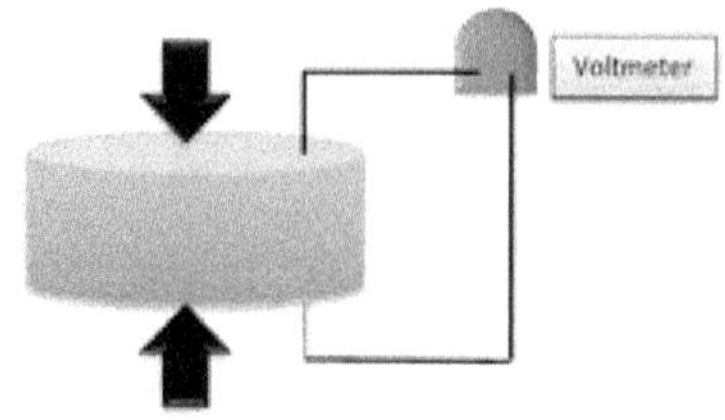

Piezoelectric converters demonstrate all of the advantages exhibited by electromagnetic converters with the additional advantage of providing useful voltages and higher practical energy densities. A comparison is summarized in Table 1. More research on vibration based energy harvesting can be found at Ooi et al (2014), Mancelos et al (2014) and Ooi & Gilbert, (2014).

Besides harvesting energy from the environment, mechanical methods such as energy harvesting from kinetic forces such as human walking and the stress of piezoelectric materials incorporated into shoes are introduced. The bending ball of the foot has been designed as an approach using piezoelectric polyvinylidene fluoride (PVDF) which is strong and rather flexible. PVDF is manufactured into thin films by stretching them to align the polymer chains, and is also electrically poled. Silver electrodes are coated on it to form as capacitor plates to collect the stressed PVDF. Another approach is by using piezo-electric, e.g., PZT (Lead Zirconate Titanate) which is rigid and the piezo-electric effect is created through the crystal structure. Energy harvesting from walking, foot strikes and finger motions is performed by deformation of the PVDF and PZT. The piezoelectric effect is described as the relative link between the mechanical stress applied and the electrical energy generated as shown in Figure. 27.

Wind Energy Harvesting

Wind energy harvesting is an approach of converting wind generated energy into electricity. The wind turbine is the basic element in generating energy using other components such as the vertical and horizontal axis of the turbine blades, the rotor and shafts. AmbiMax (Park et al. 2005) on Eco node is one of the implementations using wind energy harvesting mechanism. A biologically-inspired mechanism (Hobbs et al. 2012) inspired by tree swaying concept is applied to wind energy harvesting. A liner array of cylinder vibration attached to the piezoelectric energy transducers is proposed in this work.

ENERGY TRANSMISSION TECHNIQUES

Wireless energy transmission has emerged as an outstanding research area due to the need for infrastructure-less environment to support IoT. The ultimate goal of IoT would be achieved when there is no more need for wireless devices to be physically plugged to power source for energy. This limits the mobility of IoT devices mainly in remote locations such as forest and underwater. Wireless energy transmission works by two principles, near-field and far-field which is distinguished based on the transmission distance between the transmitting source and the receiving source (Balasubramaniam, 2013).

Mascareñas et al. (2006) explores a new alternative to the conventional battery constrained sensor nodes by exploiting mobile host. In this architecture (Figure. 28), energy is transferred wirelessly to sen-

Figure 28. An Architecture of Wireless Energy Transmission

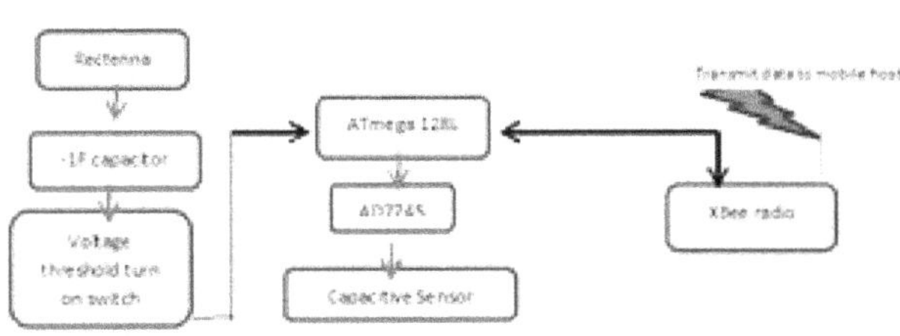

Figure 29. Wireless Energy Transmission Between The Antennas

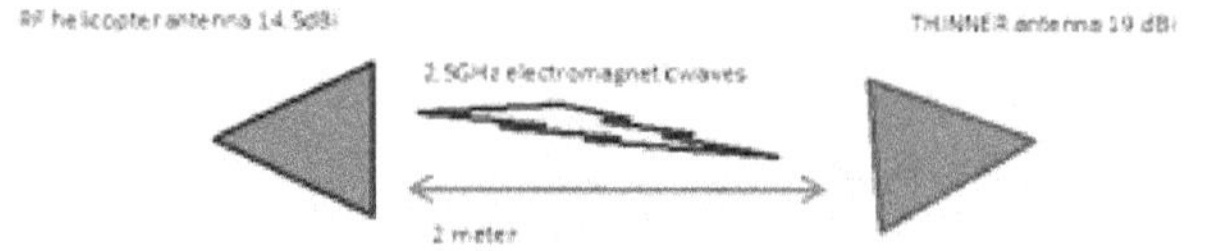

Figure 30. Power Monitoring Device(Lee, V. C., 2012).

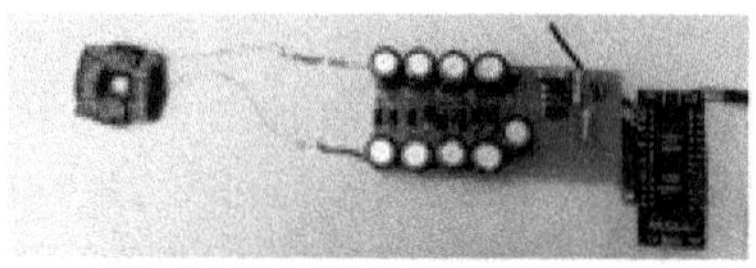

sor nodes on an as-needed basis. The model used is similar to RF energy transfer to passive RFID tags. Besides, the mobile host is also responsible for collecting sensor data from nodes to perform centralized data aggregation and processing. The mobile host in this architecture is modified to be a remote controlled (RC) helicopter. The sensor nodes in this work are called THINNER in which its antenna is tuned to the RC helicopter antenna to receive the wireless energy as well as to send sensor data. The mechanism in this approach works by mobile host transmitting a 2.5 Ghz signal through a 14.5dB antenna and the 19dB patch antenna at the sensor nodes receive these signals. The RF-to-DC conversion and multiplication is performed before it is fed into a capacitor for storage purpose. THINNER is designed specifically to be powered by wireless energy transfer by a mobile host and employs a capacitance-to-digital converter for the need to store peak displacement values. The Xbee radio is engaged in order to communicate with unmanned aerial vehicle (UAV) and also to communicate with other sensors.

One of the main constraints of this approach is the high energy consumption of THINNER node as it is operated by power being transferred wirelessly due to free space loss and requirement for power at other components such as the capacitor and storage capacitor. Typically it takes approximately about 76-3-5s for the mobile node to be within the proximity of a sensor in order to transfer energy wirelessly for the nodes to operate properly. The RF energy is transferred at 2.5 GHz and this figure is chosen for the cost factor to reduce the hardware cost while increasing the cost to purchase high gain antennas. The distance between the THINNER antenna and the mobile host antenna is approximated at 2 meter so that the RF helicopter would be able to reach the antenna of the THINNER as shown in Figure.29.

This work has contributed in the development of a novel capacitance-based wireless sensor node and peak displacement sensor. Through development and results obtained, it is feasible to deploy such mobile host for data collection and powering sensor nodes using wireless energy transmission. Moreover,

Figure 31. Energy Harvesting for Wireless Sensor Network (Lee, V. C., 2012).

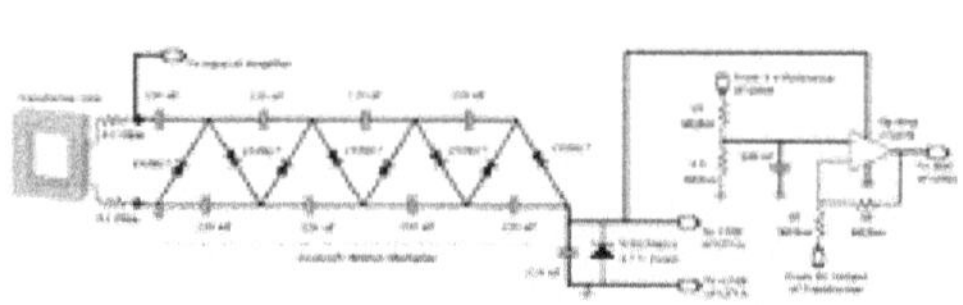

this work can contribute to post-disaster applications such as earth quake and terrorist attack for rapid assessment results in life safety and economic reasons.

Separately, Aristeidis et al. (2008) analyses the ways in which wireless non-radiative mid-range energy is transferred. Rapid development of autonomous devices such as laptops, cellular phone, house-hold devices brings into attention the need to revisit the concept of wireless energy transfer. Some non-radiative mechanisms using magnetic induction is possible but restricted by distance and low power energy transfers. The authors investigate the feasibility of using long-lived oscillatory resonant electromagnetic modes with slowly-evanescent field patterns wireless energy transfer. This method works by the principle of resonant coupling in which two same-frequency resonant objects couple each other resulting in that energy can be coupled between objects in near field.

Lee, (2012) has conducted a research on energy harvesting for wireless sensor network. The author proposed a novel, plug through power monitoring system for commercial and home use. The device use current sense transformer to detect any appliance power consumption which wirelessly couples magnetic energy from appliance to output an electromotive force voltage. Then, the voltage signal will be relayed to the analog-to-digital converter of GINA radio mote, which transmit the data to laptop or mobile device wirelessly. It allows user to view real-time power usage by using handheld device. The device is 10 times lower cost and 45 times smaller compared to other power monitoring devices in the present market. A power monitoring device based on energy harvesting is shown in Figure 30 and the schematic circuit view is shown in Figure.31.

ENERGY HARVESTING APPLICATIONS

IoT applications are developed to create a smarter surrounding as smart environment/city, smart water, smart metering, eHealth, home automation and more. These applications are classified as wireless sensor network, body area network (wearables), and wireless electronics/networks (Figure. 32) although sensor plays key role in IoT. One key aspect for these applications to be unique is to provide autonomic capabilities for energy management. The following section highlights some of the energy harvesting applications for IoT which provides autonomic capabilities in terms of energy management.

Figure 32. IoT Application Domains

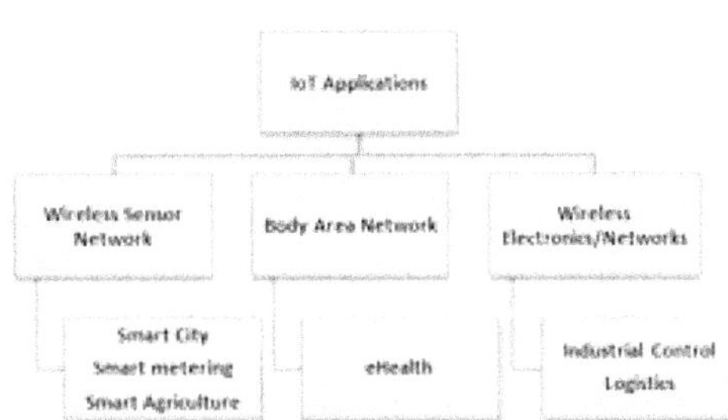

Table 2. Wireless Sensor Networks Powered by Ambient Energy Harvesting (WSN-HEAP) Seah et al. (2009)

Technology	Power Density (μ W/cm²)	Energy Harvesting Rate (mW)	Duty Cycle (%)
Vibration - electromagnetic	4.0	0.04	0.05
Vibration - piezoelectric	500	5	6
Vibration – electrostatic	3.8	0.038	0.05
Thermoelectric	60	0.6	0.72
Solar – direct sunlight	3700	37	45
Solar – indoor	3.2	0.032	0.04

Energy Harvesting in Wireless Sensor Network

Alippi & Galperti (2008) has conducted a research on adaptive system for optimal solar energy harvesting in wireless sensor network nodes. They feel that it was finite resources for wireless sensor network that is limited by energy supply which can only be provided using battery. Therefore, the idea of energy harvesting mechanism has been taken for long time operational life with solar energy. They propose a low power maximum power point tracker (MPPT) circuit that is designed for wireless sensor nodes. The circuit will convey solar energy into rechargeable batteries in every situation even cloudy or under low light condition. The design is considered effective as it works in critical weather conditions where traditional solutions do not work. MPPT circuit was designed from power converter to digital control system and work in low power operation independently from nature of solar panel and battery type.

In another research carried out by Kansal & Srivastava (2003), a distributed framework for sensor network is presented to learn the energy environment and localized algorithm to use the information for sharing between nodes. This framework enables system to exploit energy resources efficiently and increase the lifetime. They use EEHF for solar energy based sensor network. It is a way towards development of optimal set of methods to completely task distribution in the network with available energy. In order to fully resolve the harvesting problem, it needs to develop distributed method to achieve maximum lifespan and requires to be tested on real-time system development.

Besides, Seah et al. (2009) had published work on Wireless Sensor Networks Powered by Ambient Energy Harvesting (WSN-HEAP). Substantial research efforts have been used for designing energy-efficient networking protocol to maximize WSN lifespan. However, there are some WSN application required sensors to operate for longer duration after the implementation. For instance, in structural health monitoring for critical infrastructures and buildings where sensors are installed at locations almost inaccessible by humans, the alternative solution is to use ambient energy harvesting. Wireless sensor network uses ambient energy harvesting as a promising technology for many sensing application in reducing the need for battery replacement. However, there are challenges of designing wireless sensor node such as transmission power control, MAC protocol, routing and reliable of data delivery. Moreover, energy harvesting from environment is highly dependent on local environmental factors, therefore it requires further study to exploit this frontier so that energy harvesting can be optimum under any condition. Table 2 presents the achievable duty cycle by micaz with 10CM2 harvesting material.

Table 3. Comparison of energy harvesting systems for wireless sensor network (Gilbert & Balouchi (2008))

Device Number	Operating Mode / Materila	Output Power (μW)	Frequency (H_z)	Amplitude (ms^{-2})	Normalised Power (μW)	Volume (mm^3)
1	Vibration - pz	2.1	80.1	2.3	0.5	125
2	Vibration – pz	210	120	2.5	28	1000
3	Vibration – pz	375	120	2.5	50	1000
4	Vibration – pz	0.6	900	9.81	0.0007	2
5	Vibration – es	3.7	30	50	0.005	750
6	Vibration – es	1052	50	8.8	27	1800
7	Vibration – es	70	50	9.2	1.7	32
8	Vibration – cm	0.3	4400	382	$4.7x10^{-8}$	5.4
9	Vibration – cm	180	322	2.7	7.7	840
10	Vibration - cm	4000	100	0.4	25000	30 000
11	PVDF shoe stave				1300	5000
12	PZT dimorph				8400	1700
13	Em shoe insert				60000	56000
14	Em shoe insert				90000	97500
15	EAP shoe insert				1 000000	50 000
16	Thermoelectrc				50	41
17	PV cell outdoor				20000	500 000
18	PV cell indoor				1500	500 000

A research on comparison of energy harvesting systems for wireless sensor network has been done by Gilbert & Balouchi (2008). In this chapter, it discuss about the characteristic and energy requirement of sensor network nodes, access range of ambient energy sources and figure out the characteristics of energy conversion devices. They proposed a way to normalize power density by comparing diverse source and conversion method. The conversion method mainly uses three energy harvesting device such as piezoelectric, electrostatic and electromagnetic. In term of considering published data, they found

Figure 33. PicoRadio Supports Ad Hoc Ultra-Low Power Wireless Networking (Rabaey et al. (2000))

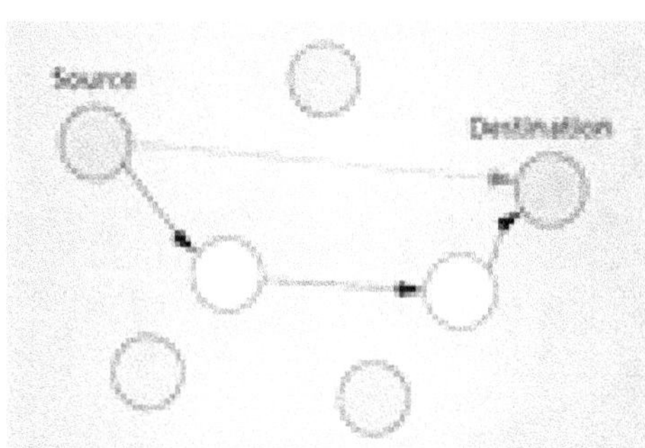

Figure 34. Energy Harvesting and Remote Powering for Implantable Biosensors (Olivio et al. (2011).

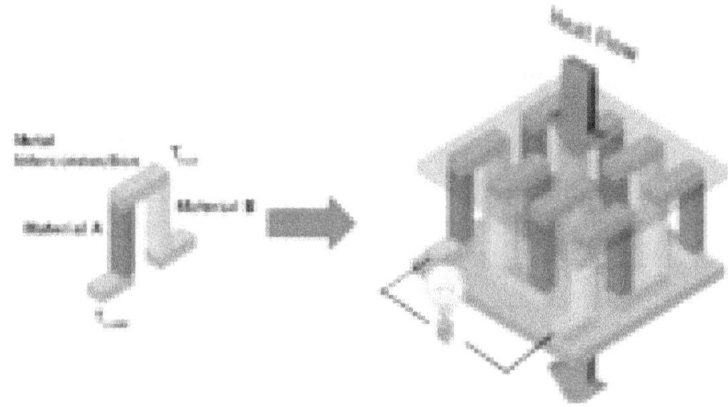

that there is large range of power levels and device volume but similar power densities are archived for best performing device in every size range. Table 3 presents a comparison of energy harvesting devices.

Park et al. (2008) has conducted a work on energy harvesting for structural health monitoring sensor networks. The work reviews the development of energy harvesting for structural health monitoring (SHM) sensing system. The system is beneficial as it can improve the reliability, safety and ownership cost engineering systems by autonomously monitoring the condition and sensing the damage before entering into critical state. However, the development of energy harvesting for large-scale source such as wind turbines exist due to high commercial products but energy harvesting for embedded sensing system is still in fancy stage. The authors argue that major drawback in the field of energy harvesting is energy generated by harvesting device is too small to direct power for most of the devices.

Kansal et al. (2007) has done research on power management in energy harvesting sensor networks. The work reviews the importance of power management in sensor network and how energy harvesting implemented on powering sensor networks. Energy harvesting has an energy neutral mode to supplement energy supply in battery powered systems. The authors discuss some challenges that is faced in this mode and modelling the characteristics of energy generation sources and loads. It presents theoretical foundations, various practical methods to optimize performance in energy neutral mode.

Figure 35. Human power-based energy harvesting strategies for mobile electronic devices (Jia & Liu (2009)

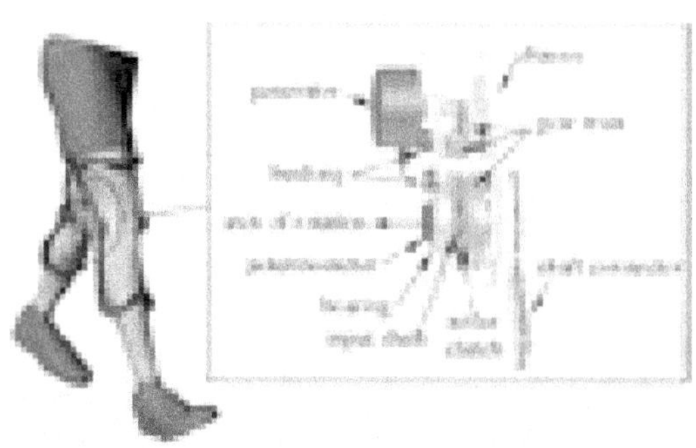

Figure 36. Energy harvesting for mobile and wireless electronic (Paradiso & Starner, 2005)

Rabaey et al. (2000) has worked on PicoRadio Supports Ad Hoc Ultra-Low Power Wireless Networking. PicoRadio application uses sensor monitoring network for environmental control in office buildings, warehouse inventory, patient monitoring and diagnostic, interactive toys etc. PicoNodes were used as self-contained and self-powered by using energy scavenging. Harvesting ambient energy will work only when applicability within environment envision for PicoNodes and size constraint of one cm chip. In the implementation of wireless ad hoc, sensor, monitor and actuator network, they found a problem which is minimizing energy consumption. The authors only proposed an ultra-low power node by optimizing all layers of protocol. The author argues that sending data through multiple intermediate hops is more energy efficient than sending through one longer hop (Figure.33).

Energy Harvesting in Body Area Networks

A study on Energy Harvesting and Remote Powering for Implantable Biosensors has been done by Olivio et al. (2011). Various popular energy harvesting techniques for implantable biosensors have been conducted. Among the areas explored were kinetic, thermoelectric, fuel cells, infrared radiation, low frequency magnetic field and inductive links. Inductive links are considered viable for implanted biosensors because communication between the internal sensors and external control devices are performed easily. High frequency inductive link produces higher data rates and have higher misalignment tolerance as compared to low frequency inductive links. The connection of multiple thermocouples placed in series (left) in electrical and placed parallel in therma forms a thermopile as shown in Figure.34. Although this can be an advantage but this technique has lower efficiency and requires higher transmission power.

Figure 37. Energy harvesting vibration sources for microsystem applications (Beeby et. al. (2006))

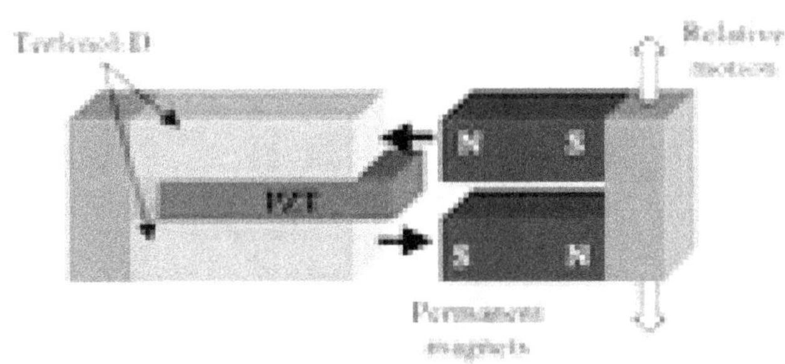

Figure 38. Powering MEMS portable devices with non-regenerative and regenerative power supply systems with special emphasis on piezoelectric energy harvesting systems (Cook et al. (2008))

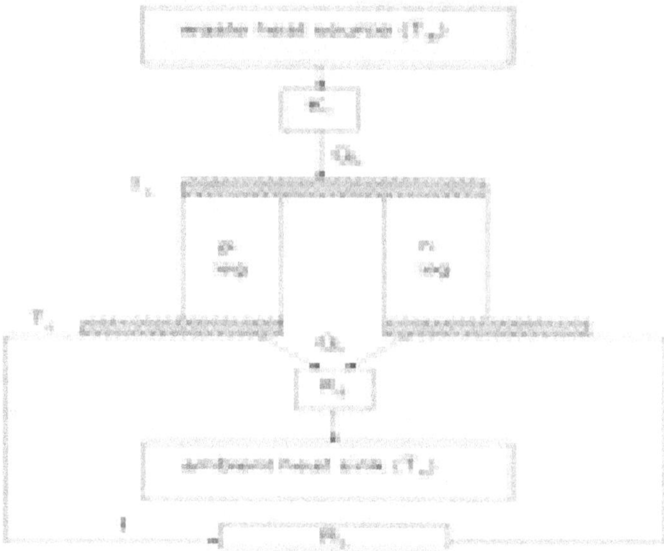

Jia & Liu (2009) conducted a research on human power-based energy harvesting strategies for mobile electronic devices. The authors proposed energy harvesting through daily activities such as joint rotation and enforcement of body weight for mobile or medical device as shown in Figure.35 . This technique seems to provide longer energy capacity since most of the mobile devices consume lesser energy. Therefore energy can be continuously harvested from human body and converted to mobile device and probably these devices no longer need recharging in the future. However, a good wearable energy harvesting system requires light weight, high power density, good regulation ability, convenient power delivery and minimum user effort. Currently, the existing technology is still unavailable to achieve the above objective. Seyedi & Sikdar (2208) have explored efficient energy transmission techniques for body sensor network using energy harvesting capabilities. An effective decision making strategy via scheduling on the transmission mode is proposed to improve the coverage quality. Three different approaches of energy transmission scheduling with varied energy levels were proposed for body sensor network with energy harvesting capability.

Figure 39. (A&B) Energy harvesting from piezoelectric material that is fully integrated into footwear (Rocha et al., 2010)

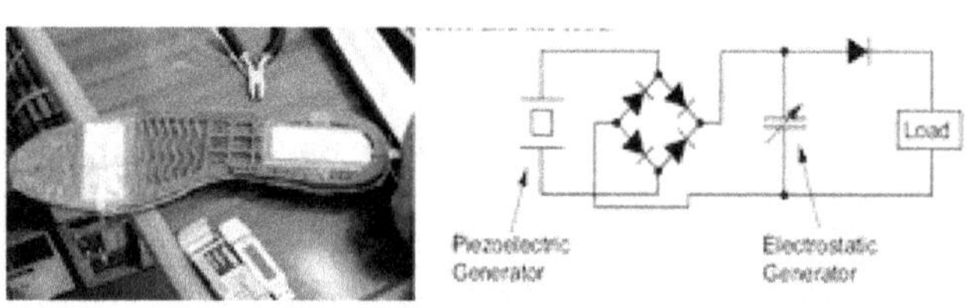

Energy Harvesting in Wireless Electronics

Energy harvesting for mobile and wireless electronic has been conducted by Paradiso & Starner (2005). The authors mentioned that battery energy is slowest trend in mobile computing in 1990. However, the revolution of battery form factor, its energy density unable to scale exponentially. To overcome the trend, on-going power management developments allow battery-powered electronics to live longer. It is not enough, as energy harvesting is the best approaches to scavenge energy from waste heat or vibration. A lot of work has been carried out on low-power electronics, wireless standards and miniaturization of devices and technology to boost the trend in wireless sensor network and mobile devices. There are several techniques to harvest energy from different environment to power electronics, and devices limit the power yield in pervasive computing. As such, Figure.36a shows the magnetic generators embedded in shoes and a strap-on overshoe managed to produce about an average of 250 mW during normal walk. Whereas Figure. 36b reveals how an assembly hosting twin motor-generators embedded into a shoe that managed to produce about 60 mW.

Beeby et. al. (2006) has worked on energy harvesting vibration sources for microsystem applications. It is state-of-the art in vibration energy harvesting for wireless and self-powered microsystems. The vibration-powered generator is similar like spring and mass system. The generator consists of three main transduction concepts which are piezoelectric, electromagnetic and electrostatic Figure.37. Vibration-powered wireless sensor is very beneficial as it can be used in ambient intelligence, medical implants and smart clothing.

A research on powering MEMS portable devices with non-regenerative and regenerative power supply systems with special emphasis on piezoelectric energy harvesting systems has been conducted by Cook et al. (2008). Their work focuses on approaches for powering MEMS using non-regenerative and regenerative power supplies. The work also reviews on the fundamental of piezoelectric energy harvesting and future trend of piezoelectric energy harvesting technology. The authors found that most of the MEMS-scaled piezoelectric energy harvesting device does not produce enough current to power the systems directly. To solve the problem, a hybrid system consisting of piezoelectric energy harvesting by improving the charge density using suitable material and a proper storage devices is proposed. Moreover, the authors propose to reduce the device size and increase the number of coupled devices in order to allow for higher frequency oscillation of piezoelectric devices to increase output power. A schematic view of the thermoelectric system is shown in Figure. 38.

Rocha et al. (2010) has conducted a research on energy harvesting from piezoelectric material that is fully integrated into footwear. Their research is about using piezoelectric polymers that is capable of harvesting energy via human walking and shoe fabrication process capable of producing and collecting energy. Figure 39a shows the first protoptype developed using two piezoelectric polymer films above the shoe sole. The researchers used electronic β-PVDF as energy harvesting element that is injected into bicolored sole together with electronic needed to improve the energy transfer and storage efficiency. The authors also included electrostatic generator to increase energy harvesting Figure 39b). It was of value by using electro-active polymers because such technology is low cost, flexible and can be easily integrated into elements such as shoes and clothes.

Mitcheson et al. (2008) has worked on energy harvesting from human and machine motion for wireless electronic devices. Their work reviews about the principle and state-of-art in motion driven tiny energy harvester and trends, existing application and future work. The motion driven energy harvester uses micro generator to apply direct use of forces and inertial forces working on proof mass. The authors

Figure 40. Future research direction towards Green IoT

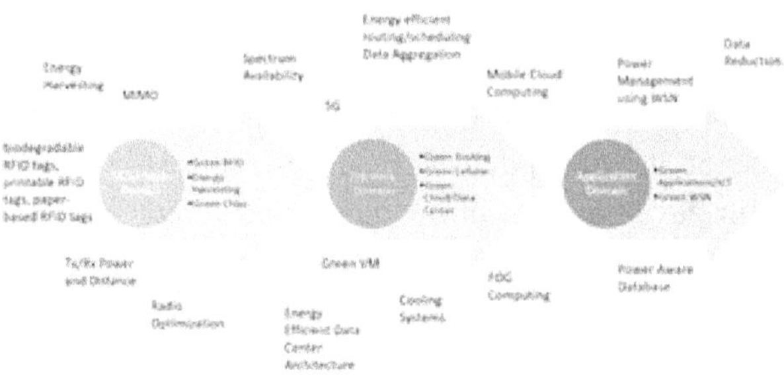

discovered that there are two existing companies offering electromagnetic harvester which is Perpetuum and Ferro Solution, it was similar to the device but their proposal has greater potential. Practical utilization of harvesting devices is depending on the efficiency of power processing and control circuitry. Lastly, an effective motion energy harvesting requires combination of complete wireless system design and power aware operation.

Discussion: Energy Harvesting

"Internet of Things via Energy Harvesting" is a famous topic and current trend that is being explored by many researchers. The main challenge for design engineers is how to fully manage the power to IoT nodes by eliminating the last wire. One such potential application is solar-powered street lighting system (Ramachandran. T et al, 2014). Energy harvesting is an ideal solution for IoT devices as it is renewable and self-sustained once it is deployed. However, such energy is not available all the time so the IoT devices should be designed in such a way to be able to be adaptable and autonomous. Therefore, mimicking biologically-inspired (Saleem et al. 2010, Dorigo et al 1996; Boonma and Suzuki, 2007) autonomous system capabilities could be a potential research in energy harvesting for IoT.

Various energy harvesting mechanisms have been examined for IoT domain in previous sections. Among the methods, solar energy harvesting is widely popular due to the availability of solar source. Solar energy could be harvested during daylight and stored in secondary storage for night or indoor use. Solar energy harvesting is limited in countries that have limited daylight such as Alaska and Sweden during certain seasons. Besides that it is limited for indoor use such as tunnel where sunlight may not be able to penetrate. But this can be overcomed by using secondary storage whereby harvested solar energy is stored in a secondary buffer for future use. However, the reliability of such approach depends on the storage capacity and energy transmission mechanism to such indoor environments. Vibration and wind energy harvesting works on the same mechanism of mechanical and physical elements of movements. Piezoelectric is widely used in these two approaches but some researchers argue that the energy generated by this material is too small to directly powering the devices. Although using secondary energy storage could solve the issue, capacitor-based storage could only provide short power surge. Sodano et al. (2004a) has proposed that rechargeable battery could be used together with piezoelectric materials to substitute capacitor.

Wireless power transmission technology has been there for some time, but it is still at its premature stage due to the challenges and limitations of such technology. Wireless power transmission is highly restricted by the distance between the antennas. Based on Choi et al. (2011), the transmission efficiency of wireless energy transmission resonators decreases proportionally over distance irrespective of the receiving coil size. Researchers are working on ways to reduce the distance whilst increasing the transmission efficiency. This is possible by improving the coupling efficiency of receiving circuit and employing magnetic resonance. Although distance can be the pitfall in energy transmission efficiency, advancements in microprocessor technology has increased power efficiency thus reducing power consumption. This can substantiate that small volume of energy generated could be able to power wireless devices. The challenges and design principles of such technologies can be found at Gungor & Hancke (2009).

FUTURE RESEARCH

As for green technology Figure 40 highlights some of the research directions that can be focused to achieve the goal. The work can be at node to node domain, network domain and application domain. In the machine to machine domain a more energy aware RFID design should be given importance with some techniques such as biodegradable RFID, printable RFID and paper-based RFID. Energy harvesting can also be deployed and transmission and receiving power should be studied to reduce energy usage. At the network various initiatives can be taken such energy efficiency in cloud, data center, utilizing FOG and mobile cloud, energy aware routing algorithms, scheduling and data aggregation as well as improving spectrum availability and utilizing 5G technology. As for the application domain, when designing applications, researchers should be cautious on how energy utilization can be reduced. Database designs should be more power aware and WSN can deployed for power management.

CONCLUSION

In this chapter, a lot of focus has been given on energy harvesting as this can be the main contributing factor towards green energy. In a smart city design the infrastructure domain, application areas as well as energy management sector should be always focusing on the green concept. Since IoT would be the main pillar in smart city projects, more emphasis should be given how to deploy green concepts in the IoT design and implementations. The green IoT Ecosystem for smart city proposed in his chapter can be guideline for future IoT designs in the smart city areas. This ecosystem can be a general framework and can also be applied to any domain rather than just emphasising on smart cities.

REFERENCES

Aarts, E., & Wichert, R. (2009). *Ambient intelligence*. Springer Berlin Heidelberg.

Abrol, A., & Jha, R. K. (2016). Power optimization in 5G networks: A step towards GrEEn communication. *IEEE Access: Practical Innovations, Open Solutions*, *4*, 1355–1374. doi:10.1109/ACCESS.2016.2549641

Albreem, M. A. (2015, April). 5G wireless communication systems: Vision and challenges. In *2015 International Conference on Computer, Communications, and Control Technology (I4CT)* (pp. 493-497). IEEE. 10.1109/I4CT.2015.7219627

Albreem, M. A., El-Saleh, A. A., Isa, M., Salah, W., Jusoh, M., Azizan, M. M., & Ali, A. (2017, November). Green internet of things (IoT): An overview. In *2017 IEEE 4th International Conference on Smart Instrumentation, Measurement and Application (ICSIMA)* (pp. 1-6). IEEE.

Alippi, C., & Galperti, C. (2008). *An Adaptive System for Optimal Solar Energy Harvesting in Wireless Sensor Network Nodes. IEEE Transactions on Circuits and Systems, 55*(6), 1742–1750. doi:10.1109/TCSI.2008.922023

Alsamhi, S., Ma, O., Ansari, M. S., & Meng, Q. (2018). *Greening internet of things for smart everythings with a green-environment life: A survey and future prospects.* arXiv preprint arXiv:1805.00844.

Alsamhi, S. H., Ma, O., Ansari, M. S., & Meng, Q. (2019). Greening internet of things for greener and smarter cities: A survey and future prospects. *Telecommunication Systems, 72*(4), 609–632. doi:10.100711235-019-00597-1

Anastasi, G., Conti, M., Di Francesco, M., & Passarella, A. (2009). Energy conservation in wireless sensor networks: A survey. *Ad Hoc Networks, 7*(3), 537–568. doi:10.1016/j.adhoc.2008.06.003

Anastasi, G., Conti, M., Di Francesco, M., Passarella, A., Denko, M., & Yang, L. (2006). How to prolong the lifetime of wireless sensor networks. *Mobile Ad Hoc and Pervasive Communications*, 1-26.

Aristeidis, K., Joannopoulos, J. D., & Marin, S. (2008). *Efficient wireless non-radiative mid-range energy transfer*. Elsevier.

Arshad, R., Zahoor, S., Shah, M. A., Wahid, A., & Yu, H. (2017). Green IoT: An investigation on energy saving practices for 2020 and beyond. *IEEE Access: Practical Innovations, Open Solutions, 5*, 15667–15681. doi:10.1109/ACCESS.2017.2686092

Ashraf, A., & Porres, I. (2018). Multi-objective dynamic virtual machine consolidation in the cloud using ant colony system. *International Journal of Parallel. Emergent and Distributed Systems, 33*(1), 103–120. doi:10.1080/17445760.2017.1278601

Azevedo, J. A. R., & Santos, F. E. S. (2012). Energy harvesting from wind and water for autonomous wireless sensor nodes. *IET Circuits, Devices & Systems, 6*(6), 413–420. doi:10.1049/iet-cds.2011.0287

Balasubramaniam, R. (2013). *Wireless energy transfer using resonant inductive coupling*. In *Proceedings of Sixth IRAJ International Conference* (pp. 44-47). Pune, India: Power Publishers.

Beeby, S. P., Tudor, M. J., & White, N. M. (2006). Energy harvesting vibration sources for microsystems applications. *Measurement Science & Technology, 17*(12), R175–R195. doi:10.1088/0957-0233/17/12/R01

Boonma, P., & Suzuki, J. (2007). BiSNET: A biologically-inspired middleware architecture for self-managing wireless sensor networks. *Computer Networks, 51*(16), 4599–4616. doi:10.1016/j.comnet.2007.06.006

Bruneo, D., Lhoas, A., Longo, F., & Puliafito, A. (2014). Modeling and evaluation of energy policies in green clouds. *IEEE Transactions on Parallel and Distributed Systems, 26*(11), 3052–3065. doi:10.1109/TPDS.2014.2364194

Burrus. (2014). *The Internet of Things Is Far Bigger Than Anyone Realizes*. Retrieved from Wired web site: https://www.wired.com/2014/11/the-internet-of-things-bigger/

Chan, C. A., Gygax, A. F., Wong, E., Leckie, C. A., Nirmalathas, A., & Kilper, D. C. (2013). Methodologies for assessing the use-phase power consumption and greenhouse gas emissions of telecommunications network services. *Environmental Science & Technology*, *47*(1), 485–492. doi:10.1021/es303384y PMID:23211093

Chean, L. T., Ponnusamy, V., & Fati, S. M. (2018, April). Authentication scheme using unique identification method with homomorphic encryption in Mobile Cloud Computing. In *2018 IEEE Symposium on Computer Applications & Industrial Electronics (ISCAIE)* (pp. 195-200). IEEE. 10.1109/ISCAIE.2018.8405469

Chih-Lin, I., Rowell, C., Han, S., Xu, Z., Li, G., & Pan, Z. (2014). Toward green and soft: A 5G perspective. *IEEE Communications Magazine*, *52*(2), 66–73. doi:10.1109/MCOM.2014.6736745

Choi, J., Cho, J. K., & Seo, C. (2011). Analysis on Transmission efficiency of wireless energy transmission resonator based on magnetic resonance. In Microwave Workshop Series on Innovative Wireless Power Transmission: Technologies, Systems, and Applications (IMWS), 2011 IEEE MTT-S International (pp. 199-202). IEEE. doi:10.2528/PIERM11050903

Chong, F. T., Heck, M. J., Ranganathan, P., Saleh, A. A., & Wassel, H. M. (2013). Data center energy efficiency: Improving energy efficiency in data centers beyond technology scaling. *IEEE Design & Test*, *31*(1), 93–104. doi:10.1109/MDAT.2013.2294466

Cook-Chennault, K. A., Thambi, N., & Sastry, A. M. (2008). Powering MEMS portable devices—A review of non-regenerative and regenerative power supply systems with special emphasis on piezoelectric energy harvesting systems. *Smart Materials and Structures*, *17*(4), 043001. doi:10.1088/0964-1726/17/4/043001

Dayarathna, M., Wen, Y., & Fan, R. (2015). Data center energy consumption modeling: A survey. *IEEE Communications Surveys and Tutorials*, *18*(1), 732–794. doi:10.1109/COMST.2015.2481183

Dorigo, M., Maniezzo, V., & Colomi, A. (1996). Ant system: Optimization by a colony of cooperating agents. *IEEE Transactions on Systems, Man, and Cybernetics. Part B, Cybernetics*, *26*(1), 29–41. doi:10.1109/3477.484436 PMID:18263004

Eu, Z. A., Tan, H. P., & Seah, W. K. (2011). Design and performance analysis of MAC schemes for wireless sensor networks powered by ambient energy harvesting. *Ad Hoc Networks*, *9*(3), 300–323. doi:10.1016/j.adhoc.2010.07.014

Eu, Z. A., Tan, H. P., & Seah, W. K. G. (2009). *Routing and relay nodeplacement in wireless sensor networks powered by ambient energy har-vesting. In Proc. of the IEEE Wireless Communications & Networking Conference*. WCNC.

Farahnakian, F., Ashraf, A., Pahikkala, T., Liljeberg, P., Plosila, J., Porres, I., & Tenhunen, H. (2014). Using ant colony system to consolidate VMs for green cloud computing. *IEEE Transactions on Services Computing*, *8*(2), 187–198. doi:10.1109/TSC.2014.2382555

Feng, W., Alshaer, H., & Elmirghani, J. M. (2010). Green information and communication technology: Energy efficiency in a motorway model. *IET Communications*, *4*(7), 850–860. doi:10.1049/iet-com.2009.0153

Gapchup, A., Wani, A., Wadghule, A., & Jadhav, S. (2017). Emerging trends of green IoT for smart world. *International Journal of Innovative Research in Computer and Communication Engineering, 5*(2), 2139–2148.

Gilbert, J. M., & Balouchi, F. (2008). Comparison of energy harvesting systems for wireless sensor networks. *International Journal of Automation and Computing, 5*(4), 334–347. doi:10.100711633-008-0334-2

Gungor, V. C. & Hancke, G. P. (2009). Industrial wireless sensor networks: Challenges, design principles, and technical approaches. *Industrial Electronics, IEEE Transactions on, 56*(10), 4258-4265.

Hande, A., Polk, T., Walker, W., & Bhatia, D. (2007). Indoor solar energyharvesting for sensor network router nodes. *Microprocessors and Microsystems, 31*(6), 420–432. doi:10.1016/j.micpro.2007.02.006

Hobbs, W. B., & Hu, D. L. (2012). Tree-inspired piezoelectric energy harvesting. [tree inspired]. *Journal of Fluids and Structures, 28*, 103–114. doi:10.1016/j.jfluidstructs.2011.08.005

Internet of Thing. (1994-2014). Retrieved from Tech Target web site: https://whatis.techtarget.com/definition/Internet-of-Things

Internet of Thing (IoT). (n.d.). Retrieved from Cisco web site: https://www.cisco.com/web/solutions/trends/iot/overview.html

Internet of Things. (2014). Retrieved from Microsoft web site: http://www.microsoft.com/windowsembedded/en-us/internet-of-things.aspx

Jia, D., & Liu, J. (2009). Human power-based energy harvesting strategies for mobile electronic devices. *Frontiers of Energy and Power Engineering in China, 3*(1), 27–46. doi:10.100711708-009-0002-4

Jiang, X., Polastre, J., & Culler, D. (2005). Perpetual Environmentally Powered Sensor Networks. *Fourth International Symposium on Information Processing in Sensor Networks, 36*(5), 463–468.

Kansal, A., Hsu, J., Zahedi, S., & Srivastava, M. B. (2007). Power management in energy harvesting sensor networks. *ACM Transactions on Embedded Computing Systems, 6*(4), 32. doi:10.1145/1274858.1274870

Kansal, A., & Srivastava, M. B. (2003). An environmental energy harvesting framework for sensor networks. In *Low Power Electronics and Design, 2003. ISLPED'03, Proceedings of the 2003 International Symposium on* (pp. 481-486). 10.1145/871506.871624

Koutitas, G. (2010). Green network planning of single frequency networks. *IEEE Transactions on Broadcasting, 56*(4), 541–550. doi:10.1109/TBC.2010.2056252

Lee, R.-G., Chen, K.-C., Chiang, S.-S., Lai, C.-C., Liu, H.-S., & Wei, M.-S. (2006). A Backup Routing with Wireless Sensor Network for Bridge Monitoring System. In *Annual Conference on Communication Networks and Services Research.* IEEE Computer Society.

Lee, V. C. (2012). *Energy Harvesting for Wireless Sensor Network* (Doctoral dissertation). University of California, Berkeley, CA.

Leung, V. (2015). Green Internet of Things for smart cities. *International workshop on smart cities and urban informatics.*

Lin, Y. H., Chou, Z. T., Yu, C. W., & Jan, R. H. (2015). Optimal and maximized configurable power saving protocols for corona-based wireless sensor networks. *IEEE Transactions on Mobile Computing*, *14*(12), 2544–2559. doi:10.1109/TMC.2015.2404796

Liu, Q., Ma, Y., Alhussein, M., Zhang, Y., & Peng, L. (2016). Green data center with IoT sensing and cloud-assisted smart temperature control system. *Computer Networks*, *101*, 104–112. doi:10.1016/j.comnet.2015.11.024

Liu, X., & Ansari, N. (2019). Toward green IoT: Energy solutions and key challenges. *IEEE Communications Magazine*, *57*(3), 104–110. doi:10.1109/MCOM.2019.1800175

Mainwaring, A., Culler, D., Polastre, J., Szewczyk, R., & Anderson, J. (2002). Wireless Sensor Networks for Habitat Monitoring. In *Proc. 1st ACM International Workshop on Wireless Sensor Networks and Applications*. ACM.

Maksimovic, M. (2017). Green Internet of Things (G-IoT) at engineering education institution: the classroom of tomorrow. *Green Internet of Things*, 270-273.

Maksimovic, M. (2017). The role of green internet of things (G-IoT) and big data in making cities smarter, safer and more sustainable. *International Journal of Computing and Digital Systems*, *6*(04), 175–184. doi:10.12785/IJCDS/060403

Maksimovic, M. (2017). Greening the future: Green Internet of Things (G-IoT) as a key technological enabler of sustainable development. In Internet of Things and Big Data Analytics toward Next Generation Intelligence. Springer.

Maksimović, M., & Omanović-Mikličanin, E. (2017). Green internet of things and green nanotechnology role in realizing smart and sustainable agriculture. In *VIII International Scientific Agriculture Symposium "AGROSYM 2017* (pp. 2290-2295). Academic Press.

Mancelos, N., Correia, J., Pires, L. M., Oliveira, L. B., & Oliveira, J. P. (2014). Piezoelectric Energy Harvester for a CMOS Wireless Sensor. In *Technological Innovation for Collective Awareness Systems* (pp. 470–477). Springer Berlin Heidelberg. doi:10.1007/978-3-642-54734-8_52

Manyika, J. (2013). *Disruptive technologies: Advance that will transform life, business and the global economy*. Retrieved from McKinsey & Company web site: https://www.mckinsey.com/insights/business_technology/disruptive_technologies

Mao, G. (2017). 5G green mobile communication networks. *China Communications*, *14*(2), 183–184. doi:10.1109/CC.2017.7868166

Mardamutu, K., Ponnusamy, V., & Zaman, N. (2018). Green energy in data centers. In Sustainable Development: Concepts, Methodologies, Tools, and Applications (pp. 386-401). IGI Global. doi:10.4018/978-1-5225-3817-2.ch019

Mascarenas, D. L. (2006). *Development of an impedance-based wireless sensor node for monitoring of bolted joint preload* (M.S. Thesis). Dept. of Structural Engineering, University of California, San Diego, CA.

Miragliotta, G., & Shrouf, F. (2012, September). Using Internet of Things to improve eco-efficiency in manufacturing: a review on available knowledge and a framework for IoT adoption. In *IFIP International Conference on Advances in Production Management Systems* (pp. 96-102). Springer.

Mitcheson, P. D., Yeatman, E. M., Rao, G. K., Holmes, A. S., & Green, T. C. (2008). Energy harvesting from human and machine motion for wireless electronic devices. *Proceedings of the IEEE*, *96*(9), 1457–1486. doi:10.1109/JPROC.2008.927494

Naeem, M., Pareek, U., Lee, D. C., & Anpalagan, A. (2013). Estimation of distribution algorithm for resource allocation in green cooperative cognitive radio sensor networks. *Sensors (Basel)*, *13*(4), 4884–4905. doi:10.3390130404884 PMID:23584119

Nandyala, C. S., & Kim, H. K. (2016). Green IoT agriculture and healthcare application (GAHA). *International Journal of Smart Home*, *10*(4), 289–300. doi:10.14257/ijsh.2016.10.4.26

Olivo, J., Carrara, S., & De Micheli, G. (2011). Energy harvesting and remote powering for implantable biosensors. *IEEE Sensors Journal*, *11*, 1573–1586.

Ooi, B., & Gilbert, J. (2014). Design of wideband vibration-based electromagnetic generator by means of dual-resonator. *Sensors and Actuators. A, Physical*, *213*(0), 9–18. doi:10.1016/j.sna.2014.03.037

Ooi, B. A., Aziz, R., & Gilbert, J. (2014). Analytical and numerical modelling for dual-resonator vibration-based electromagnetic generator. In *5th International Conference on Intelligent and Advanced Systems (ICIAS)* (pp. 1-5). Kuala Lumpur: IEEE. 10.1109/ICIAS.2014.6869443

Pacis, D. M. M., Subido, E. D., Jr., & Bugtai, N. T. (2017). Research on the Application of Internet of Things (IoT) Technology Towards A Green Manufacturing Industry: A Literature Review. In *DLSU Research Congress 2017* (pp. 1-11). Academic Press.

Paradiso, J. A., & Starner, T. (2005). Energy scavenging for mobile and wireless electronics. *Pervasive Computing, IEEE*, *4*(1), 18–27.

Park, C., & Chou, P.H. (2007). AmbiMax: Autonomous Energy Harvesting Platform for Multi-Supply Wireless Sensor Nodes. *3rd Annual IEEE Communications Society on Sensor and Ad Hoc Communications and Networks*, *1*, 168–177.

Park, G., Rosing, T., Todd, M. D., Farrar, C. R., & Hodgkiss, W. (2008). Energy harvesting for structural health monitoring sensor networks. *Journal of Infrastructure Systems*, *14*(1), 64–79. doi:10.1061/(ASCE)1076-0342(2008)14:1(64)

Pazowski, P. (2015). Green computing: latest practices and technologies for ICT sustainability. In *Managing Intellectual Capital and Innovation for Sustainable and Inclusive Society, Joint International Conference, Bari, Italy* (pp. 1853-1860). Academic Press.

Peoples, C., Parr, G., McClean, S., Scotney, B., & Morrow, P. (2013). Performance evaluation of green data centre management supporting sustainable growth of the internet of things. *Simulation Modelling Practice and Theory*, *34*, 221–242. doi:10.1016/j.simpat.2012.12.008

Ponnusamy, V. (2014). Energy analysis in Wireless Sensor Network: A comparison. *International Journal of Computer Networks and Communications Security*, *2*(9), 328–338.

Ponnusamy, V. (Ed.). (2016). *Biologically-inspired Energy Harvesting Through Wireless Sensor Technologies*. IGI Global. doi:10.4018/978-1-4666-9792-8

Ponnusamy, V., Abdullah, A., & Downe, A. G. (2012). Energy efficient routing protocols in wireless sensor networks: A survey. In Wireless Sensor Networks and Energy Efficiency: Protocols, Routing and Management (pp. 237-261). IGI Global. doi:10.4018/978-1-4666-0101-7.ch010

Ponnusamy, V., Low, T. J., & Amin, A. H. M. (2014). Biologically inspired mobile agent–based sensor network (BIMAS). *International Journal of Mobile Communications*, *12*(5), 449–476. doi:10.1504/IJMC.2014.064593

Ponnusamy, V., Tay, Y. P., Lee, L. H., Low, T. J., & Zhao, C. W. (2020). Energy harvesting methods for Internet of Things. In Securing the Internet of Things: Concepts, Methodologies, Tools, and Applications (pp. 956-976). IGI Global. doi:10.4018/978-1-5225-9866-4.ch042

Popa, D., Popa, D. D., & Codescu, M. M. (2017). Reliability for a green internet of things. *Buletinul AGIR*, (1).

Rabaey, J. M., Ammer, M. J., da Silva, J. L. Jr, Patel, D., & Roundy, S. (2000). PicoRadio supports ad hoc ultra-low power wireless networking. *Computer*, *33*(7), 42–48. doi:10.1109/2.869369

Ramachandran, T., Jung, L. T., Ponnusamy, V., & Hudaya, A. (2014, June). ZigBee wireless communication for monitoring renewable street light system. In *Computer and Information Sciences (ICCOINS), 2014 International Conference on* (pp. 1-6). IEEE.

Rault, T., Bouabdallah, A., & Challal, Y. (2014). Energy efficiency in wireless sensor networks: A top-down survey. *Computer Networks*, *67*, 104–122. doi:10.1016/j.comnet.2014.03.027

Rekha, R. V., & Sekar, J. R. (2016). An unified deployment framework for realization of green Internet of Things (GIoT). *Middle East Journal of Scientific Research*, *24*(2), 187–196.

Rocha, J. G. Goncalves, L. M. Rocha, P. F. Silva, M. P. & Lanceros-Mendez, S. (2008). Energy harvesting from piezoelectric materials fully integrated in footwear. *Industrial Electronics, IEEE Transactions on, 57*(3), 813-819.

Roundy, S. J. (2003). *Energy scavenging for wireless sensor nodes with a focus on vibration to electricity conversion* (Doctoral dissertation). University of California, Berkeley, CA.

Roy, A., Datta, A., Siddiquee, J., Poddar, B., Biswas, B., Saha, S., & Sarkar, P. (2016, October). Energy-efficient Data Centers and smart temperature control system with IoT sensing. In *2016 IEEE 7th Annual Information Technology, Electronics and Mobile Communication Conference (IEMCON)* (pp. 1-4). IEEE. 10.1109/IEMCON.2016.7746251

Saleem, M., Di Caro, G. A., & Farooq, M. (2011). Swarm intelligence based routing protocol for wireless sensor networks: Survey and future directions. *Information Sciences*, *181*(20), 4597–4624. doi:10.1016/j.ins.2010.07.005

Sander & Saulny. (2007). *Bridge collapse in Minneapolis Kills at least 7*. Retrieved from New York Times web site: https://www.nytimes.com/2007/08/02/us/02bridge.html?_r=0

Sarkar, S., & Misra, S. (2016). Theoretical modelling of fog computing: A green computing paradigm to support IoT applications. *Iet Networks*, *5*(2), 23–29. doi:10.1049/iet-net.2015.0034

Seyedi, A., & Sikdar, B. (2008). Energy Efficient Transmission Strategies for Body Sensor Networks with Energy Harvesting. *Proc. of the42nd Annual Conference on Information Sciences and Systems (CISS)*, 704–709. 10.1109/CISS.2008.4558613

Shaikh, F. K., & Zeadally, S. (2016). Energy harvesting in wireless sensor networks: A comprehensive review. *Renewable & Sustainable Energy Reviews*, *55*, 1041–1054. doi:10.1016/j.rser.2015.11.010

Shaikh, F. K., Zeadally, S., & Exposito, E. (2015). Enabling technologies for green internet of things. *IEEE Systems Journal*, *11*(2), 983–994. doi:10.1109/JSYST.2015.2415194

Shuja, J., Bilal, K., Madani, S. A., Othman, M., Ranjan, R., Balaji, P., & Khan, S. U. (2014). Survey of techniques and architectures for designing energy-efficient data centers. *IEEE Systems Journal*, *10*(2), 507–519. doi:10.1109/JSYST.2014.2315823

Simjee, F., Pai, H., & Chou, P. H. (2006). Everlast: Long-life, Supercapacitor operated Wireless Sensor Node. In *Proc 2006 International Symposium on Low Power Electronics and Design*. ACM. 10.1145/1165573.1165619

Simjee, F. I., & Chou, P. H. (2008). Efficient Charging of Supercapacitors forExtended Lifetime of Wireless Sensor Nodes. *IEEE Transactions on Power Electronics*, *23*(3), 1526–1536. doi:10.1109/TPEL.2008.921078

Sodano, H. A., Inman, D. J., & Park, G. (2005). Generation and storage of electricity from power harvesting devices. *Journal of Intelligent Material Systems and Structures*, *16*(1), 67–75. doi:10.1177/1045389X05047210

Song, W., Xiao, Z., Chen, Q., & Luo, H. (2013). Adaptive resource provisioning for the cloud using online bin packing. *IEEE Transactions on Computers*, *63*(11), 2647–2660. doi:10.1109/TC.2013.148

Sun, K., & Ryoo, I. (2015, October). A study on medium access control scheme for energy efficiency in wireless smart sensor networks. In *2015 International Conference on Information and Communication Technology Convergence (ICTC)* (pp. 623-625). IEEE. 10.1109/ICTC.2015.7354625

Tan, H. P., Eu, Z. A., & Seah, W. K. (2009). Impact of Power Control in Wireless Sensor Networks Powered by Ambient Energy Harvesting for Railroad Health Monitoring. *Proc. of the 2nd Intl Workshop on Applications of Ad hoc and Sensor Networks (AASNET)*. 10.1109/WAINA.2009.113

Tushar, W., Wijerathne, N., Li, W. T., Yuen, C., Poor, H. V., Saha, T. K., & Wood, K. L. (2018). Internet of things for green building management: Disruptive innovations through low-cost sensor technology and artificial intelligence. *IEEE Signal Processing Magazine*, *35*(5), 100–110. doi:10.1109/MSP.2018.2842096

Wang, H. I. (2014). Constructing the green campus within the internet of things architecture. *International Journal of Distributed Sensor Networks*, *10*(3), 804627. doi:10.1155/2014/804627

Wang, T., Xia, Y., Muppala, J., Hamdi, M., & Foufou, S. (2014, December). A general framework for performance guaranteed green data center networking. In *2014 IEEE Global Communications Conference* (pp. 2510-2515). IEEE. 10.1109/GLOCOM.2014.7037185

Werner-Allen, G., Lorincz, K., Ruiz, M., Marcillo, O., Johnson, J., Lees, J., & Welsh, M. (2006, March-April). Deploying a Wireless Sensor Network on an Active Volcano. *IEEE Internet Computing*, *10*(2), 18–25. doi:10.1109/MIC.2006.26

Wu, Q., Li, G. Y., Chen, W., Ng, D. W. K., & Schober, R. (2017). An overview of sustainable green 5G networks. *IEEE Wireless Communications*, *24*(4), 72–80. doi:10.1109/MWC.2017.1600343

Xiong, N., Han, W., & Vandenberg, A. (2012). Green cloud computing schemes based on networks: A survey. *IET Communications*, *6*(18), 3294–3300. doi:10.1049/iet-com.2011.0293

Ye, W., Heidemann, J., & Estrin, D. (2002, June). An energy-efficient MAC protocol for wireless sensor networks. In *Proceedings. Twenty-First Annual Joint Conference of the IEEE Computer and Communications Societies* (Vol. 3, pp. 1567-1576). IEEE.

Zhu, C., Leung, V. C., Shu, L., & Ngai, E. C. H. (2015). Green internet of things for smart world. *IEEE Access: Practical Innovations, Open Solutions*, *3*, 2151–2162. doi:10.1109/ACCESS.2015.2497312

Chapter 2
IoT Underwater Wireless Sensor Network Monitoring

Low Tang Jung
Universiti Teknologi PETRONAS, Malaysia

ABSTRACT

Water is a crucial resource for all life on earth, and it is fast becoming one of the limited natural resources to humankind, especially clean drinking water and water for agricultural uses. Sensor technologies and wireless communications (both terrestrial and underwater) have been seriously investigated by researchers to find ways to integrate these technologies for a novel data-sensing and data-collecting network for long-term water pollution monitoring purposes. This chapter describes an IoT-based underwater wireless sensor network (UWSN) which is believed to have a huge potential for monitoring the health of river, lake, reservoir, and marine environment. The sensed data from IoT sensors are communicated wirelessly via acoustic channels to a data collection center for further processing and interpretation. It is foreseeing that judicious deployment of IoT-based UWSN is a promising solution for long-term water quality surveillance.

INTRODUCTION

Water is a crucial resource for all life on earth and it is fast becoming one of the limited natural resources to humankind. It can be foreseen in near future that water, especially clean drinking water and water for agricultural purposes, is realistically becoming a scarce resource. Clean drinking water is already a revenue generating commodity since the past decade. It is therefore been observed in many countries (Water Framework Directive WFD, Council Directive 2000/60/EC) that water quality monitoring in a long-term continuous mode is beginning to catch on as an essential component in the environmental pollution monitoring and control systems. Thanks to the worrying realistic effect from industry waste and climate change, the sensor technologies, and wireless communications (both terrestrial and underwater) are being seriously investigated by many researchers to find ways to integrate these technologies for a novel data-sensing and data-collecting network for long-term pollution monitoring systems.

DOI: 10.4018/978-1-7998-6709-8.ch002

The application of wireless sensor networks in underwater domain has a huge potential for monitoring the health of river, lake, reservoir, and marine environment. By deploying the in-situ sensors for continuous sampling of the environmental parameters or indicators offers the advantage of reducing the operation costs and to provide real-time information on pollutant fluctuations. Essentially, the UWSN monitoring system comprises of a network of underwater sensors deployed at key locations for a time frame of months in a continuous operation mode. The sensed data from the sensors will be communicated by wireless means via an acoustic channel to a data collection center for further processing and interpretation. It is believed that judicious deployment of underwater sensor network is a promising solution for long-term water quality surveillance.

BACKGROUND

A handful of underwater wireless sensor network (UWSN) have been deployed for water quality or pollution monitoring and two prominent works in this area are briefly mentioned here. Smart-Coast (Regan, Lawler, & McCarthy, 2009) was a project aimed to develop a wireless sensor network with a distinct "plug-and-play" feature that incorporates novel sensor nodes and low power consumption. This system was based on Zigbee communications standard. The "plug-and-play" platform was designed to sense pH level, temperature, conductivity, depth, and turbidity.

The Fraser River Water Quality System by Ethier & Bedard (2007) was a project initiated for monitoring water quality and meteorological parameters in real-time mode all year round. A moored buoy platform was used for station location and in-situ water sampling. A 3m tall Oceanographic Data Acquisition System (ODAS) buoy was designed for this purpose. The monitoring operation was scheduled in continuous mode with a biweekly sampling. ODAS was claimed to be able to distinguish tidally driven events to initiate sampling of organic contaminates.

Practically there are three general network scenarios for UWSN deployment: static two-dimensional UWSN for underwater bottom environment monitoring, static three-dimensional UWSN for underwater column monitoring, and three-dimensional mobile network with autonomous underwater vehicles (Akyildiz, Pompili, & Melodia, 2005). However, in terms of aquatic applications, UWSN can be classified into two categories: long-term non-time critical aquatic monitoring, and short-term time critical exploration (Cui, Kong, Gerla, & Zhou, 2006). The use case study in this chapter falls in the first category. The use case in this chapter is about a long-term water pollution monitoring application. Other applications fall into the first category may include marine biology, oceanography, ocean floor seismic monitoring, etc. As for the second category, the applications may include setting up of ad-hoc UWSN at the site of a shipwreck for liquid toxic leakage monitoring, radiation detection, etc.

This chapter starts by describing the possible UWSN architecture for underwater pollution monitoring application which includes the description of a basic architecture and an extended architecture. It should be mentioned here that the use case study in this chapter is based on the extended architecture. Then, the details of data transmission and data acquisition scheduling process for the proposed UWSN monitoring is described which leads to a proposed scheduling algorithm. Next, the sink node battery power capacity issue is discussed, and a battery power capacity estimation method is proposed. A conclusion is provided at the end of this chapter.

UWSN ARCHITECTURE

Basic UWSN Architecture

UWSN is basically an underwater network utilizing sensor nodes equipped with sensors and acoustic modems for communications (Sozer, Stojanovic, & Proakis, 2000). Figure 1 shows a basic 2D static architecture that may be used for underwater environment monitoring. All sensor nodes and the sink node(s) are stationed at their respective fix/static locations at the water bottom for data collection purposes. These nodes can be connected in a star-, a tree-, or a hybrid of star-tree-topology. All the nodes can communicate with each other in sending their own data, and to forward data from other nodes, to a sink node in a multi-hop manner. It follows that the sink node that communicates directly with the surface gateway would need a high-capacity power supply for the fact that it must relay data packets from all the other nodes to the surface gateway.

Figure 1. Basic UWSN architecture for underwater environment monitoring

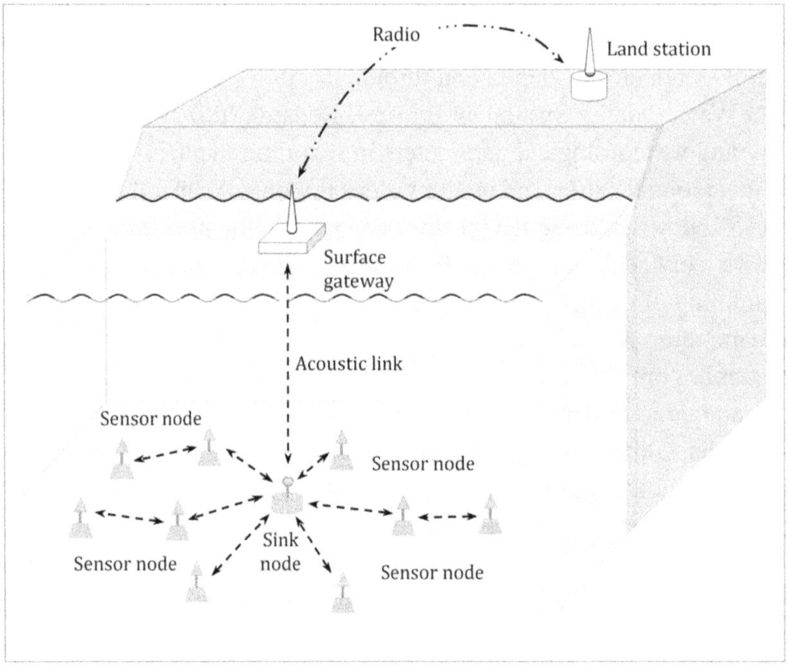

Two acoustic modems are available on the sink node. One is used to communicate with the sensor nodes in the network and the other one is to communicate with a moored gateway placed on the water surface. The gateway is equipped with a radio link to relay data packets to a land station. The land station shall provide monitoring and control of the data acquisition of all sensor nodes in ad-hoc real-time mode or in long-term time scheduled mode (Akyildiz, Pompili, & Melodia, 2006).

UWSN Architecture for Water Column Monitoring

With reference to the basic UWSN architecture in Figure 1 and the relevant works in the literature reviewed, the architecture for this use case study is shown in Figure 2. This architecture is aimed for shallow water deployment such as at coastal areas, lakes, reservoirs, etc for a long-term non-time critical water pollution or water quality monitoring. The overall architecture represents a static 3D network topology, and it is seen as the vertical extension to the basic static 2D topology in Figure 1.

Matter of fact, static 3D architecture is foreseen as one of the most suitable setups for long-term environmental monitoring application (Kong, Cui, Wu, & Gerla, 2005). The sensor nodes and the sink nodes are practically arranged, by some means of anchoring mechanism, into tiers forming a column of sensing network in a body of water. All nodes are non-mobile or having limited mobility of about 3 to 5 knots (1 to 1.5 m/s) due to typical underwater current (Kong et al., 2005).

The number of tiers in the topology depends on the depth of the water body to be monitored and the pollution monitoring resolution requirements. Naturally, higher resolution demands more tiers and more sensor nodes per tier. Take note that Figure 2 shows an architecture with each tier represents a disk-like star topology with a centralized sink node.

Typically, the top tier is placed about 30 to 50 meters away from the water surface to avoid acoustic wave reflection complications near the water surface. This precaution is also applicable to the tier of sensor/sink nodes one layer above the ground tier to avoid wave reflection due to ground surface.

Figure 2. UWSN architecture for underwater pollution monitoring

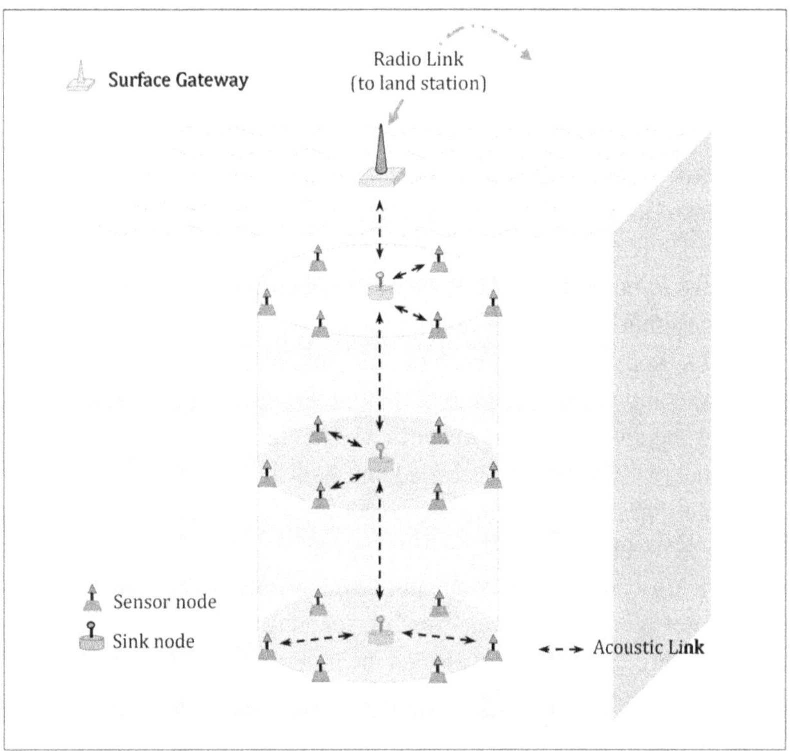

At each tier, the distance between the sink node and all the sensor nodes is arranged in such a way that it is for a one hop data transmission. This is to be in consistence to the works found in the reviewed literature. In practice, the actual distance/range shall depend on the transmission range of the sensor nodes and the acoustic modem deployed. The sink nodes are also placed at a distance for one hop transmission. Again, this range is to be in line with the research works found in the literature. The total number of sensor nodes to be deployed at each tier is very much dependent on the network deployment budget for underwater sensor/sink nodes are, very often, costly items.

As illustrated in Figure 2, each tier consists of a centralized star topology whereby a sink node at the center is surrounded by several sensor nodes. All the nodes at each tier can be quantified as a set of nodes represented by N = {*n1, n2, ... ns*} with n_s acting as a static sink node. If the transmission range of each sensor node is l and each node is of omnidirectional type then a one hop link can exist between a sensor node and the sink node if and only if $|L_{is}|$ £ l, where L_{is} is the distance between a sensor node n_i and the sink node n_s. The overall network can therefore be viewed as a 3D-column having an approximated volume of $2\pi(L_{is})^2H$, where H is the height measured from the sink node at the bottom tier to the surface gateway.

Figure 3. Data link layer general packet format

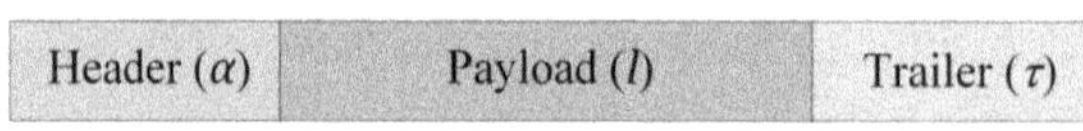

For this use case study, it is assumed a 3-tier network. In each tier the distance between the sensor nodes and the sink node is to be 50m. The sink node to sink node distance is also set to 50m and the surface gateway is 50m away from the top tier sink node. In such arrangement, each tier can cover approximately 8000m^2 of a disk-like area and for a 3-tier setup the column height would be 150m. This setup is therefore able to monitor a water-column having a volume of approximately 1.2 million m^3. Note that all the distances in this case study are chosen to be practically applicable in a shallow water environment.

It should be mentioned here that for a more inclusive and practical network setup the cluster of sensor nodes in a tier do not have to be anchored at the same height as the sink node. This is to avoid forming a disk-like shape topology that has a more restricted water-body volume and in return offers a lower monitoring resolution. The sensor nodes at each tier can in fact be anchored at different height to cover a larger monitored volume and yet keeping intact the one hop transmission range within the sink node coverage. Take note that the single column network architecture in Figure 2 can be duplicated several times over should the area to be monitored is broad and wide.

Take note also that the types of sensor node to be deployed in the proposed monitoring network and the types of pollutant to be monitored by the sensors are not the scope of this case study. They belong to the field of sensor technology and pollutant sensing or measurement. This use case study is emphasizing more on the underwater one hop data transmission scheduling for the sensed data (in relation to the proposed network architecture/topology) and also on the integration of the proposed 2Q algorithm (*Data Packet Size Optimization Algorithm With Performance Metric of Throughput Efficiency and Energy Efficiency Under Different Bit-Error-Rate* or in short simply as "*2Q Optimization Algorithm*" whereby the optimum packet size is qualified by two performance metrics i.e. throughput efficiency and energy efficiency in the data transmission scheduling process. In this use case study, it is therefore assumed that all sensor types are appropriately chosen per their sensing specifications for sensing the relevant pollutants and can generate data packets having the generic data link layer packet format as shown in Figure 3.

An addressing scheme using the address byte format shown in Figure 4 is hereby suggested to uniquely identify the nodes in the proposed network. The address byte shall comprise of three fields: T, S, and N field. These fields literally stand for Tier, Sink, and Node. The two bits in the *T* field are for the network tiers addressing and identification. The two bits in the *S* field are for the sink nodes addressing and identification, and the four bits in *N* field are for the sensor nodes addressing and identification.

Take note that the address byte shown in Figure 4 can be used to address or access a maximum of 4 tiers, a maximum of 4 sink nodes, and a maximum of 16 sensor nodes. It should be mentioned here that based on digital system numbering sequence convention, the numbering of tiers, sink nodes, and sensor nodes starts from '0'. For instance, in Figure 2, Tier 0 is the ground tier and Tier 2 refers to the tier nearest to the surface gateway. Note that the suggested addressing scheme can be easily expanded into a generic form to accommodate more address bytes for accessing larger number of nodes/sinks in a large network.

Figure 4. Address byte format

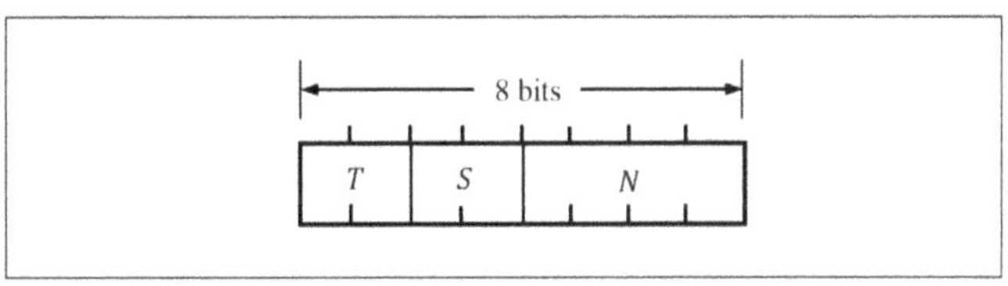

Symbolically the nodes in Figure 2 can be accessed with an address byte having the form of $<T_i:S_j:N_k>$, where *i,j,k* are integers starting from '0'. T_i represents the tier to be accessed, S_j the sink node to be accessed, and N_k the sensor node to be accessed. Taking the network in Figure 2 as an example, $<T_1:S_1:N_6>$ is a valid address that belongs to sensor node 6 which has a one hop link with sink node 1 at tier 1. Thus, the address byte would have these bits: 01010110. For address byte: 01100110 which translate to $<T_1:S_2:N_6>$ would be considered an invalid address because there is only one sink node at tier 1 in Figure 2. However, this address is valid if there exists an additional similar column in the network.

DATA TRANSMISSION AND DATA ACQUISITION

As stated above, the proposed UWSN is aimed for a long-term non-time-critical water pollution monitoring. It means this UWSN is not to be an ad-hoc network and its deployment duration can be defined for a minimum of 3 months to a maximum of one year. It is typically between 3 to 6 months depending on the water quality when the nodes need to be taken out of water for occlusion maintenances. For non-time-critical monitoring applications, data acquisition is normally time scheduled rather than event triggered. This implies that data transmission would be in time scheduled mode as well. Data transmission can be invoked immediately when a sensor has acquired a data or alternatively, to conserve node energy, the data acquired may be aggregated into a group and send as a data group to the sink at some specific scheduled time slot. An efficient dynamic frame aggregation scheme proposed by Lin et al., (2009) may be adopted for this approach.

For this use case study, the author adopted one of the more commonly used approach for time scheduled data transmission and acquisition – the Time Division Multiple Access (TDMA) method. It should be mentioned here that Frequency Division Multiple Access (FDMA) is not applicable in underwater acoustic channel due to its severe bandwidth limitation.

Fundamentally TDMA is a mechanism that allows a cyclic assignment of a time slot for each of the network node to transmit data or to acquire data. In TDMA scheduling some forms of time synchronization procedure may be necessary to minimize the 'timing drift' problem due to the nodes' hardware-based timing requirements. A generic timing diagram of TDMA scheduling adopted from (Penteado, Costa, & Pedroza, 2010) is shown in Figure 5.

This timing diagram shows the time slot allocated for each node to transmit its data packets to the sink node in sequential manner. At each allocated time slot, the acoustic channel/link is reserved for one particular sensor/sink node only. This is one of the TDMA properties in eliminating channel contention among the sensor/sink nodes for preventing data collision. Due to the relatively slow speed of acoustic wave (nominally accepted as 1500m/s), which is in the order of 5 times slower than RF wave, time guard (TG) intervals need to be inserted in between the node transmission time slots to prevent transmissions overlapping and interfering with transmissions in other time slots.

Figure 5. Generic timing diagram of TDMA scheduling

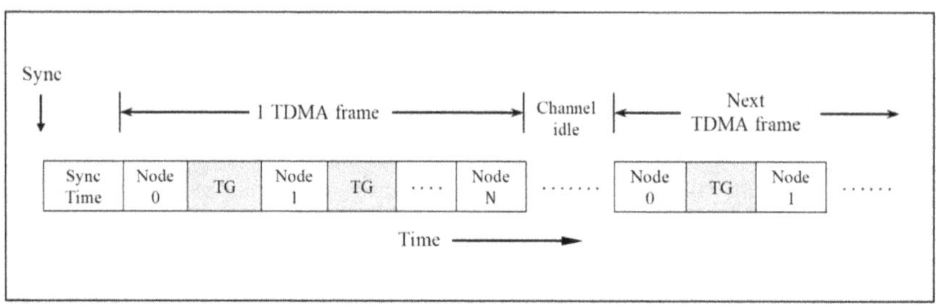

In this use case study, the basic implementation concept of TDMA time-scheduling scheme for data transmission shall first be explained with reference to a single tier network topology. After which the basic concept is extended to the multi-tier topology of Figure 2. A prerequisite for the implementation of this scheme is that all sensor nodes in a particular tier must have completed their data acquisition cycle i.e. already holding the sensed data in their respective data queue before the sequence of data transmission for that particular tier is kicked-off.

The whole TDMA scheduling sequence starts off with a Sync signal being broadcast from the sink node at a particular tier to all sensor nodes in that particular tier. The sink node initiates the broadcasting of Sync signal upon receiving a data collection command from the time scheduler located at the land station via the surface gateway. Alternatively, data collection command can be initiated by the surface

Figure 6. TDMA scheduling for multi-tier single column UWSN monitoring setup

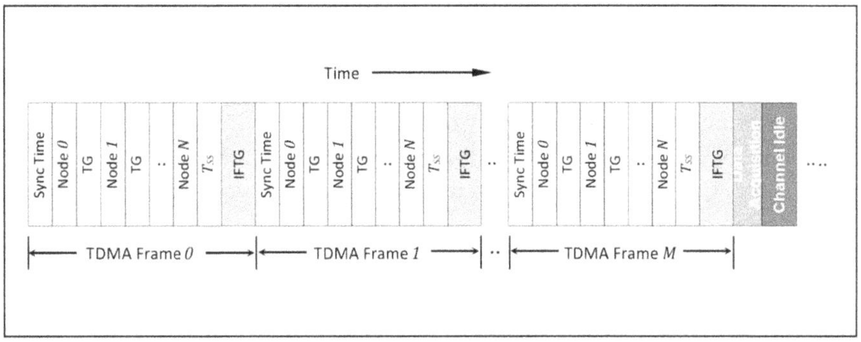

gateway itself. Since all sensor nodes are just one hop away from the sink therefore this Sync signal can synchronize simultaneously the timing requirement in each sensor node.

Upon receiving the Sync signal, each sensor node will take a duration of Sync Time to complete their respective time synchronization process. It is worth mentioned here that the actual Sync Time duration is device dependent. It is therefore best to have homogeneous modem for all nodes in the entire network. After synchronization, Node 0 is given a specific time slot to exclusively use the acoustic channel to transmit its data to the sink. A Time Guard (TG) interval is inserted at the end of Node 0 transmission. After which Node 1 will have the exclusive right to use the channel to transfer its data to the same sink. This procedure is repeated until the last sensor node in the tier, Node N, completed its data transfer. At this juncture 1 TDMA time frame is consumed and one scheduling sequence (or cycle) is over. The channel shall then go into idle state until the next sequence/cycle is initiated. At the idle state all nodes shall go into sleep mode to preserve energy or shall wake up periodically to perform data acquisition tasks. For information, the computation of a time slot duration and the TG duration is explained in the next section.

For long-term non-time-critical underwater pollution monitoring application, it is common to have the sensor nodes configured with data acquisition rate in the range of tens of minutes per acquisition. It is also common to have an acquired data being sent to the assigned sink node as soon as it is made available in the sensor node. This simply means the data transmission rate is normally in sync with the data acquisition rate. So, the consequence of configuring the data transmission/acquisition rate to tens of minutes per data sampling is that the channel idle time will be very much longer than the TDMA time frame. Hence a portion of this channel idle time can be utilized by the sink node to transfer the data that it has collected, from all the sensor nodes under its custody, to the surface gateway.

The sink node in such a situation is, of course, expected to have storage capacity large enough to aggregate the data packets from all nodes in one TDMA time frame. Take notice that multi-hop transmission is necessary for the bottom sink to transmit its data packets via some intermediate sinks to the surface gateway unless the UWSN is of a 2D static architecture where the sink node is directly linked to surface gateway.

The basic one-tier TDMA scheduling scheme explained above can be extended to suit the multi-tier UWSN architecture. In its simplicity, the multi-tier architecture simply needs a multi-frame TDMA scheduling. That is, it is the cyclic assignment of TDMA time frame (refer to Figure 5) to each of the tier in the multi-tier network. The timing diagram in Figure 6 can be used to illustrate this concept. This timing diagram shows the scheduling in a network with M tiers and therefore there are M TDMA frames.

An inter-frame time guard (IFTG) is needed in between two TDMA frames and keep in mind that inside each TDMA frame there are intra-frame time guards (TG). For multi-tier network all sink nodes, except the topmost sink, need to do multi-hop transmission to relay data packets up to the surface gateway. Therefore, it is necessary to have IFTG as a reserved time slot for multi-hop data delivery. For static 3D deployment, a network with M tiers would need an IFTG of M*$thop$ units time where $thop$ is the time for one-hop transmission based on the specification that the sink nodes are placed one-hop away from each other. TSS in each TDMA frame is the time slot reserved for the sink node to transfer the aggregated data packets to the sink node one tier or one-hop away.

With reference to Figure 6, it can be seen that the whole cycle of multi-frame TDMA scheduling starts off from TDMA Frame *0*. This time frame allows all sensor nodes at tier 0 to be appropriately scheduled to transmit their data packets to the sink node at tier 0. After collecting the data packets from each of the sensor node this sink node is then allocated a time slot of T_{SS} to transfer these packets to the adjacent sink node i.e., the sink node just one tier above it (one-hop away).

Next, an IFTG time slot is allocated to this sink node to ensure the aggregated packets can reach the surface gateway via a multi-hop relaying process. The objective of having multi-hop data transmission for the sink-to-sink route is to better conserves energy in the sink nodes. Multi-hop transmission is known to be more energy efficient than long range direct transmission approach. It is not difficult to see this advantage especially in considering the power needed by the bottom sink to transfer data packets directly to surface gateway instead of using the multi-hop route. At the end of IFTG time slot the scheduling of all nodes in tier 0 is considered complete at this point of time. The whole sequence is now repeated for TDMA Frame *1* to take care for the scheduling of the sensor nodes and the sink node placed at tier *1*. The scheduling process goes on in this manner until the Mth tier is being served.

At the end of the last (M^{th}) IFTG time slot the surface gateway shall send a data acquisition signal to all the sink nodes, again, via multi-hop transmission. Upon receiving the signal from the surface gateway, all sink nodes will broadcast this data acquisition signal to all the sensor nodes within its range to initiate sensor's data acquisition processes, respectively. A form of time scheduling is needed here for the sinks to broadcast the data acquisition signal to their respective sensor nodes to prevent signal collision. Finally, the channel goes into idle state waiting for the next cycle of scheduling to be initiated by the command from the scheduler at the land station.

It should be highlighted here that there is an inherent issue in time synchronization to ensure smooth scheduling of data acquisition and data transmission. At certain occasions some form of synchronization

Figure 7. TDMA scheduling for UWSN with multiple tiers and columns

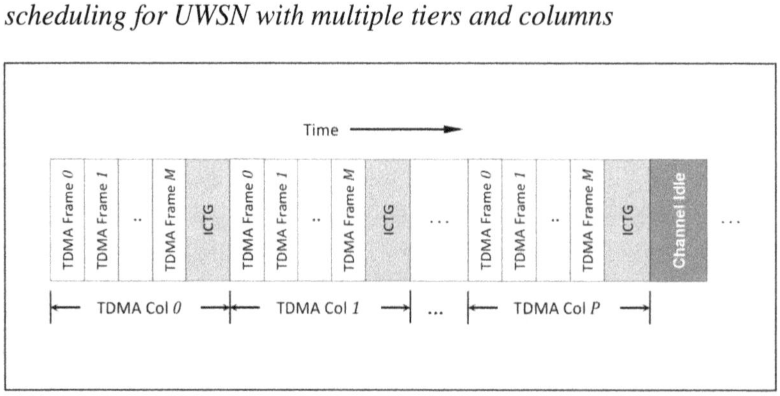

protocols may be needed to keep most of the nodes in the sleep mode so that the active nodes at any time are sparsely distributed to reduce the probability of channel contention and data collisions. However, in this use case study this issue does not pose a serious problem because there is only one active node at a time busy transferring data packet.

The whole cycle of multi-frame TDMA scheduling described above is applicable to the one column multi-tier architecture depicted in Figure 2. By extending this concept one step further, this one column multi-tier scheduling scheme can be applied to a multi-column multi-tier UWSN architecture. The fundamental timing diagram for such implementation is presented in Figure 7 above. However, the details and/or the variant of this timing diagram and its related TDMA scheduling sequence are left as a further research direction for the interested readers. The ICTG label in Figure 7 is the acronym for Inter-Column Time Guard. The timing diagram in Figure 7 assumes there are *P* columns in an arbitrary UWSN.

SCHEDULING ALGORITHM AND TIME SLOT CALCULATION

Scheduling Algorithm

The multi-frame TDMA scheduling algorithm for the proposed single column multi-tier UWSN architecture shown in Figure 2 is presented here. The algorithm is to base on the timing diagram shown in Figure 6. Some of the essential prerequisites for the algorithm are also given. These prerequisites are needed to fulfill the requirements of the components (time slot) in the TDMA time frames. For information, this algorithm is by no means exhaustive. Its variants can be derived from the user's intended or desired network architecture.

Prerequisites

- All nodes are of homogeneous type so that all sensor modems have same hardware configurations and specifications to support constant Sync Time. That is, the Sync Time is derived from these homogeneous specifications.
- The sensor nodes have completed their respective data acquisition process and are holding the latest acquired/sensed data in their data queue.
- The time slot for each sensor node to send its data packet to the sink node has been properly computed (see sample calculation in next subsection).
- Duration for TG, T_{SS}, IFTG time slots have been computed accordingly fulfilling their respective requirements (refer to the sample calculations in the next subsection).
- Each TDMA frame is referenced with the general format of *TDMA_frame_i* where $i = 0,1,2 \ldots M$ with *TDMA_frame_0* belongs to the scheduling time frame for the cluster in the lowest tier (furthest away from the surface gateway), and *TDMA_frame_M* for cluster in the highest tier (nearest to the surface gateway).
- There are *M* tiers in the water body column to be monitored.
- There are *N* sensors in each tier.
- By default, one cycle of scheduling sequence starts from *TDMA_frame_0*.
- The BER (*p*) of the link has been predetermined. BER is Bit-Error-Rate.
- The distance (*l*) between sensor-sink and sink-sink pair has been predetermined.

- Data transfer rate (*R*) has been predetermined.

Algorithm

```
01:     Surface gateway: receive(Sync) signal from land station scheduler;
02:     Surface gateway: send(Sync) via multi-sink route to sink_0 at lowest tier;
03:     for (i = 0 ; i <= M ; i++) /*for TDMA frame scheduling control*/
04:     {TDMA_frame_i:
05:     sink_i : broadcast(Sync) to sensor_0 .. sensor_N ; /*for tier_i*/
06:     sensor_0 .. sensor_N : synchronize(timer);
07:     for (j = 0 ; j <= N ; j++) /*for tier_i sensors scheduling control */
08:     {node_j : invoke 2Q algorithm for optimal(packet_size);
09:             node_j : send(data_packet) to sink_i;
10:             allocate TG time slot for (data_packet) to reach sink_i;
11:             sink_i: store(data_packet);
11:     }/*repeat from step 8*/
12:     sink_i: invoke 2Q algorithm for optimal(aggregated_data);
13:     allocate T_SS time for sink_i : send(aggregated_data) to sink_i+1;
14:     allocate IFTG time slot for (aggregated_data) to reach surface gateway;
15:     surface gateway: send(aggregated_data) to land station;
15:     }/*repeat from step 4*/
16:     surface gateway: send(data_acquisition) signal to all sinks via multi-hop transmission;
17:     all sinks: broadcast(data_acquisition) signal to all sensor nodes to initiate data acquisition process based on predefined broadcasting scheduling;
18:     channel idle: all nodes go into sleep mode.
19:     TDMA scheduling cycle repeat from step 1 on next command initiated from land station scheduler to the surface gateway.
```

Sample Time Slot Calculation

Samples of time slot calculation are hereby given in response to the timing diagram shown in Figure 6 and the prerequisites listed in the preceding subsection.

Time slot for a sensor node and its TG duration:

Parameters needed:

B_n : Total data bits (payload) sent by a node.

l : Distance between the source-sink pair in meter.

c : Nominal speed of sound in underwater (1500 m/s).

B_h : Header (sensor node address + sink node address) bits.

R : transmission bit rate in bps.

Total bits send out by a node = $B_n + B_h$

∴ Time needed for a node = $(B_p + B_h)/R$ s

$TG = 2l/c$ (Note: $2l$ is for worst case latency)

Example: A sensor transmits 3 bytes of sensed data at a rate of 1000 bps to a sink 50m away using the address format shown in Figure 6.3.

Time needed by a node = (3×8 + 8 + 8)/1000 = 40 ms.

TG = 100/1500 = 66.67 ms.

Time slot for a sink node (T_{SS}):

Parameters needed:

NB_n : Total data bits (payload) aggregated from N sensors nodes.

l : Distance between the sink-sink pair in meter.

c : Nominal speed of sound in underwater (1500 m/s).

B_h : Header ($sink_i$ address + $sink_{i+1}$ address) bits.

R : Transmission bit rate in bps.

Example: A sink transfers data bytes aggregated from 8 sensor nodes (with each sensor producing 3 bytes of sensed data) at a rate of 1000 bps to a sink 50m away using the address format shown in Figure 6.3.

Time needed by a sink node = (8×3×8 + 8 + 8)/1000 = 208 ms.

TG = 100/1500 = 66.67 ms.

∴ T_{SS} = 208 + 66.67 = 274.67 ms (Note: T_{SS} is a one-hop time)

Time slot for IFTG:

Parameters needed:

T_{SS} : Time for one hop.

M : Number of tiers in the network.

$IFTG = MT_{SS}$ in seconds = 824 ms for 3 tiers network.

Note: All calculations are based on the worst-case scenario. Other variants do exist, and it is left as a topic for possible further research to the interested readers.

BATTERY POWER CAPACITY ESTIMATION

The finite battery power source in the sensor/sink nodes is always an important issue in UWSN (in fact in WSN too). Battery energy conservation hence becomes an important factor in enhancing the life span of the entire network. The sensible strategy is to have the nodes stay in idle/inactive mode more often than in the active mode to conserve energy. The power consumption in some of the commercial acoustic modems shown in Table 1 (EvoLogics (Germany), LinkQuest Inc. (USA), Woods Hole Oceanographic Institution (USA), aptly manifests the reason for this preference. One of the solutions to enhance network lifespan is to have the power capacity of the battery be sufficiently large enough to support a long-term network deployment to avoid frequent battery recharging or replacement.

Table 1. Acoustic modem power consumption

State	LinkQuest	EvoLogics	WHOI
Transmit	1 – 12W	2.5 – 40 W	50 – 100W
Receive	0.8W	5 – 50 mW	0.1 – 2 W
Idle	8 mW	3 mW	0.1 – 0.25W

This section presents a network battery power capacity estimation method specifically for long-term non-time critical UWSN applications based on the architecture illustrated in Figure 2 and the scheduling algorithm explained in subsection 4.1. The estimation method described here would be general enough for adopting in other types of UWSN architecture.

In general, the topology in Figure 2 shows that each tier has a cluster of sensor nodes monitoring their surrounding conditions to periodically send the acquired (and possible processed) data to a central sink at the same tier. Subsequently, the sink shall aggregate the processed data packets and send them to a central surface gateway. The main power consumption at each node is when the node is transmitting data packets and when it is receiving data packets.

Technically, the node's sensing and processing power consumptions are assumed to be negligible (Bhardwaj, Garnett, & Chandrakasan, 2001; Kalpakis, Dasgupta, & Namjoshi, 2002). This statement holds true for this case study because the data acquisition frequency/rate in the proposed UWSN is in the order of tens of minutes per sampling (Boehm et al., 2002; Holman, Stanley, & Ozkan-Haller, 2003). The power consumption in data sensing and data processing in the nodes are therefore not considered a major issue.

A closer look at Figure 2 reveals that the water-body column represents an architecture of 3 tiers having 3 chained clusters. In the context of data transmission, the sensor nodes at each tier formed a cluster in star topology (with a centralized sink node) where the sink nodes formed a chained topology. Collectively the nodes in the water column represents a linear chained network. This implies that the sink node at the uppermost tier is the candidate of bottleneck in terms of its battery lifetime and thus is the main factor affecting the whole network lifespan.

It is not difficult to see the reason for this by recalling the TDMA scheduling scheme explained in section 4. The uppermost sink node practically carries the burden of transmitting the data packets of all other nodes in the whole network to the surface gateway. This node would certainly deplete its energy quickest, and worst, if it is downed or failed the whole network is downed with it. Putting the probability of the node's technical failure aside, estimating the battery capacity of this particular node is of special interest for estimating the UWSN lifespan. In a nutshell, if the battery capacity of the uppermost sink can be determined then the lifespan of the whole network can be quite easily estimated. It follows that the battery capacity (or lifetime) of all other nodes can be readily derived from the lifetime of the uppermost sink node.

It should be mentioned here that the battery capacity and the lifetime of the surface gateway is not an issue. This is because this gateway can be conveniently powered by a solar source and/or complement with a rechargeable battery. Moreover, changing or replacing the power supply unit in this node is not a difficult task at all. It is therefore safe to assume technically that the surface gateway has infinite power supply. It follows that the estimation method presented in this section will not include this gateway.

Three important parameters have been identified from the topology in Figure 2 for estimating the battery power capacity of a node. They include:

1. The sensor node data acquisition frequency – It is understood that higher data acquisition frequency/rate causes the nodes to stay active more frequent thus consuming more power. This parameter is denoted as D_f. The data acquisition frequency in this case study is made with reference to the works in (Boehm et al., 2002; Holman et al., 2003).
2. The signal transmission range between nodes – Transmission process consumed the most power in a node (cf Table 1) and unfortunately higher power is needed for long range data transmission. This parameter is denoted as S_R. By the way, keep in mind that for this case study the distances between nodes are kept at a constant of 50m.
3. The number of sensor nodes in a cluster – More sensor nodes aptly means more energy would be drained off from the cluster sink since more data packets need to be transferred to/through it. This parameter is denoted as N. Note that N is related to D_f. It is not difficult to see how power consumption in the sink node can quickly becoming an issue when N is large and D_f is high. As far as power consumption is concerned the choice of N is considered an important network design factor.

The battery capacity of a node in the network shown in Figure 2 can be derived from the basic transmitter power equation given in Urick, (1975). It is restated here as:

$$P_T = 2\pi \times 1\text{m} \times Dw \times IT$$

where,

P_T is the transmitter power to achieve an intensity of I_T at 1m in the direction of the receiver node measured in Watts.
D_w is water depth measured in meter.
I_T is power intensity at 1m from the source in unit of W/m^2.

However, the power intensity at 1m from the source can be written as:

$$I_T = 10^{SL/10} \times 0.67 \times 10^{-18}$$

where,

SL is the source signal level in *dB re μPa* and I_T is converted into W/m^2 with the conversion factor of 1 *μPa* = 0.67×10^{-18} (Dahl, Miller, Cato, & Andrew, 2007; "Reson, "Support > Self-help > Essential Acoustic Formulae",").

The source signal level (*SL*) can be derived from the basic passive sonar equation found in ("Reson, "Support > Self-help > Essential Acoustic Formulae"," ; Urick, 1975) as:

$$SL = SNR + TL + (NL - DI)$$

where,

SNR is signal-to-noise ratio (figure of merit) at the receiver node.
TL is transmission loss.
NL is underwater ambient noise level.
DI is source signal directivity index.
Note: All the parameters are quantified in *dB re μPa.*

By taking the following considerations for shallow water environment:

(i) With the nodes transmitting signal in omnidirectional, $DI = 0$.
(ii) Recommended nominal value of NL = 70 *dB re μPa* (Dahl et al., 2007; Urick, 1975).
(iii) Recommended nominal value of SNR = 30 *dB re μPa* ("Nat. Phy. Lab. (NPL), UK," ; Urick, 1975).

The *SL* expression can now be simplified into:

$$SL = TL + 100\ dB\ re\ \mu\text{Pa}$$

In shallow water acoustic communications, the acoustic signal is considered to propagate in "cylindrical spreading" mode and the transmission loss (*TL*) in *dB re μPa* for cylindrical propagation mode has been derived in ("Nat. Phy. Lab. (NPL), UK," ; Urick, 1975) as:

$$TL = 10\log S_R + \alpha SR \times 10^{-3}$$

where,

S_R is the range between a source node and a sink node in meter.
α is the frequency dependent medium absorption coefficient in *dB*/km.

The frequency dependent absorption medium coefficient (α) can be evaluated with reference to the works of (Fisher & Simmons, 1977; Jurdak, Lopes, & Baldi, 2004) as below where α is in *dB*/km and *f* in kHz:

$$\alpha = \begin{cases} 6.01 \times f^{0.8552} \times 10^{-2} & 1 \le f \le 6 \\ 9.7888 \times f^{1.7885} \times 10^{-3} & 7 \le f \le 20 \end{cases}$$

Note: (i) α is valid for temperatures between 4°C and 20°C.

(ii) Variants of α can be found in (Ainslie & McColm, 1998; Kinsler, Frey, Coppens, & Sanders, 1999).

Now let's look at an example. For a source-to-sink distance (S_R) of 50m with transmission frequency (*f*) of 8 kHz (as a typical voice range frequency adopted in underwater acoustic transmission) the following parameters can be obtained:

$\alpha = 0.404\ dB$/km
$TL = 17\ dB\ re\ \mu Pa$
$SL = 117\ dB\ re\ \mu Pa$
$I_T = 0.336 \times 10^{-6}$ W/m^2

Therefore, for a node placed at a depth of 50m, the power needed to transmit a signal such that a *SNR* of 30 *dB re μPa* is to be desired at the sink, can be calculated as:

$P_T = 2\pi \times 50 \times 50 \times 0.336 \times 10^{-6} = 5.3$m

Then let the number of sensor nodes in a cluster (in a particular tier) be *N* and each sensor node in this cluster will send *B* bytes of data acquired per one data sampling process to the sink node (at the same tier). So, the sink node would have to receive *NB* bytes (or 8*NB* bits) of data and then transmit all these bits to the next sink node which is S_R meters away. This whole event is considered a "receive-and-transmit" session for a sink node.

For a conservative analysis, 8*NB* bits can be translated approximately into 8*NB* signals in a sink node transmission. This means the power requirement at the sink node per data transmission can be derived as $8NBP_T$ watts. Do bear in mind however that the sink node will also consume power when receiving data bits. The receive-power consumption is typically around one-fifth of the transmit-power in various commercially available acoustic modems ("EvoLogics (Germany), LinkQuest Inc. (USA), Woods Hole Oceanographic Institution, (USA)."; Jurdak et al., 2004). The total power consumption per data reception and transmission at the sink node can thus be written as:

$P_S = 8NBP_T + 8/5(NBP_T) = 9.6NBP_T$ watts.

If the sensor node data acquisition frequency (or rate) is D_f, where D_f is quantified in samples/day, then the power consumption of a sink node in a day would be:

$P_D = D_f P_S$ watts.

At this juncture the power consumption per day in a sink node in a particular tier has been derived.

As mentioned earlier, the lifespan of the network shown in Figure 2 is practically depending on the battery power capacity of the uppermost sink node. It is thus important to know the power requirement of this node. Notice that the uppermost sink node is the last node in a chain of sink nodes in terms of data transmission. Essentially this sink node is responsible to relay data packets to the surface gateway forwarded by the sink nodes in other tiers plus data packets from its own cluster nodes. Hence for an M tiers network the power requirement for the uppermost sink node can be computed as:

$P_U = MP_D$ Watts/day ; P_U means the power for uppermost sink

With the value of P_U known, the battery power capacity needed by this node to support the network for a duration of say, D_y days, can be readily obtained by multiplying D_y with P_U. It should be pointed out at this point that the power capacity estimation presented above did not take into consideration the packet header bits. The header bits may vary from one application to the others so is not included in this estimation analysis. Power capacity estimation with header bits could be a further research direction for the interested readers. A practical example is now given below to illustrate the calculation of power capacity estimation.

Let's make some assumptions based on the network in Figure 2:

- There are 8 sensor nodes in a cluster ($N = 8$).
- Each sensor node needs 3 bytes to hold a sample of acquired data ($B = 3$). Note: 24-bit resolution is acceptable for a data sample in most cases.
- Sensor node is scheduled to perform 1 data sampling every 30 minutes throughout a day (24 hours).
- The power requirement for a sink to transmit a signal is taken from the example above i.e. $P_T = 5.3$mW. Note: S_R is 50m.
- It is a three tiers network ($M = 3$).
- It is one hop transmission for data packets.

The calculation:

The power requirement per data "reception-and-transmission" session for the uppermost sink node is,

$P_S = 9.6 \times 8 \times 3 \times 5.3 \times 10^{-3} = 1.2$ W.

For 1 data sampling per 30 minutes, the sampling frequency per day would therefore be,

$D_f = 48$ samples per day.

The total power consumption for a sink node per day is then,

$P_D = 48 \times 1.2 = 57.6$ W.

Hence the power requirement for the uppermost sink node per day with respect to a 3-tier arrangement would be,

$P_U = 3 \times 57.6 = 172.8$ W/day

So, if the network is expected to operate, say for 90 days, before the battery in the uppermost sink node is replaced, then this node should be supplied with a power capacity of at least 15.55 kW. Subsequently the power requirement for other sink nodes in the network can be estimated as:

2nd tier sink node $= 2 \times 57.6 \times 90 = 10.368$ kW
Bottom tier sink node $= 57.6 \times 90 = 5.184$ kW

Since all the sensor nodes are transmitting data packets only to the sink node at their respective tier in one hop manner therefore each sensor node may have a power requirement equivalent to that in a sink node i.e. 5.184 kW.

Let's have some feel of 'reality' on the value of P_U calculated above. P_U shows the power needed in a day which is 172.8W per 24 hours. This translates to 7.2W per hour (for simplicity a linear power consumption is assumed). If the node modem is powered by a typical 12V car battery then the current drained per hour would be 7.2/12 which is 0.6A. Based on a simplistic linear estimation, for 90 days (2160 hours) the current drained would be 1,296AH (Ampere-Hour).

Now, a car battery normally has a rating of 100AH. The uppermost sink node will therefore need at least 13 (i.e. 1296/100) car batteries to support the network for a 90-day operation. By the way, these batteries are to be connected in parallel configuration. If the modem can be operated with a 24V 100AH source, then the number of batteries would be halved. For instance, the UWM1000 modem from Link-Quest Inc. (USA) can be powered from 9V to 24V. Readers may refer to the reference link in (W.E. Johns, Notes on batteries) to gain further details on battery lifetime calculation/estimation.

CONCLUSION

This chapter presents a use case study to show how the proposed 2Q algorithm may be integrated into data transmission process in an IoT-based UWSN for long-term water pollution monitoring application. The UWSN adopted in this case study (cf Figure 2) is a static 3D architecture deployed in the form of a single chained multi-tier network with each tier having a cluster of nodes arranged in a centralized star topology. This chained-star topology appropriately formed a column of nodes to monitor a "column" of water body for pollutants in a long-term non-time critical mode.

An algorithm for data packets transmission literally termed as multi-tier TDMA scheduling algorithm is presented to show the possibility of allowing all the nodes in the network to transmit data packets without collision and no channel contention. The proposed 2Q data packet size optimization algorithm was integrated into this multi-tier TDMA timing scheme. This scheme however is by no means exhaustive i.e. there are possible to evolve variants out of it. For instance, the 2Q optimization can be integrated at the end of each TDMA frame instead of at the end of each sensor node transmission. Basic TDMA time slots calculation is explained and illustrated with numerical examples.

Since the proposed 2Q optimization algorithm was developed with respect to energy efficiency therefore the sensor/sink node battery power capacity issue is also being considered and discussed in this chapter. A method to estimate the battery power capacity for the sensor/sink node for the proposed IoT-based UWSN architecture is described. A numerical example is given based on the requirements of the uppermost sink node and this example is extended to illustrate how the power capacity of other nodes can be derived. Among all the nodes in the proposed network, the uppermost sink node is considered the principle node that shall determine the lifespan of the whole network. Therefore, its power capacity of uttermost importance.

A few topics/directions are highlighted in this chapter that could be of interest to those interested readers or researchers for further research or investigations. For instance, enhancing power capacity estimation method, variants of data transmission time calculation, transmission scheme based on different network architecture/topology, etc. just to name a few.

REFERENCES

Ainslie, M. A., & McColm, J. G. (1998). A simplified formula for viscous and chemical absorption in sea water. *The Journal of the Acoustical Society of America*, *103*(3), 1671–1672. doi:10.1121/1.421258

Akyildiz, I. F., Pompili, D., & Melodia, T. (2005). Underwater acoustic sensor networks: Research challenges. *Ad Hoc Networks*, *3*(3), 257–279. doi:10.1016/j.adhoc.2005.01.004

Akyildiz, I. F., Pompili, D., & Melodia, T. (2006). State-of-the-art in protocol research for underwater acoustic sensor networks. *Proceedings of the 1st ACM international workshop on Underwater networks.* 10.1145/1161039.1161043

Bhardwaj, M., Garnett, T., & Chandrakasan, A. P. (2001). *Upper bounds on the lifetime of sensor networks.* Paper presented at the ICC 2001. IEEE International Conference on Communications. Conference Record (Cat. No. 01CH37240). 10.1109/ICC.2001.937346

Boehm, A. B., Grant, S., Kim, J., Mowbray, S., McGee, C., Clark, C., Foley, D. M., & Wellman, D. (2002). Decadal and shorter period variability of surf zone water quality at Huntington Beach, California. *Environmental Science & Technology*, *36*(18), 3885–3892. doi:10.1021/es020524u PMID:12269739

Cui, J.-H., Kong, J., Gerla, M., & Zhou, S. (2006). The challenges of building mobile underwater wireless networks for aquatic applications. *IEEE Network*, *20*(3), 12–18. doi:10.1109/MNET.2006.1637927

Dahl, P. H., Miller, J. H., Cato, D. H., & Andrew, R. K. (2007). Underwater ambient noise. *Acoustics Today*, *3*(1), 23–33. doi:10.1121/1.2961145

Ethier, A., & Bedard, J. (2007). *Development of a real-time water quality buoy for the fraser river estuary.* Paper presented at the OCEANS 2007. 10.1109/OCEANS.2007.4449424

Fisher, F., & Simmons, V. (1977). Sound absorption in sea water. *The Journal of the Acoustical Society of America*, *62*(3), 558–564. doi:10.1121/1.381574

Holman, R., Stanley, J., & Ozkan-Haller, T. (2003). Applying video sensor networks to nearshore environment monitoring. *IEEE Pervasive Computing*, *2*(4), 14–21. doi:10.1109/MPRV.2003.1251165

Jurdak, R., Lopes, C. V., & Baldi, P. (2004). Battery lifetime estimation and optimization for underwater sensor networks. *IEEE Sensor Network Operations*, *2006*, 397–420.

Kalpakis, K., Dasgupta, K., & Namjoshi, P. (2002). Maximum lifetime data gathering and aggregation in wireless sensor networks. In *Networks* (pp. 685–696). World Scientific. doi:10.1142/9789812776730_0055

Kinsler, L. E., Frey, A. R., Coppens, A. B., & Sanders, J. V. (1999). *Fundamentals of acoustics*. Academic Press.

Kong, J., Cui, J.-h., Wu, D., & Gerla, M. (2005). *Building underwater ad-hoc networks and sensor networks for large scale real-time aquatic applications*. Paper presented at the MILCOM 2005-2005 IEEE Military Communications Conference.

Lin, Y. D., Yeh, J. H., Yang, T. H., Ku, C. Y., Tsao, S. L., & Lai, Y. C. (2009). Efficient dynamic frame aggregation in IEEE 802.11 s mesh networks. *International Journal of Communication Systems*, *22*(10), 1319–1338. doi:10.1002/dac.1028

Nat. Phy. Lab. (NPL), UK. (n.d.). Retrieved from http://resource.npl.co.uk/acoustics/techguides/concepts/sonarequ.html

Penteado, D., Costa, L., & Pedroza, A. (2010). *Deep-ocean data acquisition using underwater sensor networks*. Paper presented at the The Twentieth International Offshore and Polar Engineering Conference.

Reason. (n.d.). *Support > Self-help > Essential Acoustic Formulae*. Retrieved from http://www.reson.com/support/self-help/essential-acoustic-formulae/

Regan, F., Lawler, A., & McCarthy, A. (2009). *SmartCoast Project- Smart Water Quality Monitoring System*. Environmental Protection Agency.

Sozer, E. M., Stojanovic, M., & Proakis, J. G. (2000). Underwater acoustic networks. *IEEE Journal of Oceanic Engineering*, *25*(1), 72–83. doi:10.1109/48.820738

Urick, R. J. (1975). *Principles of underwater sound-2*. W.E. Johns, Notes on batteries. Retrieved from http://www.gizmology.net/batteries.htm

Water Framework Directive, W. F. D. (n.d.). *Council Directive 2000/60/EC*. Retrieved from http://www.defra.gov.uk/environment/water/wfd/

ADDITIONAL READING

Alam, S. S., Islam, A. J., Hassan, M. M., Rafid, M. N. M., Chakma, N., & Imtiaz, M. N. (2018). Design and development of a low-cost IoT based environmental pollution monitoring system. In *Proceedings International Conference on Electrical Engineering and Information & Communication Technology (iCEEiCT)*, pp. 652-656. 10.1109/CEEICT.2018.8628053

Chowdury, M. S. U., Emran, S. U. T., Ghosh, S., Pathak, A., Alam, M. M., Absar, N., Anderson, K., & Hossain, M. S. (2019). IoT based real-time river water quality monitoring system. *Procedia Computer Science*, *155*, 161–168. doi:10.1016/j.procs.2019.08.025

Daigavane, V. V., & Gaikwad, M. A. (2017). Water quality monitoring system based on IOT. *Advances in wireless and mobile communications. Research India Publications*, *10*(5), 1107–1116.

Islam, M. M., Arefin, M. S., Khatun, S., Mokarrama, M. J., & Mahi, A. M. (2020). Developing an IoT based water pollution monitoring system. *Proceedings of Image Processing and Capsule Networks, Advances in Intelligent Systems and Computing, Springer, 1200*, 561–573. doi:10.1007/978-3-030-51859-2_51

Mourvika, S., Monika, A., Jyoti, B., & Trupti, M. (2018). IOT based water quality monitoring system. *International Journal of Scientific Research in Computer Science. Engineering and Information Technology, 3*(1), 1423–1428.

Shafi, U., Mumtaz, R., Anwar, H., Qamar, A. M., & Khurshid, H. (2018). Surface water pollution detection using Internet of Things. In *Proceedings 15th International Conference on Smart Cities: Improving Quality of Life Using ICT & IoT (HONET-ICT)*, pp. 92-96. 10.1109/HONET.2018.8551341

Soundarya, P., Prathyusha, V., Niharika, A. V., Karthick, T., & Suresh, S. (2017). IoT based water quality monitoring system. *International Journal of Applied Engineering Research: IJAER, 12*(16), 5447–5454.

KEY TERMS AND DEFINITIONS

Acoustic Channel: Is a communication channel or a link for underwater communication using acoustic wave for data or information transmission.

Battery Power Estimation: A technique to systematically estimate the power of a battery to be used in a node in UWSN.

IoT: The internet of things (or in short IoT) is a term referring to the millions or billions of physical devices that are connected to the internet for collecting and sharing data. As many as 100 billion of connected IoT devices is forecasted by 2025.]

IoT Node: Is a sensor node deployed in the UWSN to perform some gathering of sensory information and able to communicate with other nodes in the network. This node can be accessed via Internet.

Long-Term Monitoring: Is referring to the time span of a UWSN node for pollution monitoring. Typically for 3 months or more of continuous monitoring operation.

Pollution Monitoring: Pollution monitoring in this chapter is referring to as a kind of environment monitoring via periodic sampling of water quality for the purpose to observe and study the impact of the pollution.

Sensor Node: Is a physical sensor deployed in the UWSN for gathering and processing sensory information for underwater environment.

Sink Node: A node in the UWSN to aggregate data from other sensor nodes deployed in the network.

UWSN: Underwater wireless sensor networks (UWSN) is a wireless network deployed in underwater environment using some forms of physical sensors with acoustic capability to perform collaborative monitoring and data collection tasks.

Chapter 3
Composite Identity of Things (CIDoT) on Permissioned Blockchain Network for Identity Management of IoT Devices

Anang Hudaya Muhamad Amin
Higher Colleges of Technology, UAE

Fred N. Kiwanuka
Higher Colleges of Technology, UAE

Nabih T. J. Abdelmajid
Higher Colleges of Technology, UAE

Saif Hamad AlKaabi
Higher Colleges of Technology, UAE

Sultan Khalid Abdulqader Rashed Ahli
Higher Colleges of Technology, UAE

ABSTRACT

Internet of things (IoT) is in the forefront of many existing smart applications, including autonomous systems and green technology. IoT devices have been commonly used in the monitoring of energy efficiency and process automation. As the application spreads across different kinds of applications and technology, a large number of IoT devices need to be managed and configured, as they are capable of generating massive amount of sensory data. Looking from this perspective, there is a need for a proper mechanism to identify each IoT devices within the system and their respective applications. Participation of these IoT devices in complex systems requires a tamper-proof identity to be generated and stored for the purpose of device identification and verification. This chapter presents a comprehensive approach on identity management of IoT devices using a composite identity of things (CIDoT) with permissioned blockchain implementation. The proposed approach described in this chapter takes into account both physical and logical domains in generating the composite identity.

DOI: 10.4018/978-1-7998-6709-8.ch003

1. INTRODUCTION

The rapid growth in the development of Internet-of-Things (IoT) has led to an increase in its utilization for smart applications. In green technology, IoT is becoming a forefront in the monitoring of energy efficiency and process automation. In a complex monitoring systems such as smart building and smart factory, large number of IoT devices need to be managed and configured. In addition, these devices are used in different kinds of integrated applications that could generate massive amount of sensory data. As such, there is a need for a proper mechanism to identify each IoT devices within the system and their respective applications. Participation of these IoT devices in complex-systems requires a tamper-proof identity to be generated and stored for the purpose of device identification. In addition, having an effective identity management mechanism of IoT devices would enable us to eliminate the possibility of identity spoofing and presence of rogue devices in the system. This perhaps could be achieved through integrating multiple information that defines the physical and logical identity of the devices.

With the advent in the field of process automation, Industrial process control and monitoring is a vital task that has been carried out by networks of Internet-of-Things (IoT). Critical event such as a surge in the electrical current or sudden increase in the temperature of boiler could be detected and monitored seamlessly. Rapid growth in the utilization of IoT in smart applications have continuously expand the IoT deployment in large-scale networks.

Sustainability is the key consideration for IoT deployments in smart applications. According to Mahadi et al. (Abu Hassan et al., 2018), there are four important factors that influence the sustainability of IoT usage in smart applications, namely performance expectancy, effort expectancy, social influence, and facilitating conditions. These four factors are derived from the Unified Theory of Acceptance and Use of Technology (UTAUT), as described in (Venkatesh et al., 2003).

There are numerous examples of IoT implementations for smart applications, including the works by Jain et. al. (Jain et al., 2019) in smart foundry and Catarinucci et. al. (Catarinucci et al., 2015) in their works on IoT-aware smart healthcare system. Apart from these initiatives, there are a number of IoT deployments focusing on enhancing and improving existing green technology applications. These can be seen from the works carried out by Garcia et. al. (Garcia et al., 2018) in wireless sensor network (WSN) based monitoring scheme for green technology, and IoT-based smart agriculture by Gondchawar and Kawitkar (Gondchawar & Kawitkar, 2016).

Our particular interest in IoT deployment for smart applications is in the smart building monitoring and management. Smart building usually comprises of different types of IoT devices and different kinds of applications. Rapid advances in different kinds of technology, including IoT, data analytics, and machine learning has risen up the demands for smart-building applications. There are different types of smart building applications ranging from smart office, smart library, smart home, and smart facilities. The benefits of having smart systems in building and infrastructure management is such that it helps in ensuring reduction in the amount of wasted energy used, as well as improved resource utilization.

Smart building applications typically consist of five components as described by Qolomany et. al (Qolomany et al., 2019). Figure 1 shows the composition of these components, which include sensors and actuators, smart control devices, software platform, networking and communication, and HVAC system. Integration of these components is essential in ensuring smooth execution of the applications.

An important aspect of smart building management is to ensure the safety and security of interconnected devices, users, and the applications. Security is an important aspect in smart building applications as its breach impact to the physical objects tend to be at a larger scale since it can directly affects our

Figure 1. Smart building components

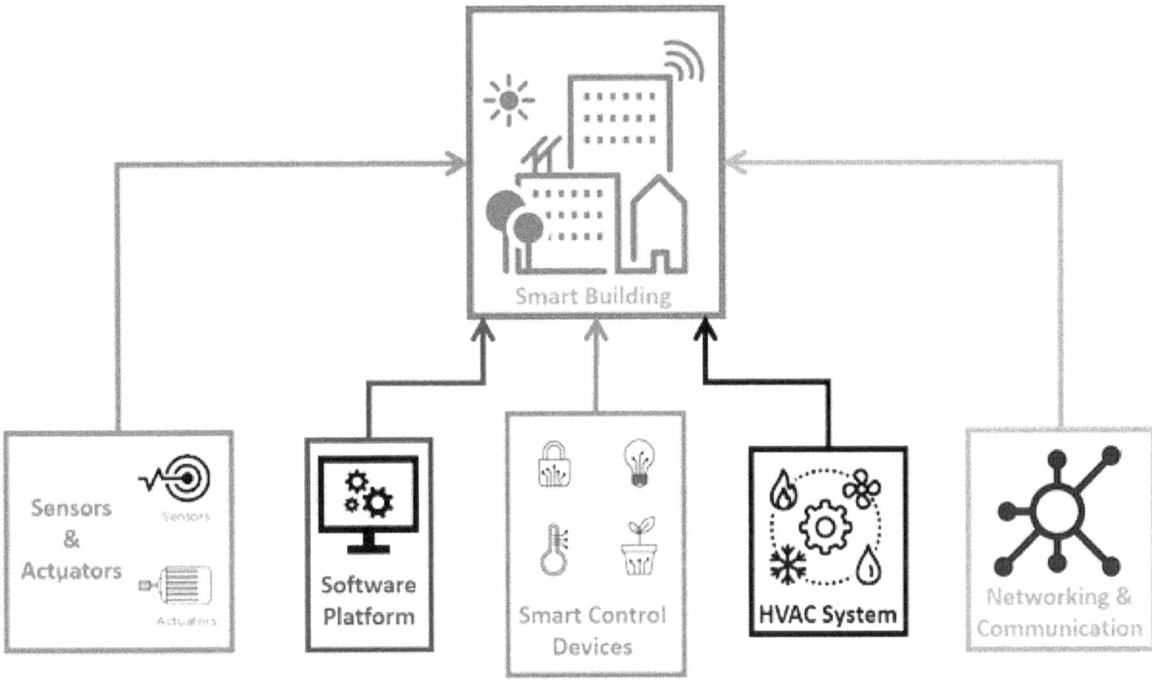

physical environment. For instance, if a hacker can hack into the smart air-conditioning device, they can simply manipulate the temperature control for the building. Some of the important works in ensuring the safety and security of IoT devices in smart building applications can be seen through the works of Bandara et al. (Bandara et al., 2016) and Hernandez-Ramos et al. (Hernández-Ramos et al., 2015) on secure access control framework for API-enabled devices.

Apart from access control, another important security measure to be considered is the identity management of IoT devices in smart building applications. Device identity plays an important role in ensuring the system is secure from identity spoofing and the existence of rouge devices as described by Zhang et al (Zhang et al., 2015). With increasing number of devices that take part in smart building systems, the need for a tamper-proof identity management is essential. In addition, our hypothesis is such that device identity management should also incorporate information related to different kinds of applications commonly accessing the devices, as well as the information on users accessing data from these devices. With this composite information, the device identity could be further strengthened by adding multiple contextual information into it.

In this chapter, we present a comprehensive approach on identity management of IoT devices using a composite Identity-of-Things (CIDoT) with permissioned blockchain implementation. Our proposed approach takes into account both physical and logical domains in generating the composite identity. The logical domain also include the respective application that utilizes the IoT device, either for sensory data acquisition or process automation. We propose the use of permissioned blockchain for identity storage and verification, which enables the identity to be immutable. For simulation purposes, we will demonstrate the use of CIDoT in the smart building application. The proposed identity management would enable the system owner to acquire a holistic view of the IoT device identity from the infrastructure and application perspectives.

The outline of this chapter is as follows: Section 2 will present some of the reviews on the current works related to IoT device identity. Section 3 will be focusing on the theoretical foundation for composite identity for devices, from the context of smart building applications. In section 4, we delve into the possibility of incorporating IoT data in a blockchain network, by looking into some existing works in IoT-blockchain implementations. We will also explore the potential of blockchain for IoT identity management. Section 5 presents the proposed framework for composite identity-of-things on IoT-Blockchain network for smart building applications. Discussions on the underlying structure of the proposed framework will be presented. Finally, Section 6 provides a conclusion for this chapter.

2. IOT AND IDENTITY

IoT network usually comprises of different types of sensors and processing devices that are connected over a communication network. A typical structure of IoT network comprises of heterogeneous devices as described by Sethi and Sarangi (Sethi & Sarangi, 2017) can be visually represented as layers as shown in Figure 2.

The sensory layer is basically the physical layer, which contains sensors used to sense and gather information about the environment. The communication layer is responsible for connecting to other smart things, network devices, and servers. Finally, the application layer deals with the process of delivering application specific services to the user.

The existence and introduction of IoT devices enables seamless integration between physical and cyber world (Wang, 2018). In this particular circumstance, it makes the physical world easier to be attacked through the cyber world. A malicious instruction or execution of code from the cyber world can cause serious damage in the physical world, and thus, IoT security issues should be handled carefully and effectively. Tackling security issues related to IoT devices requires specific measures, due to several distinct properties such as resource-constrained end-devices, high privacy requirements, heterogeneous devices and communications, and trust management.

An important security aspect in relations to IoT devices is their identity. With common heterogeneity and mobility features of IoT devices, identity requires a more dynamic management approach. According to Mahalle et. al. (Mahalle et al., 2010), with IoT devices, addressing identity problem requires changes to the architecture for naming, addressing, and discovery. These changes are essential as to be capable of managing different forms of communications and systems integration. The term Identity-of-Things (IDoT) as discussed by Lam and Chi (Lam & Chi, 2016), basically extends the functionality of existing identity and access management (IAM) models by considering logical and physical identity of devices based upon several distinctive attributes and different kinds of authentication approaches.

2.1. Attacks on IoT Device Identity

Generally, IoT devices are mostly prone to the network-based attacks. Common attacks on IoT device identity includes distributed denial-of-service (DDoS) and IP spoofing. These types of attack basically take advantage of the vulnerabilities related to the network configurations of ioT devices, including their identity.

Figure 2. Layered representation of IoT Network

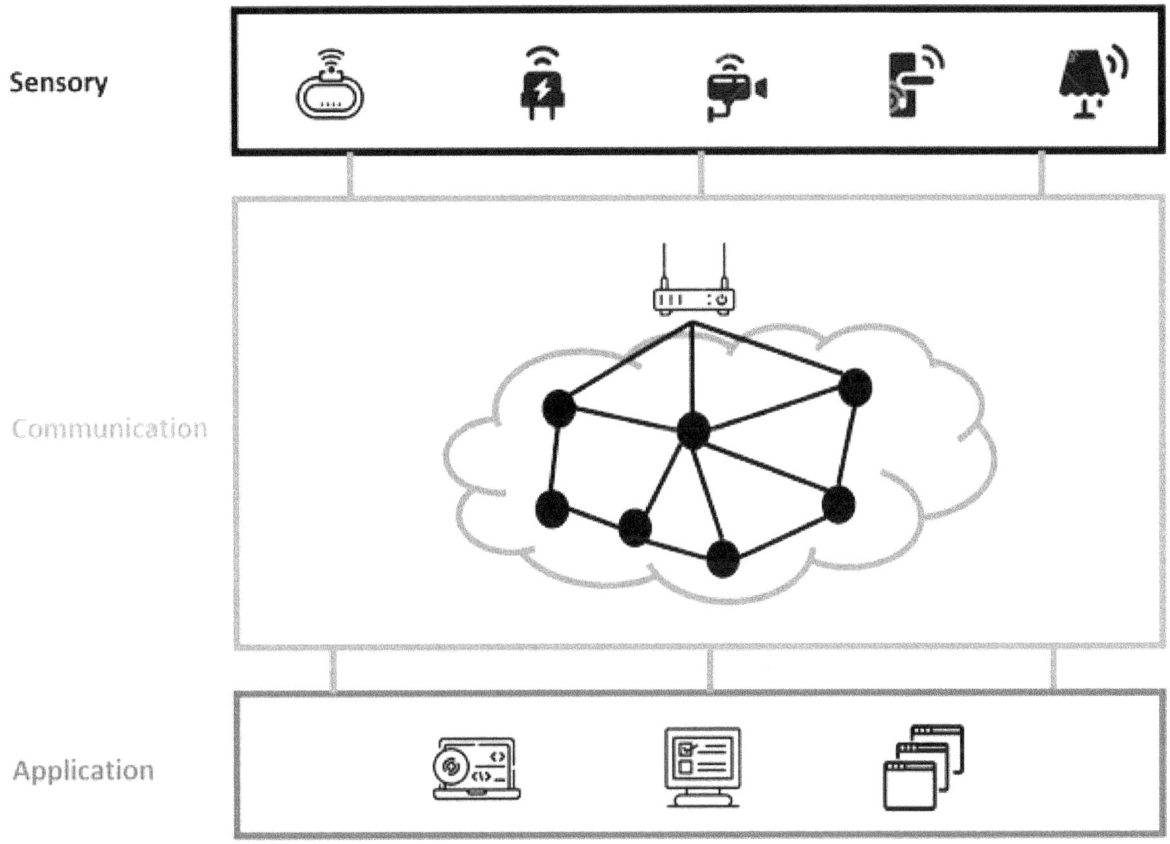

Consider a scenario of an IoT infrastructure as shown in Figure 3. The IoT devices in this network are connected to a gateway that enables different applications to access the data generated by these devices. Each application may require access to the data generated by one or more of these devices. These applications in return, can be used by multiple different users.

This particular scenario as shown in Figure 3 could raise two important questions: Firstly, how users could verify that data used by the application comes from a legitimate IoT devices within the large-scale distributed network? Secondly, is there any mechanism to check the validity and authenticity of the application accessing data from the IoT devices? These questions need to be addressed in order to ensure the security and safety of all the components of the network, including the devices, applications, and users.

Figure 4 shows an extended scenario whereby a rogue IoT device and malicious application exist, which attempt to access data from the IoT device. This scenario may happen due to the fact that the identity of each entity within the system is currently managed in silos, in which the identity of the devices are managed separately from the applications and users that utilizing and accessing these devices. In this particular situation, the gateway should have a complete identity information of all the participating entities within the system.

One of the mostly known DDoS attacks on IoT devices is the Mirai attack, as described by Kolias et al (Kolias et al., 2017). Mirai exploits the insecure IoT devices through their open Telnet ports. It scanned

Figure 3. Common infrastructure for IoT system with connected devices, applications, and users

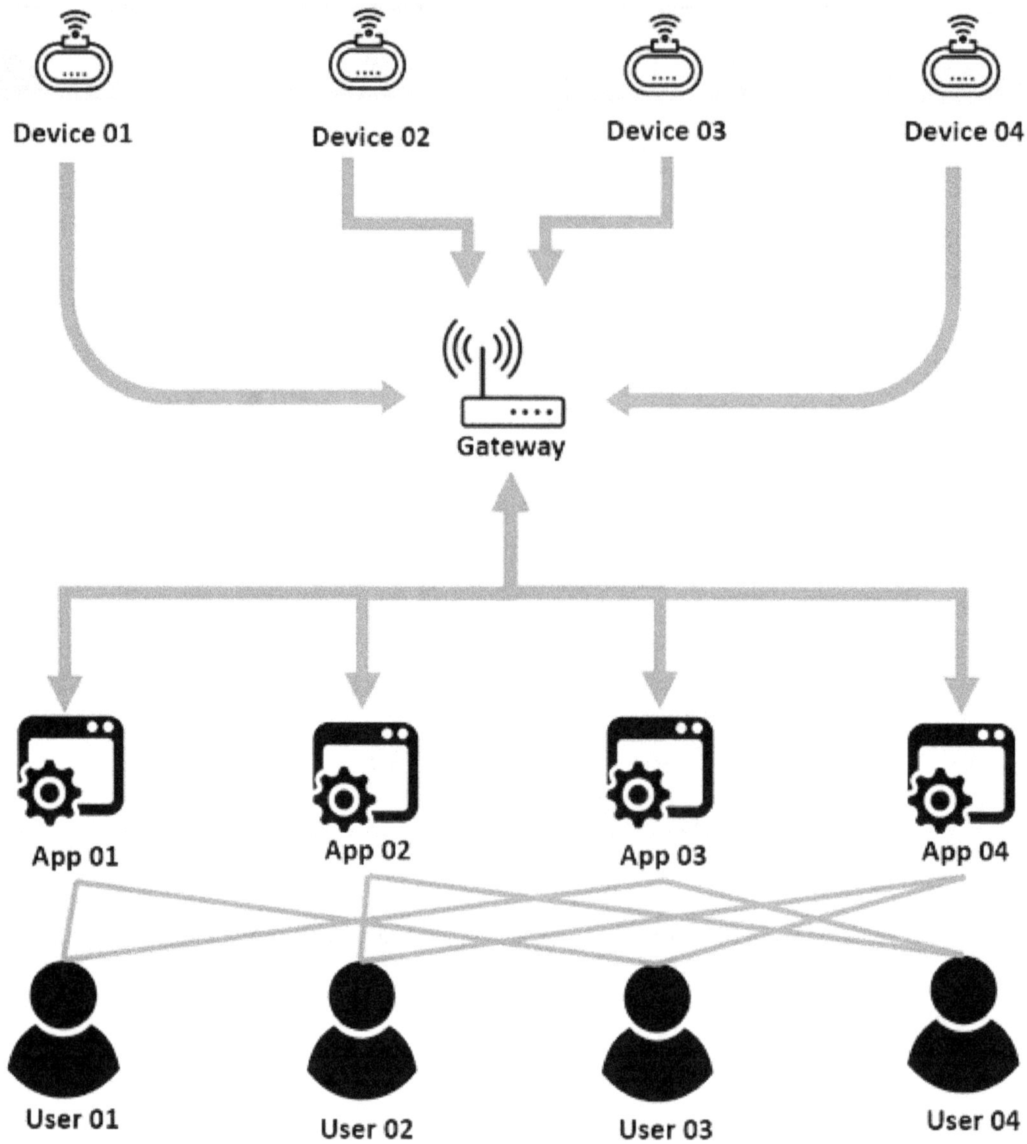

big blocks of the internet for open Telnet ports, then uses default username/password combinations in attempt to log in. The attack was able to amass a botnet army. Figure 5 shows the Mirai attack mechanism.

In this attack, IoT devices act as the intermediary, used by the attacker to launch massive attack on a particular target. Mirai's first big wave of attacks came on September 19, 2016, and was used against the French host OVH (Fruhlinger, 2018). The OVH hosted a tool that is used by Minecraft server hosts to fight against the DDoS attacks.

Another form of attack on IoT device identity can be seen in the Sybil attack. A Sybil attack as described by Jan et. al (Jan et al., 2015) whereby an adversary (attacker) attempts to forge multiple identities at a given time to mislead legitimate IoT devices into believing that they are having many neighbors.

Figure 4. Presence of rouge device and malicious application that can compromise the IoT network and related applications

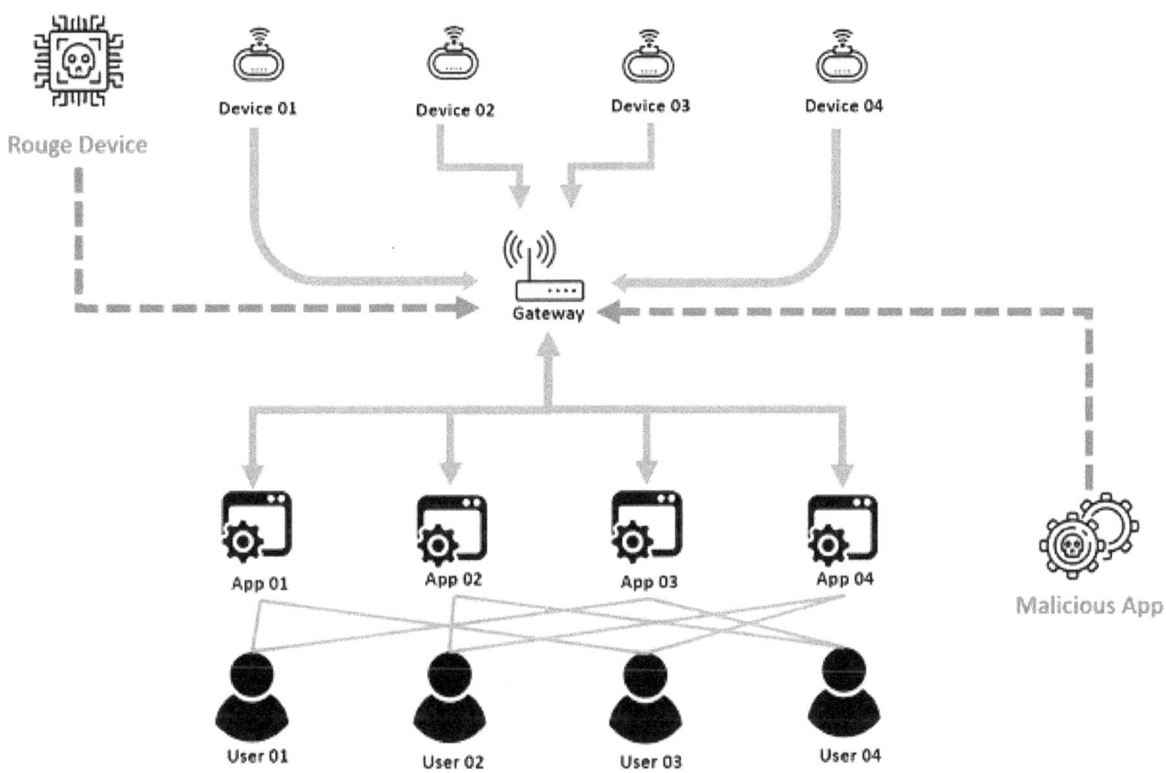

This kind of attacks usually happened as a results of device-to-device communications over error-prone wireless channels.

2.2. IoT Device Identity Management Models

There are a number of different approaches in strengthening the identity management for IoT devices, as well as attack detection mechanisms involving identity spoofing that have been published in the literature. Jan et al (Jan et al., 2018) proposed the used of two-tier detection scheme for detecting Sybil nodes and their forged identities. In this particular work, the detection involves identifying nodes that exert high energy as potential Sybil nodes. On the other related works, Guo and Heidemann (Guo & Heidemann, 2018) proposed the use of device detection mechanism using information retrieved from the IP traffic flow. The proposed scheme utilizes IP-based identity for both public IoT devices, as well as other devices that are located within a similar network. The main issue of this kind of detection mechanisms is such that it relies on a singular form of identity that could easily be tampered with.

In different implementation, Trnka et al (Trnka et al., 2018) proposed a centralized identity repository that hold the records of all IoT devices and their roles. The OAuth 2.0 token mechanism was used and governed by the repository for the purpose of communication initiation and access to all the services offered within the network. The work of Bello and Mahadevan (Bello & Mahadevan, 2019) is also focusing on centralized conceptual cloud-based identity management framework, which is based on a cloud management model for IoT device identity.

An identity management model that incorporates both logical (application/user) and physical (device) identities can be seen in the works of Bernabe et al (Bernabé et al., 2017) on holistic identity management using a claims-based approach. The proposed approach allows partial identity to be used for verification, allowing personally identifiable information (PII) of users and devices to remain confidential. A limitation of such approach is that it still relies on a centralized identity credential issuer. Hence, dependency on a single entity in managing identity is still exist, and it is prone to single point of failure.

Apart from the centralized approach in managing the identity of IoT devices in a distributed network, there are several approaches that look into integrating multiple information as identity for IoT devices. The additional information such as the application accessing the device, as well as the user/owner of the device can be incorporated as part of the identity information. Zhu et al (Zhu et al., 2017) proposed a Blockchain-based Identity Framework for IoT (BIFIT). The proposed scheme introduces identity self-management approach that can be performed by end users. The proposed approach basically extracts the device signatures and creates blockchain-based identifiers for their device owners. It also correlates device signature (low level identities) and owner identity in order to use them in authentication credentials and to make sure that any IoT entity is behaving normally.

In this chapter, we will explore the possibility of creating and using a composite identity as the identity model for IoT devices in a distributed network. The hypothesis is such that adding more information will basically strengthen the identity credentials for IoT devices. In the following section, we will describe the theoretical foundation for the proposed composite identity of IoT devices.

3. COMPOSITE IDENTITY: A THEORETICAL PERSPECTIVE

The notion of composite identity is increasingly adapted by different kinds of applications that usually being deployed in a distributed network. For instance, Singi et al (Singi et al., 2019) introduces the composite identity framework for global distributed software delivery. The common hypothesis of the use of composite identity in identifying object or element is such that it strengthens the identity against potential attacks like identity spoofing and forged identity.

Our definition of composite identity is a composition of multiple identity elements that builds up a single main identity of a particular IoT device. Composite identity basically incorporates multiple identifiable information that distinctively represent a device and its related applications or operations. Having composite identity enables clear verification and authentication of devices, similar to how individuals are verified using multiple identity elements (ID card and fingerprint). Figure 5 shows the composite identity for IoT devices.

The composite identity *ID* for a given device *x* can be represented in the following equation:

$$ID_x = \{\alpha n_j; 1£n£m \quad (1)$$

Whereby αn is an identity element within the m composition.

Collectively, for a system with *X* number of devices x_i where $1£i£X$, there exists (n) composite identities each with a unique identifier ID_x. The composite identity is defined by a set of *m* identity elements αn as defined in Equation 2.

Figure 5. Identity elements that comprise the composite identity for IoT device.

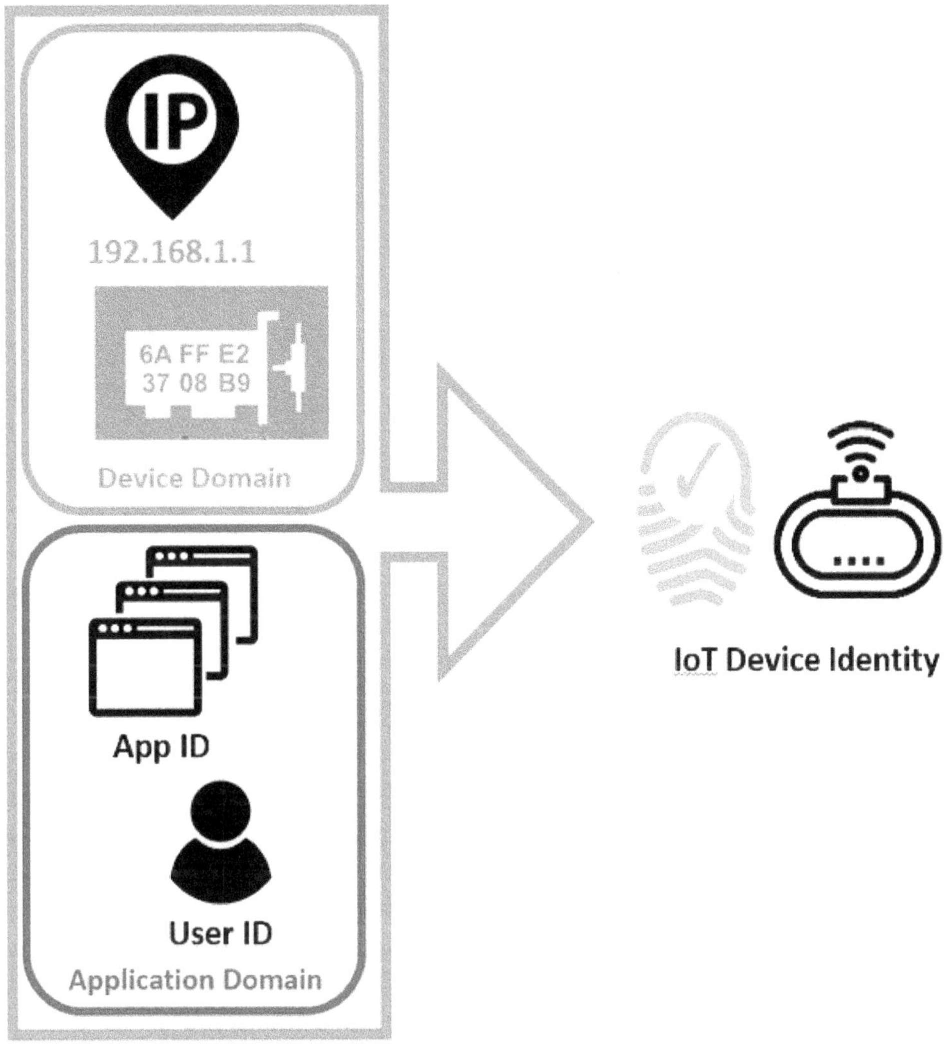

$$x_i \leftarrow ID_i \leftarrow \alpha_{in};\ for\ 1 \leq n \leq m;\ 1 \leq i \leq X \qquad (2)$$

From this representation, we can observe that there are two important factors that contribute to the size of the composite identity, namely the number of IoT devices and the number of identity elements that are incorporated into the device identity. Figure 6 and 7 shows the projected growth in terms of the size of the identity with increasing number of identity elements and the number of IoT devices within the network.

A definition of composite identity is basically a combination of different identity elements that builds up a single main identity of a particular IoT device. Composite identity can provide a complete view of known identity including its applications and operations by incorporating multiple identifiable information that clearly represent a device. By implementing composite identity, we enables clear verification

Figure 6. Growth trend in the number of identity element data per network for increasing number of devices in a network

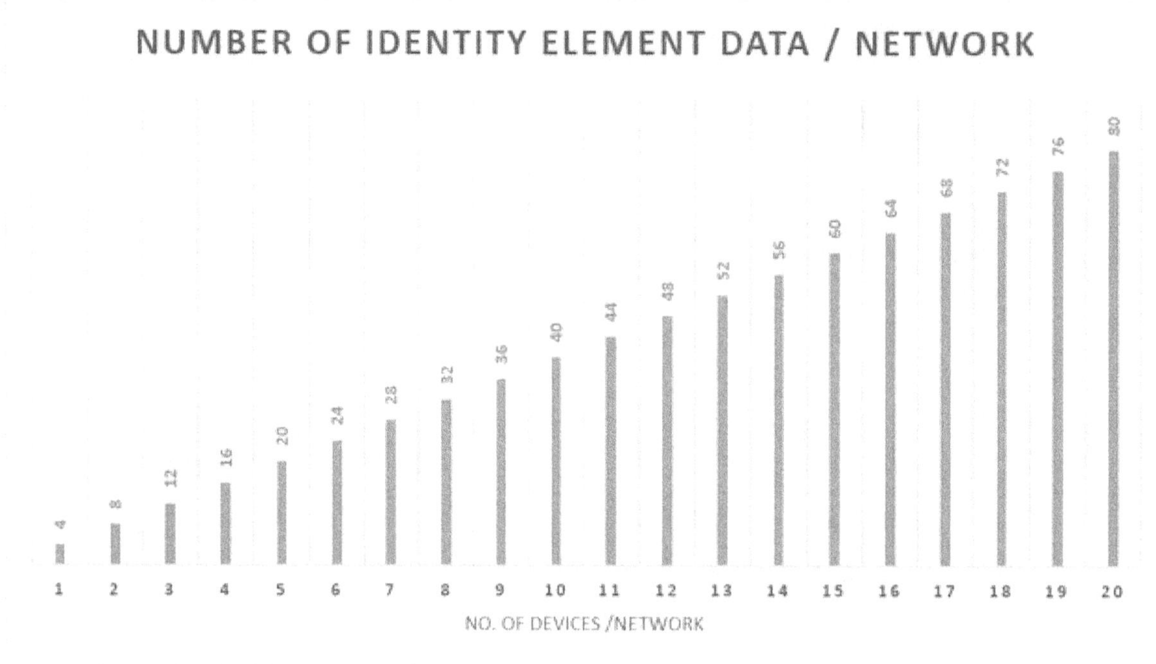

of devices, in a similar way on how individuals are verified using multiple identity elements (ID card and fingerprint).

In considering the identity elements for composite identity, two important domains are taken into account as previously shown in Figure 5, namely device and application domains. Each domain consists of a number of identity elements. Note that both logical and physical identity are being considered. For example, in the device domain, we considered both device addresses (IP address – logical) and MAC address – physical) as identity elements in the composite identity. On the application domain, the application identifier, such as the software license and the user ID can be used as the identity elements in the composition.

Based on the projected trends, one of the key important issue to be considered in implementing composite identity for IoT devices is the network storage capacity to accommodate huge amount of identity elements for all the devices within the network. To address this particular issue, we propose the use of distributed storage in a form of distributed ledger within a permissioned blockchain network.

Blockchain is a distributed ledger technology that enables secure and tamper-proof data transactions through its smart contract implementations. In the following section, we will further discuss the implementation of blockchain network for distributed data storage in IoT network. We will delve into the possibility of implementing a permissioned blockchain scheme for IoT device identity management.

Figure 7. Growth trend for the number of identity elements per network with increasing number of identity elements per identity for the entire network

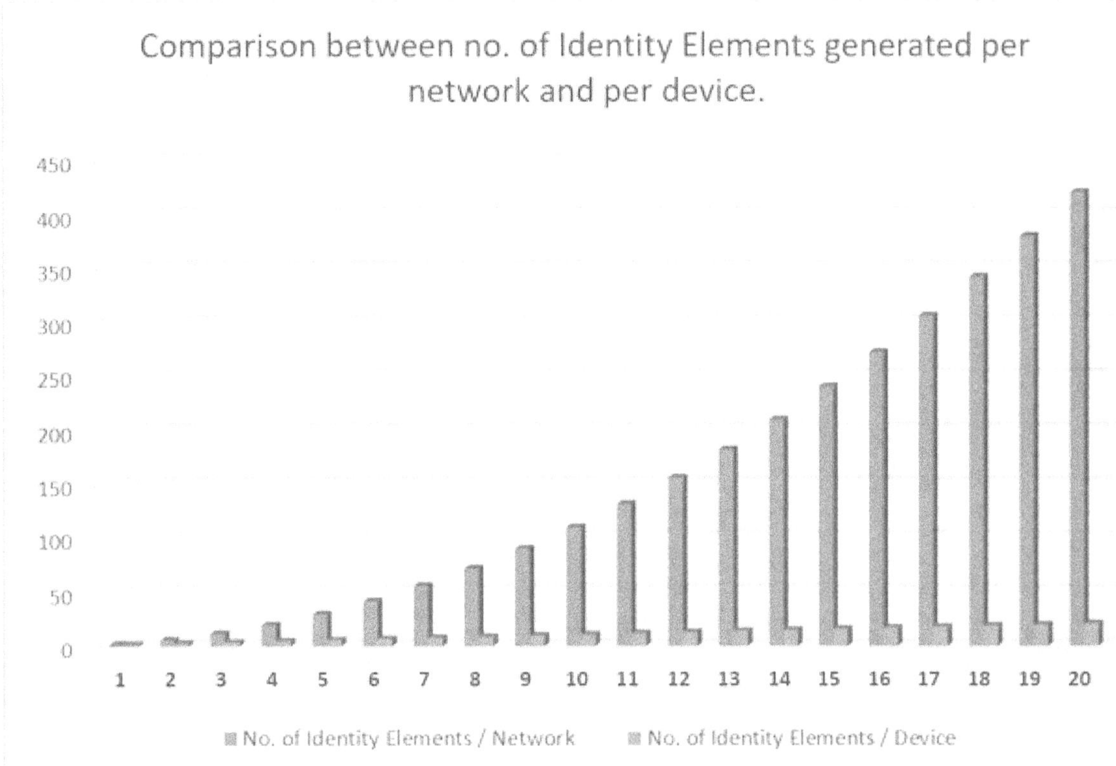

4. IOT AND BLOCKCHAIN

One of the limitations of data storage and processing in IoT network is such that it is commonly being conducted in a centralized manner. A typical IoT network infrastructure usually consists of IoT devices, a gateway, and a centralized repository for data processing and storage. With this particular system design, it is prone to single point-of-failure. For instance, a centralized event monitoring application, such as the works of Jan et al. (Jan et al., 2018) suffer from several drawbacks when applied to the IoT-related applications. With massive number of IoT devices in a network, the amount of transmitted data to a centralized repository can be significantly increased. As a result connection bottleneck from IoT network to the central server could cause the system performance to be deteriorated. Another potential problem with the existing setup is affecting the integrity of event data obtained from the IoT devices, as described in the works by Chen et al (Chen et al., 2018).

4.1. Blockchain for Distributed Data Storage and Verification

In a typical setup, there are two types of blockchain models namely public and permissioned blockchain. Public blockchain consists of a distributed ledger that can be accessible by anyone, without any prior authorization and verification. Permissioned blockchain on the other hand, is a form of blockchain model

Figure 8. Smart contract defines the access to data that resides on the distributed ledger within the blockchain network

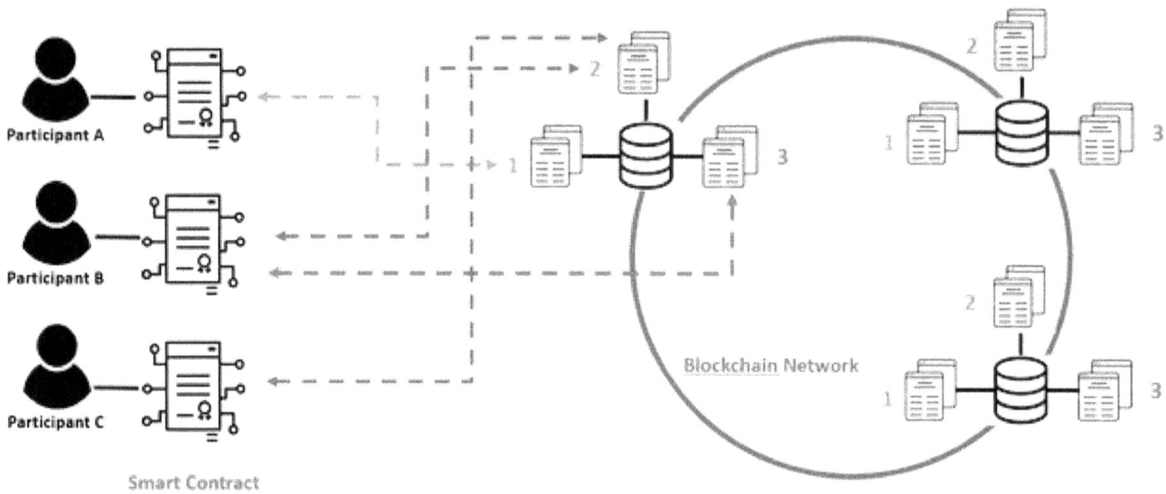

comprises of authorized participants that can collaboratively access and share data within a secure data network. The benefit of implementing permissioned blockchain for distributed data storage and verification is that it allows data to be shared and replicated in a secure form with consensus, provenance, immutability, and finality (Andolfatto, 2018). Consequently, this model can be useful in securing critical information including identities such as IoT device identity.

In a permissioned blockchain network, access to data within the distributed ledger are confined by the rules and policies created within a smart contract. A smart contract can be customized according to different data access level required. Figure 8 shows how smart contract defines the access level for three different participants accessing data in the distributed ledger. Note that Participant A could only access data that is related to Component 1 but not Components 2 and 3. This is achieved through the customization done on the smart contract.

The blockchain network also allows participants to verify and access the data through predefined set of queries in the smart contract implementation. This feature enhances the capability of blockchain network as a mean for distributed data storage and verification mechanism.

4.2. Blockchain For IoT Device Identity Management

A number of IoT identity management models have implemented blockchain for its data storage mechanism. These include the works by Omar and Basir (Omar & Basir, 2018) on semi-decentralized identity management model using blockchain for IoT devices. Their proposed model creates and stores encrypted device identity within a blockchain network. The proposed model includes both identity creation as well as transfer of ownership features within its implementation. In a related work, Dittmann and Jelitto(Dittmann & Jelitto, 2019) has also proposed an identity management model based on blockchain proxy concept that stores the PKI-encrypted identity of IoT devices.

Beside identity management, implementation of blockchain can also be seen in the access control management for IoT devices. Xu et al (Xu et al., 2018) has proposed a blockchain-enabled decentralized capability-based access control for IoT devices. The implementation model being proposed includes a robust identity-based capability token management strategy that utilizes a smart contract for registration, propagation and revocation of the access authorization. In another related work, Guin et al (Guin et al., 2018) has also proposed the use of blockchain network to authenticate resource-constrained IoT devices using identities generated by SRAM-based physically unclonable functions (PUFs) in order to minimize the impact of the attacks conducted by rogue devices.

The related works discussed in this section have so far been focusing on the IoT device identity management, without extending its consideration on the applications and the users that utilizing these devices. It is also important to take into account these factors as they are highly related to the IoT devices being used in the network. In addition, both logical and physical identity of IoT devices, incorporating the

User and/or application information that are linked to these devices should be considered. With such approach, we expand the spectrum of identity information of IoT devices, to include both device and application domains.

In the next section, the composite identity-of-things with an integrated blockchain network for IoT device identity management will be further discussed and explored. A typical setup for identity management using permissioned blockchain on smart home application will be reviewed.

5. COMPOSITE IDENTITY-OF-THINGS (CIDOT) FRAMEWORK ON IOT-BLOCKCHAIN NETWORK

Identity of IoT device typically being set up using either the logical or physical identification of the particular device. The dependency for single identification element for IoT devices as discussed in Section 2, may leads towards potential breach of confidentiality and eventually capable of bringing down the entire device network, through the network attacks such as the Sybil and Mirai attacks (See Section 2.1).

Composite identity of IoT device is mainly composed of a set of logical and physical identity elements, including MAC address, IP address, application ID, device type, and device owner. The composition is solely rely upon the network provider or owner, in defining which identity elements should be taken into consideration. In implementation within an integrated IoT-blockchain network, this composite identity is consequently stored as a data block in a shared distributed ledger of the blockchain network.

The CIDoT identity management model for IoT devices as described in the previous section basically utilizes a permissioned blockchain network for identity storage and verification. A permissioned blockchain network can be implemented in different kinds of platform including Hyperledger Fabric (Androulaki et al., 2018), Quorum (Baliga et al., 2018), and Multichain (Greenspan, 2015). In this chapter, the Hyperledger Fabric implementation has been considered. Hyperledger Fabric is a modular and extensible open-source system for permissioned blockchain deployment. It is one of the Hyperledger projects hosted by the Linux Foundation (www.hyperledger.org).

There are a number of different components that made up the hyperledger fabric. Table 1 shows the components and their respective functions.

The permissioned blockchain model for distributed identity management requires the use of blockchain proxy that handles the device identity records to be stored and verified on blockchain network, as proposed in (Dittmann & Jelitto, 2019). Figure 9 shows the system view of the identity management model for IoT devices. In ensuring the authenticity of data stored in the blockchain network, a PKI certificate authority (CA) is registered with the blockchain that acts as identity provider trusted by the

Table 1. Hyperledger Fabric components and their functions

Component	Function
Distributed Ledger	Distributed database that contains data and related transactions.
Peers	Network services that maintain the ledger and execute smart contracts.
Ordering Service	Manage transaction sequence and distributes blocks to peers.
Smart Contract	Transaction logic whose output agreed upon by the peers.
Consensus	The process by which agreement is obtained on the peer network.

participating peers. The CA is responsible to issue an identity certificate to each blockchain proxy which uses the certified private key to sign the blockchain transactions. The choice of CA is depending upon the decision made by the network owner/provider. Hyperledger Fabric also come with its own CA as the certificate authority that issues certificates to the blockchain proxy. The blockchain proxy could represent the network owner/provider that owns the IoT devices. Also note that the identity elements that made up the composite identity come from two main sources, namely the IoT devices, and the related application systems that linked up to the network.

In Figure 9, it is also shown that the blockchain proxy will hold its security credential in its wallet, which is used to verify its identity to the blockchain network. The proxy has a responsibility to submit the transaction to the blockchain peer. The blockchain transaction in this system model is implemented in the form of smart contract, which can be executed by the blockchain peer, in order to commit it to the ledger.

5.1. Smart Contract for Identity Data Access and Verification

An important component in Hyperledger Fabric implementation is chaincode. Chaincode is a program written in programming language such as javascript, java, and Go, which contain the list of smart contract transactions for storing and updating data on distributed ledger. Figure 10 shows a chaincode, consists of a number of smart contract code written in Javascript that can be used to access the composite identity data in the CIDoT implementation.

The chaincode as shown in Figure 10, includes two smart contracts, namely **queryDev()** that is used to query for an identity information of a particular device; and **createDev()**, that is used to create a device identity information. In order to access these smart contracts, the user application requires access to them. Figure 11 shows a Javascript code used to access the smart contracts.

The Javascript code shown in Figure 11, basically used to execute the smart contract **queryDev()** with input parameter: **DEV4**. Also note that in order to execute this smart contract, the user is verified against the credentials stored in the wallet.

5.2. System Flow

Figure 12 illustrates the process flow for the composite identity creation and storage in a permissioned blockchain network.

The composite identity is created by the blockchain proxy, based on the received identity elements from the IoT device and the organization/user that runs the application. This identity will then be stored

Figure 9. System view of the CIDoT identity management model for IoT devices with permissioned blockchain network

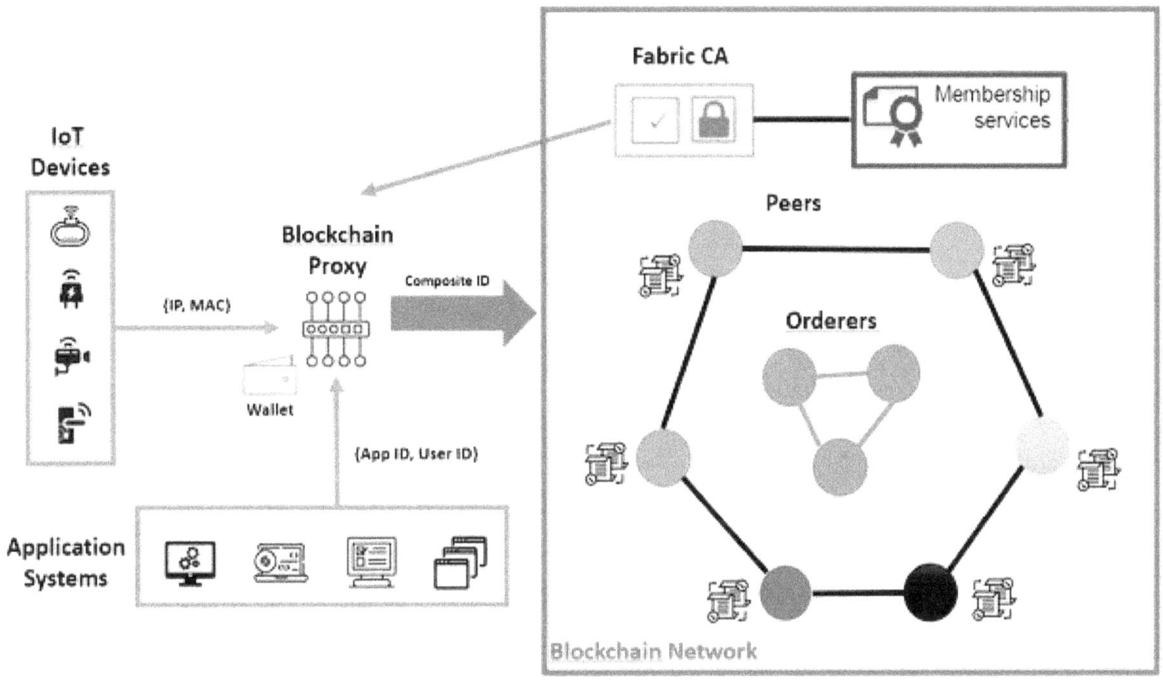

as data block in Blockchain network. This is achieved via smart contract configuration made by the blockchain proxy.

5.3. Application Design

Smart building application as described in the Introduction section, typically consists of a number of IoT devices that are used to monitor the conditions of the building. The implementation of composite identity management (CIDoT) of the IoT devices basically follows the system model as shown in the previous section. Figure 13 shows a typical composite identity data stored in JSON format within the Hyperledger Fabric permissioned blockchain.

With implementation of data format such as JSON, data can be retrieved by different applications more dynamically and being independent of the programming platform.

6. CONCLUSION

This chapter presents an alternative approach towards enhancing the identity of IoT devices in a distributed network using a composite identity design within an integrated IoT-permissioned blockchain network. IoT device is an important component that builds up smart applications, including smart building management system and other green systems. The composite Identity-of-Things (CIDoT) model could help in providing another layer of defense for the IoT network against the network-based attacks such as the

Figure 10. A snapshot of the CIDoT chaincode that consists of two smart contract codes to access data stored in the blockchain network

```
/*
 * SPDX-License-Identifier: Apache-2.0
 * This is the chaincode code written in Node.js programming language for the CIdOT Application.

 * Last revision: 13th of March 2020
 */

'use strict';

const { Contract } = require('fabric-contract-api');

class CidotData extends Contract {

    async queryDev(ctx, devNumber) {
        const devAsBytes = await ctx.stub.getState(devNumber); // get the device from chaincode state
        if (!devAsBytes || devAsBytes.length === 0) {
            throw new Error(`${devNumber} does not exist`);
        }
        console.log(devAsBytes.toString());
        return devAsBytes.toString();
    }

    async createDev(ctx, devNumber, macAdd, app, ipAdd, owner) {
        console.info('============= START : Create Device ===========');

        const dev = {
            ipAdd,
            docType: 'dev',
            macAdd,
            app,
            owner,
        };

        await ctx.stub.putState(devNumber, Buffer.from(JSON.stringify(dev)));
        console.info('============= END : Create Device ===========');
    }
}

module.exports = CidotData;
```

presence of rogue devices, as well as the botnet attacks. Two important enhancement towards existing identity management model for IoT devices have been presented. These are the increased complexity of the device identity through composition, and secure and verified data access and storage in a distributed network using permissioned blockchain infrastructure. In this chapter, we also reviewed a number of different identity management models for IoT devices.

The CIDoT model for IoT device identity management is still considered in its infancy. More detail works need to be carried out in order to enhance its functionality and efficiency. Some of the important aspects to consider is the involvement of certificate authority (CA) to delegate and verify the proxy certificate to access data in the blockchain. In addition, access customization is also need to be further explored, in order to create a more dynamic data query and access.

Figure 11. A snapshot of the CIDoT user application code written in Javascript to execute the smart contracts in Figure 10

```
/*
 * SPDX-License-Identifier: Apache-2.0
 * This is the application code written in Node.js programming language for the CIdOT Application.
 * to access the smart contract included in the chaincode.
 * Date created: 12th of March 2020
 * Last revision: 13th of March 2020
 *
 */
'use strict';

const { Gateway, Wallets } = require('fabric-network');
const path = require('path');
const fs = require('fs');

async function main() {
    try {
        // load the network configuration
        const ccpPath = path.resolve(__dirname, '..', '..', 'first-network', 'connection-org1.json');
        const ccp = JSON.parse(fs.readFileSync(ccpPath, 'utf8'));

        // Create a new file system based wallet for managing identities.
        const walletPath = path.join(process.cwd(), 'wallet');
        const wallet = await Wallets.newFileSystemWallet(walletPath);
        console.log(`Wallet path: ${walletPath}`);
        const identity = await wallet.get('user1');
        if (!identity) {
            console.log('An identity for the user "user1" does not exist in the wallet');
            console.log('Run the registerUser.js application before retrying');
            return;
        }

        // Create a new gateway for connecting to our peer node.
        const gateway = new Gateway();
        await gateway.connect(ccp, { wallet, identity: 'user1', discovery: { enabled: true, asLocalhost: true } });

        // Get the network (channel) our contract is deployed to.
        const network = await gateway.getNetwork('mychannel');

        // Get the contract from the network.
        const contract = network.getContract('cidot');

        // Evaluate the specified transaction.
        const result = await contract.evaluateTransaction('queryDev', 'DEV4');
        console.log(`Transaction has been evaluated, result is: ${result.toString()}`);

    } catch (error) {
        console.error(`Failed to evaluate transaction: ${error}`);
        process.exit(1);
    }
}

main();
```

Figure 12. System flow for the composite identity creation and storage on the permissioned blockchain network

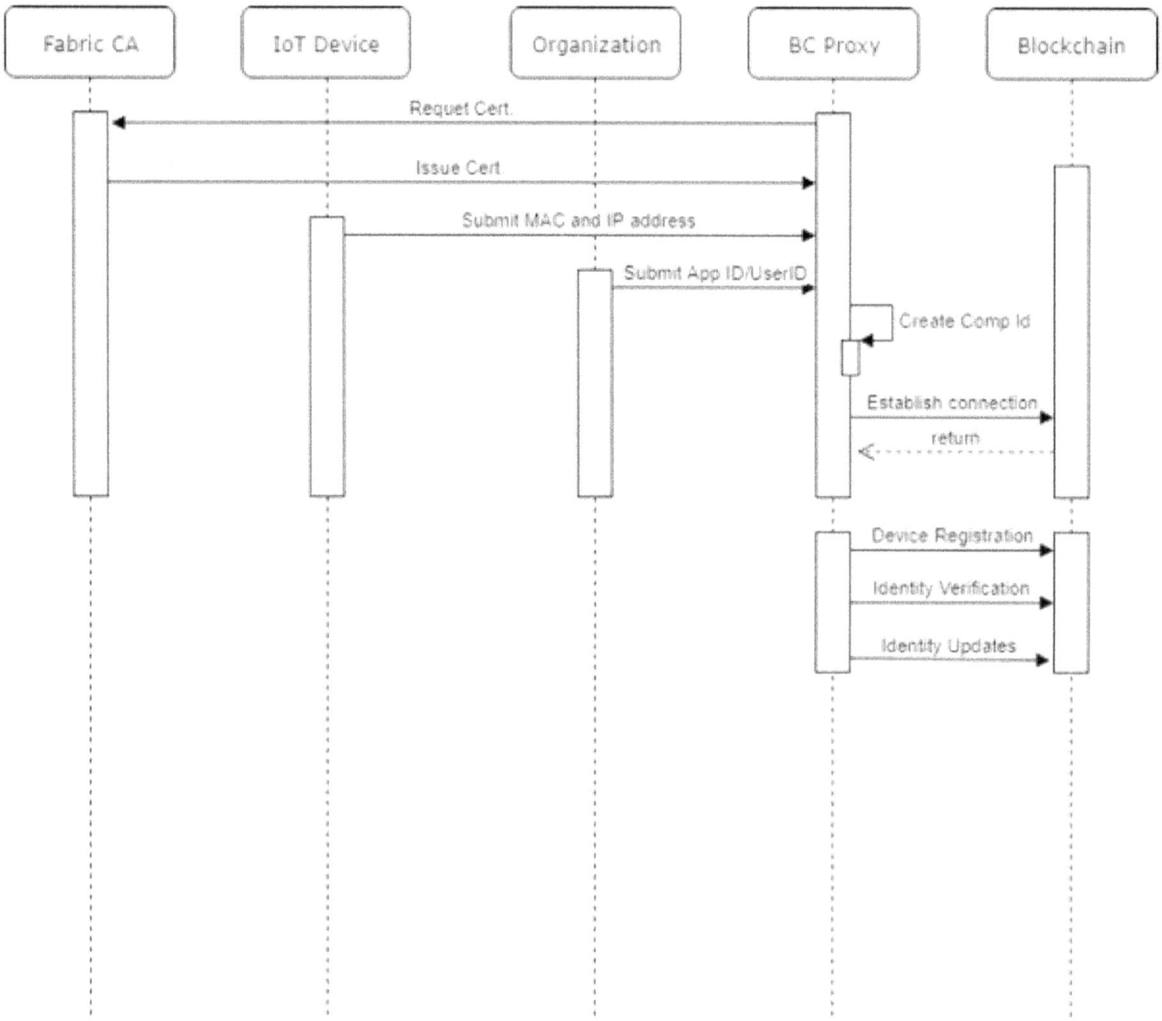

Figure 13. Composite identity data defined in the JSON format

```
const devapp = [
    {
        ipAdd: '192.168.1.1',
        macAdd: 'FA:DF:0F:D5:3E:1A',
        app: 'Matlab',
        owner: 'Tomoko',
    },
    {
        ipAdd: '192.168.1.2',
        macAdd: 'EB:E0:1E:C4:4F:2B',
        app: 'SensorAnalytics',
        owner: 'Brad',
    },
    {
        ipAdd: '192.168.1.3',
        macAdd: 'DC:F1:2D:B3:50:3C',
        app: 'TensorFlow',
        owner: 'Jin Soo',
    },
    {
        ipAdd: '192.168.1.4',
        macAdd: 'CD:02:3C:A2:61:4D',
        app: 'Matlab',
        owner: 'Max',
    },
    {
        ipAdd: '192.168.1.5',
        macAdd: 'BE:13:4B:91:72:5E',
        app: 'SensorAnalytics',
        owner: 'Adriana',
    },
    {
        ipAdd: '192.168.1.6',
        macAdd: 'AF:24:5A:80:83:6F',
        app: 'TensorFlow',
        owner: 'Michel',
    },
    {
        ipAdd: '192.168.1.7',
        macAdd: '90:35:69:7F:94:70',
        app: 'Matlab',
        owner: 'Aarav',
    },
    {
        ipAdd: '192.168.1.10',
        macAdd: '63:68:96:4C:C7:A3',
        app: 'TensorFlow',
        owner: 'Shotaro',
    },
];
```

REFERENCES

Abu Hassan, M., Kudus, N., Sidek, S., Mohamed, S., & Aripin, M. A. (2018). Factors That Influence The Sustainability Of IoT Usage For Smart Living. *International Journal of Pure and Applied Mathematics*, *119*, 573–587.

Andolfatto, D. (2018). Blockchain: What it is, what it does, and why you probably don't need one. *Review - Federal Reserve Bank of St. Louis*, *100*(2), 87–95. doi:10.20955/r.2018.87-95

Androulaki, E., Barger, A., Bortnikov, V., Cachin, C., Christidis, K., De Caro, A., Enyeart, D., Ferris, C., Laventman, G., Manevich, Y., & Muralidharan, S. (2018, April). Hyperledger fabric: a distributed operating system for permissioned blockchains. In *Proceedings of the Thirteenth EuroSys Conference* (pp. 1-15). 10.1145/3190508.3190538

Baliga, A., Subhod, I., Kamat, P., & Chatterjee, S. (2018). *Performance evaluation of the quorum blockchain platform*. arXiv preprint arXiv:1809.03421.

Bandara, S., Yashiro, T., Koshizuka, N., & Sakamura, K. (2016, August). Access control framework for api-enabled devices in smart buildings. In *2016 22nd Asia-Pacific Conference on Communications (APCC)* (pp. 210-217). IEEE. 10.1109/APCC.2016.7581479

Bello, A. & Mahadevan, V. (2019). A cloud based conceptual identity management model for secured Internet of Things operation. *Journal of Cyber Security and Mobility*, 53-74.

Bernabé, J. B., Ramos, J. L. H., & Gómez-Skarmeta, A. F. (2017). Holistic Privacy-Preserving Identity Management System for the Internet of Things. *Mobile Information Systems*, *2017*, 6384186–1. doi:10.1155/2017/6384186

Catarinucci, L., De Donno, D., Mainetti, L., Palano, L., Patrono, L., Stefanizzi, M. L., & Tarricone, L. (2015). An IoT-aware architecture for smart healthcare systems. *IEEE Internet of Things Journal*, *2*(6), 515–526. doi:10.1109/JIOT.2015.2417684

Chen, Y. J., Wang, L. C., & Wang, S. (2018). Stochastic Blockchain for IoT Data Integrity. *IEEE Transactions on Network Science and Engineering*.

Dittmann, G., & Jelitto, J. (2019, June). A Blockchain Proxy for Lightweight IoT Devices. In *2019 Crypto Valley Conference on Blockchain Technology (CVCBT)* (pp. 82-85). IEEE. 10.1109/CVCBT.2019.00015

Fruhlinger, J. (2018). *The Mirai botnet explained: How teen scammers and CCTV cameras almost brought down the internet*. CSO Online. Disponible en: https://www. csoonline. com/article/3258748/security/the-mirai-botnet-explainedhow-teen-scammers-and-cctv-cameras-almost-brought-down-the-internet. html

Garcia, G. T., Sanchez, V. M., Marin, C. N. L., Cortez, J. I., Acevedo, C. A. R., Gonzalez, G. S. A., Ameca, J. L. H., & Garcia, M. D. C. M. (2018). Wireless sensor network for monitoring physical variables applied to green technology (IoT green technology). *European Journal of Electrical Engineering and Computer Science*, *2*(2). Advance online publication. doi:10.24018/ejece.2018.2.2.15

Gondchawar, N., & Kawitkar, R. S. (2016). IoT based smart agriculture. *International Journal of Advanced Research in Computer and Communication Engineering*, *5*(6), 838–842.

Greenspan, G. (2015). *Multichain private blockchain-white paper.*

Guin, U., Cui, P., & Skjellum, A. (2018, July). Ensuring proof-of-authenticity of IoT edge devices using blockchain technology. In *2018 IEEE International Conference on Internet of Things (iThings) and IEEE Green Computing and Communications (GreenCom) and IEEE Cyber, Physical and Social Computing (CPSCom) and IEEE Smart Data (SmartData)* (pp. 1042-1049). IEEE. 10.1109/Cybermatics_2018.2018.00193

Guo, H., & Heidemann, J. (2018, August). IP-based IoT device detection. In *Proceedings of the 2018 Workshop on IoT Security and Privacy* (pp. 36-42). 10.1145/3229565.3229572

Hernández-Ramos, J. L., Moreno, M. V., Bernabé, J. B., Carrillo, D. G., & Skarmeta, A. F. (2015). SAFIR: Secure access framework for IoT-enabled services on smart buildings. *Journal of Computer and System Sciences*, *81*(8), 1452–1463. doi:10.1016/j.jcss.2014.12.021

Jain, R. K., Banerjee, P., Baksi, D., & Samanta, S. K. (2019, July). IoT Based Interface Device for Automatic Molding Machine towards SMART FOUNDRY-2020. In *2019 10th International Conference on Computing, Communication and Networking Technologies (ICCCNT)* (pp. 1-6). IEEE.

Jan, M.A., Nanda, P., He, X., & Liu, R.P. (2015, August). A sybil attack detection scheme for a centralized clustering-based hierarchical network. In *2015 IEEE Trustcom/BigDataSE/ISPA* (Vol. 1, pp. 318-325). IEEE.

Jan, M. A., Nanda, P., He, X., & Liu, R. P. (2018). A Sybil attack detection scheme for a forest wildfire monitoring application. *Future Generation Computer Systems*, *80*, 613–626. doi:10.1016/j.future.2016.05.034

Kolias, C., Kambourakis, G., Stavrou, A., & Voas, J. (2017). DDoS in the IoT: Mirai and other botnets. *Computer*, *50*(7), 80–84. doi:10.1109/MC.2017.201

Lam, K. Y., & Chi, C. H. (2016, November). Identity in the Internet-of-Things (IoT): New challenges and opportunities. In *International Conference on Information and Communications Security* (pp. 18-26). Springer. 10.1007/978-3-319-50011-9_2

Mahalle, P., Babar, S., Prasad, N. R., & Prasad, R. (2010, July). Identity management framework towards internet of things (IoT): Roadmap and key challenges. In *International Conference on Network Security and Applications* (pp. 430-439). Springer. 10.1007/978-3-642-14478-3_43

Omar, A. S., & Basir, O. (2018, July). Identity management in IoT networks using Blockchain and Smart Contracts. In *2018 IEEE International Conference on Internet of Things (iThings) and IEEE Green Computing and Communications (GreenCom) and IEEE Cyber, Physical and Social Computing (CPSCom) and IEEE Smart Data (SmartData)* (pp. 994-1000). IEEE. 10.1109/Cybermatics_2018.2018.00187

Qolomany, B., Al-Fuqaha, A., Gupta, A., Benhaddou, D., Alwajidi, S., Qadir, J., & Fong, A. C. (2019). Leveraging machine learning and big data for smart buildings: A comprehensive survey. *IEEE Access: Practical Innovations, Open Solutions*, *7*, 90316–90356. doi:10.1109/ACCESS.2019.2926642

Sethi, P., & Sarangi, S. R. (2017). Internet of things: Architectures, protocols, and applications. *Journal of Electrical and Computer Engineering*, *2017*, 2017. doi:10.1155/2017/9324035

Singi, K., Kaulgud, V., Bose, R. J. C., & Podder, S. (2019, May). ShIFt-Software Identity Framework for Global Software Delivery. In *2019 ACM/IEEE 14th International Conference on Global Software Engineering (ICGSE)* (pp. 122-128). IEEE. 10.1109/ICGSE.2019.00032

Trnka, M., Cerny, T., & Stickney, N. (2018). Survey of Authentication and Authorization for the Internet of Things. *Security and Communication Networks, 2018*. doi:10.1155/2018/4351603

Venkatesh, V., Morris, M. G., Gordon, B., & Davis, F. D. (2003, September). User acceptance of information technology: Toward a unified view. *Management Information Systems Quarterly*, *27*(3), 425–478. doi:10.2307/30036540

Wang, Z. (2018). A privacy-preserving and accountable authentication protocol for IoT end-devices with weaker identity. *Future Generation Computer Systems*, *82*, 342–348. doi:10.1016/j.future.2017.09.042

Xu, R., Chen, Y., Blasch, E., & Chen, G. (2018, July). Blendcac: A blockchain-enabled decentralized capability-based access control for iots. In *2018 IEEE International Conference on Internet of Things (iThings) and IEEE Green Computing and Communications (GreenCom) and IEEE Cyber, Physical and Social Computing (CPSCom) and IEEE Smart Data (SmartData)* (pp. 1027-1034). IEEE. 10.1109/Cybermatics_2018.2018.00191

Zhang, Z. K., Cho, M. C. Y., & Shieh, S. (2015, April). Emerging security threats and countermeasures in IoT. In *Proceedings of the 10th ACM Symposium on Information, Computer and Communications Security* (pp. 1-6). 10.1145/2714576.2737091

Zhu, X., Badr, Y., Pacheco, J., & Hariri, S. (2017, September). Autonomic identity framework for the internet of things. In *2017 International Conference on Cloud and Autonomic Computing (ICCAC)* (pp. 69-79). IEEE. 10.1109/ICCAC.2017.14

Chapter 4
Applications of Vibration-Based Energy Harvesting (VEH) Devices

Muhammad Faruq Foong
Aeronautics Laboratory, Universiti Teknologi Malaysia, Malaysia

Chung Ket Thein
https://orcid.org/0000-0001-8600-8561
School of Aerospace, University of Nottingham, Ningbo, China

Beng Lee Ooi
Intel Microelectronics (M) Sdn. Bhd., Malaysia

ABSTRACT

This chapter reviews present usage of vibration-based energy harvesting (VEH) devices and their applications. The evolution of energy resources and advancement in electronic technologies has resulted in the need for a self-sustainable wireless/portable electronic device in current modern society. Batteries are non-beneficial in the miniaturization process of electronic designing, and alternative power supplies are desperately needed to drive the advance of the wireless/portable development further. VEH has emerged as one of the most promising alternatives to replace conventional batteries and as the solution for the bottleneck. Consideration of creating an optimal vibration energy harvester is suggested through an analytical model of a mechanical transducer, including a relatively new method defined as triboelectricity. Useful applications and usages of VEH are presented, and some suggestions for improvement are also given. Lastly, the trend of energy harvesting is annotated and commented in-line with the demand of electronic sensors market.

ENERGY HARVESTING FOR WIRELESS SENSOR NETWORKS (WSN)

Due to the advancement in electronic technology, the size and power consumption of CMOS technology has led to wide research activities on wireless/portable devices. By removing the wires from the devices, there is a potential for embedding sensors in previously inaccessible locations, such as roof

DOI: 10.4018/978-1-7998-6709-8.ch004

tops, underneath the raised-access floor panel, implant into the concrete wall and etc. Presently, most of the remote electronic devices are still powered by batteries with finite lifespan (Wang et al., 2016). This not only limits the miniaturization process of the device size, but also creates environmental pollution caused by the improper disposal of used batteries (NHDES, 2017). Not to mention the cost of labors and parts involved during the batteries replacement process, especially in an enormous electronic device system. Energy harvesting can be used as an alternative long-lasting energy sources for the sensor devices, particularly in applications where the replacement of batteries is practically unfeasible or costly.

Energy harvesting is generally defined as the process of extracting available energy from the ambient surrounding (otherwise known as ambient energy) and converting it into useful electrical power. Ambient energies can come in the form of light, wind, kinetic, heat and many others form of waves. Hence, this makes energy harvesting a promising alternative as the replacement for the batteries. Applications of energy harvesting tend to focus on this aspect (battery replacement in low-powered electronics) because energy that is harvested from the ambient surroundings are usually in the order of micro- or milliwatts, hence limiting its practical usage. However, in the past decade, research have shown that it is possible to harvest energy to the order of Watts, allowing broader applications usages for larger-powered components such as railway signal lights (Dotti and Sosa, 2019). The main advantage of energy harvesting is that ambient energy resources are redundant and have ideally an infinite lifetime. Besides, they are also considered as a green energy resources, with very little to no environmental pollution impact.

The idea of providing a sustainable source of power for wireless sensor nodes (WSN) through energy harvesting has been actively discussed by Roundy (2003) in the past decade. WSN plays an important role in realizing the concept of Internet-of Things (IoT). This concept envisions a stage where humans and machines can transfer data over a wireless network without the need of physical interactions. IoT is a crucial component towards digitalizing the manufacturing industry, creating a cyber-physical environment where human and machines can communicate and exchange data in real- time globally. This sort of environment is commonly known as a 'smart factory' and it represents the main goal of the fourth industrial revolution (Industry 4.0).

Figure 1 shows the schematic of an IoT system powered by a sustainable energy harvesting source. IoT is generally a platform which transmits the collected data from the sensors to a central control station via the cloud computing system. The processes of data collection and transmission are executed using sensors, which explains the main purpose of WSN in an IoT system. Ideally, these sensors are required to be powered by a sustainable energy source to reduce maintenance cost and machine downtime. However, till date many WSN are still powered by batteries which lead to an issue in energy sustainability as these batteries have finite lifespan. Different techniques have been imposed to reduce the energy dissipation effect and prolong the battery life. One of the methods is called the duty-cycle operation. This method periodically places the sensor nodes in 'sleep' mode to save the energy, or in other words, the cycle of operation of the sensor's nodes operate intermittently rather than continuously. However, this comes with the trade-off on the data latency, since during the 'sleep' mode, no data can be transmitted or received by either ends. On top of that, batteries are also prone to leakage, as it will continuously deplete over time due to energy leakage even if the devices are periodically placed in the 'sleep' mode.

Another technique to prolong battery life was demonstrated by Chamanian et al. (2014) as depicted in Figure 2. In this design, a WSN device (MicaZ Mote) is integrated with an energy harvesting circuit and two rechargeable batteries. The energy harvester (which produces 169.0 μW of power) was used to charge the two rechargeable batteries, which then channeled to power the MicaZ Mote device that required 186μW in operation mode. With that, theoretically only net amount of 17μW will be directly

Figure 1. Schematic of an IoT platform integrated with a sustainable energy source system

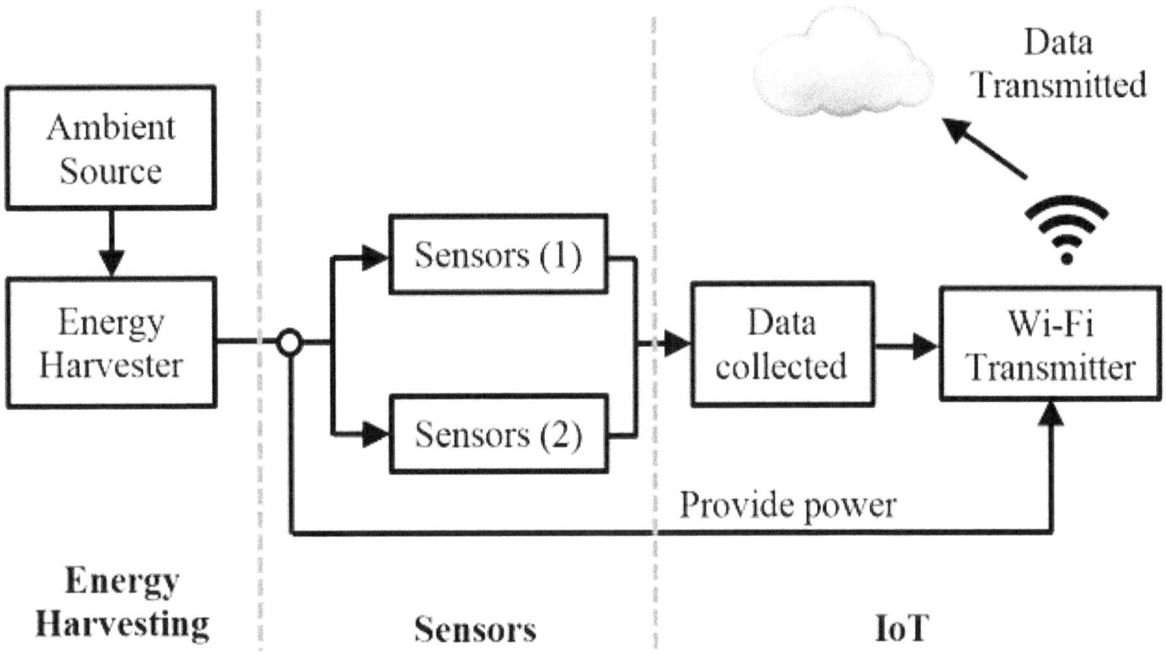

drained out from the batteries, hence prolong the battery life. This specific circuit was anticipated to increase the battery life for more than 10 times of its usual lifespan. However, the increase in battery life may be overestimated since it was determined based on the average current flowing into the batteries and disregards other factors such as electrode corrosion, energy leakage, rectification conversion losses and operating temperature (Prauzek et al., 2018). Furthermore, based on practical usage experiences, rechargeable batteries can only last up to a few thousand charging and discharging cycles under optimum operating conditions. Hence, it may be better to replace batteries with a fast charging/discharging super-capacitor as they maintain a longer cycle lifetime (up to hundreds of thousands of times with minimal change in performance). Despite the great cycle lifetime, supercapacitors are low in energy density, thus they cannot be used as a continuous power source. However, this can be compensated by having multiple supercapacitor cells connected in series to achieve a greater energy density and well-controlled duty-cycle operation system to minimize the power usage of the circuit without compromising the system performance. It may also be noted that this not only increases the efficiency of the power system, but also significantly reduces the device sizes as capacitors are relatively small in size as compared to batteries.

VIBRATION ENERGY HARVESTING (VEH)

Among the promising ambient resources for a self-powered WSN device are solar, wind, thermal, and vibrations. Solar energy harvesting usually adopts the use of materials that exhibits the photovoltaic (PV) effect to convert light energy into electrical energy. However, Fathabadi (2019) proposed the use of an electrochemical device instead which is lighter in weight, does not require a maximum power point tracking controller and has a higher efficiency compared to PV cells. Wind energy can be harvested

Figure 2. Example of an energy harvesting circuit powering a WSN (Chamanian et al., 2014)

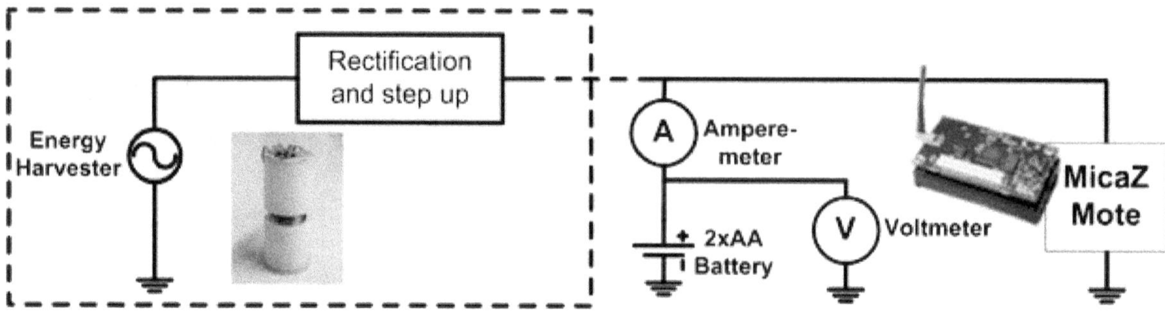

using two different methods, either by converting the wind energy into rotational energy to drive a generator such as using a turbine or using the wind to induce oscillations onto a certain structure which is then converted into electrical power. For a smaller scale system, the latter method is often preferred due to its flexibility in sizes. The drawback of this method is that it lacks in efficiency as the wind energy must first be converted into an oscillating mechanical motion which is then transformed into electrical power using a transducer, in contrast to directly driving a generator (like in wind turbines). However, by arranging oscillating structures in a certain manner (such as the in-line tandem configuration) as to advantage of the vortex shedding effect, the efficiency and power output of the harvester can be significantly increased (Lee et al., 2019). Thermal energy harvesting strongly relies on the temperature gradient between two surfaces of a thermoelectric material. These materials create an electrical potential when heat travels from the hotter surface to the cooler side. The performance of a thermal energy harvester is often evaluated based on its Carnot efficiency, which defines the ratio of the temperature difference between two surfaces to the larger temperature of the surfaces. Over the past decade, many researches have been conducted on thermal energy harvesting through pyroelectricity due to its ability to generate a greater current output in low thermal applications as compared to thermoelectricity (Cuadras et al., 2010; Bowen et al., 2014; Zhang et al., 2020a).

Aforementioned energy resources have some limitations like in case of solar energy, the effectiveness of the harvester will drop at location where there is an absence of sunlight. In the case of wind energy, wind energy harvesters cannot be used in lower altitude or urban areas, at where there is an inconsistent and weak wind density. For thermal generators, the efficiency of the generator is strongly dependent on the temperature distribution on the plates. An uneven distribution will decrease the harvester's performance, however, to control the temperature distribution may prove to be challenging (Tang et al., 2015). Among these energy resources, vibration sources claimed to be one of the most attractive approach, apart from their abundance availability in many targeted wireless sensor or portable electronic device applications, they also claimed to exhibit a greater power density and efficiency (Du Toit, 2005). Vibration sources may vary considerably in amplitude and dominant frequency on different architectural structures and geographical locations. Some examples of mechanical excitations are moving vehicles, rotations (Kubba and Jiang, 2014), ocean tides and waves (Viet et al., 2016), civil structures, household appliances and even human motion (Li et al., 2018). Table 1 gives the approximate frequency ranges and amplitude of acceleration levels for several vibration resources. It is shown that the fundamental frequencies of the measured resources fall between the range from 0.1 Hz to around 150 Hz and acceleration amplitudes ranges from less than 0.03 g to 40 g. Yet, the vibration present in most environments is not made up of a single frequency but is typically made up of number of fundamental frequencies and their harmonics. For example the vibration data that shown in Figure 3, was taken by (Ooi, 2010) from a desktop computer case when a CD room running, indicates a fundamental frequency of 120 Hz with an acceleration amplitude of 0.54 ms^{-2} with the second and higher harmonics present at lower amplitudes. Since most of the vibration-based conversion devices have a relatively narrow range of operating frequencies (high Quality factor), it is important that the nature of the source be understood first so that the generator can be designed to resonate at an appropriate frequency peak in the vibration spectrum, thus optimal power can be harvested (Thein et al., 2019a). While many energy resources look promising to be used as an alternative power source for portable/wireless electronic devices, this chapter will only focus on energy harvesting from a vibration source. This is not to suggest that vibration-based power sources are the best energy scavenging solution, but only recommend as one of the most appropriate energy resources for the self-powered electronic systems.

Figure 3. Vibration spectra for desktop computer casing with CD is running (Ooi, 2010)

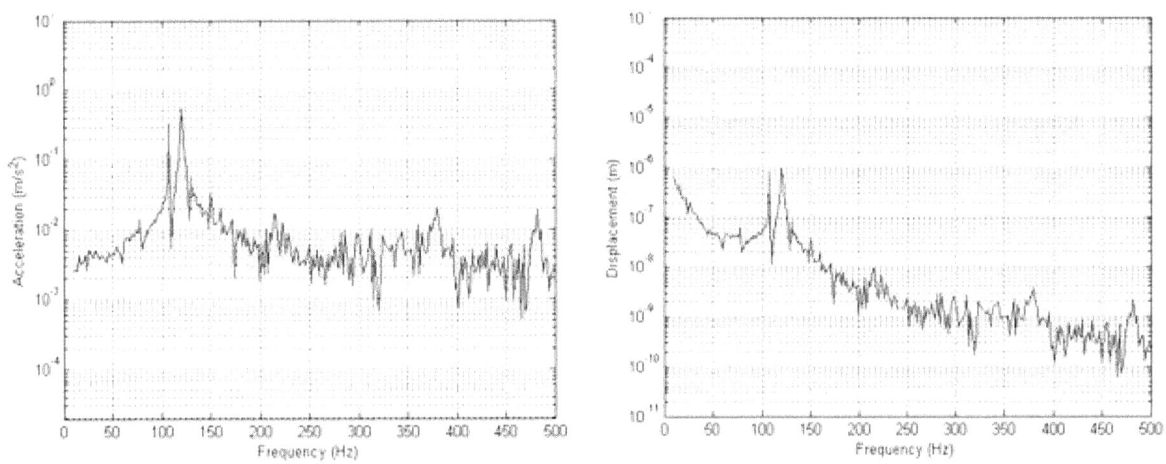

MODELING OF A GENERAL VEH CONVERTER SYSTEM

A simple model based on the schematic in Figure 4 has been proposed by Williams and Yates, 1996) to model common inertia vibration-based converter system. The model consists of a seismic mass, m, a spring of stiffness, k, the damping coefficient, c, the time varying position, $z(t)$, of the seismic mass relative to the support casing that is driven by an external time varying vibration force, $y(t)$. When the converter model is subjected to an external vibration force, the mass moves out of phase with the converter housing, so that there is a net movement between the mass and the housing, which cause loss of mechanical energy that is represented transformation of energy into electrical form. The motion equation for this system may be described as

$$m\ddot{z}(t)+c\dot{z}(t)+kz(t)=-m\ddot{y}(t) \tag{1}$$

Table 1. Frequency ranges and acceleration levels for some vibration resources

Vibration source	Frequency range (Hz)	Acceleration (g)	Source of Information
Civil structures	< 10	< 1	(Rhimi and Lainef, 2012)
Ocean waves	0.1 – 10	-	(Shirai et al., 2020)
Human motion	2 – 10	0.05 – 0.7	(Green et al., 2013)
Train carriage	20 – 40	0.1 – 2	(Bradai et al., 2018)
Mining	10 – 30	0.05 – 3	(ReVibe Energy, 2020)
Bridges	1 – 40	0.01 – 0.5	(Khan and Iqbal, 2018)
Railway track	10 – 150	10 - 40	(Dotti and Sosa., 2019)
Human heartbeat	< 50	< 0.03	(Karami and Ansari, 2017)
Household appliances	25 – 120	0.06 – 3.5	(Roundy, 2003)

For a sinusoidal external excitation vibration force where, $y(t) = Y\sin(\omega t)$, the steady-state solution for z(t) *is*

$$z(t)=\left[\frac{Y\omega^2}{\sqrt{\left(\omega_o^{\,2}-\omega^2\right)^2+\left(2\omega_o\omega\zeta_F\right)^2}}\right]sin(\omega t-\phi) \tag{2}$$

and the phase angle ϕ is

$$\phi=\tan^{-1}\left(\frac{2\omega_o\omega\zeta_F}{\omega_o^{\,2}-\omega^2}\right) \tag{3}$$

where $\omega_o=\sqrt{k/m}$ is the natural frequency of the converter system, $\zeta_F=c/\left(2m\omega_o\right)$ is the full damping factor and Y is the amplitude of external excitation force. If the damping factor of the model is the sum of electrical induced $\left(\zeta_e\right)$ and mechanical loss $\left(\zeta_m\right)$ damping, $\zeta_F=\zeta_e+\zeta_m$, the average power delivered to the electrical domain, P_{avg}, can be written as

$$P_{avg}=\frac{m\zeta_e Y^2(\frac{\omega}{\omega_o})^3\omega^3}{[1-\left(\frac{\omega}{\omega_o}\right)^2]^2+2\zeta_F\frac{\omega}{\omega_o}^2} \tag{4}$$

Under an ideal condition where there is no power loss due to the converter's internal impedance, then the theoretical maximum average power delivered ($P_{avg(max)}$) occur when the external vibrating force frequency is matched with the resonant frequency, $\omega = \omega O$, and when $\zeta_e=\zeta_m$ which yield

Figure 4. Model of a common VEH converter system (Williams and Yates, 1996)

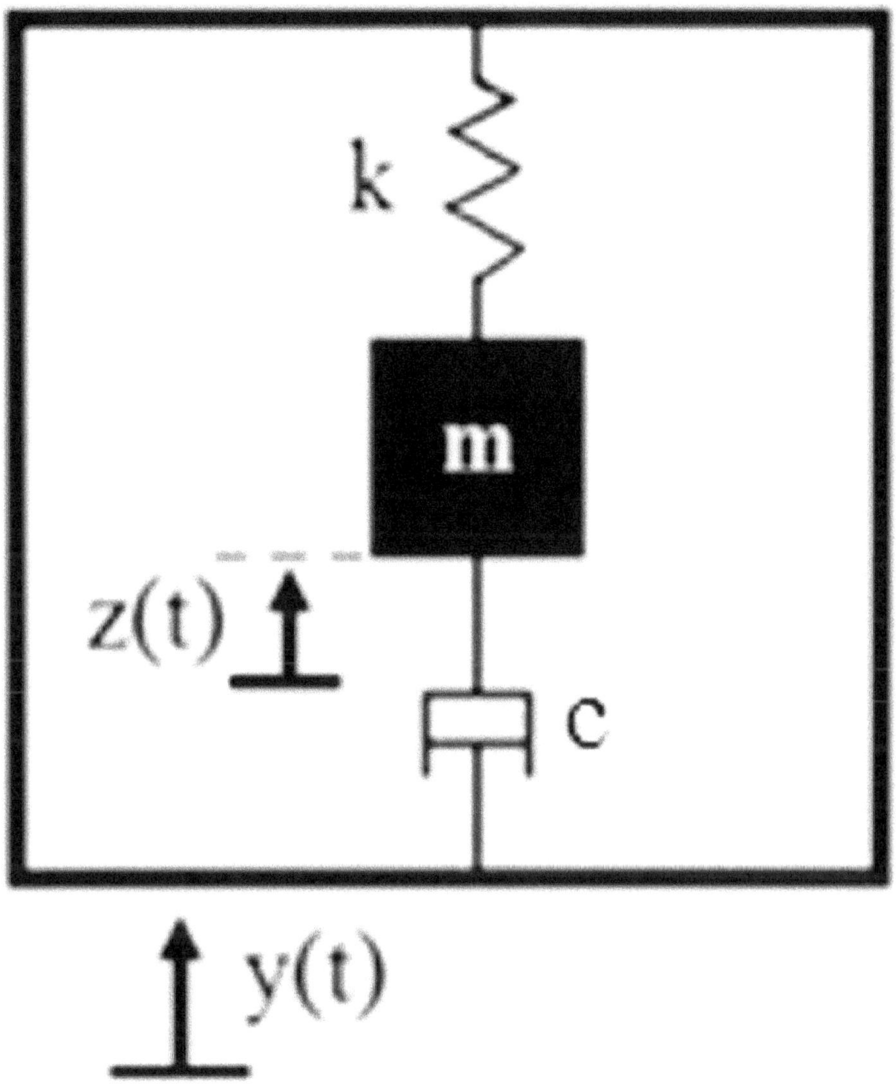

$$P_{avg(\max)} = \frac{mA^2}{16\omega\zeta_m} \tag{5}$$

where $A = Y\omega 2$ is the amplitude of external excitation force. From equation (5), is can be seen that the power is inversely proportional to frequency. Therefore, if the acceleration amplitude of the vibrations is constant or decreasing with frequency, then the vibration-based converter should be designed to resonate at the possible lowest fundamental frequency and highest acceleration.

The relationship of the full damping ratio, external driving frequency and average power in equation (4) is plotted in Figure 5 by assuming that the proof mass, *m* and amplitude, *Y*, are constant with unity gain that the resonant frequency is 100 Hz and input vibration frequencies vary from 10 to 1000 Hz. By assuming that $\zeta_m = \zeta_e$, three different damping ratios (ζ_F) output response are plotted. It can be seen

Figure 5. Output response for systems with ζ_F = 0.01, 0.05 and 0.1 (Ooi, 2010)

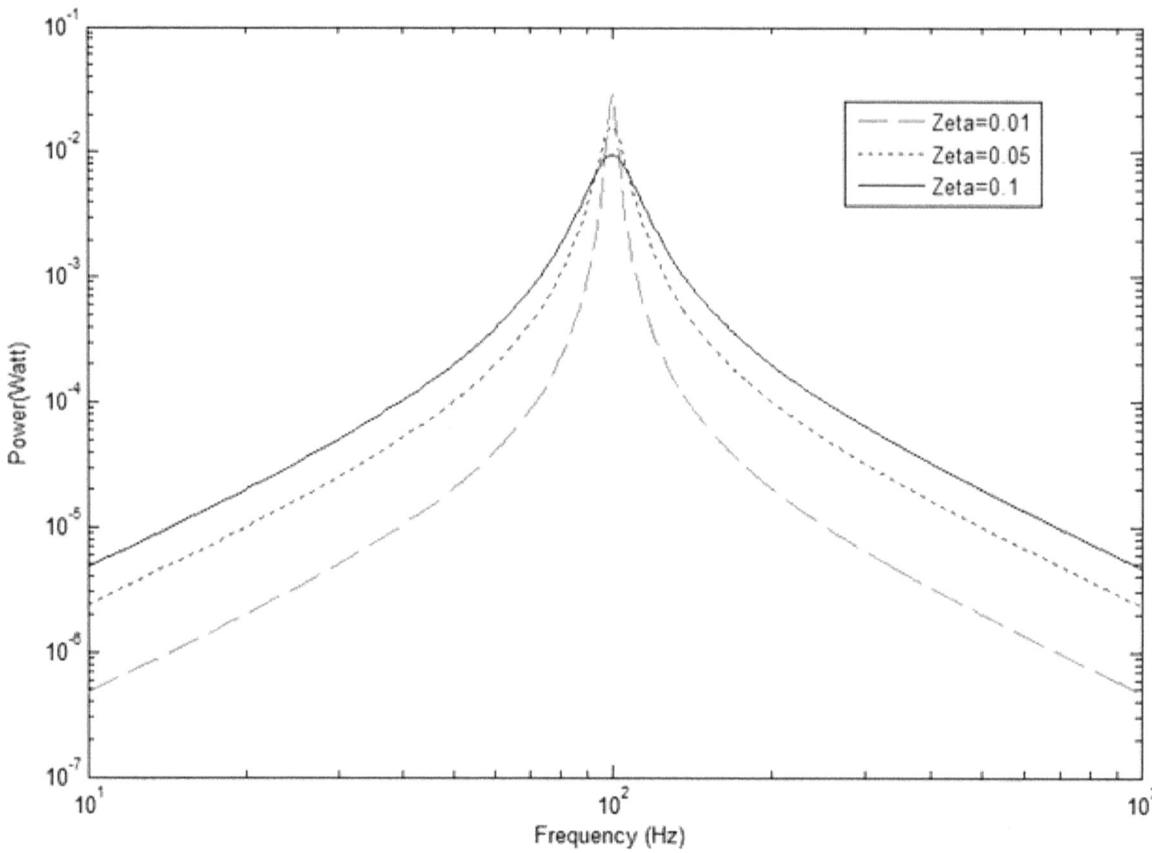

that a converter will operate at its optimum condition when the vibration frequency is matched with the system resonant frequency ($\omega = \omega O$). It may also be noted that increasing the damping, ζ_F has the effect of reducing the peak output power and Quality factor but increasing the operational bandwidth. To create a highly efficient converter, the converter should be designed to resonate at the lowest frequency peak in the vibration spectrum provided that higher frequency peaks do not have higher acceleration amplitude. Or more specifically, higher value of $Y2'\omega$. If the case in Figure 7 is used as an example, for a system with proof mass (m) of 1.0 grams and full damping factor (ζ_F) of 0.02 as suggested in Kumar et al. (2014), according to equation (5), the theoretical maximum average power of ~2.42μW could be delivered from the vibration-based energy harvester, which is sufficient to drive most of the ultra-low-power sensors (European Editors, 2013) and this raise the potential for vibration-based energy harvesting in ultra-low-powered electronic systems.

To further enhance the performance, the converter needs to be designed at the minimal value of damping factor to provide a greater operational output at the resonant frequency. Equation (5) indicates that the generated power will tend to be infinity as ζ_m approaches to zero, but in practice it is not possible to reduce the ζ_m to an arbitrarily small value. This implies that the generated power will always be a finite value and depends on the design of the converter. The amplitude is, $z = Y/2\zeta_F$, and so reducing the damping factor implies increasing the displacement range of the mass. The maximum movement range of the mass can move is limited by the size and the design of the resonator. Therefore, the damping ratio must be designed to be large enough to keep the range of movement within the limits which the resonator is capable of making without any physical damage. Thus, if the vibration source has a fixed frequency then a low level of damping (high Quality factor) is preferable but if the source frequency varies, it may give better average power to use a higher damping factor (low Quality factor) resonator. On the contrary, one of the main penalties for this solution is the power output is still less than what it would be if the resonance frequency matched the driving frequency. To date, frequency matching is a common issue for all the linear vibration-to-electricity converters and in order to resolve this, many alternative solutions have been proposed in the literature. These approaches are non-linear oscillation (Lai et al., 2019; Friswell et al., 2013, Cottone et al., 2009), frequency-up-conversion (Dechant et al., 2017), dual-resonator cantilevers (Ooi and Gilbert, 2014), multiple harvesters (Thein et al., 2019b; Wang et al., 2019; Cui et al., 2015; Zhang and Afzalul, 2014), mechanical stoppers (Zhou et al., 2020; Maeguchi et al., 2015) and magnetic bi-stable effect (Zhang et al., 2020b; Zhou et al., 2019; Lai et al., 2019; Wang et al., 2017; dos Santos et al., 2016; Mann and Owens, 2010). Every aforementioned approach has its own advantages and drawbacks in term of device size, output density, design cost and so on. However, it is worth to mention that some of the works are focus on energy harvesting from non-resonant frequencies as the nature of the interested input excitation are irregular, have large amplitudes and at lower frequencies (Halim and Park, 2014; Li et al., 2013a; Yang and Lee, 2010). Due to the large amplitude excitations, significant power can still be attained by the harvesters even at non-resonant frequencies.

TYPES OF MECHANICAL-TO-ELECTRICAL CONVERSION METHODS FOR VEH DEVICES

There are many known methods of converting mechanical energy (vibration) into electrical power. In this chapter, the three most common methods (piezoelectric, electromagnetic and electrostatic) are discussed and compared. Furthermore, a relative new conversion method (triboelectric) is also covered.

Piezoelectric VEH

Piezoelectric materials such as lead zirconate titanate and zinc oxide can generate an electrical charge when the materials are physically deformed or via versa. This is due to the shift in positive and negative charges within the material crystals when under the right condition, would result in an external electric field. The open voltage output of a piezoelectric transducer, V_{PZ}, can generally be described by the following equation (Roundy, 2003)

$$V_{PZ} = \frac{-\partial h}{\varepsilon}\sigma \tag{6}$$

where ε, σ and h *a*re the dielectric constant, bending stress and thickness of the piezoelectric material. The power output of a piezoelectric harvester can then simply be determined using Ohm's law by dividing the square of the voltage output with the magnitude of the device resistance. Equation (6) suggests

Figure 6. A piezoelectric VEH cymbal design proposed by Daniels et al. (2013)

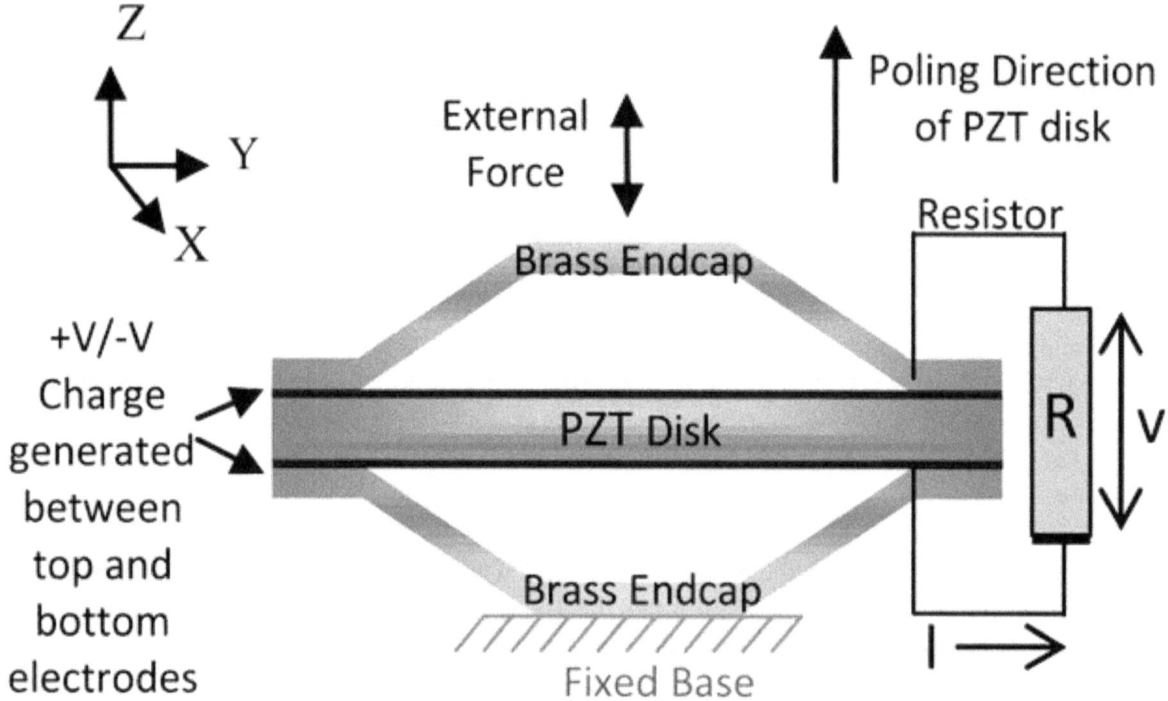

the power output of this harvester is dependent on the type and thickness of the piezoelectric material and the stress induced in it. The output of a piezoelectric transducer is strongly dependent on the volume of the piezoelectric material that is subjected to stress, hence to produce a higher output, it is advisable that high average stress is applied across the entire piezoelectric volume.

Since the power generated by piezoelectric materials is depends on the stress generated in the material itself and it is more appropriate to be used in cantilever beam designs form, as cantilever structure exhibits beam resonance frequency at lower frequencies range. However, other designs such as the cymbal structure design in Figure 6 (Daniels et al., 2013) and the diaphragm design (Yang et al., 2017) have been explored using this method. Though these designs typically have a lower power output than cantilever designs, their physical size and shape ease the device to be placed in small awkward locations such as in shoe heels. Piezoelectric transducer is attractive as it is not only can be easily implemented for micro-scale designs at lower frequencies applications, but also exhibits a higher efficiency conversion rate, power density and does not require a separate voltage source to operate.

ELECTROMAGNETIC VEH

Based on Faraday's law of electromagnetism, electrical power can also be generated when a conducting material causes a disturbance in a magnetic field. The principle of electromagnetic vibration energy harvesting is simple and direct, in where amplified vibrations are used to cause variations in the magnetic field of a permanent magnet, hence inducing a current flow in the conductor. The voltage induced in an electromagnetic harvester, V_{EM}, is

$$V_{EM} = Kv \tag{7}$$

where K represents the electromagnetic transduction factor derived from the properties of the coil and the magnet components and v is the relative velocity between the two said components.

Depending on application, one can choose to vibrate the magnet or the conductor while fixing the position of the other. In some cases, it may be desirable to vibrate both components (Foong et al., 2019; Klein and Zuo, 2017; Ooi and Gilbert, 2014) to generate a greater output power or to increase the operational frequency bandwidth. For example, the design proposed by Foong et al. (2019) in Figure 7 applies the concept of anti-phase resonance. This design causes both the coil and the magnet components to vibrate but in the opposite direction with respect to each other. This would increase the relative velocity between the coil and the magnet components, hence results in an increase of voltage and output power.

Like in the piezoelectric transduction method, it does not require a separate voltage source to operate. Another advantage is that they are robust and proven to have a longer lifespan compared to other methods (Wang et al., 2016). However, the drawback of this method is that the use of permanent magnets can be bulky, which leads to challenge of integrating them into micro-scale applications. Additionally, the voltage output of an electromagnetic harvester can be quite low due to its low impedance. Kim et al. (2012) reported that under a critical harvester volume of approximate 0.5 cm^3, piezoelectric conversion is more favorable as the electromagnetic power transformation factor drastically drops below this volume. However, the output power of electromagnetic harvesters is observed to has surpass not only piezoelectric harvesters, but also other types of VEH devices at the macro-scale volumes. Beeby et al. (2007) demonstrated that through detailed optimization, a cantilever beam-based electromagnetic energy harvester is capable of achieving a higher power density (power per unit volume) as compared to piezoelectric and electrostatic harvesters even at small volume.

ELECTROSTATIC VEH

The concept of electrostatic conversion involves the use of a variable capacitor that consists of two conducting plates separated by an air, vacuum medium or a dielectric material. When the two conductors move relatively to each other, the energy stored in the capacitor varies resulting in an unbalanced electrical charge. This concept makes it easier for electrostatic harvesters to adopt a micro-size structure (Naito and Uenishi, 2019; Dorzhiev et al., 2015; Mitcheson and Green, 2012), though some researches have also designed macro-sized electrostatic VEH (Zhang et al., 2016). Assuming that the capacitor consists of two parallel rectangular electrodes, the voltage generated across the electrodes, V_{ES}, is (Boisseau et al., 2011)

$$V_{ES} = \frac{QD}{\varepsilon_0 Lw} \tag{8}$$

where Q is the charge in the capacitor, D is the distance between the two parallel electrodes, $\varepsilon 0$ is the dielectric permittivity of free space and L and w are the length and width of the electrodes. The electrostatic conversion method can generally be divided into two categories, which are the electret-free and the electret-based electrostatic conversion. The electret-free converter operates based on an external electri-

Figure 7. An antiphase electromagnetic VEH design proposed by Foong et al. (2019)

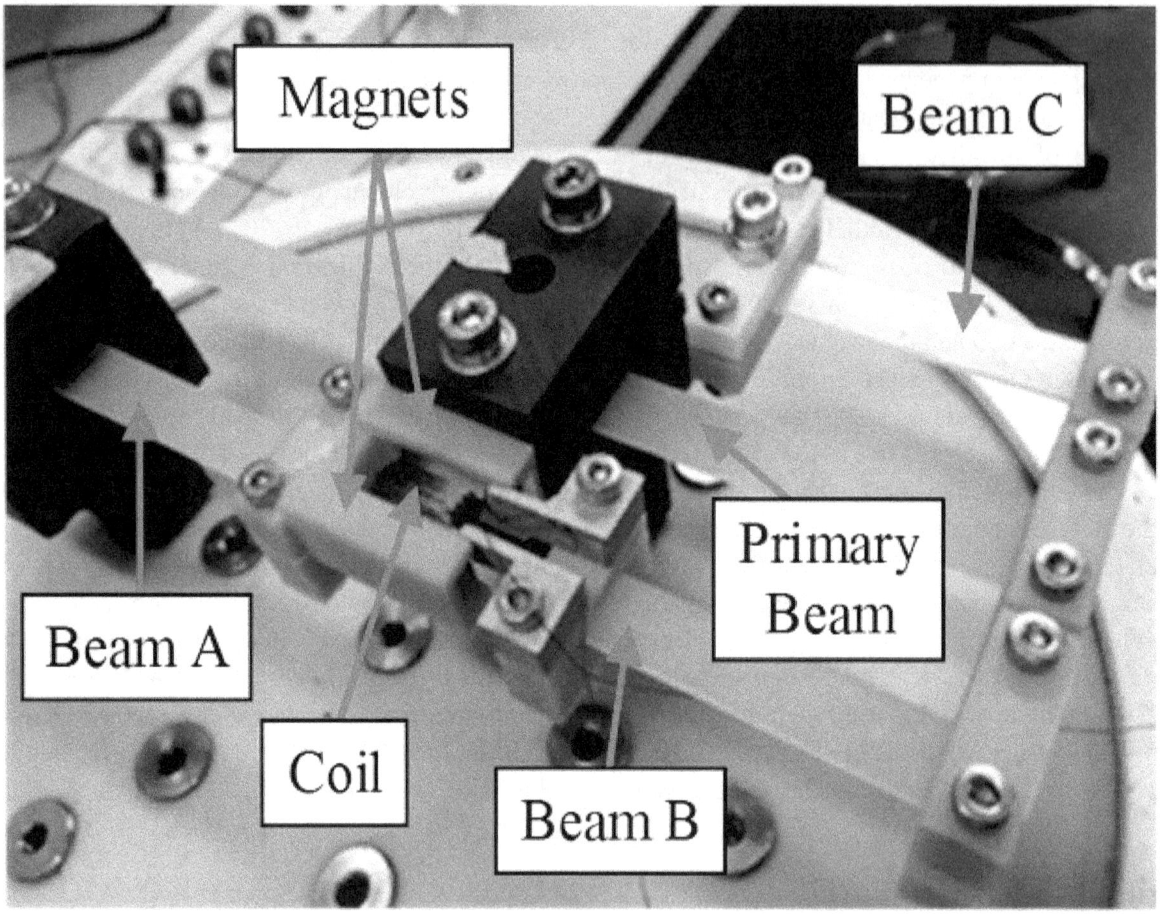

cal input which charges the capacitor up to a certain voltage. The stored energy can then be increased by varying the relative distance between the two conducting plates inside the capacitor using either the charge-constrained method or the voltage-constrained method (Torres and Rincon-Mora, 2009). This however highlights one of the main drawbacks of the electret-free converter in that a separate voltage source is required to initially charge the capacitor.

On the other hand, the conducting plates inside the capacitor of an electret-based converter (See Figure 8) are covered with a layer of electrets (quasi-permanent polarized dielectric material) to provide a constant charge supply. Due to this, the electret-based converter does not require an outside voltage source to operate. However, the electrets have a limited lifespan and its efficiency degrade over time (Du Toit, 2005). The main benefits of an electrostatic harvester are that they have a high voltage generation and can easily be integrated into a micro-scale level since the manufacturing process for electrostatic transducers are very similar to the manufacturing standards of micro-electromechanical systems (MEMS).

On top of that, there is a separate type of electrostatic energy harvesting method called triboelectric energy harvesting, which was introduced and was actively discussed by Fan et al. (2012). The triboelectric effect is a type of contact electrification that occurs when a material becomes electrically charged after coming into frictional contact with a dissimilar material. Static electricity that is generated when

Figure 8. (a) An electret-based converter with its equivalent circuit representation (b) with load R and (c) with capacitances (Boisseau et. al., 2013)

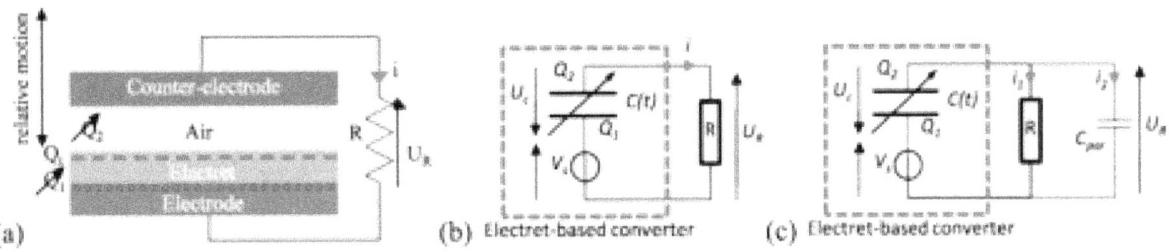

two objects are rubbed against each other is an example of the triboelectric effect. Since the triboelectric effect also applies the concept as electrostatic, triboelectricity energy harvesting is categorised as a subset of electrostatic energy harvesting. As the definition states, the electrical power produced from the triboelectric effect is strongly dependent on the type of material. Zou et al. (2019) developed a new method to quantify materials based on their normalised triboelectric charge density (TECD), with polytetrafluoroethylene as the reference TECD. The TECD of a material indicates the capability of a material to obtain or lose electrons after contact electrification. Generally, materials with a higher negative TECD value would generate a large static electricity. Due to its similarity with the electrostatic harvester, the voltage output of triboelectric harvesters can be also be calculated using equation (8) by replacing the capacitor charge term with the TECD value of the contact material (Zhu et al. 2015). In terms of VEH, vibration-based triboelectric harvesters have made remarkable progress over the past decade (Fu et al., 2019; Hamid and Butler, 2019; Yang et al., 2014; Yang et al., 2013; Chen et al., 2013; Hu et al., 2013). In some examples such as in Figure 9, the triboelectric harvester was able to generate sufficient energy from human walking activities to simultaneously power 40 commercial LED bulbs.

The output of a triboelectric harvester is similar to piezoelectric and electrostatic harvester, in where a large voltage output can easily be obtained, making it a promising new electrical conversion method. However, the main concern of a triboelectric harvester is in its durability of the material for long-term continuous applications. The constant friction contact between two materials to produce the triboelectric effect gives rise to material abrasion and heat generation which can reduce the efficiency of the harvester,

Figure 9. Broadband triboelectric VEH design proposed by Yang et al. (2014)

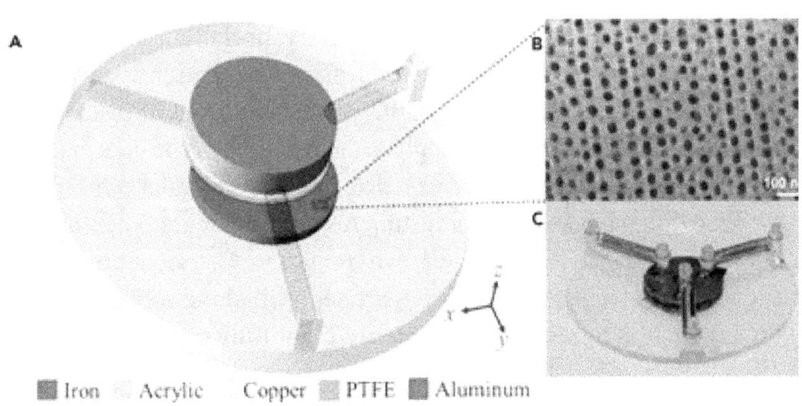

especially at higher frequencies (Chen and Wang, 2017). Additionally, friction may also introduce undesirable damping into the harvester.

SUMMARY ON MECHANICAL-TO-ELECTRICAL CONVERSION METHODS

Table 2 summarizes the comparison between the three mechanical-to-electrical conversion methods discussed in this chapter. Generally, piezoelectric and electrostatic conversion methods have a similar characteristic in where these methods result in a high voltage and a low current output. On the other hand, the electromagnetic conversion method generates a low voltage but a high current output. Due to this, piezoelectric and electrostatic harvesters tend to produce relatively larger impedance compared to electromagnetic harvesters, which offsets the power generated from these harvesters. However, most studies have shown that piezoelectric harvesters are generally better at micro-scale volume (Roundy, 2003) whereas electromagnetic harvesters are usually preferred at larger volume scales where its output power can be several times larger (Hendijanizadeh, 2013; Kim et al., 2012).

Table 2. Comparison between four different electrical conversion methods

Conversion method	Concept	Advantages	Disadvantages
Piezoelectric	Piezoelectric effect	High voltage, high efficiency	Low current, expensive, material factor
Electromagnetic	Faraday's law	High current, long life span, best choice at macro-size.	Low voltage, bulky.
Electrostatic	Static electricity	High voltage, easy to integrate into micro-scale.	Low current, may require external voltage source, durability and friction concerns for triboelectric harvesters.

APPLICATIONS OF VEH

Recall that the biggest challenge in portable systems is the batteries limited lifespan. Typical battery-powered vibration sensor lifetime is around three years (Weddell, 2010) due to environmental uncertainty the lifetime period is expected to be shorter. Many portable systems are installed at physically inaccessible location or implant into the human body (for medical micro-sensors), considering the cost of ownership, changing the system batteries periodically is not an option, especially for a ubiquitous system. Not to mention that batteries are not practically applicable in many applications, such as automotive engine sensors that operate in high thermal range. Hence, this raise the interest of research activities in energy harvesting to seek for an alternative solution to replace the batteries or the concept of "Integrated Power unit" – combination of batteries and harvester as a primary and secondary energy sources to prolong the battery lifespan. There are many energy harvesting resources that are potentially feasible to be used as a replacement, however each resource has their own limitation. Consequently, this chapter only focus on the applications of VEH and any other resources are set beyond the scope for exploration. Added to the above, the following section in rest of the chapter will not fully cover all the applications of VEH, but to the best knowledge of the author of this chapter, only those that are impactful and insightful will be further analyzed.

VEH FROM TRAFFIC FLOW

VEH has been extended to many applications for different purposes. One of the most common application VEH is applied on is to harvest the traffic-induced vibration from road surfaces or bridges. Lucy (2010) in association with Technion Israel Institute of Technology developed a piezoelectric VEH which converts the mechanical energy exerted by the weight of passing vehicles into electrical energy. The piezoelectric harvester was embedded along a two-wheel footprint as in Figure 10. In this work, the electrical storage system was integrated in the roads and did not take up any public space, not to mentioned it is able to function in all weather conditions. This demonstration claims that provided there are 600 heavy vehicles travelling at speed of 72 km/h through the interval per hour on average, an average of 200 kW of electrical power can be harvested from generators along a single traffic lane for a distance of one kilometer, which is sufficient to supply for the average electricity consumption in 200 – 300 households. Apart from supplying the harvested electricity to various roadside applications, the system can also be used to monitor the vehicle's speed and weight effectively. For example, it can be used as a weigh bridge to measure the load of the vehicle by monitoring the average output from the generator. This will not reduce the turnaround time for a vehicle to entering or leaving the place, but also diversify the cost of ownership of road surface embedded generator. However, the drawback for this parallel roadway harvesters is the alignment issue, since it is very difficult to ensure that the vehicle wheels are always perpendicularly on top of the harvester footprints. Not to mention the inconsistent driving attitudes such as overtaking, right/left turning etc., which will further reduce the available electrical power that can be harvested.

Ding et al. (2020) performed a similar study by considering the piezoelectric VEH as a double plate layer. The model based on the plate theory displayed a good agreement with experimental results and showed that larger power can be generated if the piezoelectric transducers are place near the load center. If the transducers strayed more than 20 cm from the load center, the power output significantly reduces to almost zero. However, in both studies, the effect of vehicle speed on the performance of the harvester as well as how the vehicle speed correlates with the vibration frequency induced of the harvester was not discussed in depth. This study may be important as it would allow for a more practical understanding in traffic flow harvesting since the input of a VEH device depends on two main factors which are the acceleration of the excitation (can be related to the vehicles weight) and its frequency (can be related to the speed of the vehicles).

Peigney and Seigert (2013) took a different approach in where energy was harvested from the traffic-induced vibrations under a road bridge to power structural health monitoring sensors installed on the bridge, which used to provide quantitative and reliable data on the real conditions of the bridge. Sensors were initially placed at different locations under the bridge to evaluate the acceleration levels. It was found that the sensors that are placed on the water pipes under the bridge (See Figure 11) recorded a 10 times larger acceleration level as compared to the acceleration levels measured at the mid span girder of the bridge, with a vibrating frequency of 14.6 Hz. Burst excitations signals were observed from the sensors when a single lorry passes the bridge. For heavier lorries, the acceleration levels generated from the burst excitations can reach up to 0.6 ms-2. A test was then conducted by placing a cantilever piezoelectric VEH device at location '0' in Figure 11. The device was tuned to achieve resonance at 14.5

Figure 10. Parallel road harvesters

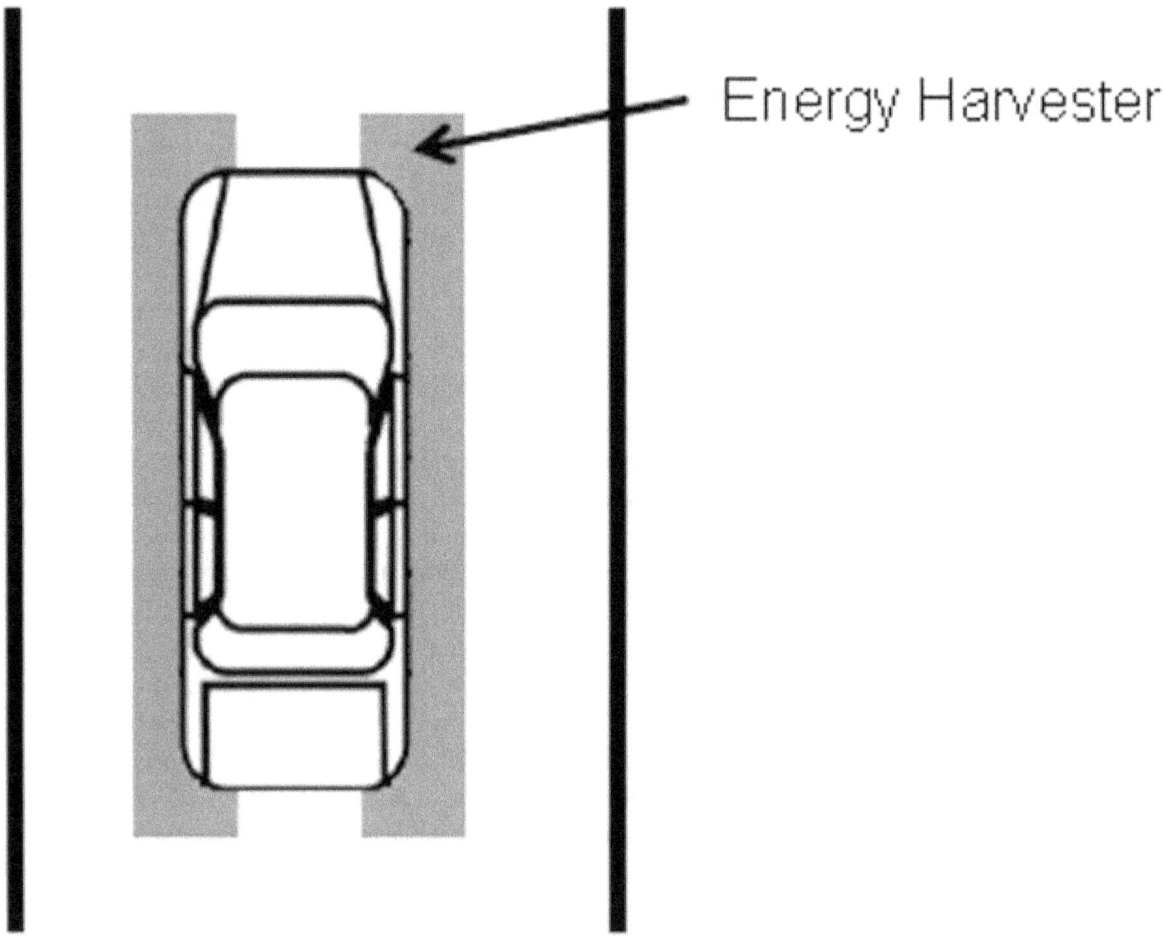

Hz. Results indicate that at peak traffic intensity, a maximum power output of 0.03mW can be obtained, which provides sufficient energy to power a structural health monitoring wireless sensor node.

VEH FOR RAILWAY APPLICATIONS

Railway transportations such as freight trains, commuters, monorail and subways have experienced rapid transformation over the past few decades, now being one of the most preferred forms of public transportation for everyday commutes. Nowadays, high-speed and efficient inter-state transport have become a necessity which have urged railway development to produce a faster, precise and more intelligent monitoring system. Currently, the Shanghai Maglev train holds the record as the world's fastest train, recording a maximum operation speed of 430.0 km/h. Regardless, the more crucial issue here is to ensure that these high-speed trains are equipped with essential railway infrastructures such as wireless monitoring sensors, signal lights and global positioning system (GPS) to facilitate its operation. How-

Figure 11. (Left) Location of harvester placement under a road bridge and (Right) A design of a piezo-electric VEH (Peigney and Seigert, 2013)

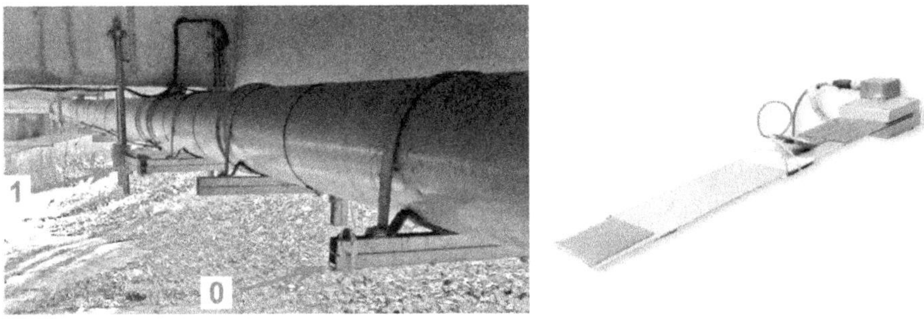

ever, large portions of the railway tracks are located in rural areas where a direct power supply is usually unavailable, which makes it necessary to find an alternative power source for the railway infrastructures.

Vibrations are induced in railway tracks when a train passes above it and causes the track to deflect vertically. These vibrations, while mostly exhibiting a pulse like signal, have significantly high acceleration levels which makes VEH a viable option to explore. Wang et al. (2012) came up with an electromagnetic VEH design which uses two one-way clutches, three shafts, four pairs of gear meshing, two racks and

Figure 12. Working principle of a simplified mechanical motion rectifier for railway track VEH (Lin et al., 2018)

a pinion to convert the vibrations of the track into a unidirectional direction generator. This method is known as the mechanical motion or vibration rectifier. The method causes the generator to rotate in a single direction with a relatively uniform speed, making full use of the pulse like signal of the track vibrations. The design was later simplified by Lin et al. (2018) by reducing the number of shafts and gears to one shaft and two gear meshing as shown in Figure 12. The improved design reduced energy loss from friction and misalignments and increased the efficiency by around 20.0%. The proposed harvester can generate a power output of 28.0 W at 1.0 Hz frequency and a 5.0 mm harmonic displacement input when tested at a location beside a railway track. This makes it possible for the harvester to target on powering major railway infrastructures such as warning signal lights and switches, which requires an average power of 10.0 W or more. In another study, Zhang et al. (2017) also applied a similar method but with a slightly smaller design. The design however applied the mechanism from Wang et al. (2012) and not the simplified design by Lin et al. (2018). Instead of placing the harvester beside the railway tracks, the harvester was designed to be placed under the tracks where the vibrations are more pronounced. A single prototype was able to generate a peak voltage output of 58.0 V at 1.0 Hz, which is nearly enough to power railway infrastructures. If the design was upgraded to the later simplified design, sufficient power may be achievable.

Bradai et al. (2018) measured the vibrations induced on the train itself instead of the railway track. The main reason for this was there was a growing need to install monitoring sensors in train carriages, but electricity is not always available for all carriages in a freight train. A data logger sensor was placed onto one of the train carriages on a freight train travelling from Leipzif to Hamburg in Germany. The data logger recorded a maximum acceleration level of 20.0 ms^{-2} as seen in Figure 13, with a frequency range between 20.0 Hz to 40.0 Hz. However, the average acceleration levels are about 3.0 ms^{-2}. By replicating the train-induced vibration profile onto an electrodynamic shaker, a conventional electromagnetic VEH harvester was then tested. The tested design has a volume of less than 2.0 cm^3 and was able to achieve a peak power output of 14.8 mW at a frequency of 27.0 Hz, which outperforms most of the other reported VEH designed for train wagon applications.

VEH FROM CIVIL STRUCTURES

Another promising area of vibration energy is harvest from the civil structures. Civil structures like tall buildings, fly-over bridges, tunnels, communication base stations, are very susceptible to external vibration. In the modern society, to provide more space on a fixed amount of land, building structures expansion are lean toward the vertical direction. Especially in the urban area, where most of the skyscrapers are built to accommodate more and more population onto an acre of land. Yet, vibration that cause by huge dynamic loadings from winds, earthquakes, traffics or even human motions has been a serious concern since the early days. Large deflection amplitude of the building structure not only cause damage to the civil component but also cause discomfort to its human occupants (Kareem et al., 1999). Hence, vibrations of a tall building are serious concern to engineers and many researches has been carried out to mitigate the harmful vibration into a useful electrical energy (Ni et al., 2011; Tang and Zuo, 2011). Instead of dissipating the vibration energy into heat waste in the conventional vibration suppression devices, the vibration energy can be converted into electricity through dual functional tuned mass damper (TMD) energy transducers (Zuo and Cui, 2013). Practically, TMD is implemented in many tall buildings, such as Taipei 101 building in Taipei (See Figure 14a). To make the 508 m tall

Figure 13. Acceleration levels on a freight train carriage travelling from Leipzig to Hamburg (Bradai et al., 2018)

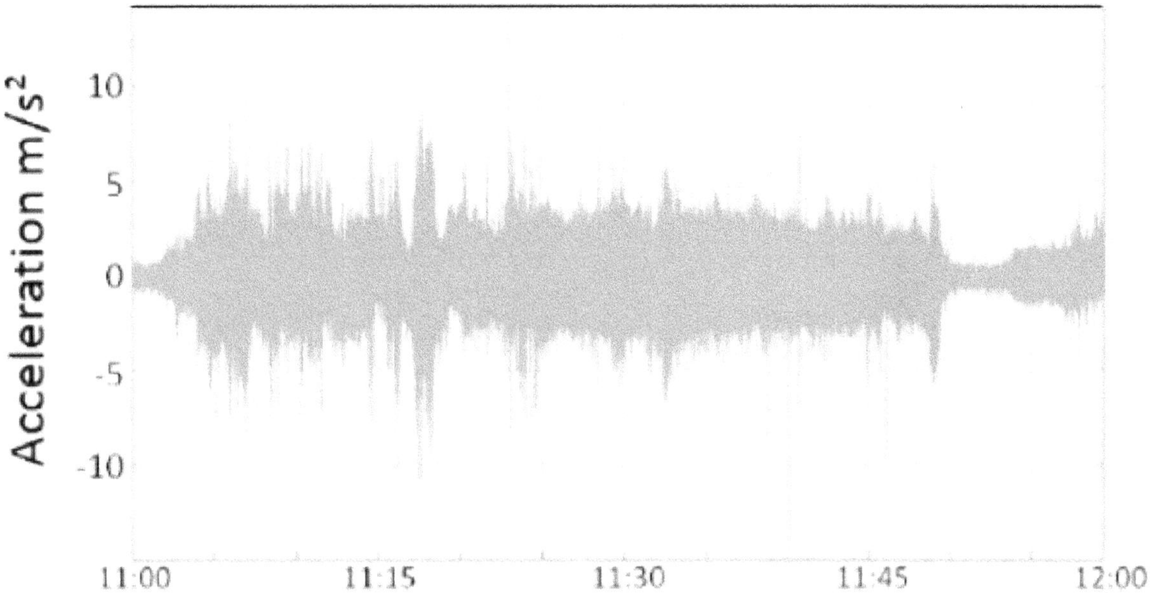

building endurable to earthquakes, typhoons and fierce winds atmospheric assaults, a 730-ton giant pendulum TMD (See Figure 14b) has been included to counteract the building movement that due to external force. The TMD uses a computer-controlled counter moving weight to adaptively move against the building sway, it is pushed in the opposite direction to the building sway to dampen the oscillation, reducing sway due to wind flow up to 36.0 cm or 40% from the initial value of 60.0 cm. The schematic of TMD modelled as a two-degree-of-freedom (2DOF) system is shown in Figure 14c, where the terms with subscript '1' representing the model of the vibrating structure (the building in this case) and the terms with subscript '2' representing the model of TMD suspension. This model is very useful for use in the modern mechanical analysis and simulation.

Ni et al. (2011) investigated the power available to be harvested from the TMD system when a 76-storey building was subjected to high wind event of 13.5 m/s at height of 10.0 m. The result shows that more than 85.0 kW of power is available for harvesting from the TMD dynamics. In this study, similar performance level of energy harvesting can also be achieved by using multiple TMDs system, constructing in series configuration with significantly smaller mass than the single TMD system, but a larger stroke or building sway was required. A year later, for the first time Tang and Zuo (2012b) demonstrated that both vibration mitigation and energy harvesting can be achieved simultaneously. In the three-story building prototype experiment, a conventional energy-dissipating mechanical suspension was replaced with an electricity-generating tuned mass damper (TMD), in which the damping force can be controlled by the duty cycle of pulse width modulation (PWM) signal that was generated according to the building sway condition. The TMD composed of a high energy density rotational motor with gears and rack-pinion mechanism. From the experiment, a power output of 60.0 mW was harvested from the proposed regenerative electromagnetic TMD (simultaneously with vibration control) under a harmonic force with amplitude of 15.0 N at 3.3 Hz. From the result, it can be seen that the harvested output is suf-

Figure 14. a) location at where the TMD system is placed on the Taipei 101 building (taken by ChungKet Thein). b) Illustration of the TMD system (Tso et al. 2013). c) TMD modelled as a 2DOF system (Tang and Zuo, 2012a)

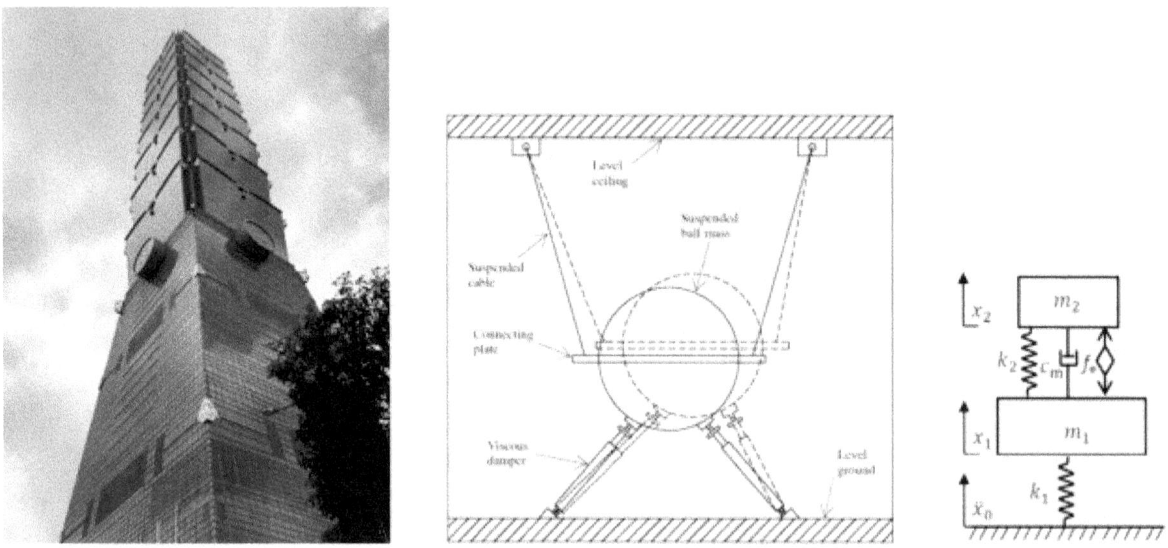

(a) (b) (c)

ficient to supply to most of the ultra-low-powered microsensors that is ubiquitous in most of the smart building, such as fire alarm sensors, temperature & humidity sensors, security sensors and etc. It may also be suggested that to obtain a true output for a real physical three-story building, one may just upscale the output according to the up-scaling size of the prototype as well as the input force and frequency.

Apart from building structures, traffic tunnels are also one of the potential civil infrastructures. Traffic surveillance and structure condition are mandatory requirements for the safety operation of the traffic tunnels. Hence, a consistent monitoring system is required to provide a systematic overview of the tunnels during regular operation to minimize human losses especially during the event of landslide, traffic accidents or even terrorist attacks. This surveillance can be achieved by implementing a highly efficient sensor network that consistently providing the real- time update information in the tunnel. However, due to the narrow civil structure of the tunnel, installation of sensor within the tunnel pose a great challenge, not to mention that the sensor node might be embedded completely into the concrete. Thus, wired connection between nodes is impractical and wireless communication is highly preferable under these circumstances. For this application, a sustainable and low maintenance power sources for the sensor nodes are needed, such as vibration energy harvesting. Traffic tunnels are frequently the bottleneck for traffic and working at full capacity most of the time, so sufficient vibration energy can be harvested from the vibration excitation caused by the traffic flow in the tunnel. Wischke et al. (2011) studied the vibration available to be harvested from two different types of tunnel: a road traffic and a fixed railway tunnel. The sensors were mounted on the tunnel concrete walls, asphalt road surface and railway track for each of the tunnel, respectively. The results show that the vibration amplitude levels at the tunnel wall from both types of tunnel are below 0.01 ms^{-2}, which is way too weak for a transducer, not to mention that the different resonant frequency that created by various vehicle/train sizes, further preventing the employment of a high-Quality Factor harvester. Added to that, the vibration amplitude obtained from the road surface were insignificant too, only maximum peak of 0.015 ms^{-2} amplitude was recorded when big truck passing by the road, which was not yield sufficient energy to power any wireless sensor devices. The weak vibration transferred from the vehicle was caused by the suspension and pneumatic tires that were implemented for passenger comfort. Not to mention that to harvest the vibration, a harvester has to be embedded into the road deck in order to protect them against the impact from the passing by vehicles, which further make the application complex and costly. In the railway tunnel, vibration amplitudes of 3.5 ms^{-2} and 1.0 ms^{-2} were recorded at the railway and at the railway sleeper, respectively. To avoid any unwanted disturbance to the train, the railway sleeper was chosen as the vibration energy harvesting position. A four cantilever-based piezoelectric energy harvester which electrically connected in parallel was mounted to the railway sleeper and average energy of 395.0 μJ was recorded per passing by train, which proved to be sufficient to power for one radio frequency (RF) protocol cycle. Other than a RF module, the harvested energy may also be used for the monitoring and analyzing purposes to prevent the failures in the civil structures other than tunnels, such as bridges, foot walks and embankment.

VEH FOR MEDICAL IMPLANTABLE DEVICES

Due to the significant advances in extremely low power and miniaturization in electronics devices, implantable biomedical devices have considerable potential for energy harvesting from the human body itself. Previously, most of the implantable biomedical devices are powered through wires (See Figure 15). This may cause discomfort, skin allergy or infections and many unwanted problems to the patients. To avoid the wires, implanted batteries may be used to power the devices, but due to its finite lifetime, this may increase the patients' surgical risk and burden on maintenance operations, especially to the elders. Hence, a battery-less and a standalone implantable device designs are highly desired. By eliminating the batteries and wires, this can further extend the devices' lifespan and resulting in a smaller, longer lasting, safer implantable biomedical devices, which provide a greater comfort to the patients as well as reducing the surgical risks and costs. Currently, one of the most frequently used bioengineering technologies is pacemakers. Due to its relation to human health, this particular implantable device has significantly grown as one of the most lucrative business. For example, as of the year 2016, it was estimated that there were 1.14 million pacemakers being used globally and is expected to rise to 1.43 million units by the year 2023. (Mikulic, 2019).

As the global population continues to grow exponentially yet advances in healthcare, this increases in the number of elderly populations. There are increasing incidences of cardiac disorders health concern even in a developed country like United States, creating the need of cardiac pacemaker to maintain proper heart rhythm of patients. In the effort to seek for the solution of a leadless pacemaker, Ouyang et al. (2019) proposed a symbiotic cardiac pacemaker (SPM) which is powered by micro triboelectric VEH. The triboelectric harvester is surgically implanted into the heart of a patient and generates power from the patient's own heartbeat as seen in Figure 16. The SPM consist of three main components which are the triboelectric energy harvester (iTENG) unit, the power management unit and the pacemaker unit. Initially, the power management unit is switched off and the energy harvested by iTENG is stored inside a capacitor. The power management unit is then switched on via a wireless switch to drive the pacemaker under controlled conditions. The iTENG harvester consisted of seven different layers which

Figure 15. Example of a pacemaker with wires/lead implanted in a patient (Deterre, 2013)

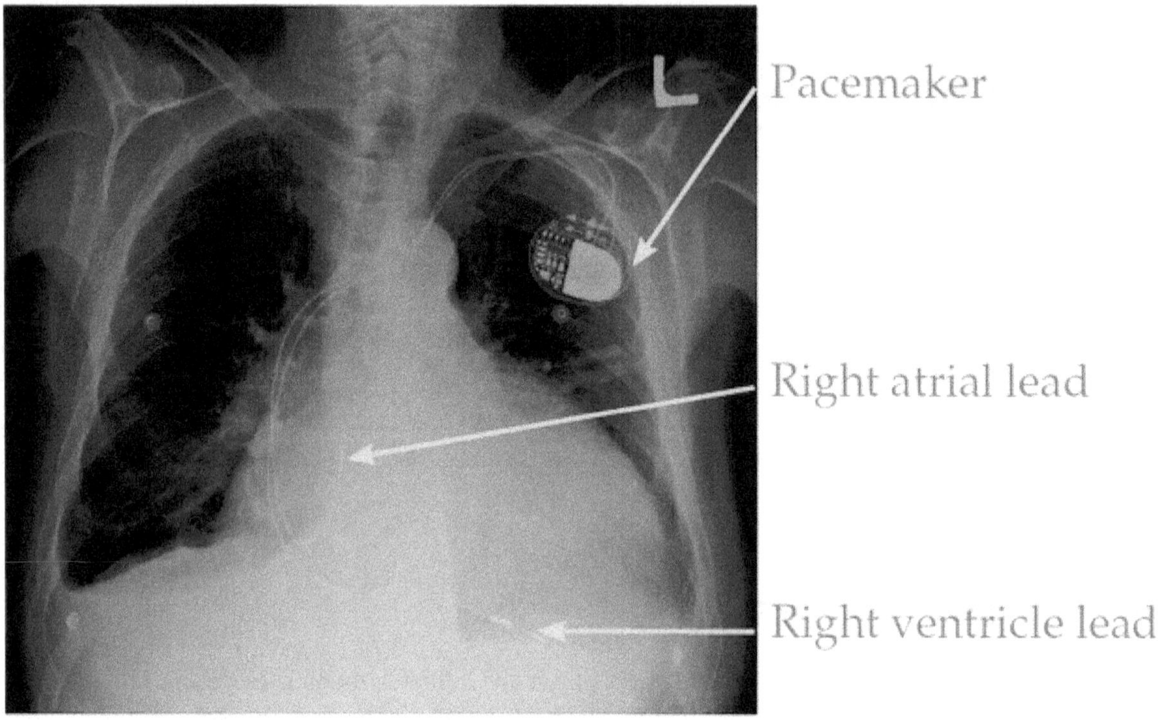

makes up its core structure, supporting structure and encapsulation layer. The core structure consists of two triboelectric layers, with one of the layers being a nanostructured thin film polytetrafluoroethylene (PTFE) material. This layer was polarized using the corona discharge method to further increase its surface charge density. Preliminary investigation showed that the triboelectric VEH can attain a peak power density of 110.0 mWm^{-2} under a load resistance of 100.0 MΩ.

A test was then conducted by implanting the SPM system into a large male Yorkshire porcine as shown in Figure 17. Prior to the implantation operation, the porcine made to fast for 12 hours due to operational procedures. A 100 μf capacitor was used to store the energy generated by iTENG as this capacitor can be charged up to 3.55 V in roughly three hours when subjected to a rate of 77 heartbeats per minute. The open circuit voltage of iTENG peaked at 65.2V and the energy generated from each cardiac cycle was 0.495 μJ, which was higher than the heart pacing threshold for humans (0.377 μJ).

Karami and Ansari (2017) studied the benefits of a fan folded piezoelectric VEH for self-powered pacemakers. The design itself is made of several cantilevered beams layers stacked on top of each other and joined together at consecutive ends with a rigid links (Figure 18). Two piezoelectric layers were attached on each beam, making each beam a bimorph piezoelectric harvester. A seismic mass was then added to the free end of the topmost beam to tune its natural frequency. Excluding the seismic mass, the device itself has a volume of approximately 1 cm^3 and was able to generate an average output power of 15.2 μW under a rate of 20 to 100 heartbeats per minute. This power is sufficient to power a typical pacemaker (< 10 μW). The harvester is also MRI (Magnetic resonance imaging) friendly since it is free of any magnetic materials. Nevertheless, the power output of 15.2 μW was only achieved when a seismic mass of 18.4 g (tungsten) was used. Visually (Figure 18), the seismic mass has almost the same

Figure 16. Schematics of a symbiotic cardiac pacemaker

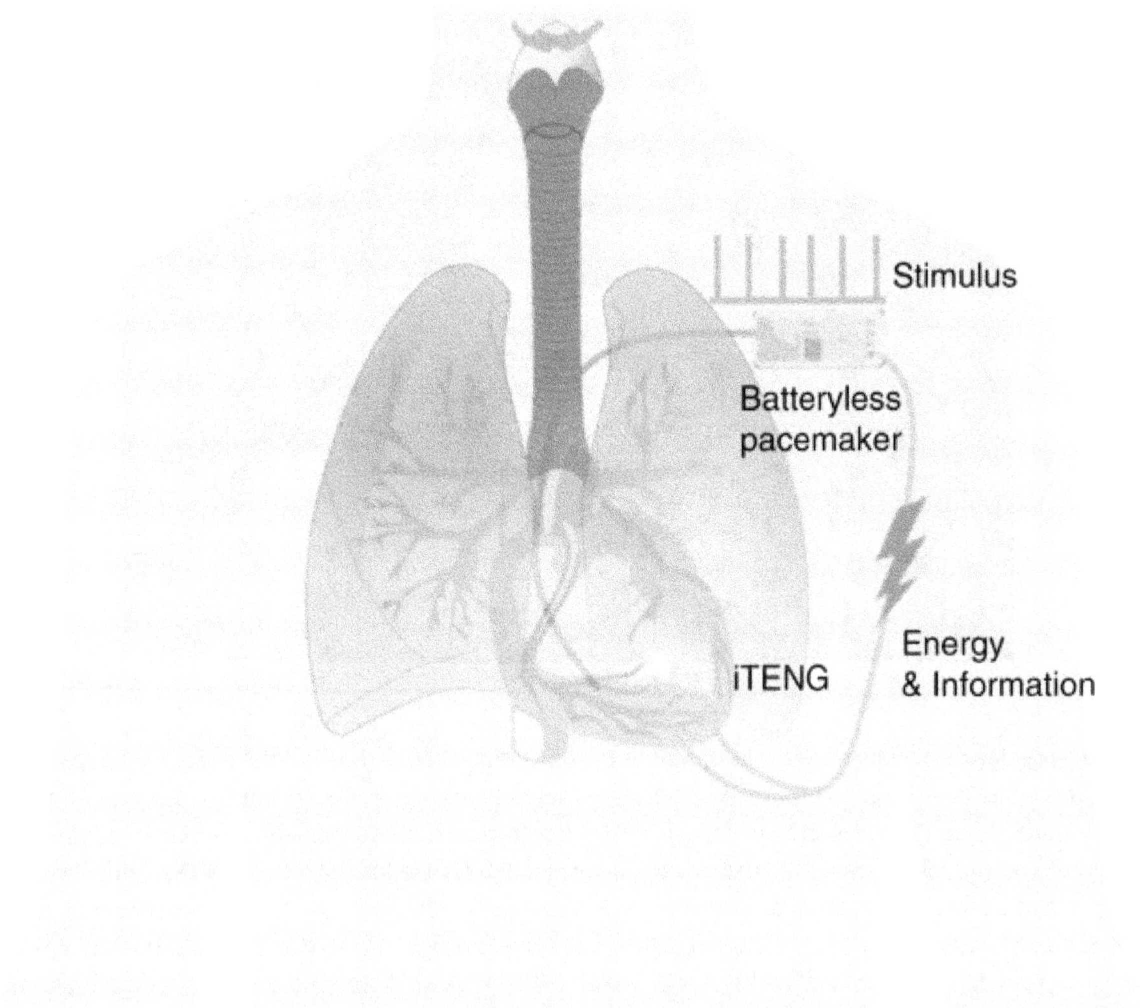

volume as the harvester device, which makes the practical volume of the actual device almost double. It was reported that the required size constraint for an energy harvesting device to be implantable must not be larger than 1 cm^3 without the seismic mass. However, there should be a limit as to how large the seismic mass can be which was not specified in the report. Hence, this makes it questionable if the whole device is practical for implantation.

VEH IN FOR AUTOMOTIVE APPLICATIONS

In the automotive industry, it is well aware that the function of a vehicle suspension system is to support the weight of vehicle chassis, to isolate the body from road disturbances and to ensure comfort of the passenger in the vehicle. However, it is believed that one of the main energy losses comes from the dissipation of vibration energy by shock absorbers in the vehicle suspension that is caused by road

Figure 17. A triboelectric VEH being implanted onto a porcine's heart (Ouyang et al., 2019)

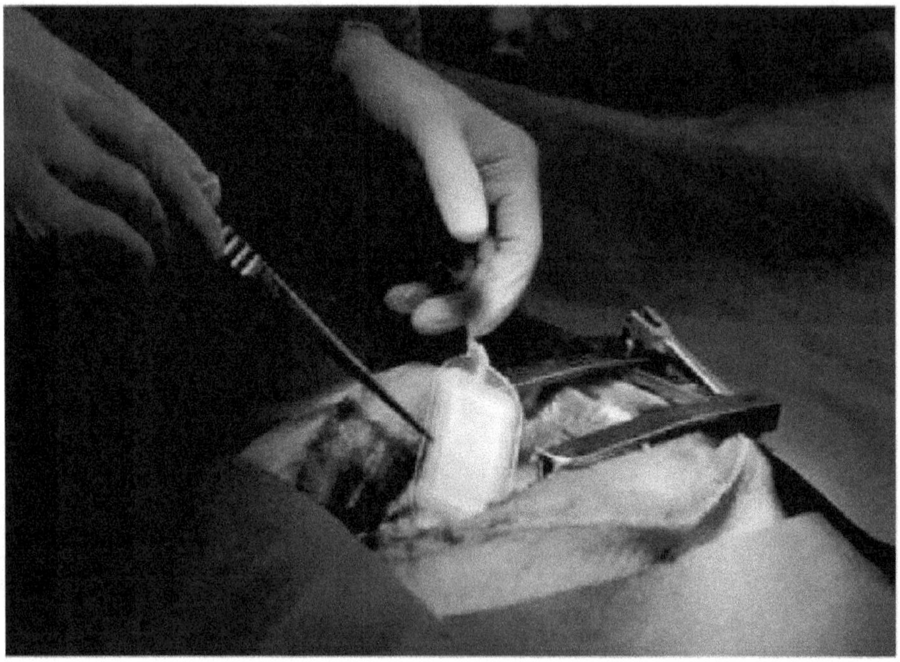

irregularity during acceleration or deceleration of the vehicle. Thus, to avoid the energy wastage, VEH mechanism can be applied to turn the kinetic energy into electricity or other potential energy and store it for later use. Since 2004, research in energy harvesting from vehicle suspension has exponentially risen to approximately 30 research publications by the year 2018 (Abdelkareem et al., 2018a). For a vehicle travelling at a speed of 100 km/h and depending on the type of vehicle class, the average potential power that can be generated from the vehicle suspension system can reach up to 0.1 kW for cars, 0.55 kW for buses, 3.5 kW from lorries and 3.0 kW for off-road vehicles (Abdelkareem et al., 2018b). Different harvesting mechanisms have been proposed for vehicle suspension energy harvesting, ranging from linear systems to rotational system. Figure 19 shows the schematic diagram of a VEH for vehicle suspensions.

In 2010, Zuo et al. (2010) proposed a linear electromagnetic VEH device for vehicle suspensions illustrated in Figure 20. The design focused on the arrangement of the coil and the materials used for the decide. Zuo et al. (2010) determined that a higher power output can be obtained by using multiple coil windings with four different phases, namely phase 0°, 90°, 180° and 270°. Each coil winding has around 250 to 300 turns. Through finite element analysis, it found that placing a material with a low magnetic permeability (like aluminum) at the center of the magnet stacks and encasing the harvester with a material of high magnetic permeability (soft steel) would increase the magnetic flux density within the coil regions. Tang et al. (2014) later improved the performance of the design by considering a double layer magnet arrangement consisting of radial and axial oriented magnets to increase the magnetic flux density. The final improved design recorded a power density of 27.2×10^4 Wm^{-3} as compared to a power density of 17.2×10^4 Wm^{-3} from the first design.

Li et al. (2013b) demonstrated that it was also possible to apply the mechanical motion rectifier method to harvest energy from vehicle suspensions due to the similarity between the motion of vehicle suspensions and railway tracks. One the benefits in doing so is that it reduces friction in the suspension

Figure 18. Fan folded piezoelectric VEH for powering pacemakers (Karami and Ansari, 2017)

Figure 19. Schematic diagram of a VEH for vehicle suspensions (Abdelkareem et al., 2018a)

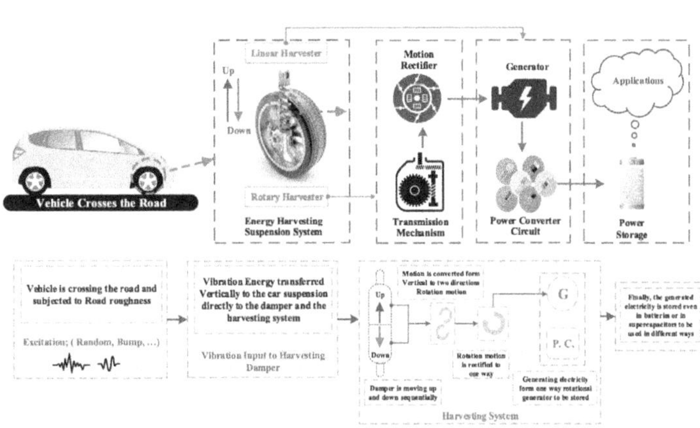

Figure 20. Linear electromagnetic VEH for vehicle suspensions consisting of four different coil phases (Zuo et al., 2010)

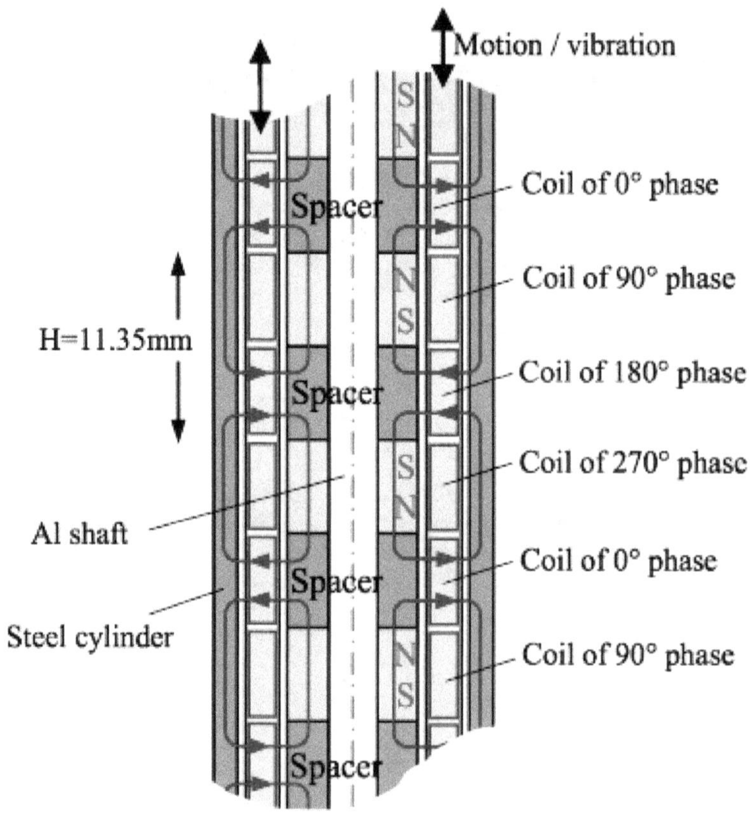

and increases energy efficiency. The design also incorporates a rack and pinion system as shown in Figure 21. Subsequently, rotational harvesters have been shown to be superior than linear harvesters in terms of power output and efficiency (Hendijanizadeh, 2013). The prototype design was fixed onto a Chevrolet Suburban SUV vehicle and achieved a power output of 15.4 W at a speed of 24.1 km/h when driven on a smooth paved road. Overall, an efficiency of 60.0% was recorded, which is higher than the maximum efficiency for linear harvesters (50.0%). Improvements to the device can be made using the same method proposed by Lin et al. (2018), that is using a simplified mechanical motion rectifier to increase efficiency and output.

Another application of VEH in automotive is as a power source for tire monitoring sensors. These sensors are often fixed onto the inner liner of vehicle tires to take advantage on the tire deformation and acceleration drop while driving (See Figure 22). This makes it almost impossible to connect wires or cables to the sensors since the tires would be rotating at very high speeds. In Figure 22, the acceleration curve describes a steady-state region (a), a region where the tire experience a sudden increase in acceleration (b), an instant increase in acceleration due to the increase of radial curvature when the tire is flat (c), the duration between two acceleration peaks where a significant change in acceleration can be observed (d) and the time when the tire is in contact with the surface (e). Normally, a VEH device would produce maximum power at the two acceleration peaks encompassing region (d). At these peaks, the stress of the tires is also maximized. Tire monitoring sensors function as a tire pressure sensor or as

Figure 21. Electromagnetic VEH incorporating the rack and pinion system (a) overall and (b) inner structure (Li et al., 2013b)

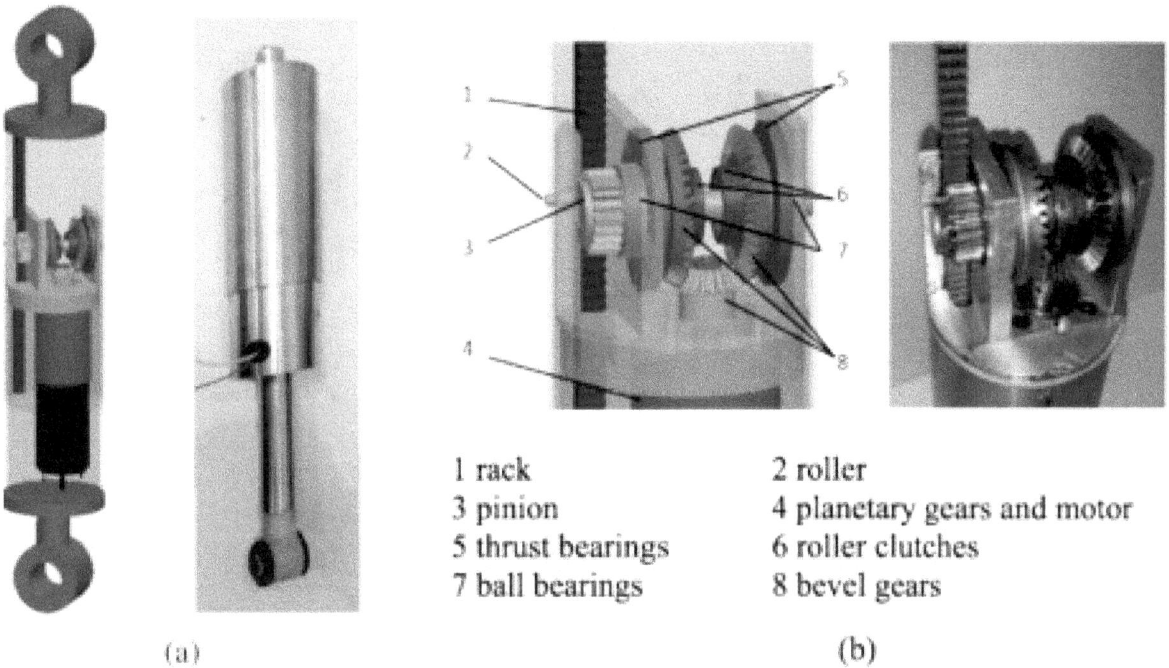

a motion indicator. On average, a rotating vehicle tire would experience peak acceleration in the order of 10^3 g at frequencies larger than 100 Hz (Lohndorf et al., 2014).

Eshghi et al. (2017) proposed a unimorph piezoelectric VEH device to convert the two maximum stress peaks into useful power. The device consists of a clamp-clamp cantilever beam with a seismic mass placed in the middle as seen in Figure 23. The design was optimized using two different methods, namely the deterministic design optimization (DDO) and the reliability-based design optimization (RBDO). Experimental results show that both optimized devices were capable of producing peak power outputs of nearly 2.0 mW when the vehicle is moving at speed of 20.0 km/h, which is sufficient to drive the tire monitoring sensors. However, a crack was detected on the piezoelectric material when the DDO device was tested at speeds higher than 20.0 km/h. The RBDO device remained crack-free even when it was tested at a speed of 60.0 km/h. Apart from that, it also showed than if the vehicle was travelling at 10.0 km/h instead, the peak power output would drop to less than half the power at 20 km/h, which suggest that the fundamental frequency of the tires changes with vehicle speed. This raises two issues; one on the durability and practicality of the device since generally, a vehicle would move at greater speeds, especially during long distance travel and the other on the efficiency of the device to produce sufficient power at different speeds. It may be more appropriate to use a cantilever beam design instead as the stresses in cantilever beams are typically eight times lower than clamp-clamped beams if considering the same beam dimensions and deflection. Although this would inevitably lead to a trade-off in terms of power, it can be compensated by using a bimorph piezoelectric harvester beam. Moreover, if several cantilever beams were arranged in an array and are properly tuned, the device would also be able to produce sufficient energy at several different speeds.

Figure 22. Radial acceleration of a moving vehicle tire (Eshghi et al., 2017)

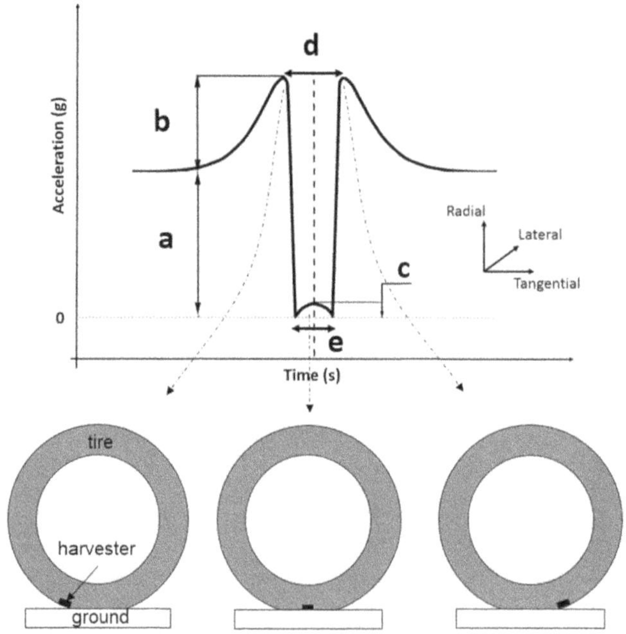

CONCLUSION AND FUTURE PROSPECTS

Figure 23. Piezoelectric VEH for tire pressure monitoring sensors (Eshghi et al., 2017)

This chapter presents the current overview of the trend of energy resources used by sensors and portable devices. The environmental impact caused by batteries is identified and various types of energy available to be harvested from the surrounding environment are also reviewed. The arguments suggested that vibration energy harvesting has the potential to be one of the most promising alternative energy sources due to its redundancy at most of the famous application locations. Common designs of vibration energy harvesting (VEH) devices were discussed and compared. A general vibration-based energy harvesting (VEH) mathematical model is also presented and design consideration toward an optimal VEH devices are discussed and analyzed. From the modelling, it is shown that the output power generated by the excitation vibration created from many surrounded environment sources is sufficient to be used for powering the sensor and portable devices. Three different mechanical-to-electrical conversion methods for VEH devices were then outlined together with theirs benefits and disadvantages of each. Lastly, some insightful and useful applications of VEH are also presented and reviewed.

A typical sensor uses Lithium Thionyl Chloride ($LiSOCl_2$) batteries that come with lifespan of roughly 4 years (Weddell, 2010). However, most of the batteries start to show uncertainty of output after 2 years, thus maintaining a ubiquitous sensor system is problematic and costly. Not to mention that $LiSOCl_2$ batteries are strictly regulated and cannot be transported on passenger planes in large packages, hence this may lead to weeks of lead time for batteries replacement. Energy harvesting system is always being compared with batteries of its higher up-front cost. However, many incidents have shown that the higher up-front cost can be offset by the reduction in maintenance when batteries in conventional devices fail in long run. On top of that, wireless technologies are also eliminating the need of wires within the system, hence saving the installation and wiring costs. Application of energy harvesting has now expanded to powering larger-powered devices, which could potentially boost the scale of manufacturing and allow wider adoption of energy harvester as a replacement of batteries, thus significantly reducing the up-front cost of ownership of energy harvesting systems. While is it true that the application and demand of electronic sensors is increasing over the years as they play a vital role in this technological era, it is believed that the higher up-front cost of an energy harvesting system will further reduced when the demand for a self-power solution increases.

ACKNOWLEDGMENT

The authors would like to acknowledge the funding body from the University of Nottingham Ningbo China for the partial support from New Researchers Grant, Faculty of Science and Engineering (FOSE) and Institute of Asia and Pacific Studies (IAPS) Research Grant from for funding this research.

REFERENCES

Abdelkareem, M. A. A., Xu, L., Ahmed Ali, M. K., Elagouz, A., Mi, J., Guo, S., Liu, Y., & Zuo, L. (2018a). Vibration energy harvesting in automotive suspension system: A detailed review. *Applied Energy*, *229*, 672–699. doi:10.1016/j.apenergy.2018.08.030

Abdelkareem, M. A. A., Xu, L., Guo, X., Ahmed Ali, M. K., Elagouz, A., Hassan, M. A., Essa, F. A., & Zou, J. (2018b). Energy harvesting sensitivity analysis and assessment of the potential power and full car dynamics for different road modes. *Mechanical Systems and Signal Processing*, *110*, 307–332. doi:10.1016/j.ymssp.2018.03.009

Ansari, M. H., & Karami, M. A. (2017). Experimental investigation of fan-folded piezoelectric energy harvesters for powering pacemakers. *Smart Materials and Structures*, *26*(6), 065001. doi:10.1088/1361-665X/aa6cfd PMID:29674807

Beeby, S. P., Torah, R. N., Tudor, M. J., Glynne-Jones, P., O'Donnell, T., Saha, C. R., & Roy, S. (2007). A micro-electromagnetic generator for vibration energy harvesting. *Journal of Micromechanics and Microengineering*, *17*(7), 1257–1265. doi:10.1088/0960-1317/17/7/007

Boisseau, S., Despesse, G., Monfray, S., Puscasu, O., & Skotnicki, T. (2013). Semi-flexible bimetal-based thermal energy harvesters. *Smart Materials and Structures*, *22*(2), 025021. doi:10.1088/0964-1726/22/2/025021

Boisseau, S., Despesse, G., Ricart, T., Defay, E., & Sylvestre, A. (2011). Cantilever-based electret energy harvesters. *Smart Materials and Structures*, *20*(10), 105013. doi:10.1088/0964-1726/20/10/105013

Bowen, C. R., Taylor, J., LeBoulbar, E., Zabek, D., Chauhan, A., & Vaish, R. (2014). Pyroelectric materials and devices for energy harvesting applications. *Energy & Environmental Science*, *7*(12), 3836–3856. doi:10.1039/C4EE01759E

Bradai, S., Naifar, S., Viehweger, C., & Kanoun, O. (2018). Electromagnetic Vibration Energy Harvesting for Railway Applications. *MATEC Web of Conference, ICoEV 2017*, *148*(8). 10.1051/matecconf/201814812004

Chamanian, S., Baghaee, S., Ulusan, H., Zorlu, O., Kulah, H., & Biyikoglu, E. U. (2014). Powering-up Wireless Sensor Nodes Utilizing Rechargeable Batteries and an Electromagnetic Vibration Energy Harvesting System. *Energies*, *7*(10), 6323–6339. doi:10.3390/en7106323

Chen, J., & Wang, Z. L. (2017). Reviving Vibration Energy Harvesting and Self-Powered Sensing by a Triboelectric Nanogenerator. *Joule*, *1*(3), 480–521. doi:10.1016/j.joule.2017.09.004

Chen, J., Zhu, G., Yang, W., Jing, Q., Bai, P., Yang, Y., Hou, T. C., & Wang, Z. L. (2013). Harmonic-Resonator-Based Triboelectric Nanogenerator as a Sustainable Power Source and a Self-Powered Active Vibration Sensor. *Advanced Materials*, *25*(42), 6094–6099. doi:10.1002/adma.201302397 PMID:23999798

Cottone, F., Vocca, H., & Gammaitoni, L. (2009). Nonlinear energy harvesting. *Physical Review Letters*, *102*(8), 080601. doi:10.1103/PhysRevLett.102.080601 PMID:19257728

Cuadras, A., Gasulla, M., & Ferarri, V. (2010). Thermal energy harvesting through pyroelectricity. *Sensors and Actuators. A, Physical*, *159*(1), 132–139. doi:10.1016/j.sna.2009.12.018

Cui, Y., Zhang, Q., Yao, M., Dong, W., & Gao, S. (2015). Vibration piezoelectric energy harvester with multi-beam. *AIP Advances*, *5*(4), 041332. doi:10.1063/1.4919049

Daniels, A., Giuliano, A., Zhu, M., & Tiwari, A. (2013). Modeling, Validation and Design Analyses of a Piezoelectric Cymbal Transducer for Non-Resonant Energy Harvesting. *IEEE International Conference on Green Computing and Communications and IEEE Internet of Things and IEEE Cyber, Physical and Social Computing*, 1665–1667. 10.1109/GreenCom-iThings-CPSCom.2013.303

Dechant, E., Fedulov, F., Chashin, D. V., Fetisov, L. Y., Fetisov, Y. K., & Shamonin, M. (2017). Low-frequency, broadband vibration energy harvester using coupled oscillators and frequency up-conversion by mechanical stoppers. *Smart Materials and Structures*, *26*(6), 065021. doi:10.1088/1361-665X/aa6e92

Deterre, M. (2013). *Towards an energy harvester for leadless pacemakers* (PhD Thesis). Paris: Universite Paris-Sud.

Ding, G., Zhao, X., Wang, J., & Xu, C. (2020). Vibration energy harvesting from roads under traffic loads. *Road Materials and Pavement Design*, *21*(3), 780–799. doi:10.1080/14680629.2018.1527719

Dorzhiev, V., Karami, A., Basset, P., Marty, F., Dragunov, V., & Galayko, D. (2015). Electret-Free Micromachined Silicon Electrostatic Vibration Energy Harvester With the Bennet's Doubler as Conditioning Circuit. *IEEE Electron Device Letters*, *36*(2), 183–185. doi:10.1109/LED.2014.2387213

dos Santos, M. P. S., Ferreira, J. A. F., Simões, J. A. O., Pascoal, R., Torrão, J., Xue, X., & Furlani, E. P. (2016). Magnetic levitation-based electromagnetic energy harvesting: A semi-analytical non-linear model for energy transduction. *Scientific Reports*, *6*(1), 18579. doi:10.1038rep18579 PMID:26725842

Dotti, F. E., & Sosa, M. D. (2019). Pendulum systems for harvesting vibration energy from railroad tracks and sleepers during the passage of a high-speed train: A feasibility evaluation. *Theoretical & Applied Mechanics Letters*, *9*(4), 229–235. doi:10.1016/j.taml.2019.03.005

Du Toit, N. E. (2005). *Modeling and Design of a MEMS Piezoelectric Vibration Energy Harvester* (MSc. thesis). Massachusetts Institute of Technology.

Eshghi, A. T., Lee, S., Sadoughi, M. K. M., Hu, C., Kim, Y. C., & Seo, J. H. (2017). Design optimization under uncertainty and speed variability for a piezoelectric energy harvester powering a tire pressure monitoring sensor. *Smart Materials and Structures*, *26*(10), 105037. doi:10.1088/1361-665X/aa8886

European Editors. (2013). *Digi-Key Corporation*. https://www.digikey.com/en/articles/techzone/2013/sep/smaller-smarter-ultralowpower-sensors-raise-potential-for-energy-harvesting-in-medical-implants

Fan, F. R., Tian, Z. Q., & Wang, Z. L. (2012). Flexible triboelectric generator. *Nano Energy*, *1*(2), 328–334. doi:10.1016/j.nanoen.2012.01.004

Fathabadi, H. (2019). Solar energy harvesting in buildings using a proposed novel electrochemical device as an alternative to PV modules. *Renewable Energy*, *133*, 118–125. doi:10.1016/j.renene.2018.10.010

Foong, F. M., Thein, C. K., Ooi, B. L., & Yurchenko, D. (2019). Increased power output of an electromagnetic vibration energy harvester through anti-phase resonance. *Mechanical Systems and Signal Processing*, *116*, 129–145. doi:10.1016/j.ymssp.2018.06.012

Friswell, M. I., Ali, S. F., Bilgen, O., Adhikari, S., Lees, A. W., & Litak, G. (2013). Non-linear piezoelectric vibration energy harvesting from a vertical cantilever beam with tip mass. *Journal of Intelligent Material Systems and Structures*, *23*(13), 1505–1521. doi:10.1177/1045389X12455722

Fu, Y., Ouyang, H., & Davis, R. B. (2019). Triboelectric energy harvesting from the vibro-impact of three cantilevered beams. *Mechanical Systems and Signal Processing*, *121*, 509–531. doi:10.1016/j.ymssp.2018.11.043

Green, P. L., Papatheou, E., & Sims, N. D. (2013). Energy harvesting from human motion and bridge vibrations: An evaluation of current nonlinear energy harvesting solutions. *Journal of Intelligent Material Systems and Structures*, *24*(12), 1494–1505. doi:10.1177/1045389X12473379

Halim, M. A., & Park, J. Y. (2014). A non-resonant, frequency up-converted electromagnetic energy harvester from human-body-induced vibration for hand-held smart system applications. *Journal of Applied Physics*, *115*(9), 094901. doi:10.1063/1.4867216

Hamid, H. M. A., & Butler, Z. C. (2019). Design and Optimization of a MEMS Triboelectric Energy Harvester for Nano-sensor Applications. *IEEE Sensors Applications Symposium. SAS 2018*, 1–6. 10.1109/SAS.2019.8706072

Hendijanizadeh, M., Sharkh, S. M., Elliott, S. J., & Torbati, M. M. (2013). Output power and efficiency of electromagnetic energy harvesting systems with constrained range of motion. *Smart Materials and Structures*, *22*(12), 125009. doi:10.1088/0964-1726/22/12/125009

Hu, Y., Yang, J., Jing, Q., Niu, S., Wu, W., & Wang, Z. L. (2013). Triboelectric Nanogenerator Built on Suspended 3D Spiral Structure as Vibration and Positioning Sensor and Wave Energy Harvester. *ACS Nano*, *7*(11), 10424–10432. doi:10.1021/nn405209u PMID:24168315

Kareem, A., Kijewski, T., & Tamura, Y. (1999). Mitigation of motions of tall buildings with specific examples of recent applications. *Wind and Structures*, *2*(3), 201–251. doi:10.12989/was.1999.2.3.201

Khan, F. U., & Iqbal, M. (2018). Electromagnetic Bridge Energy Harvester Utilizing Bridge's Vibrations and Ambient Wind for Wireless Sensor Node Application. *Journal of Sensors*, *2018*, 3849683. doi:10.1155/2018/3849683

Kim, S. G., Priya, S., & Kanno, I. (2012). Piezoelectric MEMS for energy harvesting. *MRS Bulletin*, *37*(11), 1039–1050. doi:10.1557/mrs.2012.275

Klein, J., & Zuo, L. (2017). A velocity-amplified electromagnetic energy harvester for small amplitude vibration. *Smart Materials and Structures*, *26*(9), 095057. doi:10.1088/1361-665X/aa7e54

Kubba, A. E., & Jiang, K. (2014). A Comprehensive Study on Technologies of Tyre Monitoring Systems and Possible Energy Solutions. *Sensors (Basel)*, *14*(6), 10306–10345. doi:10.3390140610306 PMID:24922457

Kumar, A., Sharma, A., Kumar, R., Vaish, R., & Chauhan, V. S. (2014). Finite element analysis of vibration energy harvesting using lead-free piezoelectric materials: A comparative study. *Journal of Asian Ceramic Societies*, *2*(2), 138–143. doi:10.1016/j.jascer.2014.02.001

Lai, S.-K., Wang, C., & Zhang, L. H. (2019). A nonlinear multi-stable piezomagnetoelastic harvester array for low-intensity, low-frequency, and broadband vibrations. *Mechanical Systems and Signal Processing*, *122*, 87–102. doi:10.1016/j.ymssp.2018.12.020

Lee, Y. J., Zhou, G., & Lua, K. B. (2019). Vortex-induced vibration wind energy harvesting by piezoelectric MEMS device in formation. *Nature*, *9*, 20404. PMID:31892701

Li, K., He, Q., Wang, J., Zhou, Z., & Li, X. (2018). Wearable energy harvesters generating electricity from low-frequency human limb movement. *Microsystems & Nanoengineering*, *4*(24), 24. doi:10.103841378-018-0024-3 PMID:31057912

Li, X., Hehn, T., Thewes, M., Kuehne, I., Frey, A., Scholl, G., & Manoli, Y. (2013a). Non-resonant electromagnetic energy harvester for car-key applications. *Journal of Physics: Conference Series*, *476*, 012096. doi:10.1088/1742-6596/476/1/012096

Li, Z., Zuo, L., Kuang, J., & Luhrs, G. (2013b). Energy-harvesting shock absorber with a mechanical motion rectifier. *Smart Materials and Structures*, *22*(2), 025008. doi:10.1088/0964-1726/22/2/025008

Lin, T., Wang, J. J., & Zuo, L. (2018). Efficient electromagnetic energy harvester for railroad transportation. *Mechatronics*, *53*, 277–286. doi:10.1016/j.mechatronics.2018.06.019

Lohndorf, M., Kvisteroy, T., Westby, E., & Halvorsen, E. (2007). Evaluation of energy harvesting concepts for tire pressure monitoring systems. Proceedings PowerMEMS 2007, 331 – 334.

Lucy, E. A. (2010). *Innowattech: Harvesting Energy and Data A stand-alone technology*. Retrieved March 2015, 2, from https://www.academia.edu/29338723/Innowattech_Harvesting_Energy_and_Data_A_stand_alone_technology

Maeguchi, T., Masuda, A., Katsumura, H., Kagata, H., & Okumura, H. (2015). Band widening of piezoelectric vibration energy harvesters by utilizing mechanical stoppers and magnets. *Journal of Physics: Conference Series*, *660*, 012099. doi:10.1088/1742-6596/660/1/012099

Mann, B. P., & Owens, B. A. (2010). Investigations of a nonlinear energy harvester with a bistable potential well. *Journal of Sound and Vibration*, *329*(9), 1215–1226. doi:10.1016/j.jsv.2009.11.034

Mikulic, M. (2019). *Global number of pacemakers in 2016 and 2023*. https://www.statista.com/statistics/800794/pacemakers-market-volume-in-units-worldwide/

Mitcheson, P. D., & Green, T. C. (2012). Maximum Effectiveness of Electrostatic Energy Harvesters When Coupled to Interface Circuits. *IEEE Transactions on Circuits and Systems*, *59*(12), 3098–3111. doi:10.1109/TCSI.2012.2206432

Naito, Y., & Uenishi, K. (2019). Electrostatic MEMS Vibration Energy Harvesters inside of Tire Treads. *Sensors (Basel)*, *19*(4), 890. doi:10.339019040890 PMID:30795502

New Hampshire Department of Environmental Services. (2017). *Universal Waste, Batteries: Management Requirements for Handlers and Transporters*. Environmental Fact Sheet, WMD-HW-18.

Ni, T., Zuo, L., & Kareem, A. (2011). Assessment of Energy Potential and Vibration Mitigation of Regenerative Tuned Mass Dampers on Wind Excited Tall Buildings. *Proceedings of the ASME 2011 International Design Engineering Technical Conferences & Computers and Information in Engineering Conference, IDETC/CIE 2011*, 333–342. 10.1115/DETC2011-48728

Ooi, B. L. (2010). *Optimisation and Frequency Tuning Concepts for a Vibration Energy Harvester* (PhD Thesis). University of Hull.

Ooi, B. L., & Gilbert, J. M. (2014). Design of wideband vibration-based electromagnetic generator bymeans of dual-resonator. *Sensors and Actuators. A, Physical, 213*, 9–18. doi:10.1016/j.sna.2014.03.037

Ouyang, H., Liu, Z., Li, N., Shi, B., Zou, Y., Xie, F., Ma, Y., Li, Z., Li, H., Zheng, Q., Qu, X., Fan, Y., Wang, Z. L., Zhang, H., & Li, Z. (2019). Symbiotic cardiac pacemaker. *Nature Communications, 10*(1), 1821. doi:10.103841467-019-09851-1 PMID:31015519

Peigney, M., & Siegert, D. (2013). Piezoelectric energy harvesting from traffic-induced bridge vibrations. *Smart Materials and Structures, 22*(9), 095019. doi:10.1088/0964-1726/22/9/095019

Prauzek, M., Konecny, J., Borova, M., Janosova, K., Hlavica, J., & Musilek, P. (2018). Energy Harvesting Sources, Storage Devices and System Topologies for Environmental Wireless Sensor Networks: A Review. *Sensors (Basel), 18*(8), 2446. doi:10.339018082446 PMID:30060513

ReVibe Energy. (2020). *The modelC*. https://revibeenergy.com/modelc/

Rhimi, M., & Lainef, R. (2012). Tunable Energy Harvesting from Ambient Vibrations in Civil Structures. *Journal of Energy Engineering, 134*(4), 185–193. doi:10.1061/(ASCE)EY.1943-7897.0000077

Roundy, S. J. (2003). *Energy Scavenging for Wireless Sensor Nodes with a Focus on Vibration to Electricity Conversion* (PhD. thesis). Berkeley, CA: University of California.

Shirai, H., Mitamura, H., Arai, N., & Moriya, K. (2020). Study of Energy Harvesting from Low-Frequency Vibration with Ferromagnetic Powder and Non-magnetic Fluid. *Plasmonics, 15*(2), 559–571. doi:10.100711468-019-01067-9

Tang, X., Lei, T., & Zuo, L. (2014). Design and Optimization of a Tubular Linear Electromagnetic Vibration Energy Harvester. *IEEE/ASME Transactions on Mechatronics, 19*(2), 615–622. doi:10.1109/TMECH.2013.2249666

Tang, X., & Zuo, L. (2011). Enhanced vibration energy harvesting using dual-mass systems. *Journal of Sound and Vibration, 330*(21), 5199–5209. doi:10.1016/j.jsv.2011.05.019

Tang, X., & Zuo, L. (2012a). Vibration energy harvesting from random force and motion excitations. *Smart Materials and Structures, 21*(7), 075025. doi:10.1088/0964-1726/21/7/075025

Tang, X., & Zuo, L. (2012b). Simultaneous energy harvesting and vibration control of structures with tuned mass dampers. *Journal of Intelligent Material Systems and Structures, 23*(18), 2117–2127. doi:10.1177/1045389X12462644

Tang, Z. B., Deng, Y. D., Su, C. Q., Shuai, W. W., & Xie, C. J. (2015). A research on thermoelectric generator's electrical performance under temperature mismatch conditions for automotive waste heat recovery system. *Case Studies in Thermal Engineering, 5*, 143–150. doi:10.1016/j.csite.2015.03.006

Thein, C. K., Foong, F. M., & Shu, Y. C. (2019a). Damping ratio and power output prediction of an electromagnetic energy harvester designed through finite element analysis. *Sensors and Actuators. A, Physical, 286*, 220–231. doi:10.1016/j.sna.2018.12.041

Thein, C. K., Foong, F. M., & Shu, Y. C. (2019b). Spring amplification and dynamic friction modelling of a 2DOF/2SDOF system in an electromagnetic vibration energy harvester – Experiment, simulation, and analytical analysis. *Mechanical Systems and Signal Processing*, *132*, 232–252. doi:10.1016/j.ymssp.2019.06.028

Torres, E. O., & Rincon-Mora, G. A. (2009). Electrostatic Energy-Harvesting and Battery-Charging CMOS System Prototype. *IEEE Transactions on Circuits and Systems. I, Regular Papers*, *56*(9), 1938–1948. doi:10.1109/TCSI.2008.2011578

Tso, M. H., Yuan, J., & Wong, W. O. (2013). Design and experimental study of a hybrid vibration absorber for global vibration control. *Engineering Structures*, *56*, 1058–1069. doi:10.1016/j.engstruct.2013.06.017

Viet, N. V., Xie, X. D., Liew, K. M., Banthia, N., & Wang, Q. (2016). Energy harvesting from ocean waves by a floating energy harvester. *Energy*, *112*, 1219–1226. doi:10.1016/j.energy.2016.07.019

Wang, J. J., Penamalli, G. P., & Zuo, L. (2012). Electromagnetic Energy Harvesting from Train Induced Railway Track Vibrations. *8th IEEE/ASME International Conference on Mechatronic and Embedded Systems and Applications*, 29–34. 10.1109/MESA.2012.6275532

Wang, L., Luo, G., Jiang, Z., Zhang, F., Zao, L., Yang, P., Lin, Q., & Maeda, R. (2019). Broadband vibration energy harvesting for wireless sensor node power supply in train container. *The Review of Scientific Instruments*, *90*(12), 125003. doi:10.1063/1.5127243 PMID:31893793

Wang, W., Cao, J., Zhang, N., Lin, J., & Liao, W. H. (2017). Magnetic-spring based energy harvesting from human motions: Design, modeling and experiments. *Energy Conversion and Management*, *132*, 189–197. doi:10.1016/j.enconman.2016.11.026

Wang, X., Liang, X., Hao, Z., Du, H., Zhang, N., & Qian, M. (2016). Comparison of electromagnetic and piezoelectric vibration energy harvesters with different interface circuits. *Mechanical Systems and Signal Processing*, *72–73*, 906–924. doi:10.1016/j.ymssp.2015.10.016

Weddell, A. (2010). *Notes from "Energy Harvesting for Wireless Automation" Conference*. Hanson Wade.

Williams, C. B., & Yates, R. B. (1996). Analysis of a micro-electric generator for microsystems. *Sensors and Actuators. A, Physical*, *52*(1-3), 8–11. doi:10.1016/0924-4247(96)80118-X

Wischke, M., Masur, M., Kroner, M., & Woias, P. (2011). Vibration harvesting in traffic tunnels to power wireless sensor nodes. *Smart Materials and Structures*, *20*(8), 085014. doi:10.1088/0964-1726/20/8/085014

Yang, B., & Lee, C. (2010). Non-resonant electromagnetic wideband energy harvesting mechanism for low frequency vibrations. *Microsystem Technologies*, *16*(6), 916–966. doi:10.100700542-010-1059-z

Yang, J., Chen, J., Yang, Y., Zhang, H., Yang, W., Bai, P., Su, Y., & Wang, Z. L. (2014). Broadband vibrational energy harvesting on a triboelectric nanogenerator. *Advanced Energy Materials*, *4*(6), 1301322. doi:10.1002/aenm.201301322

Yang, W. Q., Chen, J., Zhu, G., Wen, X., Bai, P., Su, Y., Lin, Y., & Wang, Z. L. (2013). Harvesting vibration energy by a triple-cantilever based triboelectric nanogenerator. *Nano Research*, *6*(12), 880–886. doi:10.100712274-013-0364-0

Yang, Y. W., Wang, S., Stein, P., Xu, B. X., & Yang, T. (2017). Vibration-based energy harvesting with a clamped piezoelectric circular diaphragm: Analysis and identification of optimal structural parameters. *Smart Materials and Structures*, *26*(4), 045011. doi:10.1088/1361-665X/aa5fda

Zhang, C., Harne, R. L., Li, B., & Wang, K. W. (2020b). Harmonic analysis and experimental validation of bistable vibration energy harvesters interfaced with rectifying electrical circuits. *Communications in Nonlinear Science and Numerical Simulation*, *82*, 105069. doi:10.1016/j.cnsns.2019.105069

Zhang, H., & Afzalul, K. (2014). Design and Analysis of a Connected Broadband Multi-Piezoelectric-Bimorph Beam Energy Harvester. *IEEE Transactions on Ultrasonics, Ferroelectrics, and Frequency Control*, *61*(6), 1016–1023. doi:10.1109/TUFFC.2014.2997 PMID:24859665

Zhang, X., Pan, H., Qi, L., Zhang, Z., Yuan, Y., & Liu, Y. (2017). A renewable energy harvesting system using a mechanical vibration rectifier (MVR) for railroads. *Applied Energy*, *204*, 1535–1543. doi:10.1016/j.apenergy.2017.04.064

Zhang, Y., Phuong, P. T. T., Roake, E., Khanbareh, H., Wang, Y., Dunn, S., & Bowen, C. (2020a). Thermal Energy Harvesting Using Pyroelectric-Electrochemical Coupling in Ferroelectric Materials. *Joule*, *4*(2), 301–309. doi:10.1016/j.joule.2019.12.019

Zhang, Y., Wang, T., Zhang, A., Peng, Z., Luo, D., Chen, R., & Wang, F. (2016). Electrostatic energy harvesting device with dual resonant structure for wideband random vibration sources at low frequency. *The Review of Scientific Instruments*, *87*(12), 125001. doi:10.1063/1.4968811 PMID:28040962

Zhou, K., Dai, H. L., Abdelkefi, A., & Ni, Q. (2020). Theoretical modeling and nonlinear analysis of piezoelectric energy harvesters with different stoppers. *International Journal of Mechanical Sciences*, *166*, 105233. doi:10.1016/j.ijmecsci.2019.105233

Zhou, Z., Qin, W., Du, W., Zhu, P., & Liu, Q. (2019). Improving energy harvesting from random excitation by nonlinear flexible bi-stable energy harvester with a variable potential energy function. *Mechanical Systems and Signal Processing*, *115*, 162–172. doi:10.1016/j.ymssp.2018.06.003

Zhu, G., Peng, B., Chen, J., Jing, Q., & Wang, Z. L. (2015). Triboelectric nanogenerators as a new energy technology: From fundamentals, devices, to applications. *Nano Energy*, *14*, 126–138. doi:10.1016/j.nanoen.2014.11.050

Zou, H., Zhang, Y., Guo, L., Wang, P., He, X., Dai, G., Zheng, H., Chen, C., Wang, A. C., Xu, C., & Wang, Z. L. (2019). Quantifying the triboelectric series. *Nature Communications*, *10*(1), 1427. doi:10.103841467-019-09461-x PMID:30926850

Zuo, L., & Cui, W. (2013). Dual-Functional Energy-Harvesting and Vibration Control: Electromagnetic Resonant Shunt Series Tuned Mass Dampers. *Journal of Vibration and Acoustics*, *135*(5), 510181–510189. doi:10.1115/1.4024095 PMID:23918165

Zuo, L., Scully, B., Shestani, J., & Zhou, Y. (2010). Design and characterization of an electromagnetic energy harvester for vehicle suspensions. *Smart Materials and Structures*, *19*(4), 045003. doi:10.1088/0964-1726/19/4/045003

Chapter 5
Energy Harvesting Models and Techniques for Green IoT:
A Review

Saira Muzafar
King Faisal University, Hofuf, Saudi Arabia

ABSTRACT

Wireless sensor networks (WSNs) have been popular due to their wide range of applications in almost all walks of life including industry controls, environmental monitoring, health, transportation, military. Usually conventional sensor networks are dedicated, and private networks have little or no communication with the outside world; hence, when connected to the external world by using internet protocol (IP), they form a network of connected devices, sensors, and systems and form an internet of things (IoT). Due to rapid development of IoT infrastructures worldwide, energy demand has been increased significantly to meet the power requirements of billions of connected devices. Since WSN is a foundation of IoT; hence, IoT also inherited with the challenge of providing consistent energy and to maintain hazard-free environments. This chapter is the extension of "Energy Harvesting Models and Techniques: A Review," which focuses on possible energy sources available in ambient environment and the technological mechanism to harvest energy for WSN and IoT that promotes green energy concept.

INTRODUCTION

WSN was explored to a great extent by researchers in past three decades due to its wide use and significance in industry, environment monitoring, military and many other applications. It comprises spatially dispersed and dedicated sensors and has limited communication with external world as it is usually a private and dedicated network of sensors. In order to exchange information with the outside world that is generated by sensor network it is needed to have specific gateways in WSN. This is achieved by using internet Protocol (IP) to connect a particular WSN with internet (Vasseur, J-P. et al., 2010). Hence, sensor devices have their own IP addresses and can be linked to the outside world and form internet of connected systems. (Khan, Z.A. et al., 2016). This technical advancement in IT sector allow the integration of both

DOI: 10.4018/978-1-7998-6709-8.ch005

physical and virtual in a new paradigm along with innovative and diversified applications in all walks of life and termed as internet of things (IoT). The term "Internet of Things" was first coined by Kevin Ashton in 1998 (Ashton, K. 2009). Later In 2001, IoT vision was introduced by MIT Auto-ID center (Brock, D. 2001) and the International Telecommunication Union (ITU) formally used the term in 2005 (ITU, 2005). IoT is a worldwide network of interconnected devices, objects, machines and systems that can be identified by unique addresses. Technical advancements in last decade increases the number of devices and subscribers all over the world exponentially, According to (IHS Markit 2017) it is expected that by 2030 there will be 125 billion IoT devices worldwide. The entire IoT is built upon four Cs, that is connect, collect, compute and create new business models which requires tremendous amount of data processing and power consumption. The result is the great amount of carbon di oxide (CO_2) emission that becomes an environmental and health hazard. Also the power resources of theses smart devices are battery depended and limited which have a great impact on their performance and usability (Jhanjhi, N. Z. et al., 2017), hence in order to overcome these problems green computing attracts the attention of many researchers. The standard IoT is an evolution of WSN and considered as a combination of Wireless Sensor Networks (WSNs) and Radio Frequency Identification Devices (RFID) (Khan, Z.A. et al., 2016). As energy and power consumption has been a major concern in WSN deployment from the beginning and since IoT is considered as the evolution of WSN hence the energy and the low power consumption is the inherited problem in IoT as well, and the demand rises with the exponential growth of digital society. The launch of recent 5G networks and the drastic increase in IoT devices along with many versatile applications imposes two major challenges to cope up with the increased demand of power consumption, maintenance in diversified and new IT infrastructures and networks such as 5G, another problem is the associated environmental risk of carbon emission (GSMA, 2015) and the electronic waste. One of the effective solution to address both the problems is to harvest environmental energy. Hence, the need of low power consumption and environmental friendly devices are the main triggers of green IoT.

Energy Harvesting

Harvesting energy from environment is not a new technology; it has been used since decades to scavenge energy from wind, sun and moving water. The new thing is to implement this idea in modern systems. It is a process of conversion of environmental energy into electrical energy. Energy is around us in many different forms such as sound, vibration, mechanical, thermal, chemical, electrical, solar, wind, RF etc. Figure1 (Hi-globe, n.d.) shows the ambient energy sources in the environment. There are many types of energy harvesters each offering differing degrees of usefulness depending on the application (Alippi C. et al., 2009).Energy harvesting or energy scavenging has an important application in areas where wiring is not practical or it is difficult to supply energy, hence it has great scope in wireless sensor networks to keep them operational for almost infinite time.

Wireless sensor networks are suitable to deploy in areas with harsh weather condition and flexible enough to embed with any system of interest. However, constant power supply is a big challenge to keep sensor node functional. To address power problem in WSN, we broadly have two solutions

Reduce Power Consumption By Using Low Power Networks

Low power consumption in WSN has been a popular research topic since past two decades for the sake of increasing sensor network life. Continuous research has been conducted to investigate and improve

Figure 1. Different sources of ambient energy

low-power MAC protocols (Lu C. et al., 2010) (Filip N. et al., 2011), routing protocols (Hua W.S. et al., 2014) (Raghunathan V. et al., 2005) transport protocols (A. B. Nacef 2011), and operating systems (S. Dash. S et al., 2012).

Extracting Energy From Ambient Environment

An alternative to power IoT devices and sensor nodes is to extract energy from surroundings.

As the total power cost of wireless sensor nodes has been reduced to low level by now (Zhou H.-Y. et al., 2008) with the development of microelectronics, low-power design, and network technology (Vullers R. J. M. et al., 2009), and this give rise to adapt energy harvesting techniques for the provision of energy for both IoT sensor networks.

Traditionally, batteries are the power source of WSN. Despite a reliable source of power supply batteries have limitation of size and limited functional time span even they are not used at the fullest. As main function of the wireless node is to sense and compute environmental energy in the form of heat, sound, pressure and to communicate neighboring nodes. Hence, it is good to extract operational energy from the surroundings in support of green environment cost and maintenance overhead. Wireless sensors are useless when their battery dies, also batteries can cause a serious environmental hazard because batteries contain heavy metals and toxic chemicals for example mercury and lead. Moreover, it is expensive to maintain and replace batteries especially for the places difficult to reach such as environmental monitoring, biomedical fields and military surveillance. Life of the network depends on the life of the sensor nodes, which requires continuous power supply. Use of batteries as a power source is appropriate for a limited time period as average life of a battery is 1-18 months only also batteries discharge 0.1 to 5% per month also constant and consistent high temperature exposure (>40C) increases the rate significantly. Hence, battery can be a good choice for the applications of wireless sensor networks if they require less energy conversion overhead.

Due to short life time of battery and other limitations, it is important to find an alternate solution for constant energy source. Energy harvesting is one such promising technique to power sensor nodes continuously at lower cost. However, it is important to consider environment, size and life of energy harvesting system.

Energy Harvesting Applications

WSN has been declared as top ten breakthrough technologies of 21stcentury that will change the world (MIT Technology Review 2003) and with the advent of IoT they find innovative applications in almost all walks of life. Different type of energy harvesters for WSN are available, their usefulness depends on the type of application. Some of the potential applications are reviewed below:

Application in Industry

Sensors application in industry requires continuous power supply and often operated in harsh environment. Industry applications usually require continuous monitoring of pressure and temperature parameters, such as oil and gas fields.

Applications in Biomedical and Fitness devices

Healthcare is one of the major problem worldwide particularly in elderly population. The technical advancement of WSN and machine to machine communication lead the world to IoT era and introduces new trends in healthcare such as remote patient monitoring, drug administration, timely and convenient diagnostics etc. The complete network of Iot devices, sensors and people generate tremendous amount of data. For timely and real time data transmission and processing increases energy consumption demands significantly.

Also, energy harvesting technique has useful and innovative applications for implantable body sensors in medical field, as they need regular battery replacement. Such as pacemakers, defibrillator, ear devices for monitoring biological activity in patient ear.

Environmental Monitoring Applications

Wireless sensors are widely used for environmental and habitat monitoring such as forests, landslides, avalanches, pollution detection, weather monitoring, agriculture, waterways, cattle monitoring, glaciers in ocean (Padhy P. et al., 2005) and many more. The biggest challenge in these applications to address power requirements to process data in harsh environmental conditions (Khan, Z.A. et al., 2016).

Military Applications

WSN has been a mature technology for battlefield activities of surveillance and security because of high spatial resolution and coverage of sensors which help in perimeter and force protection, logistic information, exchange of critical information, command, intelligence, intrusion, monitoring of sea and land area.

The evolutionary paradigm of WSN in the form of IoT is extremely useful and has many strong applications to meet military and battlefield requirements. It is used to connect different types of military equipment and vehicles such as tanks, ships, drones, planes along with soldiers and military bases to form a consistent and cohesive network of systems people and things to provide effective situational awareness. Combination of different technologies under IoT umbrella such as biometrics, sensors for quick data transmission allow soldiers and military personnel to handle dangerous situations successfully. This Internet of Military Things comprises a network of sensors, wearables, and other IoT devices as shown in Fig 2. (IEEE Computer Society, 2017).

Figure 2. Internet of Battlefield Things (IoBT)

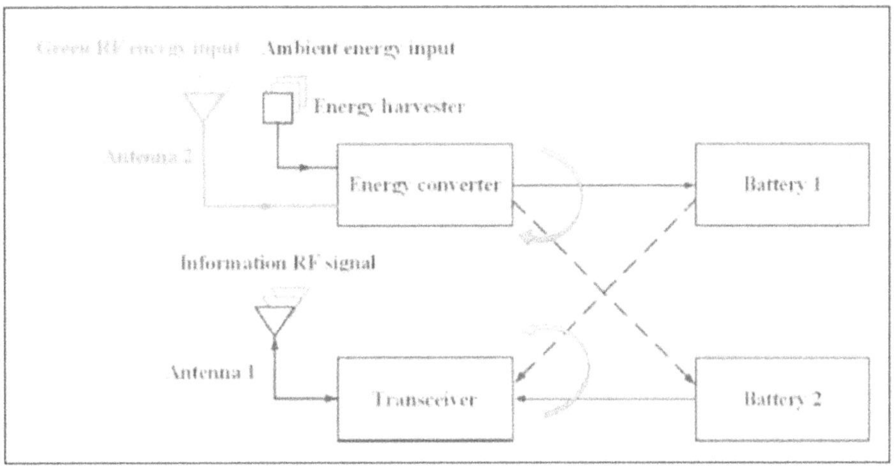

Applications in Structural Maintenance

Structures such as buildings and bridges require regular inspection and maintenance in order to avoid any harmful incident. Sensors can indicate problems in advance this will reduce the cost of maintenance and failure.

Applications in Transport Management

Traffic congestion is a major problem in big cities all over the world. Intelligent transport management system consists of sensors help to prevent traffic congestion by synchronizing traffic lights and measuring vehicle flow.

Material Recycling

Safe disposal of e-waste and to collect recyclable material can be done by using RFID tags. The IoT infrastructure can be installed for efficient processing of collected recycled material.

Incorporation of IoT and WSN technology along with appropriate sensors can be used to contribute to greener environment (Khan, Z.A. et al., 2016).

Smart Water Monitoring

These applications involves to monitor water for different chemicals levels and check distribution of water for leakage etc. Monitoring of water level can play a very important role in flood prevention in certain situations [75]. The challenge is to maintain continuous monitoring, processing and communication of gathered data with real-time constraints. These challenges can be realized using under water sensor networks in conjunction with IoT technologies (Khan, Z.A. et al., 2016).

Smart City and Smart Home

Smart city is one of the biggest application of IoT, the IoT can optimize traffic control systems, power grids and vehicle parking etc. Smart home is one of the great application of IoT and WSN. The concept of smart homes using IoTs is extremely beneficial for resource optimization such as electricity and also helps carbon reduction. The main objective is to control, monitor and automate lighting, temperature, ventilation in the building in order to optimize energy consumption in smart grid (Hubert, T., et al, 2011). The sensors are main contributors that monitor the consumption and detect the user need then decision could be taken for controlling the resources (Khan, Z.A. et al., 2016). However, concerning with the green IoT, where we focus mainly to the energy harvesting and the alternative resources to save the energy for the IoT. The vital group of IoT applications are linked with the smart homes and smart cities (Almusaylim et al., 2019) (Humayun, M. et al., 2020) (Khan, A. et al., 2020). To manage the security and privacy of the internet of things, researchers presented several approaches including (Almulhim. M, et al., 2019) (Jhanjhi, N. Z., et al., 2015). Since the security and privacy requires further computation and resources to apply different encryption techniques, which requires more energy from IoT networks, in other words, the security and privacy is inversely proportional to the IoT energy related resources. The researchers are paying maximum attention to design different lightweight/energy efficient routing techniques and light weight schemes (Ponnusamy, V., et al., 2017) (Diro, A., et al., 2020) to address this issue and to enhance the IoT energy resources. In addition, researches are in (Hussain, S. J., et al., 2019) (Jhanjhi, N. Z., et al., 2018) focused on the several techniques to enhance to energy harvesting of smart application running on the gadgets, and to elaborate their cyber security concerns. Green IoT has multiple vectors to consider, we need to consider its several vectors to keep the IoT green including, harvesting energy, lightweight/energy efficient routing protocol, smart applications with smart gadgets, etc. and to reduce the cyber security threats.

Energy Harvesting Sources for Green IoT

According to Law of Energy Conservation, energy can neither be created nor destroyed; it only transforms from one form to another As our physical world is embedded with different energies such as wind, solar, kinetic etc. As there are different methods are available to extract free energy from the environment, it is important to find a suitable harvesting technique that can effectively produce a free and green energy. Table 1 (Zhou G et al., 2014) presents the comparison of power density of some of the potential energy scavenging technologies.

The power density expresses the captured energy per unit volume, area, or mass.

Common known energy sources include solar, magnetic, mechanical vibrations, thermal, wind and water. The captured energy can directly be used or stored in a storage device i.e. a capacitor or a battery depends on the output power impedance and magnitude, and requirement of the application.

Table 1. Power density comparison from various energy scavenging technologies

Energy harvesting Method	Power Density
Solar energy-Outdoor Solar energy-Indoor	15m 15mW/cm3—bright sunny day 0.15mW/cm3—cloudy day 10–100 μW/cm2
Vibrations (piezoelectric) Vibrations (electrostatic conversion) Vibrations (electromagnetic conversion)	*330 μW/cm3* *0.021 μW/mm3—105Hz* *306 μW/cm3—52Hz*
Thermoelectric	5∘C gradient 40 μW/cm3
Wind flow	16.2 μW/cm3—5m/s
Acoustic noise	3 nW/cm3—75 dB 960 nW/cm3—100 dB
Magnetic field energy	*130 μW/cm3—200 μT, 60Hz*

Potential energy harvesting methods are discussed below

1. Photonic Method

Generally, two methods are used for solar power extraction. First is solar thermal power generation which is suitable for large engineering systems and not applicable in WSNs for it is not available on small or nano scale level (Tabbakh S. R. K. et al., 2010). However the second method is to use photovoltaic solar cells and is suitable for WSNs. The energy harvesting by sun is cheaper and nonpolluting, but it is not very efficient for dark areas, as this method require continuous light source. Therefore, it is not suitable for networks deployed in buildings and overcast areas. The other light such as fluorescent and infrared can also be used as the power source for solar cells (Tabbakh S. R. K. et al., 2010).

Photonic process is based on conversion of photons coming from indoor or outdoor light source. Photovoltaic cells are used to harness energy from solar or artificial light. Photovoltaic cells are solid state electrical device capable of converting solar/light energy into electrical energy by photovoltaic effect having the efficiency rate up to 30% (Stojcev M. K. et al., 2009).

Following points should be considered for solar power based wireless sensor networks

- power supply requirements of the application
- Environmental parameters affecting the solar cells efficiency, such as solar radiation intensity, dense cloudy and snow weather, seasonal and climatically changes, latitude etc
- Angle and position at which the solar panel installed
- characteristics and power features of the solar panel chemistry and capacity of energy storage components used to store the harvested energy (Chalasani S. et al., 2008) (Morais R. et al., 2008) (Stojcev M. K. et al., 2009).

Broadly, we can divide solar cells into two categories

a. Crystalline Solar cells

Monocrystalline Silicon Solar Cells

Monocrystalline Silicon solar cells are made up of pure silicon and cell volume is considered as most efficient of all solar cells up to 30%. Most suitable and economical type for residential solar panels.

Polycrystalline Solar Cells

The very first solar cells were polycrystalline or multi-crystalline that introduced to industry in 1981. Each cell consists of several crystals that prohibit free electron movement consequently makes them less efficient than monocrystalline operated at 13-16% efficiency. Also, due to lower heat tolerance their efficiency decreased in high temperature.

b. Thin Film Solar cells

Thin film solar cells are generally suitable for smaller-scale applications such as solar powered pocket calculators, travel lights, and camping gear used in remote locations as they are less efficient than Crystalline Solar cells. Thin- film solar cells consists of thin layers made of semiconductor material such as amorphous silicon (a-Si) and copper indium gallium deselenide (CIGS).The layers of semiconductor material used such as a-Si must be thin in order to absorb solar energy efficiently. On the other hand, sun exposure cause significant degradation in power output of these cells. This problem can be addressed by thinning a-Si layers, resulting less absorption of sunlight. Hence, they are not suitable for larger-scale applications.

Photovoltaic energy harvesting is a mature technology with the limitation of light availability for power generation. Hence, may not be a good choice if a network is deployed in overcast area.

Thermal Energy Harvesting

This method uses two techniques

a. Thermo Energy Harvesting

This method use thermoelectric generators and produce electricity by temperature difference.

Thermoelectric generators based on Seebeck effect can generate electricity from heat flux, they are used for temperature sensing and to harvest waste heat. The performance of thermoelectric material is based on its electrical and thermal conductivity and the potential to produce voltage difference based on Seeback effect. There are many sources of waste heat which can be used around us, such as geothermal (N. Hubbell 2009), industrial waste heat, engine exhaust and the heat of sun (Elancheziyan A. et al., 2012).Energy can be harvested as long as difference in temperature is maintained and the temperature gradient is proportional to the efficiency of thermoelectric generator. Because it is difficult to maintain a

Table 2. Energy Harvesting Methods corresponding to different vibration sources

Form of vibration Source	Energy Harvesting Method
Force Excitation	Piezoelectric Energy Harvesting Method
Velocity Excitation	Electromagnetic Energy Harvesting Method
Displacement Excitation	Piezoelectric Electrostatic Energy Harvesting Method

significant temperature gradient on a small device (Yick J. et al., 2008), the small available temperature difference between the surfaces of theTEG, commonly no higher than 10◦C (Yong Y.-T. et al., 2009), and the micro size of the devices (A. B. Nacef 2011)generate less output power resulting limited efficiency. Current range of conversion efficiency of thermoelectric generators is 6% to 11% and or the temperature gradient less than 40C, the thermoelectric generators are less efficient i.e. hardly 1%.Therefore, wireless sensor nodes based on thermoelectric harvesters are suitable in surroundings where temperature is high such as industry. They have wide application of converting waste heat energy of industrial processes into electricity. Using waste heat energy for energy scavenging helps to reduce pollution in environment, however due to lower rate of energy conversion and high cost their use is limited. Presently, TEGs are mainly used in high-tech fields such as medical treatment, military, submarines, and cosmic space (Zhou, G. et al., 2014).

b. Pyroelectric Energy Harvesting

Pyroelectric energy harvesting is efficient than thermoelectric energy harvesting. But this method requires a varying temperature input which limits their application. In this process voltage is generated by cooling or heating pyroelectric materials. Charge can only be generated when the temperature of the material is changing; if temperature is constant charge will not be produced.

Energy Harvesting through Mechanical Vibration

Moving objects generate mechanical, vibrational and kinetic energy and is commonly available all the time. These energies can be converted into electrical energy by using following mechanism.

1. Piezoelectric
2. Electrostatic
3. Electromagnetic

Depending on the type of force excitation vibration energy harvesting system converts the mechanical energy into electrical energy.Table2 (Zhou, G. et al., 2014) shows the vibration source and the corresponding energy harvesting method.

Figure 3. Illustration of (a) d31 mode; (b) d33 mode

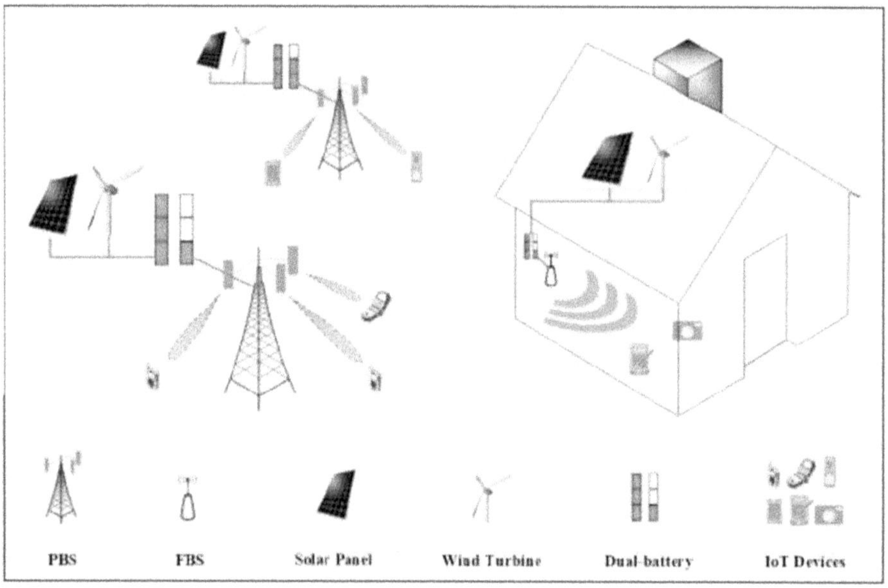

1. Energy Harvesting Through Piezo

Piezo is from Greek origin "piezein" that means press. Mechanical energy generated from vibration, force or pressure can be converted into electricity by using piezoelectric materials. Piezoelectric effect can be direct or indirect. For energy harvesting direct piezoelectric effect is used.

Generally piezoelectric materials are crystals and constructed of semiconductor materials such as Silicon and Germanium with added elements. These materials are capable of producing electricity when vibrated or compressed. Vibration based energy harvesting system requires to match the frequency of source and harvesting device. When a piezoelectric material is subject to mechanical stress it produces an electric charge. Piezoelectric devices have minimum moving parts and generate energy without an additional voltage source. The main disadvantage is piezoelectric materials are fragile and sometimes they allow current leakage.

Several piezoelectric materials are available such as piezoelectric mono crystal, piezoelectric ceramics PZT, piezoelectric polymers, and piezoelectric composites out of which piezoelectric ceramic PZT is most commonly used material because of its low cost, high efficiency rate and large electromechanical coupling constants. However, it is not capable of sustaining high strain, hence not suitable for the applications of high-frequency periodic load. An alternative is poly vinylidene fluoride (PVDF), although this piezoelectric material has smaller electromechanical coupling but offers more flexibility and higher mechanical strength which makes it suitable for high-frequency cyclic load.Figure3 (Zhou, G. et al., 2014) shows two coupling modes of operation for piezoelectric materials In thed33mode, the vibration force is applied in the same direction as the poling direction, while in thed31mode the force is applied perpendicular to the poling direction (Saadon S. et al., 2011).The d33 mode has a higher coupling coefficient than d31 mode, but d31 mode is able to produce a larger strain in a small external force at very low voltage source and small-sized devices compared to d33 mode, and it allows piezoelectric material to be strained on a bending cantilever (Lu F. et al., 2004), hence they are suitable for energy harvesting in wireless sensor networks.

Piezoelectric transducer is one of the key structures in piezoelectric energy harvesting system and the selection of it had better be consistent with the environmental vibration so as to maximize the harvesting of mechanical energy. According to the different structures, piezoelectric transducers can be divided into unimorph, bimorph, rainbow type, cymbal type, moonie type, stack type, and so on (Erturk A. et al., 2009) (Erturk A. et al., 2010) (Saadon S. et al., 2011) (Karami M.A. et al., 2011) (Li X. et al., 2011) (Palosaari J. et al., 2012).Table 3(Zhou, G. et al., 2014) shows the characteristics of different piezoelectric transducers.

Table 3. Characteristics of different piezoelectric transducers.

Piezoelectric transducers	Characteristics
Unimorph	A highest power can be generated under lower excitation frequencies and load resistances, widely used in cantilever beam structure
Bimorph	Variable range of resonant frequency: 5Hz~100Hz, can be used in various configuration such as cantilever beam and end-end clamped beam
Rigid bimorph	Similar to bimorph but easier mounting, suitable for wide bandwidth, and widely used in cantilever beam structure
Macrofiber composite (MFC)	Flexible, both d31 and d33 modes practicable, suitable for low strain and high frequency application, large area coverage, and can be used as a bimorph component
Stack type	Suitable for the application of low frequency (~10Hz) and large uniaxial stress, easy Mounting
Rainbow type	Ability to generate a higher output voltage in the given stress level, can be stacked to amplify output voltage
Moonie and Cymbal type	Ability to withstand high stress, high output charge level, and adjustable resonance frequency

The major advantages of piezoelectric energy harvesting devices are simple structure, low cost, no fever, no electromagnetic interference, easiness to manufacture, relatively high output voltage level, electromechanical conversion efficiency, and so forth (Wang X. 2012).At present, the piezoelectric harvesters are capable of producing voltage from 2 to 10 V. In order to obtain better performance and improve the efficiency as well as practicability of piezoelectric energy harvester, the ongoing research mainly focus to improve the existing piezoelectric materials property, researching new piezoelectric materials with large piezoelectric constant strain, high electromechanical coupling coefficient and low loss, and improving the structure of piezoelectric energy harvester to increase the output power.

2. Electrostatic (Capacitive)Energy Harvesting

The electrostatic or capacitive energy harvester based on the technique of changing the capacitance of a vibration dependent variable capacitor. Electrical charge stored on the parallel plates of the capacitor causes electrostatic effect between the plates and energy will be harvested by fixing one plate and move the other through mechanical motion to change variable capacitor parameters (Miranda, M. 2004). According to electrostatics principle this action coverts the mechanical energy into electrical energy. Electrostatic harvesters and microelectronic devices can be integrated however capacitor needs voltage source for initial charging.

3. Electromagnetic Energy Harvesting

This technique is based on Faraday's law of electromagnetic induction and generates electrical energy from mechanical energy by using inductive spring mass system. In this method there is no mechanical contact between the parts of a separate voltage source, which results improvement in the reliability and reduction in the mechanical damping. However, they are not suitable for sensor nodes because of their larger size.

Wireless Energy Harvesting

Wireless energy harvesting can be divided broadly into two categories:

a. RF energy harvesting

Electromagnetic radiation is commonly available in environment, and is a good source of continuous low power energy harvesting. In this method electromagnetic signals can be converted into electric signals by using rectenna. RF energy harvesting devices can completely replace the battery from sensor node. It has a wide application in military, agriculture, industry etc.

Although other options of scavenging energy such as solar, wind, vibration etc also available to produce green energy for IoT devices however they are not reliable techniques due to high possibility of variable amount of energy production. (Sun, H. et al. 2018).

Further RF energy harvesting technique allow to carry and process information concurrently along with energy production (Nasir et al. 2013; Zhou et al. 2013). Which is a quite useful aspect in context of IoT.

(Shah et al. 2016) scrutinize a two-way amplify-and-forward relay network in which two nodes communicate and transmit information simultaneously to a relay node. The constrained relay node first harvest the energy from the received signal and then amplify and forward the signal to its destination.

A RF powered energy harvesting sensor (RFEH) for high voltage electrical equipment monitoring is presented by (Xiao et al. 2016) that can be a low cost alternative of surface acoustic wave (SAW) sensor. The RFEH system consists of several nodes with a dual channel reader that can hold out against short burst of 25 kV electrical discharge for 60 s.

b. Resonant energy harvesting

This technique based on resonant inductive coupling in which electrical energy is transferred and captured between two highly resonant coils at the same frequency. Generally, two implementations are possible for resonant inductive coupling, weak and strong inductive coupling. In weak inductive coupling coils are distance apart by few centimeters only, if the receiving and external powered coil are matched strong coupling is possible between electromagnetic resonant devices for longer distance. Since both the coils have no physical connection, resonant inductive coupling is considered as a wireless energy harvesting technique. Some implementations for WSNs are discussed in (Reinisch H. et al., 2011) (Mandal S. et al., 2010) (Heer H. et al., 2010).

Dynamic Fluid Energy Harvesting

Dynamic fluid energy refers to the power of air and flowing water. Two methods can be used to harvest kinetic energy of wind or flowing water. The first method uses mechanical parts i.e. miniature wind turbines (Fei F. et al., 2012) and the second is based on non mechanical parts that works on the similar principal but induces mechanical vibration which can be converted into electricity by piezoelectric (Wang D.-A. et al., 2010) (Pobering S. et al., 2009), electrostatic (Bu L. et al., 2011), or electromagnetic principles (Wang D.-A. et al., 2010).

Harvesting Wind Energy

At present, four methods are in use to harvest wind energy, miniature wind turbines, piezoelectric wind harvesters, miniature wind belt and electromagnetic wind generators. Traditional wind turbines are the mature technique for harvesting wind energy, even though large-scale wind turbines are efficient; however small-scale wind turbines are less efficient because of high viscous drag on the blades at low Reynolds numbers (Matova S.P. et al., 2011).The micro wind belt generator has a higher output power at high wind speeds (>5.5m/s), but its output power reduces significantly at low wind speeds (<3.5 m/s),and the wind belt generator can be extremely noisy during Operation(Zhu D. et al., 2010).Wind energy harvesters made of piezoelectric materials are efficient, smaller in size, light weight, start instantly without any dead time, having compact structure, sensitive to low speed of wind. They almost have no heat dissipation and showing extremely low magnetic permeability (Tan Y. K. et al., 2007). However, they may generate very low output power due to low strain coefficient. Another limitation is piezoelectric material must vibrate at their resonant frequency for the generation of maximum power. The electromagnetic wind generators are reliable and have small mechanical damping, but the heavy magnet will result in the generator being not able to operate at low wind speed (Ramasur D. et al., 2012).Tan and Panda presented a piezoelectric wind harvester and utilized the bimorph piezoelectric actuator for a low power autonomous wind speed sensor (Tan Y. K. et al., 2007). Comparing with other wind mills, this harvester has smaller size and it is lighter in weight, but the harvesting power is of lesser magnitude and also the network is not capable of handling unexpected events as the duty cycle of sensor is too small. Hence, it is suitable for WSNs having insensitive delay and handling smaller amounts of information. Some examples of wireless sensor networks implemented with wind energy harvesting systems can be found in (Fei F. et al.,2012) (Sardini E. et al., 2011) (Tan Y. K. et al., 2011).Wind energy is infinitely available everywhere, it is non polluting and renewable. However, having low energy density, regional differences, and unstable factor for the wind speed is fluctuant and unpredictable which will resulting in equable electricity all the time (Ucar A. et al., 2009).Moreover, the wind harvester can be noisy as it has moving mechanical part (Wan Z. G. et al., 2011). Therefore, research is needed to find techniques for maximizing wind energy capture.

Harvesting Energy from Flowing Water

The flowing water contains kinetic energy due to the water pressure fluctuation, and it can be converted into electrical energy by energy harvesters (Wang D.-A. et al., 2011). Like wind, flowing water is also renewable, and can produce pollution-free, continuous, and dependable energy source for wireless sensor nodes (Pobering S. et al., 2004). For hydropower can be developed in any size and any scale, it is applicable in WSNs (Tabbakh S. R. K. et al., 2010).Energy harvesting from flowing water is used for the WSNs implemented to monitor the water quality and hydraulic state (Ye G. et al., 2012)of rivers, canals and other water distribution systems. Flowing water is capable to generate enough power for WSNs, however the water turbine bigger in size is quite challenging to install it in pipelines and integrate with the sensor nodes. While the flow-induced vibration energy harvester has a micro size, it needs further research to increase the conversion efficiency and optimize the generated power to be appropriate for powering wireless sensor nodes (Mohamed M. I. et al., 2011).

Acoustic Energy Harvesting

Sound is available and generated profusely in environment such as noise of jet engines, train station; loud subways, during industrial processes, hydraulic systems, machine shops and even a sound produced by a crying baby can be harvested to power the sensor nodes. To achieve this, an acoustic energy device can be used to convert the noise into electrical energy. Acoustic Energy Harvesting device comprises of three main components a resonator, which oscillates at resonating frequency to produce mechanical oscillations (sound), a membrane capturing energy is attached to the one end of the resonator and a storage device. The membrane is connected to a piezoelectric material that converts the vibration caused by sound into electrical energy which is then stored by a storage device with signal conditioning circuit (Zhou, G. et al., 2014).An acoustic energy harvesting will not only satisfy the power demand of wireless sensors, but also improve the environmental conditions by reducing the noise (Filip N. et al., 2011) (Hua W.S. et al., 2014).

Hybrid Energy Harvesting

The amount of energy from a single energy harvester is typically small (Seah W. K. G. et al. 2009) and highly variable with time, its location, and working conditions of load (Li L. et al., 2008) (Li. Y et al., 2008), hence depending on a one energy source for some particular application may not help to keep the node functional in case of short supply of energy from environment. Many types of energy sources are available in the environment simultaneously. In order to make each sensor node obtain the required energy from environment as much as possible, we may need to design such a kind of energy harvesting system that it cannot only harvest energy from one type of power source (Shang Y. et al., 2007).Therefore, it is necessary to integrate a variety of energy harvesting modules in one sensor node (Silva R. et al., 2011).

How Energy Harvesting works

Energy is available profusely in ambient environment; usually this available free energy reserve is remaining unused or wasted. Such as mechanical energy generating from vibration, stress and strain, thermal energy due to heat escaped from furnaces, combustion engines, biological, solar energy from all types of light sources; electromagnetic energy captured via inductors, coils and transformers; wind and fluid energy resulting from air and liquid flow; chemical energy from naturally recurring or biological processes; and RF energy due to radio transmitters and television broadcasting. The available sources provide energy in a form of very small packets and have unique characteristics depending on the application .Energy-harvesting devices scavenge and store these small energy packets and generate electricity for sensor nodes. Current energy harvesting circuits in use are designed particularly for the requirements of specific application.

However, in general a good harvesting system should possess following characteristics

- Harvesting circuit power consumption should be less than the amount of energy extracted from the surroundings.
- It must be in permanent active mode to capture available energy and capable of providing proper conditioned output as per application requirement.
- Harvester circuit must be capable to retain harvested energy for long periods of time with minimal leakage or loss.
- The energy management must be well-defined — and tolerate a wide range of voltage, current and waveform inputs, including over-voltage, overcharge and other irregular input conditions.

The energy harvesting system consists of two main parts:

- Conversion of the ambient energy from the environment and surroundings
- Store harvested or captured energy for later use

Typically, energy-harvesting systems contain similar components and are ideal for sensor network applications. An MCU is responsible for the processing, sensing, and communication activities. The MCU can be integrated to any number of sensors to collect data from surroundings and will usually send or receive data via a wireless transceiver. A typical block diagram is shown in Figure 4.When system is in a state of collecting energy, the MCU, sensors and communication systems are put in a standby mode which is helpful to minimize their drain on the available power.

An energy buffer is used to store the power for the rest of the application. Type and nature of application defines the size and technical specification of the buffer. A large buffer is required if the application is discontinue to access available energy source for a long period of time. However, a small buffer will be enough if the application is adjacent to energy source and does not need to be active frequently especially in case of low duty-cycle applications. In ideal situation energy buffer must have efficient energy acceptance and delivery, negligible leakage and volume with unlimited capacity.

A typical energy harvester circuit consists of at least one transducer, a power conditioning circuit, and a unit for energy storage. Together all these components work to extract energy from surroundings and generate power for the sensor node. In general, transducer is a device which converts energy from one form to another, when it is used to convert any measureable quantity such as sound, pressure etc.

Figure 4. Block Diagram of Energy Harvesting System

to an electrical current or voltage it is termed as a sensor. Because of uncontrolled and random input energy, transducer may generate irregular output and some sensors may be very sensitive to the change of voltage supply, then the output voltage needs to be regulated by DC-DC converter (Raghunathan V. et al., 2005). Hence the power conditioning circuit is crucial part of the system. A specialized DC-DC converter microchip takes in power from the transducer and outputs typically 3 to 5V, which can then be stored or used. Nano power comparators are used to support power consumption; they control and monitor the minimum and maximum limits of the input energy and the energy storage. The additional regulation may also be required between sensor and the device that stores energy to ensure that the sensor node working with a stable and safe voltage (Zhou, G et al., 2014).

Thus energy harvesting process can be summarized as

1. Detect energy source and harvest energy
2. Harvested energy converted into electrical energy and saved in a storage device for example capacitor
3. Saved electrical energy used by MCU and sensor to perform sensing, processing and communication activities
4. Data delivered from the sensor through wireless transceiver

Basic Elements of Energy Harvesting System

Every energy harvesting unit comprising different sub-systems that work parallel out of which power generating sub-systems paly a very important role to manage control strategies (Afghan, S. A . et al., 2018). Since last couple of years, there are continuous and significant advancements to address the need of low powered devices work on harvested energy sources from ambient environment such as solar, thermal, wind, sound, radio frequency etc. (Lu, C. et al., 2010). Fig 5 shows the basic building block of energy harvesting module (Afghan, S. A . et al., 2018).

Figure 5. Building Blocks of Energy Harvesting System

Source	Source Power	Harvested Power
	Ambient Light	
Indoor	0.1 mW/cm^2	10 μW/cm^2
Outdoor	100 mW/cm^2	10 mW/cm^2
	Vibration/Motion	
Human	0.5 m at 1 Hz 1 m/s^2 at 50 Hz	4 μW/cm
Industrial	1 m at 5 Hz 10 m/s^2 at 1 kHz	100 μW/cm^2
	Thermal Energy	
Human	20 mW/cm^2	30 μW/cm^2
Industrial	100 mW/cm^2	1–10 mW/cm^2
	Radio Frequency	
GSM Base Station	0.3 μW/cm^2	0.1 μW/cm^2

A. Energy Harvesting Transducers

Transducers are the key component of EH system that generate electrical energy from source of ambient energy (Afghan, S. A . et al., 2018). Every transducer used for harvesting energy has specific electrical characteristics and addresses the requirements of the system uniquely. The ambient energy sources for IoT applications has been attributed in 4 different sources as shown in Table I (Vullers, R. J. M, et al., 2010).

In order to design any IoT device it is important to select appropriate transducer that harvest the energy as much as possible from the available source and to configure the power generating sub-system compatible to the particular energy source. Based on these requirements many techniques have been used to generate the utmost energy such as MPPT algorithms for scavenging solar energy

(Sojan, S., et al., 2016), (Park, S., et al., 2013). Also to maximize the efficiency of thermoelectric generators, piezo-thermo mechanical and piezo-thermo fluidic has been researched (Monfray, S., et al., 2012)

Figure 6. Energy Harvesting Sources Along With Harvested Power for IoT

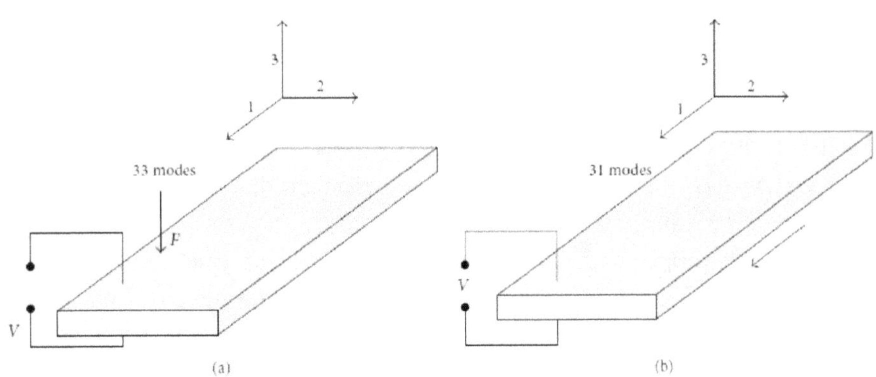

Figure 7. Dual-battery green energy harvesting architecture for IoT device

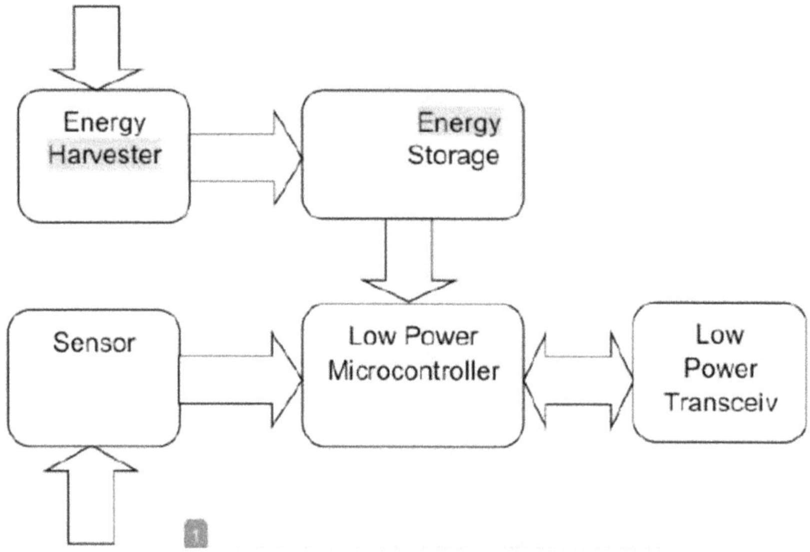

Challenges Harvesting Energy for Green IoT

Scavenging energy from surrounding to meet the power need of IoT devices is a good idea however, following are the main challenges (Liu, X., et al., 2019)

- Ambient energy generation rely on different natural and unnatural factors and may not available all the time.
- Due to unpredictable availability of green energy, it is quite difficult to schedule smooth functionalities of IoT devices.
- Another major issue is ambient energy scavenging may not be sufficient for the smooth function of IoT device even though we consider the ideal case of constant availability of ambient energy the current harvesters are not efficient enough to harvest sufficient amount of ambient energy.
- Due to directional flow of electric current, simultaneous charging and discharging is not possible in a single battery. Hence, IoT devices with traditional single battery may not able be to synchronize Iot and energy harvesting functionalities.

Dual-Battery Green Energy Harvesting Architecture for IoT Devices

In order to address these challenges (Liu, X., et al., 2018) proposes a dual-battery green energy harvesting architecture for IoT devices. Fig 6 (Liu, X., et al., 2019) shows the proposed architecture according to that each IoT device consists of an energy harvesting mechanism that scavenge energy from surroundings, an energy converter, a transceiver, dual battery and antennas.

One IoT device can be equipped with multiple energy harvesters, and multiple types of ambient energy can be harvested by single IoT device. In the proposed architecture Antenna 1 and transceiver are used for signal transmission and reception, However, Antenna 2 is to harvest green RF energy.

Since energy harvesting (EH) is the process of converting different ambient energy into electrical energy. Hence after scavenging ambient energy, the energy converter converts the harvested energy into direct current (DC) for battery charging and battery 1 is used to store that energy and simultaneously battery 2 discharges to supply power for IoT device. This process can be explained as battery cycle, in current cycle if battery 1 stores the harvested energy and battery 2 discharge the power for IoT device and in the next cycle battery 2 will store the harvested energy and battery1 will meet the power demand of IoT device. This is how the alternate role of both the batteries make sure the constant power supply to IoT device (Liu, X., et al., 2019).

Dual-battery architecture addresses below mentioned energy harvesting issues in IoT devices (Liu, X., et al., 2019):

- Use of dual-battery removes the need of prediction for available ambient energy along with the time domain. Because the amount of available harvested energy in the discharging battery is known. Hence, IoT devices can schedule their functionalities effectively based on the amount of residual energy along the time domain.
- Use of dual batteries help to accumulate the harvested energy in order to attain a certain energy level in the charging battery and used to power the IoT device in the next battery cycle. Without storing energy the instantaneous harvested energy may not be enough for IoT device proper functioning.
- Simultaneous charging and discharging is not possible in single battery, hence dual battery supports both IoT functionalities and energy harvesting from ambient environment at the same time. One battery is reserved for harvesting ambient energy while the other is used to provide power for the IoT device in each battery cycle.

Green Energy Wireless Charging

Wireless charging was not popular in past decade due to relatively low charging efficiency and the brown energy source; as huge amount of brown energy is consumed by the transmitter however very less can be harvested by the receiver (Shinohara, N., 2014).

Wireless charging can be another option if the ambient energy source is not sufficient or available, and also in the case when rate of scavenging ambient energy is fairly slow. In that scenario requisition for energy replenishment will be sent to nearby green BSs by IoT device. Wireless charging even though is not considerably efficient however the constant RF energy emission from base station (BS) can charge the IoT devices in close proximity. Antenna 2 shown in Fig. 6 (Liu, X., et al., 2019) . receives green RF energy from BS and transmits upon single or multiple public dedicated wireless charging spectrum bands, this way Antenna 2 can be a single multi-band antenna (Liu, X., et al., 2019).

Fig 7 (Liu, X., et al., 2019) shows a green femto base station (FBS) equipped with dual-batteries deployed for indoor environment, in that case dual battery can be connected to a solar panel or a wind turbine via an electric cable. WiFi access points or Low Power Wide Area (LPWA) transmitters can be modified and used as green FBSs (Liu, X., et al., 2018)

For the outdoor environment the cellular pico base stations (PBSs) PBSs are equipped with dual-batteries and powered by green energy as shown in Fig 7 (Liu, X., et al., 2019). The function of the dual-battery for the BS is the same as of the previously mentioned IoT devices.

Figure 8. Green energy wireless charging architecture

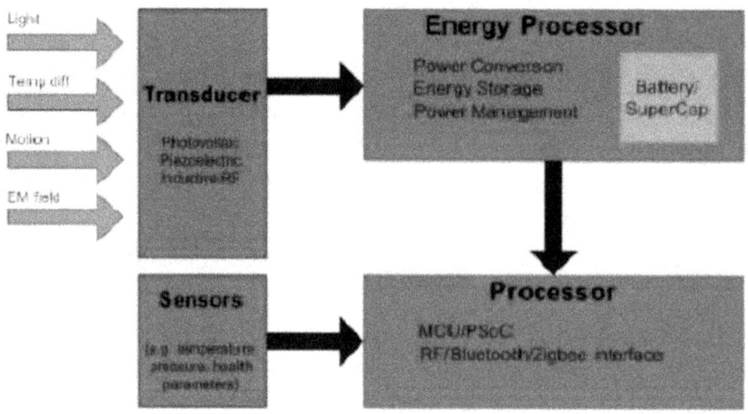

CONCLUSION AND FUTURE DIRECTIONS FOR ENERGY HARVESTING IN 5G NETWORKS

Wireless sensor networks are the backbone of IoT, to fulfil the power requirements of increased number of IoT devices green energy solutions are discussed in this chapter which are based on principal of harvesting ambient energy to make IoT devices self-powered. A dual battery architecture to harvest green energy for IoT devices continuous and smooth functionalities is also presented in the situation ambient energy is not sufficient or not available. Wireless Sensor Networks are the essential element of IoT infrastructure and already proved its significance in many commercial and industrial based applications, specifically the ones suitable in hard conditions where human interference and involvement is dangerous or difficult. Since the power is the main need of WSN to keep sensor nodes alive and functional and the technical advancement in low power electronics addresses the power challenge to some extent. However, the use of conventional batteries that usually have limited time span is costly, impose environmental hazards and also battery replacement and maintenance may not possible in many situations. Therefore, energy extracting from surroundings and ambient environment is an effective alternative both in terms of power and cost. Energy harvesting from surroundings is the promising technology to prolong the life of sensor node and most likely in future will replace the batteries to make WSNs independent of power supply. This chapter gives a brief introduction to various possible ambient energy harvesting technologies and the selection of which should be suitable for the particular applications and working environments of WSNs and eventually for IoT. Besides of that it also enlightens different mechanism of harvesting energy to achieve the ultimate goal of permanent use of wireless sensors and expand and support the use of WSN and IoT applications.

Since IoT is the backbone of 5G networks that are more conscious for energy saving due to diversified applications such as augmented reality, smart cars, smart homes etc. Harvesting ambient energy can be promising technology for 5G networks as they are deployed low powered radio access nodes that can work on both licensed and unlicensed frequency band resulting the significant energy consumption in the entire infrastructure (Imran M., et al., 2019).

Energy efficiency is the critical issue in 5G networks, hence research is needed for the selection of energy harvesting techniques in context of 5G networks, also the mobile devices likely face the ambient noise that can greatly affect energy efficiency.

REFERENCES

Afghan, S. A., Husam, A., & Husi, G. (2018). Towards the self-powered Internet of Things (IoT) by energy harvesting: Trends and technologies for green IoT. *Proc. 2nd Int. Symp. Small-Scale Intell. Manuf. Syst. (SIMS),* 1–5.

Alippi, C., Anastasi, G., Di Francesco, M., & Roveri, M. (2009). Energy Management in Wireless Sensor Networks with Energy-hugry Sensors. *IEEE Instrumentation & Measurement Magazine*, *12*(2), 16–23. doi:10.1109/MIM.2009.4811133

Almulhim, M., Islam, N., & Jhanjhi, N. Z. (2019). A Lightweight and Secure Authentication Scheme for IoT Based E-Health Applications. *International Journal of Computer Science and Network Security*, *19*(1), 107–120.

Almusaylim, Z. A., & Jhanjhi, N. Z. (2019). A review on smart home present state and challenges: Linked to context-awareness internet of things (IoT). *Wireless Networks*, *25*(6), 3193–3204. doi:10.100711276-018-1712-5

Ashton, K. (2009). *That "Internet of Things" thing in the real world, things matter more than ideas.* RFID Journal.

Brock, D. (2001). *The electronic product code (epc) a naming scheme for physical objects.* Auto-ID Center, White Paper.

Bu, L., Wu, X. M., Wang, X. H., & Liu, L. T. (2011). Liquid encapsulated energy harvester for low frequency vibrations. *Proceedings of the 16th International Solid-State Sensors,Actuators and Microsystems Conference (TRANSDUCERS '11)*, 1673–1676. 10.1109/TRANSDUCERS.2011.5969863

Chalasani, S., & Conrad, J. M. (2008). A survey of energy harvesting sources for embedded systems. *Proceedings of IEEE Southeastcon*, 442–447.

Dash, S., Swain, A. R., & Ajay, A. (2012). Reliable energy aware multi token based MAC protocol for WSN. *Proceedings of the26th IEEE International Conference on Advanced Information Networking and Applications (AINA '12)*, 144–151. 10.1109/AINA.2012.61

Diro, A., Reda, H., Chilamkurti, N., Mahmood, A., Jhanjhi, N. Z., & Nam, Y. (2020). Lightweight Authenticated-Encryption Scheme for Internet of Things Based on Publish-Subscribe Communication. *IEEE Access: Practical Innovations, Open Solutions*, *8*, 60539–60551. doi:10.1109/ACCESS.2020.2983117

Erturk, A., & Inman, D. J. (2009). An experimentally validated bimorph cantilever model for piezoelectric energy harvesting from base excitations. Iopscience Smart Materials and Structures, 18(2). doi:10.1088/0964-1726/18/2/025009

Erturk, A., Vieira, W. G. R., DeMarqui, C. Jr, & Inman, D. J. (2010). On the energy harvesting potential of piezoaeroelastic systems. *Applied Physics Letters*, *96*(18), 184103. doi:10.1063/1.3427405

Fei F., Mai J. D., & Li W. J. (2012). A wind-utter energy converter for powering wireless sensors. *Sensors and Actuators A: Physical, 173*(1), 163-171.

Filip, N., & Simu, I. (2011). About the noise energy conversion from agriculture tractor engines. *Proceedings of the 39thInternational Symposium on Agricultural Engineering*, 29–37.

GSMA-Green Power for Mobile. (2015). *The Global Telecom Tower ESCO Market*. Technical Report. Retrieved June 5, 2020 from https://www.gsma.com/mobilefordevelopment/wp-content/uploads/2015/01/140617-GSMA-report-draft-vF-KR-v7.pdf

Heer, R., Wissenwasser, J., Milnera, M., & Farmer, L. (2010). Wireless powered electronic sensors for biological applications. Proceedings of IEEE EMBC 2010, 700-703.

Hi-globe. (n.d.). Retrieved September 15, 2015, from http://hi-globe.com/?page_id=85

Hua, W. S., Li-Dong, D., De-Yi, K., Hao-Yue, P., Zhen, F., & Zhan, Z. (2014). Hybrid device for acoustic noise reduction and energy harvesting based on a silicon micro-perforated panel structure. *Chinese Physics B*, *23*(4), 044302. doi:10.1088/1674-1056/23/4/044302

Hubert, T., & Grijalva, S. (2011). Realizing smart grid benefits requires energy optimization algorithms at residential level. *Innovative Smart Grid Technologies (ISGT)*, 1–8.

Humayun, M., Jhanjhi, N. Z., Malak, Z., & Khan, A. (2020). Smart Cities and Digital Governance. In *Employing Recent Technologies for Improved Digital Governance*. IGI Global. doi:10.4018/978-1-7998-1851-9.ch005

IEEE Computer Society. (2017). Retrieved June 5, 2020 from https://www.computer.org/publications/tech-news/research/internet-of-military-battlefield-things-iomt-iobt

Imran, M., Khan, L. U., Yaqoob, I., Qureshi, M. A., & Ahmed, A. (2019). Energy harvesting in 5G networks: Taxonomy, requirements, challenges and future directions. *IEEE Wirel. Commun*. https://arxiv.org/pdf/1910.00785.pdf

International Telecommunication Union. (2005). *ITU internet report 2005: the internet of things*. International Telecommunication Union, *Workshop Report.*

Jhanjhi, N. Z., Almusali, F. A., Brohi, S. N., & Abdullah, A. (2018). Middleware Power Saving Scheme for Mobile Applications. *IEEE Fourth International Conference on Advances in Computing, Communication & Automation (ICACCA),* 1-6.

Jhanjhi, N. Z., & Almusalli, F. A. (2017). Review: Smartphones power consumption & energy saving techniques. *2017 International Conference on Innovations in Electrical Engineering and Computational Technologies (ICIEECT)*, 1-7. 10.1109/ICIEECT.2017.7916593

Karami, M. A., Bilgen, O., Inman, D. J., & Friswell, M. I. (2011). Experimental and analytical parametric study of single-crystal unimorph beams for vibration energy harvesting. *IEEE Transactions on Ultrasonics, Ferroelectrics, and Frequency Control, 58*(7), 1508–1520. doi:10.1109/TUFFC.2011.1969 PMID:21768034

Khan, A., Jhanjhi, N. Z., & Humayun, M. (2020). *Secure Smart and Remote Multipurpose Attendance Monitoring System, EW*. EAI. doi:10.4108/eai.13-7-2018.164583

Khan, A., Jhanjhi, N. Z., Humayun, M., & Ahmad, M. (2020). The Role of IoT in Digital Governance. In *Employing Recent Technologies for Improved Digital Governance*. IGI Global. . doi:10.4018/978-1-7998-1851-9.ch007

Khan, Z. A., & Abbasi, U. (2016). Evolution of wireless sensor networks toward internet of things. Emerging Communication Technologies Based on Wireless Sensor Networks: Current Research and Future Applications, 179–200. doi:10.1201/b13092

Li, L., Zhang, Y., San, H., Guo, Y. I., & Chen, X. (2008). MEMS for vibration energy harvesting. *MEMS/MOEMS Technologies and Applications III*, vol. 6836 *of Proceedings of the SPIE.*

Li, X., Guo, M., & Dong, S. (2011). A flex-compressive-mode piezoelectric transducer for mechanical vibration/strain energy harvesting. *IEEE Transactions on Ultrasonics, Ferroelectrics, and Frequency Control, 58*(4), 698–703. doi:10.1109/TUFFC.2011.1862 PMID:21507747

Li, Y., Yu, H., Su, B., & Shang, Y. (2008). Hybrid micro power source for wireless sensor network. *IEEE Sensors Journal, 8*(6), 678–681. doi:10.1109/JSEN.2008.922692

Liu, X., & Ansari, N. (2018). Dual-Battery Enabled Green Proximal M2M Communications in LPWA for IoT. *Proc. IEEE Int. Conf. Commun. (ICC)*, 1–6. 10.1109/ICC.2018.8422203

Liu, X., & Ansari, N. (2019). Toward green IoT: Energy solutions and key challenges. *IEEE Communications Magazine, 57*(3), 104–110. doi:10.1109/MCOM.2019.1800175

Lu, C., Park, S. P., Raghunathan, V., & Roy, K. (2010). Analysis and design of ultra low power thermoelectric energy harvesting systems. *Proceedings of the ACM/IEEE International Symposium on Low-Power Electronics and Design (ISLPED '10)*, 183–188. 10.1145/1840845.1840882

Lu, C., Raghunathan, V., & Roy, K. (2010). Micro-scale energy harvesting: A system design perspective. *Design and Automation Conference (ASPDAC) IEEE*, 89 – 94. 10.1109/ASPDAC.2010.5419913

Lu, F., Lee, H. P., & Lim, S. P. (2004). Modeling and analysis of micro piezoelectric power generators for micro-electromechanical systems applications. *Smart Materials and Structures, 13*(1), 57–63. doi:10.1088/0964-1726/13/1/007

Mandal, S., Turicchia, L., & Sarpeshkar, R. (2010). A low-power, battery-free tag for body sensor networks. *IEEE Pervasive Computing, 9*(1), 71-77.

Markit, I. H. S. (2017). *Number of Connected IoT Devices Will Surge to 125 Billion by 2030*. Retrieved June 5, 2020 from https://news.ihsmarkit.com/prviewer/release_only/slug/number-connected-iot-devices-will-surge-125-billion-2030-ihs-markit-says

Matova, S. P., Elfrink, R., Vullers, R. J. M., &Schaijk, R. V. (2011). Harvesting energy from airow with a michromachined piezoelectric harvester inside a Helmholtzresonator. *Journal of Micromechanics and Microengineering, 21*(10), 1-6.

Miranda, M. (2004). *Electrostatic vibration-to-electric energy conversion* (PhD thesis). Massachusetts Institute of Technology.

MIT Technology Review. (2003). *10 Emerging Technologies That Will Change the World.* Retrieved September 15, 2015 from http://www2.technologyreview.com/featured-story/401775/10-emerging-technologies-that-will-change-the/2/

Mohamed, M. I., Wu, W. Y., & Moniri, M. (2011). Power harvesting for smart sensor networks in monitoring water distribution system. *Proceedings of the International Conference onNetworking, Sensing and Control (ICNSC '11)*, 393–398. 10.1109/ICNSC.2011.5874897

Monfray, S., Puscasu, O., Savelli, G., & Sapremanian, U. (2012). Innovative thermal energy harvesting for zero power electronics. In *Silicon Nano electronics Workshop (SNW)* (pp. 1–4). IEEE. doi:10.1109/SNW.2012.6243313

Morais, R., Matos, S. G., Fernandes, M. A., Valente, A. L. G., Soares, S. F. S. P., Ferreirac, P. J. S. G., & Reis, M. J. C. S. (2008). Sun, wind and water flow as energy supply for small stationary data acquisition platforms. *Computers and Electronics in Agriculture, 64*(2), 120–132. doi:10.1016/j.compag.2008.04.005

Nacef, A. B., Senouci, S.-M., Doudane, Y. G., & Beylot, A.-L. (2011). A cooperative low power Mac protocol for wireless sensor networks. *Proceedings of the IEEE InternationalConference on Communications (ICC '11)*, 1–6. 10.1109/icc.2011.5962416

Nasir, A. A., Zhou, X., Durrani, S., & Kennedy, R. A. (2013). Relaying protocols for wireless energy harvesting and information processing. *IEEE Transactions on Wireless Communications, 12*(7), 3622–3636. doi:10.1109/TWC.2013.062413.122042

Padhy, P., Martinez, K., Riddoch, A., Hart, J., & Ong, H. (2005). Glacial environment monitoring using sensor networks. *Proceedings of the Conference on Real-World Wireless Sensor Networks.*

Palosaari, J., Leinonen, M., Hannu, J., Juuti, J., & Jantunen, H. (2012). Energy harvesting with a cymbal type piezoelectric transducer from low frequency compression. *Journal of Electroceramics, 28*(4), 214–219. doi:10.100710832-012-9713-8

Park, S., Koh, B., Wang, Y., Jaemin, K., Kim, Y., Pedram, M., & Chang, N. (2013). Maximum power transfer tracking in a solar USB charger for smartphones. *Symposium on low power electronics and design, IEEE 2013*, 88 – 93. 10.1109/ISLPED.2013.6629272

Pobering, S., & Schwesinger, N. (2004). A novel hydropower harvesting device. *Proceedings of the International Conference onMEMS, NANO and Smart Systems (ICMENS '04)*, 480–485.

Pobering, S., & Schwesinger, N. (2009). Power supply for wireless sensor systems. *Proceedings of the IEEE Sensors Conference(SENSORS '08)*, 685–688.

Ponnusamy, V., Ramachandran, T., Jung, L. T., & Jhanjhi, N. Z. (2017). Bio-Inspired Energy Scavenging in Wireless Ad Hoc Network. In *IEEE Innovations in Electrical Engineering and Computational Technologies*. ICIEET. doi:10.1109/ICIEECT.2017.7916600

Raghunathan, V., Kansal, A., Hsu, J., Friedman, J., & Srivastava, M. (2005). Design considerations for solar energy harvesting wireless embedded systems. *Proceedings of the 4th InternationalSymposium on Information Processing in Sensor Networks(IPSN '05)*, 457–462. 10.1109/IPSN.2005.1440973

Ramasur, D., & Hancke, G. P. (2012). A wind energy harvester for low power wireless sensor networks. *Proceedings of the IEEE International Instrumentation and Measurement Technology Conference (I2MTC '12)*, 2623–2627. 10.1109/I2MTC.2012.6229698

Reinisch, H., Gruber, S., Unterassinger, H., Wiessecker, M., Hofer, G., Pribyl, W., & Holweg, G. (2011). An electro-magnetic energy harvesting system with 190 nW idle mode power consumption for a BAW based wireless sensor node. *IEEE Journal of Solid-State Circuits, 46*(7), 1728-1741.

Saadon, S., & Sidek, O. (2011). Vibration-based MEMS piezoelectric energy harvester (VMPEH) modeling and analysis for green energy source. *Proceedings of the 4th International Conference on Developments in eSystems Engineering (DeSE '11)*, 527–531. 10.1109/DeSE.2011.85

Saadon, S., & Sidek, O. (2011). A review of vibration-based MEMS piezoelectric energy harvesters. *Energy Conversion and Management*, *52*(1), 500–504. doi:10.1016/j.enconman.2010.07.024

Sardini, E., &Serpelloni, M. (2011). Self-powered wireless sensor for air temperature and velocity measurements with energy harvesting capability. *IEEE Transactions on Instrumentation and Measurement, 60*(5), 1838-1844.

Seah, W. K. G., Zhi, A. E., & Tan, H.-P. (2009). Wireless Sensor Networks Powered by Ambient Energy Harvesting (WSNHEAP)—survey and challenges. *Proceedings of the 1stInternational Conference on Wireless Communication, Vehicular Technology, Information Theory and Aerospace and Electronic Systems Technology, Wireless (VITAE '09)*, 1–5.

Shah, S. T., Choi, K. W., Hasan, S. F., & Chung, M. Y. (2016). Throughput analysis of two-way relay networks with wireless energy harvesting capabilities. *Ad Hoc Networks*, *53*, 123–131. doi:10.1016/j.adhoc.2016.09.024

Shang, Y., Li, Y., Yu, H., Sun, H., & Su, B. (2007). Experimental investigation on thermoelectric generator of micro hybrid power source. *Proceedings of the 3rd International Symposium on Advanced Optical Manufacturing and Testing Technologies: Design, Manufacturing, and Testing of Micro- and Nano-Optical Devices and Systems*, 6724. 10.1117/12.782827

Shinohara, N. (2014). *Wireless Power Transfer via Radio waves*. Wiley Press.

Silva, R., Farinholt, K., & Park, G. (2011). Hybrid energy sources for embedded sensor nodes. *Proceedings of the Industrial and Commercial Applications of Smart Structures Technologies*. 10.1117/12.882010

Smoker, J., Nouh, M., Aldraihem, O., & Baz, A. (2012). Energy harvesting from a standing wave thermo acoustic-piezoelectric resonator. *Journal of Applied Physics*, *111*(10), 104901. doi:10.1063/1.4712630

Sojan, S., & Kulkarni, R.K. (2016). A comprehensive review of Energy Harvesting techniques and its potential application. *International Journal of Computer Applications, 139*(3), 14– 19.

Stojcev, M. K., Kosanovic, M. R., & Golubovic, L. R. (2009). Power management and energy harvesting techniques for wireless sensor nodes. *Proceedings of the 9th International Conference on Telecommunication in Modern Satellite, Cable, and Broadcasting Services (TELSIKS '09)*, 65–72. 10.1109/TELSKS.2009.5339410

Stokes, C. D., Duff, A., Mantini, M. J., Grant, B. A., & Venkatasubramanian, R. (2010). Nanostructured thermoelectric material and device technology for energy harvesting applications. *Proceedings of the 4th IEEE Nanotechnology Materials and Devices Conference (NMDC '10)*, 154–159. 10.1109/NMDC.2010.5652061

Sun, H., Yin, M., Wei, W., Li, J., Wang, H., & Jin, X. (2018). MEMS based energy harvesting for the Internet of Things: A survey. *Microsystem Technologies, 24*(7), 2853–2869. doi:10.100700542-018-3763-z

Tabbakh, S. R. K., Maarefdoust, R., Kyun, N. C., & Ali, B. M. (2010). Environmental taxonomy of power scavenging techniques for autonomous self powered wireless sensors. *Proceedings of the Asia Pacific Conference on Circuit and System (APCCAS '10)*, 1031–1034.

Tan, Y. K., & Panda, S. K. (2007). A novel piezoelectric based wind energy harvester for low-power autonomous wind speed sensor. *Proceedings of the 33rd Annual Conference of the IEEE Industrial Electronics Society (IECON '07)*, 2175–2180. 10.1109/IECON.2007.4460120

Tan, Y. K., & Panda, S. K. (2011a). Energy harvesting from hybrid indoor ambient light and thermal energy sources for enhanced performance of wireless sensor nodes. *IEEE Transactions on Industrial Electronics, 58*(9), 4424-4435.

Tan, Y. K., & Panda, S. K. (2011b). Self-autonomous wireless sensor nodes with wind energy harvesting for remote sensing of wind-driven wild re spread. *IEEE Transactions on Instrumentation and Measurement, 60*(4), 1367-1377.

Ucar, A., & Balo, F. (2009). Investigation of wind characteristics and assessment of wind generation potentiality in Uluda˘g-Bursa, Turkey. *Applied Energy, 86*(3), 333–339. doi:10.1016/j.apenergy.2008.05.001

Vasseur, J.-P., & Dunkels, A. (2010). *Interconnecting Smart Objects with IP - The Next Internet*. Morgan Kaufmann.

Vullers, R. J. M., Schaijk, R., Visser, H. J., Penders, J., & Hoof, C. V. (2010). Energy Harvesting for autonomous wireless sensor networks. *IEEE Solid-State Circuits Magazine, 2*(2), 29–38. doi:10.1109/MSSC.2010.936667

Vullers, R. J. M., Schaijk, R. V., Doms, I., Hoof, C. V., & Mertens, R. (2009). Micropower energy harvesting. *Solid-State Electronics, 53*(7), 684–693. doi:10.1016/j.sse.2008.12.011

Wan, Z. G., Tan, Y. K., & Yuen, C. (2011). Review on energy harvesting and energy management for sustainable wirelesssensor networks. *Proceedings of the IEEE 13th International Conference on Communication Technology (ICCT '11)*, 362–367.

Wang, D.-A., & Chang, K.-H. (2010). Electromagnetic energy harvesting from flow induced vibration. *Microelectronics Journal, 41*(6), 356–364. doi:10.1016/j.mejo.2010.04.005

Wang, D.-A., & Ko, H.-H. (2010). Piezoelectric energy harvesting from flow-induced vibration. *Journal of Micromechanics and Microengineering, 20*(2), ID025019. doi:10.1088/0960-1317/20/2/025019

Wang, D.-A., & Liu, N.-Z. (2011). A shear mode piezoelectric energy harvester based on a pressurized water flow. *Sensors and Actuators, A. Physical, 167*(2), 449–458.

Wang, X. (2012). Piezoelectric nanogenerators-Harvesting ambient mechanical energy at the nanometer scale. *Nano Energy, 1*(1), 13–24. doi:10.1016/j.nanoen.2011.09.001

Xiao, L., Qian, J. J., Lian, Z., Long, Z., Tang, X., Chen, T., Du, X., & Cao, L. (2016). RF energy powered wireless temperature sensor for monitoring electrical equipment. *Sensors and Actuators. A, Physical, 249*, 276–28. doi:10.1016/j.sna.2016.08.022

Ye, G., & Soga, K. (2012). Energy harvesting from water distribution systems. *Journal of Energy Engineering, 138*(1), 7–17. doi:10.1061/(ASCE)EY.1943-7897.0000057

Yick, J., Mukherjee, B., & Ghosal, D. (2008). Wireless sensor network survey. *Computer Networks, 52*(12), 2292–2330. doi:10.1016/j.comnet.2008.04.002

Yong, Y.-T., Chow, C.-O., Kanesan, J., & Ishii, H. (2009). A survey on design of self-powered wireless sensor network. *Proceedings of the International Conference for Technical Postgraduates (TECHPOS '09)*, 1–5. 10.1109/TECHPOS.2009.5412044

Zhou, G., Huang, L., Li, W., & Zhu, Z. (2014). Harvesting Ambient Environmental Energy for Wireless Sensor Networks: A Survey. *Journal of Sensors, 2014*, 815467. doi:10.1155/2014/815467

Zhou, H.-Y., Wu, F., & Hou, K.-M. (2008). An event-driven multi-threading real-time operating system dedicated to wireless sensor networks. *Proceedings of the International Conferenceon Embedded Software and Systems (ICESS '08)*, 3–12. 10.1109/ICESS.2008.58

Zhou, X., Zhang, R., & Ho, C. K. (2013). Wireless information and power transfer: Architecture design and rate-energy tradeoff. *IEEE Transactions on Communications, 61*(11), 4754–4767. doi:10.1109/TCOMM.2013.13.120855

Zhu, D., Beeby, S., Tudor, J., White, N., & Harris, N. (2010). A novel miniature wind generator for wireless sensing applications. *Proceedings of the 9th IEEE Sensors Conference (SENSORS '10)*, 1415–1418.

Chapter 6
Realizing the Benefits of Energy Harvesting for IoT

Shakeel Ahmed
https://orcid.org/0000-0003-1155-0991
King Faisal University, Saudi Arabia

ABSTRACT

Different types of energy which generally fulfill the requirements of computing are mostly from thermal, mechanical, solar, wind, acoustic, and wave. Typically, IoT devices are powered by batteries that have limited lifetime, and thus these IoT devices need to be self-powered or require supportive energy sources that uninterruptedly power IoT devices. Energy harvesting is one of the techniques that can be applied to power these devices, which is a procedure of apprehending energy from lone or more energy from renewable sources in the proximate atmosphere known as environmental energy which can be renovated into usable electrical energy. Numerous researches are being carried out to harvest energy. This chapter is the extended version of the previous work carried out and analyses the present works on the application of IoT in energy harvesting systems and extant different research works carried out by the investigators to classify them.

1. INTRODUCTION

Internet of Things (IoT) sometimes referred to as Internet of objects are inter-connection of commonly used networked objects (Humayun et al., 2020). IoT are physical devices or virtual devices connected over the internet using 4G/5G supported phones wirelessly. These devices have inimitable ID which enables them to be inimitably identified. These devices are not only connecting devices but are also used for direct or indirect communications with one another by sending or receiving data. These devices excerpt information via applying the process of filtering followed by processing and data condensing based on which the information or knowledge can be inferred for performing the respective operations.

IoT devices are dynamic in nature, they are self-adoptable and self-configurable devices. Vermesan et al. (Vermesan et al., 2011) define the Internet of Things as simply an interaction between the physical and digital worlds. These IoT devices are power-driven by batteries which is a main constraint on

DOI: 10.4018/978-1-7998-6709-8.ch006

such devices as the batteries capacity is finite. Also, a vast amount of energy is spent when IoT devices connect and communicate with one another, owing to this devices function for partial period, as long as battery persists. One of the feasible options to this energy issue can be solved by using the replaceable batteries which may be operative for smaller IoT systems whereas for huge IoT systems it's not operative and also the price of upholding and substituting billions of batteries is pretty more. In IoT devices consuming Batteries are not continuously helpful as they require human contribution to interchange the batteries. Henceforth, the main concern is for the procurement of electrical power which is preferred to function these devices.

To overwhelm this alternative kind of energy source to traditional batteries are needed to be elegantly devise. Utilization of Energy increase on a day-by-day basis rapidly and henceforth innovative, capable energy resources are required to ascertain. Investigators are realizing novel methods of exploiting profuse renewable energies from varied variety of sources which is a continuing research focus. A favorable answer for such kinds of problematic is energy harvesting which is defined as the process in which energies are captivated from one or more additional ecological sources which include from Solar, Radio frequency, Wind, etc. (Jiao et al., 2017) additional energy sources from heated bodies, foot strikes and finger strokes, etc., collecting and translating them into electrical energy which is operational. The energy harvested powers electrical energy of these devices and surge lifespan of IoT system. Two such energy harvester edifices are presented below:

2. ENERGY HARVESTING ARCHITECTURES

There are two classifications of energy harvesting and they differ from each other in their architectures: (Sudevalayam & Kulkarni, 2011) (a) Harvest-Use: Which is defined as the Energy harvested just-in-time and are directly used (b) Harvest-Store-Use: The Energy which is harvested every time possible and which is stowed for later usage.

2.1. Harvest-Use Architecture

The harvesting system right away powers the sensor node and the sensor node is operational; the power output produced from the harvesting systems need to be uninterruptedly beyond the least operating point. If passable energy is not obtainable, the node gets deactivated Figure 1(a) below depicts the Harvest-Use architecture. Unexpected variations in harvesting capacity near to the least power point caused the sensor nodes to fluctuate in ON/OFF conditions. This kind of system uses mechanical energy bases like pedaling, walking and pushing keys/buttons, etc. For instance, impulse of a keys/buttons deforms piezoelectric material, thus producing electrical energy to direct short wireless messages. Likewise, while walking and running the piezoelectric materials tactically placed in the shoes may deform to different extents which transmits or communicate using RFID signals and to keep track of the shoe-wearer.

2.2. Harvest-Store-Use Architecture

Harvest-Store-Use architecture which comprises of storage components to store the harvested energy and power the sensor nodes as illustrated in Figure 1(b) (Sudevalayam & Kulkarni, 2011). Energy storage is valuable when harvested energy is existing and surplus than to the current usage. Excess energy is

Figure 1. Energy harvesting architectures (Sudevalayam & Kulkarni, 2011)

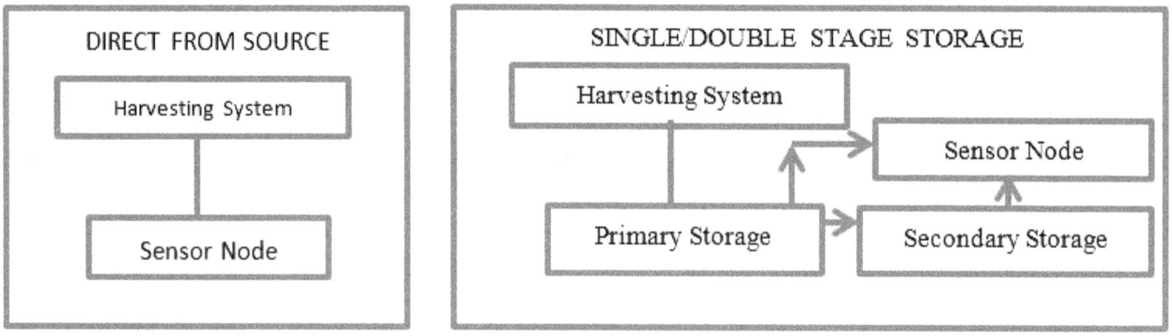

(a) Harvest-Use (b) Harvest-Store-Use

stowed for future usage for improvement and to increase the competence and performance factors when the harvesting chance doesn't exist or energy usage of the sensor node. The storage modules might be single-stage or double-stage where secondary storage is used as backup store for conditions whenever Primary storages gets drained. For instance, Harvest-Store-Use uses unrestrained, expectable energy source like solar energy. Throughout the day-time, the energy is consumed for work and as well as stowed for future use. The stowed energy is conventionally used during night-time to power the sensor nodes.

3. SOURCES OF ENERGY HARVESTING

Solar energy, Mechanical energy, Thermal energy and Wind-energy are the main four ambient sources of energy harvesting (Cao & Li, 2017). The energies which can be acquired using varied energy source by choosing light source, mechanical source, or thermal energy sources available from adjacent locations. The procedure provides and supports unlimited energy for the lifespan of these electronically enabled devices. Henceforth, the procedure of mining energies from ambient settings and changing into dispensable electrical energy is recognized as power scavenging or energy harvesting. Few of the most well know energies are produced from wind farm, energy produced from solar and energy produced from farms of tidal energy, geothermal farms etc. (Ahmed, 2016). Development of technologies has enhanced the ways where these sources can be used to power up the electronic system. The method of classifying energy harvesting is based on the methods being used for energy scavenging.

Renewable sources are categorized based on affluence of source of energy included from Solar, wind energy, thermal energy and tidal energy. The energy is harvested using photovoltaic cells, wind turbines, thermoelectric generator and mechanical vibration device such as piezoelectric device and electromagnetic device. The well recognized and identified common energy harvesting mechanisms have led in the expansion of self-powered systems which need very less cost to maintain and they can be positioned in unreachable sites wherever effortlessly. The foremost target of harvesting energy is to power the self-powered systems. The densities of power for the above mentioned techniques are presented in Table1 (Roundy, 2003).

Nevertheless, varied approaches imposes different limitations on the functioning of different electronic devices created on the category of energy they fit in the atmosphere and according to the source

Table 1. Energy harvesting technologies Power outputs (Roundy, 2003)

Energy Kind	Harvesting Methods	Power Densities
Solar	Solar Energy-outdoors Solar energy-indoors	15 mW/cm³ bright sunny-day 0.15 mW/cm³- cloudy-day. 10-100μW/cm ²
Mechanical	Vibrations (piezoelectric-Shoe insert) Vibrations (electrostatic conversions) Vibrations (electromagnetic conversions)	330μW/cm³ 0.021μW/mm³-105Hz 184μW/cm²-10Hz 306μW/cm³-52Hz
Thermal	Thermoelctric-5°C Gradient	40 μW/cm ³
Wind	Wind flow	16.2μW /cm ³ -5m/s

of energy they gain power from contrasting all such sources of energies, mechanical energy can be found in instances where other energies become insufficient. Different vibrations, motion, heat, flow induced vibrations, and noise, etc. will enable to in care the Mechanical energy from. Further, Power electronic system can be powered up by harvesting energy generated by mechanical energy. Different materials like Piezoelectric, pyro electric, electrostatic material which alter the mechanical energy into convenient energy forms (Lallart et al., 2012).

3.1. Solar Energy Harvesting

Solar energy is well-defined as the energy which is derived from the Sun and solar energy is the utmost resourceful base of environmental energy which provides utmost of the energy with very high efficacy (King et al., 2007). Solar cells collects the sunlight in solar energy harvesting and unswervingly converts sunlight into operational electricity which is joined by several technologies. Sunlight is garnered via photo voltaic cells or solar which are made using semiconductor material i.e. solar thermal collectors and silicon. Photovoltaic (PV) or concentrated solar power (CSP) are used in generating electricity directly or indirectly. Under optimal conditions solar panels are well-regarded at 15% - 20% efficacy which differs in optimal conditions.

Solar panel decreases efficacy of transformation in cloudy and rainy atmospheres. Sun is greatest opulent of renewable energy basis from the biosphere. Photo voltaic cell is the convenient solution for locations where accessibility of light is certain to be great. Specialized PV material are used for indoor applications which are enhanced and suitable for dim lights. To overcome the problem of large amounts of energy is lost when the energy is transmitted from the harvester, Maximum power point tracker (MPPT) circuits is being recommended and these MPPT competently transmits harvested solar energy to rechargeable battery from harvester. Sunlight is well-thought-out as the most suitable energy source for collecting the energy amongst all other sources the maintenance cost of such solar cells is very low-slung. It is valued to provide 1.4 kW/m2 (Sudevalayam & Kulkarni, 2011). Practical efficiency of PV cell and the theoretical efficiency of PV cell is 40% and 90% respectively (Atallah et al., 2016).

3.2. Photovoltaic Energy Harvesting

In the photovoltaic procedure, semiconductor exhibiting photoelectric outcomes are in use for obtaining direct current electricity by using solar radiations. Photovoltaic cell or solar cells are used in generating Photoelectric effect which converts light into electric current. As the process used in photosynthesis in plants using chlorophyll to absorb light, PV Dye Sensitized solar cell are used in harvesting energy as dyes and photovoltaic cell are the devices that are used in converting light energy into electrical energy (Juang et al., 2019). This type of energy typically is light energy increased mostly from sunlight.

Photovoltaic cells are suitable elucidation for spaces where availability of light and usages of batteries and added means of power supply is not conceivable and affluent. Roadway signs are an examples. The most widespread photovoltaic cells which are extra elusive to light and effortlessly obtainable and bid accurate prices to performance ratio are Silicon-based cells.

3.3. Concentrated Solar Power Harvesting

The Sunlight is focused into small beam from a big area thru the convention of lenses or mirrors, which can be moreover used as source of heat aimed at conventional power plants. Several mechanism like Stirling dish, Fresnel reflector and solar power towers are used to keep track the sun light and focus the light. The working fluid in Concentrated Solar Power (CSP) system prepares solar power heating by using the aid of focused sunlight (Ferrara et al., 2014).

The solar energy received by the earth in an hour is larger than the energy expended in a year which marks the photovoltaic material as one which is the utmost substantial harvesters of alternative energy. Energy harvesting techniques includes: Piezoelectric, thermoelectric, Photovoltaic, necessitates added mechanisms which surge the circuit convolution and price of these device, that's the reason why the wireless image sensors with on-chip array are used (Ahmed et al., 2016) and as mentioned in the paper it focuses on design, operations of power management system along with pixel arrays. Operation of CMOS - Active Pixel Sensor (APS) is based on the photoconductive mode during the process of imaging as well as photovoltaic mode for harvesting.

Power management which is accomplished by single inductor boosts or buck regulators and boosts using maximum power to charge onboard sensor point tracking storage regulator to harvests energy. High power density for video monitoring application is harvested from Energy harvesting power management unit which generates regulated voltage of 1V from the external photovoltaic harvester by using super capacitor. To control illumination and temperature variation the approach of consuming maximum power point tracking (MPPT) method is applied (Bol et al., 2015). The specific impedance tuning method is used instead of using frequency modulation scheme the effect which they have attained 20 μW and efficacy of 89%. An intelligent energy harvesting system (Nechibvute et al., 2012) which is built on MPPT for sensor nodes are used in IoT to prolong lifespan of lithium batteries by using RS trigger and make the lithium battery to charge only when the battery voltage is lesser than specific values. Additional paper presents an algorithm for energy prediction which uses the light intensity of fluorescent lamp in indoor environments (Li & Shi, 2015).

A single source solar energy harvesting system in (Shin & Inwhee, 2016). Two NiMH batteries and Maximum power point tracking (MPPT) for impedance matching are used (Taneja et al., 2008) platform which effects in higher energy harvesting efficacy. The output from solar panel is 276mW at voltage 3.11V. One of the drawbacks of consuming NiMH battery is the extraordinary self-discharge rate due to which lifespan of (Shin & Inwhee, 2016) systems is low and NiMH battery is little input/output efficacy. Sunlight in the range of 30 minutes is required to fulfill the charge NiMH batteries to meet the requirements of the application.

Heliomote (Polastre et al., 2005) (Lin et al., 2005) (Kansal et al., 2007) is additional solo source of solar energy used for harvesting the energy the system uses Mica2 (Raghunathan et al., 2005) (Hill & Culler, 2002) platform and two NiMH batteries and it doesn't use MPPT. To hold the large amount of energy for longer period of time with minimum leakage current. Outputs of solar panel is 198mW at 3.3V. 80-84% of efficiency is achieved using Heliomote. Prometheus (Mica, n.d.) is yet another single source of solar energy harvesting uses the TelosB platform.

In Fleck (Jiang et al., 2005) two NiMH batteries are considered which is another single source of solar energy used as harvesting system which deliberates since it doesn't need complex charging circuits and MPPT is not being used in Integrated Fleck1. The output of the solar panel is 2100mWh in a single day. The system is designed to work in a full solar light but it doesn't work in dim sunlight.

4. MECHANICAL ENERGY HARVESTING

The Mechanical energy harvesting which is moreover named as waste energy harvesting and low power generator. Mechanical energy is influenced by object either through its positions or through motion and which is classified into two types, which are kinetic energy and potential energy where Kinetic energy is associated to motion and energy is associated to work is potential energy done based on object or position of the object. Vibration, motion, pressure and human activity are the bases of kinetic energy. The mechanism which is used for harvesting mechanical energy is by using piezoelectric transductions, electrostatic & electromagnetic and piezoelectric transductions are used extensively for the reason that it uses piezoelectric material which consumes large power densities (Corke et al., 2007).

Piezoelectric transducers are used for harvesting the energy capability to transform from mechanical energy to electrical energy when mechanical load is applied on piezoelectric transducers mechanical strain into electric current or voltage is converted by the direct piezoelectric effect (Corke et al., 2007). An inertial mass that be used to generate movement and this measure can be adapted to electrical energy by a devices when it is set to tremor. Several of these are naturally occurred materials such as quartz, Polycrystalline ceramic which is a well-known piezoelectric material, Lead Zirconate Titanate (PZT) is being well-thought-out as it displays a high efficiency of mechanical to electrical energy conversions (Beeby et al., 2006) Diverse possessions of piezoelectric material are well-thought-out by investigators in improving various piezoelectric harvesters application in order to power diverse applications.

4.1. Energy Harvesting from Machinery Vibrations

Linear generator is designed from machinery vibrations which transforms kinetic energy to electrical energy to harvest energy (Joseph, 2005). Several techniques are used such as piezoelectric transducers, electromagnetic and electrostatic for transduction. One of the three transducers have maximum power density which is electromagnetic energy harvesters also they are not appropriate for MEMS application. Both the power density harvester and piezoelectric energy have same electromagnetic energy and have simple structure. Low power density is appropriate for MEMS applications out of the three Electrostatic energy harvesters.

4.2. Energy Harvesting from Flow Induced Vibrations

Various investigators have absorbed their attentions on Energy harvester as flow-induced vibrations (Basset et al., 2009). Prevailing device are accomplished to produce energy, the start flow speed for them is been condensed with minimum possible as 2.5 m.s-1. Most of the devices which are normally huge in volume are adept to harvest beneficial energy and relate them for the further obtainable vibration energy harvester. Therefore, it is big challenge and hard to integrate wireless sensor nodes easily with other available wireless electronic systems. Further the work ought mostly stresses on shrinking energy harvesters though these devices have to uphold current power level. In addition, researchers need to reflect to decrease the start flow speediness and let the technologies be available for widely used applications.

4.3. Energy Harvesting from Human Movement

The devices that are favored to harvest energy from human movements is Feet and this is the focal part of the human body which is concentrated which produces force due to the rigorous movement which includes Random, non-resonant or broadband non-linear devices (Safaei et al., 2019). The movement of feet mechanics can be useful for producing energy for portable electronic devices using the technique of energy harvester. Further care need be taken in demand to circumvent uneasiness in human body owing to energy harvest.

5. THERMAL ENERGY HARVESTING

Technique of renovating thermal gradients through thermoelectric generators (TEGs) into electrical energy which comes with ceramic plates on one side is warmer and the other side colder, which produces electrical energy by converting the heat flux into electrical energy with very high efficacy by the process identified as thermionic conversion.

As the thermal energy harvesting has protrusive results researches are working for many long time to progress advantageous thermionic generators. The energy is harvested by generating the steam where the turbines spin using this generated steam and which drives the generators and harvest electricity. The setup requires realistic size equipment as well as space which is common available. The slight temperature difference between the two surfaces is used as a benefit for Thermal Energy (TE) energy-harvesting. Thermoelectricity is the conception of generating electric potential with a temperature variance and vice-versa. Thermal energy is foraged to power the electronic devices by acquiring electrical energy. Thermoelectric devices are mostly used in earthly and space applications (Li et al., 2018).

6. WIND ENERGY HARVESTING

For long period of time wind-power has remained and perceived as feasible basis for "free" energy. Wind turbines are very significant player since early 1980s as renewable energy. Wind energy is extracted from mechanical turbines. Wind power covers extensive varieties of application and joined with small domestic wind turbines provide 3kW to a home and by large wind turbine farms provide up to 800MW of power.

Mechanical energy is converted into electrical energy using wind turbine generators and these are a bit rare, equated to additional producing unit usually attached to the electrical grids in which the generators works with the power source and supplies changing mechanical power (Jaziri et al., 2019). Two or three propeller like blades are used in these turbines to capture wind energy and the turbines are high towers which captures stronger and less turbulent winds. The speed of wind resolves the effectiveness and reduced of the wind energy applications. Wind is initiated due to rough heating of the earth, anomalies of earth's terrain and also from earth revolution. The electricity is produced by the wind which rotates the blades and spin the shaft of the turbines through generator. Pacing wind turbines at the higher hub heights because wind speed are higher at in height altitudes and are also less turbulent. Normally the power which the turbine creates is a function of the cube of the winds speed.

7. BIO-ENERGY HARVESTING

Renewable energy resulting from materials of various biological sources is called Bio-energy (Singh et al., 2013). These naturally-occurring Bioenergy existences assist as power source in environmental sensing network for wild life surveillance and climatic changes in zones with inadequate approachability, wherever battery replacement is not feasible. Biomass is an organic material which is a form of Chemical energy and is an alternative of bio-Energy harvesting by means of trees that can be used to power circuits, as explored the voltage is generated by these trees are small to charge batteries.

8. TIDAL ENERGY HARVESTING

The optimum source of potential energy is considered from Oceans. The energy transferred in the form tides, waves and currents which definite quantity of energy be utilized to accomplish work. Tidal power from the tides which are gained by captivating the energy from the moving water masses because of the movement of tide and the tide can generate energy in different types and forms. Kinetic and potential energy are two kinds of tidal energy which are also common and can be extracted by current between receding and rolling tide of current between low tides and high tides immediately (Taylan et al., 2018). Constructing energy from tides is easy and potentially adequate, which analogous to generating energy through water via construction of diverse dams and various machineries. Presently tidal Energy Harvesting from the source is well-thought-out as inexpensive and common source, many countries everywhere in the world coastal sites are reviewing the appropriateness in harvesting the tidal energy. The Power based on tidal is added dependable as well as foreseeable.

9. ELECTROMAGNETIC ENERGY HARVESTING

Electromagnetic induction is well-defined in which the voltage in the conductor is produced by altering the magnetic field round the conductor. The principle of electromagnetic induction can be applied to produce Electromagnetic Energy Harvesting. Electromagnetic radiation (light) is electromagnetic waves that propagate through space.

Table 2. Comparative Analysis (Atallah et al., 2016)

Energy	Mechanisms	Drawbacks
Light	Photo Voltaic cell	Connected in day-time & sunny
Wind	Wind turbine generator	Wind and Noise can't be predicted
Wave	Ocean thermal energy conversion	Variable based on the intensity of sea waves and loss to the power generating units.
Heat	Thermo electric devices	Spoils environment and atmosphere owing to release of smoke and fumes
Mechanical	Electromagnetic, electrostatic, piezoelectric transducers	Likely only in machines with high power

The above Table 2 compares the various forms of Harvesting Energy and process involved along with disadvantage with these kinds of energy.

There are many types of these waves depending on the frequency (X-rays, Ultraviolet, Visible light, and Infrared etc.). Radiant energy is the energy of electromagnetic waves. Radiant energy is in the form of radio frequencies which are produced from network router, televisions, microwave, base stations, smartphones and ovens. In electromagnetic procedure with the assistance of large opening power receiving antennas this radio frequency signals are collected by renovating the attained waves into functional DC power the enactment is dependent on RF to DC adaptation efficiency and quantity of power received by the antenna. Several investigators use RF signals for energy harvesting (Ben Elghali et al., 2007) (Aminov & Agrawal, 2014). As measured the practical proficiency of solar panel is 85% and that of theoretical proficiency is 100% (Kamalinejad et al., 2015).

10. CONCLUSION

Billions of devices are deployed in Internet of things and there are several forms of energies which includes energies from solar, wind, mechanical, thermal, acoustic, and for different applications and devices that can satisfy the requirements of computing. One of the main source of energy for various devices in IoT systems are Batteries. To circumvent the replacement of batteries and its cost, the IoT's require energy harvesting. The Energy harvesting harvests or scavenges energy from various ambient energy sources and transforms into electrical energy to recharge batteries in IoT devices. Energy harvesting permits on-site charging of rechargeable batteries, that can be cycled hundreds of times before their performance worsens. With proper hardware's and energy management techniques, lifetime of the batteries can be extended practically indefinitely. Investigators are working to address the issues of various Energy harvesting techniques and mechanisms based on environment are being well-thought-out and work out for possible feasible prospect to switch the current power supplies for energy constrained in embedded systems and also towards creating new approaches to overwhelm few limitations such as lesser amount of power generated using power harvesters. The proposed methods would help in retaining the energy harvesters as one of the optimum bases to power moveable devices in the arena of IoT technology.

REFERENCES

Ahmed, K. Z., Amir, M. F., Ko, J. H., & Mukhopadhyay, S. (2016). Reconfigurable 96x128 Active Pixel Sensor with 2.1 µW/mm2 Power Generation and Regulated MultiDomain Power Delivery for Self-Powered Imaging. *42nd European Solid-State Circuits Conference*, 507 - 510.

Ahmed, S. (2016). Exploring Current Trends of Energy Harvesting. In *Biologically-Inspired Energy Harvesting through Wireless Sensor Technologies* (pp. 268–278). IGI Global. doi:10.4018/978-1-4666-9792-8.ch014

Aminov, P., & Agrawal, J. P. (2014, May). RF energy harvesting. In *Electronic Components and Technology Conference (ECTC), 2014 IEEE 64th* (pp. 1838-1841). IEEE. 10.1109/ECTC.2014.6897549

Atallah, R., Khabbaz, M., & Assi, C. (2016). Energy harvesting in vehicular networks: A contemporary survey. *IEEE Wireless Communications, 23*(2), 70–77. doi:10.1109/MWC.2016.7462487

Basset, P., Galayko, D., Paracha, A. M., Marty, F., Dudka, A., & Bourouina, T. (2009). A batch-fabricated and electret-free silicon electrostatic vibration energy harvester. *Journal of Micromechanics and Microengineering, 19*(11), 115025. doi:10.1088/0960-1317/19/11/115025

Beeby, S. P., Tudor, M. J., & White, N. M. (2006). Energy harvesting vibration sources for microsystems applications. *Measurement Science & Technology, 17*(12), R175–R195. doi:10.1088/0957-0233/17/12/R01

Ben Elghali, S. E., Benbouzid, M. E. H., & Charpentier, J. F. (2007). Marine tidal current electric power generation technology: state of the art and current status. IEEE international electric machines & drives conference, 1407–1412. doi:10.1109/IEMDC.2007.383635

Bol, Boufouss, Flandre, & De Vos. (2015). A 0.48mm2 - 5µW-10mW Indoor/Outdoor PV Energy-Harvesting Management Unit in a 65nm SoC based on a Single Bidirectional Multi-Gain/Multi-Mode Switched-Cap Converter with Supercap Storage. *41st European Solid-State Circuits Conference*, 241 - 244.

Cao, S., & Li, J. (2017). A survey on ambient energy sources and harvesting methods for structural health monitoring applications. *Advances in Mechanical Engineering, 9*(4), 1–14. doi:10.1177/1687814017696210

Corke, P., Valencia, P., Sikka, P., Wark, T., & Overs, L. (2007, June). Long-duration solar-powered wireless sensor networks. In *Proceedings of the 4th workshop on Embedded networked sensors* (pp. 33-37). ACM. 10.1145/1278972.1278980

Ferrara, F., Gimelli, A., & Luongo, A. (2014). Small-scale Concentrated Solar Power (CSP) Plant: ORCs Comparison for Different Organic Fluids. *Energy Procedia, 45*, 217–226.

Hill, J. L., & Culler, D. E. (2002). Mica: A wireless platform for deeply embedded networks. *IEEE Micro, 22*(6), 12–24. doi:10.1109/MM.2002.1134340

Humayun, Jhanjhi, Alsayat, & Ponnusamy. (2020). Internet of things and ransomware: Evolution, mitigation and prevention. *Egyptian Informatics Journal.*

Jaziri, N., Boughamoura, A., Müller, J., Mezghani, B., Tounsi, F., & Ismail, M. (2019). A comprehensive review of thermoelectric generators: Technologies and common applications. *Energy Reports*. Advance online publication. doi:10.1016/j.egyr.2019.12.011

Jiang, X., Polastre, J., & Culler, D. (2005, April). Perpetual environmentally powered sensor networks. In *Proceedings of the 4th international symposium on Information processing in sensor networks* (p. 65). IEEE Press.

Jiao, P., Borchani, W., Hasni, H., & Lajnef, N. (2017). Enhancement of quasi-static strain energy harvesters using non-uniform cross-section post-buckled beams. *Smart Materials and Structures*, *26*(8), 085045. doi:10.1088/1361-665X/aa746e

Joseph, A. D. (2005). Works in Progress-Energy Harvesting Projects. *IEEE Pervasive Computing*, *4*(1), 69–71. doi:10.1109/MPRV.2005.8

Juang, S. S.-Y., Lin, P.-Y., Lin, Y.-C., Chen, Y.-S., Shen, P.-S., Guo, Y.-L., Wu, Y.-C., & Chen, P. (2019). Energy Harvesting Under Dim-Light Condition With Dye-Sensitized and Perovskite Solar Cells. *Frontiers in Chemistry*, *7*, 209. doi:10.3389/fchem.2019.00209 PMID:31024895

Kamalinejad, P., Mahapatra, C., Sheng, Z., Mirabbasi, S., Leung, V. C., & Guan, Y. L. (2015). Wireless energy harvesting for the internet of things. *IEEE Communications Magazine*, *53*(6), 102–108. doi:10.1109/MCOM.2015.7120024

Kansal, A., Hsu, J., Zahedi, S., & Srivastava, M. B. (2007). Power management in energy harvesting sensor networks. *ACM Transactions on Embedded Computing Systems*, *6*(4), 32. doi:10.1145/1274858.1274870

King, R. R., Law, D. C., Edmondson, K. M., Fetzer, C. M., Kinsey, G. S., Yoon, H., Sherif, R. A., & Karam, N. H. (2007). 40% efficient metamorphic GaInP/GaInAs/Ge multijunction solar cells. *Applied Physics Letters*, *90*(18), 3. doi:10.1063/1.2734507

Li, K. L., He, Q. S., Wang, J. C., Zhou, Z. G., & Li, X. X. (2018). Wearable energy harvesters generating electricity from low-frequency human limb movement. *Microsystems & Nanoengineering*, *4*(1), 4. doi:10.103841378-018-0024-3 PMID:31057912

Li, Y., & Shi, R. (2015). An intelligent solar energy-harvesting system for wireless sensor networks. *EURASIP Journal on Wireless Communications and Networking*, *2015*(1), 179. doi:10.118613638-015-0414-2

Lin, K., Yu, J., Hsu, J., Zahedi, S., Lee, D., Friedman, J., ... Srivastava, M. (2005, November). Heliomote: enabling long-lived sensor networks through solar energy harvesting. In *Proceedings of the 3rd international conference on Embedded networked sensor systems* (pp. 309-309). ACM. 10.1145/1098918.1098974

Liu, Y.-P., & Vasic, D. (2012). Self-Powered Electronics for Piezoelectric Energy Harvesting Devices. In M. Lallart (Ed.), *Small Scale Energy Harvesting* (p. 327). InTech. doi:10.5772/51211

Mica, Mica2, Mica2Dot, MicaZ, Telos. (n.d.). http://www.xbow.com/products

Nechibvute, A., Chawanda, A., & Luhanga, P. (2012). Piezoelectric Energy Harvesting Devices: An Alternative Energy Source for Wireless Sensors. *Smart Materials Research*, 1–13.

Polastre, J., Szewczyk, R., & Culler, D. (2005, April). Telos: enabling ultra-low power wireless research. In *Proceedings of the 4th international symposium on Information processing in sensor networks* (p. 48). IEEE Press.

Pozo, B., Garate, J. I., Araujo, J. Á., & Ferreiro, S. (2019). Energy Harvesting Technologies and Equivalent Electronic Structural Models - Review. *Electronics (Basel)*, *8*(5), 486. doi:10.3390/electronics8050486

Raghunathan, V., Kansal, A., Hsu, J., Friedman, J., & Srivastava, M. (2005, April). Design considerations for solar energy harvesting wireless embedded systems. In *Proceedings of the 4th international symposium on Information processing in sensor networks* (p. 64). IEEE Press 10.1109/IPSN.2005.1440973

Roundy, S. J. (2003). *Energy scavenging for wireless sensor nodes with a focus on vibration to electricity conversion* [Ph.D. dissertation]. Department of Mechanical Engineering, University of California, Berkeley, CA.

Safaei, M., Sodano, H. A., & Anton, S. R. (2019). A review of energy harvesting using piezoelectric materials: State-of-the-art a decade later (2008–2018). *Smart Materials and Structures*, *28*(11), 113001. doi:10.1088/1361-665X/ab36e4

Shin, & Inwhee. (2016). Energy management algorithm for solar-powered energy harvesting wireless sensor node for Internet of things. *IET Journals the Institution of Engineering and Technology, 10*(122), 1508–1521.

Singh, M., Muljadi, E., & Jonkman, J. (2013). Hybrid electro-mechanical simulation tool for wind turbine generators. *Proc. of IEEE Green Technologies Conf.*, 266–270. 10.1109/GreenTech.2013.48

Sudevalayam, S., & Kulkarni, P. (2011). Energy harvesting sensor nodes: Survey and implications. *IEEE Communications Surveys and Tutorials*, *13*(3), 443–461. doi:10.1109/SURV.2011.060710.00094

Taneja, J., Jeong, J., & Culler, D. (2008, April). Design, modeling, and capacity planning for micro-solar power sensor networks. In *Proceedings of the 7th international conference on Information processing in sensor networks* (pp. 407-418). IEEE Computer Society. 10.1109/IPSN.2008.67

Taylan, O., Kaya, D., Bakhsh, A. A., & Demirbas, A. (2018). Bioenergy life cycle assessment and management in energy generation. *Energy Explor. Exploit.*, *36*(1), 166 181. doi:10.1177/0144598717725871

Vermesan, O., Friess, P., & Guillemin, P. (2011). Internet of things strategic research roadmap. *Global Technological and Societal Trends*, *1*, 9–52.

Chapter 7
Internet of Things for Green Building Management:
A Survey

Muhammad Waqas Nadeem
Department of Computer Science, Lahore Garrison University, Lahore, Pakistan

Hock Guan Goh
Faculty of Information and Communication Technology (FICT), Universiti Tunku Abdul Rahman, Malaysia

Mudassar Hussain
Department of Computer Science, Lahore Garrison University, Lahore, Pakistan

Vasaki a/p Ponnusamy
Faculty of Information and Communication Technology (FICT), Universiti of Tunku Abdul Rahman, Malaysia

Muzammil Hussain
School of Systems and Technology, University of Management and Technology, Pakistan

Muhammad Adnan Khan
Department of Computer Science, Lahore Garrison University, Lahore, Pakistan, Pakistan

ABSTRACT

The expansion and development of advanced technology by mankind have a great substantial impact on the natural environment. The buildings consume a large number of natural resources during their construction, design, and operation process. Due to the manipulation of natural resources by the buildings on a large scale, there is a need of better-designed buildings for effective and efficient use of resources. The concept of "green buildings" is an innovative solution for the aforementioned issues and promotes eco-friendly activities. Internet of things (IoT) transforms the idea of green building into real. Considering a large number of IoT applications, the objective of this chapter is to survey the major IoT concepts for green building management. The survey is conducted by summarizing a large number of scientific contributions to the field (i.e., internet of things for green buildings) and presented in this study. This chapter also describes the open research challenges for future work in this emergent area.

DOI: 10.4018/978-1-7998-6709-8.ch007

INTRODUCTION

The Internet of Things (IoT) is entering the daily operations into an automatic manner of several applications and industries such as but not limited to smart homes, smart buildings, green buildings smart grids, smart cities, e-health, physical security, FinTech, asset management, logistics, etc. IoT technologies and devices provide new types of services to improve daily life. Similarly, this new technology recently developed other new types of technologies such as Cloud Computing(Armbrust et al., 2010; Mell & Grance, 2011), Big Data (Chen, Mao, & Liu, 2014), FinTech (Mehrban et al., 2020) and Energy Monitoring (Al-Turjman, Altrjman, Din, & Paul, 2019; Fulk et al., 2019; Gray, 2020; L. Han & Wanhui, 2019; Hu et al., 2019; Kowalska-Pyzalska & Byrka, 2019; Middei et al., 2019; Yang, Chen, Liu, Liu, & Chang, 2020). The concept of the smart city is emerging in several continents, where enhanced infrastructure monitoring, street lighting controls, physical security, surveillance, and public safety, meter reading, gunshot detection, and transportation analysis and optimization systems are being deployed on a city-wide scale. The concept of green buildings is related to cost-effective at user-level that begins from IoT based applications and devices (Ghosh & Ghosh, 2020; Jain, Goel, Rekhi, & Alzubi, 2020).

Buildings have a great impact on the environment in various stages. Starting from their construction, revamp, occupancy, devastation, and repurposing, buildings use raw materials, water, and energy, and produce waste, and release toxic atmospheric constituents. Thus propagates impose positive impacts on the natural environment, surroundings, and weather conditions to improve the quality of life of the residents of the buildings. The building industry consumes energy significantly. The buildings do not consume energy at construction time, but also consume high energy during whole life as well for daily life activities such as air-conditioning and lighting. This is a challenging task for civil engineers, architects, and constructors to plan resource-efficient and environment-friendly constructions. These challenges are triggered by the creation of green buildings that main goal to reduce the impact of buildings on the natural environment through sustainable design. The design of green buildings is mindful and real-world for the environment. Because green buildings generate less waste, consume less water, conserves the precise amount of energy, and use natural resources to provide a healthier and comfortable living environment to human beings. In normal construction, the method pays very little attention to the impact of toxic and non-renewable materials that are used for the construction of the building. The green buildings consider the sustainability of the environment at every stage starting from design, implementation, operation, and maintenance. The serious environmental issues and problems, the demand for the improvement of air quality constantly increase that encourages the development of green buildings at a rapid level. Firstly, to make efficient use of non-renewable resources like water, gas, and energy. To overcome this problem, the green building promotes the concept of alternate renewable energy resources such as solar energy. Several organizations, standardize the concept on green buildings at the global level, namely Green Buildings Index (GBI) of Malaysian Government, World Green Building Council (WGBC), British Building Research Establish Environment Assessment Method (BREEAM), US Green Building Council, etc. The main objectives of these standards and organizations are to facilitate durability, sustainability, economy, and comfort through the development and maintenance of green buildings. The IoT technologies perform excellently to control and maintain buildings. According to the research report of ABI, the IoT based devices and systems for green buildings is expected to grow over 64 million by 2021. The concept of green buildings with IoT evolved as connected to everything soon that present in the context of the building. Internet technologies are used for the intercommunication for IoT devices. The IoT based green buildings make sure the improvement in the Quality of Occupant Life (QOL) in the green building environment. Some of the prominent areas where IoT being implemented is like energy modeling and monitoring, management of green buildings, lighting management is discussed in this chapter.

Recently, the study of the concept "Industry 4.0" becomes the main focus of building industries to automate the buildings. Industry 4.0 is a combination of digital information and the Internet of Things (IoT) technologies that get information from different actuators and sensors sources with locations (Guoqiang, Yanming, Chao, & Yanxu, 2013; Kim, Lee, Son, & Park, 2015). In building sector industry 4.0 improve the safety ratio and reduce the consumption of energy in the appraisal of existing technology without compromising the comfort level. The IoT devices are embedded with several actuators and sensors for numerous applications that are present in the context of green buildings. The different green buildings applications are cooling, heating, air quality, load control, lighting, ventilation, natural daylighting, energy management, cooking gas management, water management, etc. Based on these, heterogeneous actuators and sensors are used in green buildings (Galinina, Mikhaylov, Andreev, Turlikov, & Koucheryavy, 2015). Furthermore, the existing controlling and sensing technologies and methodologies for green buildings are reported in (Hemant Ghayvat et al., 2015). The IoT based research for green buildings is beneficial and effective in reducing the greenhouse effect that is global warming for all continents.

The concept of green buildings is based on the Internet of Things (IoT). The IoT devices are used for controlling and monitoring of home appliances that are connected by a complex network. The wireless medium is used by the IoT devices to communicate and share adequate information among different electronic devices in the area of the building. The green building management by IoT devices and applications is an important research area. IoT devices present efficient and cost-effective solutions for the management of green buildings. The Building Management and Information System (BMIS) and IoT work together in parallels manner that is collectively called Building Internet of Things (BIoT). The BIoT controls every indoor physical device from any place of the building. The building with BIoT becomes energy-efficient, intelligent, green, and sustainable (Gaikwad, Gabhane, & Golait, 2015; Samuel, 2016).

Green building management becomes a key challenge for researchers and academia. Furthermore, based on the analysis, in recent research, various papers are presents by the authors to solve the issues that are related to IoT based green buildings management. However, a survey is needed for the existing body of knowledge related to green buildings management using IoT. The main theme of this chapter is to studies the IoT for green building management through a survey. This chapter also helps the researchers and academia to understand the several research option are available in the field of green building management using IoT. The strategies and architectures also discussed in this survey that is proposed by the researchers to improve the IoT for green building management. However, the main contribution of this chapter is to:

- The introduction of the green building concept.
- IoT to overcome and achieve several issues of green building management.
- Discuss different IoT frameworks and architectures that are existing in literature for green building management.
- IoT based energy monitoring and modeling architectures for green building management.

Green Buildings

The green buildings have various automatic operations such as ventilation, lighting, heating, etc. In earlier, these facilities are functioned individually however is smart and green buildings building their usage and action are consistent, which gives the chance to make the subsystems more accurate, reliable, and safe. It is the greatest worldwide market of IoT and almost produces a return of $8.5 billion in 2020 (Pan et al., 2015). According to a survey in the U.S, buildings are liable 38% of carbon dioxide releases, 71% electrical energy intake, 39% total energy usage.12% water intake and 40% non-industrial excess. The green building make an important effect on environmental protection, financial reserves, and energy maintenance (Pan et al., 2015)(Khanna, Arora, Chhabra, Bhardwaj, & Sharma, 2019).

Components of Green Buildings

Occupancy Sensors: The motion sensors are used to make sure of the authentic use of resources. Resources like energy, lights, and other utilities are used according to requirements with the help of sensors to enhance the savings.

Light Sensors: The light sensors manage the lights in the room according to their requirement in the context of green building. The light sensors identify a suitable light for the rooms.

Thermal Sensors: The appropriate value of the temperature is set according to the climate situation with the help of thermal sensors. Furthermore, the thermal sensors are also helpful to measure the facts and figures about the time zone to take effective decisions.

CO_2 Sensors: The surplus amount of carbon dioxide in the air is ineffective for health. The level of CO_2 is measured with the help of CO_2 sensors in residential areas.

Actuators: The actuators are used for the transformation of energy in the context of green buildings.

Medical Sensors: In the healthcare domain the proper checking and measurement of the medical parameters are necessary. So, the medical sensors are important components of green building because the timely checking of medical parameters is important for the health of residents.

Benefits

The Green Building benefits are as follows:

Improved Efficiency: The sensors that are used in the context of a building may help to explore the facts and figures about the environmental situation. The sensors also monitor the consumption and overutilization of energy and other building resources. The proper monitoring of buildings resources increases the efficiency of the building.

Easier Maintenance: The active evaluation and controlling of the modules and systems in the green buildings easily tracked and maintained. Furthermore, detectors also applied in the context of building to detect the reasons behind the malfunctioning of systems parts.

Savings: The smart control systems detect the areas of a building, where resources are underused and overused that help to maintain the best climate environment. So, the proper management of resources and energy is an important factor to save resources and energy in green buildings. (Chhabra, Vashishth, & Sharma, 2017).

Safety: The green buildings provide a safe environment for their residents. Different security features and protocols are used in green building environment to tackle emergencies such as fires, floods, earthquakes, etc.

Enable IoT Technologies For Green Buildings Management

This section thoroughly presents the latest IoT technologies that are implemented for the management of green buildings. Specifically, this chapter gives an overview of the latest IoT and green buildings based technologies, architectures, and platforms that can be applied to move the concept of green buildings in the forward direction.

Architectures For Management

There are numerous architectures and platforms are proposed by the researcher to manage the activities of the green building. Such as most of them use wireless sensors network as the main component for the infrastructure of IoT based green buildings. For instance, Syndesi is a framework that is proposed by (Evangelatos, Samarasinghe, & Rolim, 2013) which uses a wireless sensor network to create a personalized smart environment. The framework is also applicable to trace the peoples and take personalized steps (such as control of electric devices) based on preferences. This framework is deployed real-world in the premises of a building using two use-case scenarios. Furthermore, the authors proposed PhoneLab in (Nandugudi et al., 2013) which is a smartphone base testbed that gives access to the users to participate in the experiments for the collection of data. A more diverse approach presented by the authors in (Nati, Gluhak, Abangar, & Headley, 2013) that deployed testbed to mainly focus on smart and green buildings because buildings are key elements of cities in the future. The presented systems are mainly focused on to combine heterogeneous IoT devices such as programmable experimentation deployed in a real-life environment that makes the experimentations are feasible with real end-users. Consequently, researchers present user-centric IoT architectures that are feasible for the management of green buildings. Smart actuation is an also important factor in green buildings using IoT devices. So, researchers started investigating how to combine the sensory capabilities of the smartphone for smart actuation. In this context, various architectures, formworks, and techniques are introduced by the researchers using the sensing capabilities of the smartphone to automate the buildings. In (Riahi, Papaioannou, Trummer, & Aberer, 2013) the authors introduce the use of Participatory Sensing (PS) to handle the data acquisition methods for multiple PS applications efficiently produce by the sensors in the buildings. Similarly, in (Agarwal, Banerjee, Chakraborty, & Mittal, 2013) authors propose a utility-driven smartphone that executes the community-driven sensing task in the buildings. They propose a middleware framework that considers the available resources and user preferences to tune the execution of tasks passively and optimistically. An IoT testbed architecture proposed by the researchers for green buildings that enable the scalable and seamless integration of smartphones and tablets for the buildings where crowd-sourced resources are available. The architecture has dual integration capabilities. First, the architecture combines the sensory capabilities of the resources with sensing capabilities of the building for efficient smart actuation. Second, the system directly interacts with the user and provide the sensory data to the user that provide by the IoT devices on the user preferences and experience. This system is also testbed in different scenarios to examine the flexibility of the system in the premises of a building e.g. the smart luminance scenario used by the authors to evaluate the capabilities of this system.

The PS principle is used to create luminance maps and adjust the luminance units in the buildings (Angelopoulos et al., 2015). The typical diagram of the system is shown in Figure-1.

Figure 1. Typical diagram of PS based architecture (Angelopoulos et al., 2015)

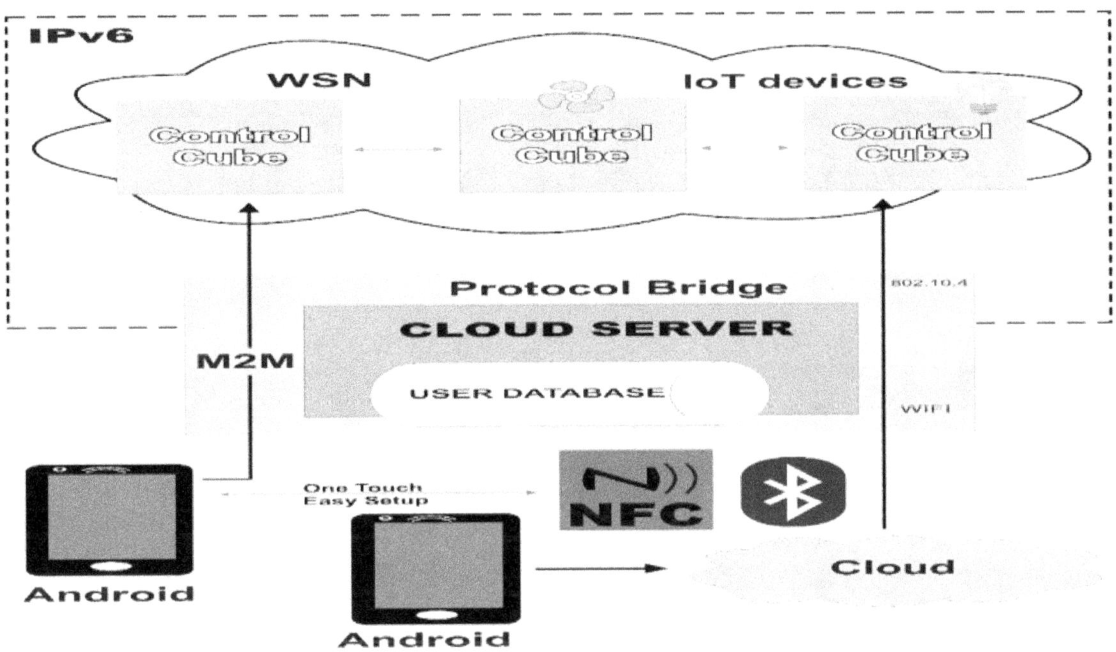

Cloud and Fog Based Technologies for Green Buildings Management

In this section, cloud and fog computing-based architecture and platforms are described briefly. The main issue of the green buildings is how these buildings operate. IoT technology has great power to automate green buildings. In this context, numerous researchers are working on automation and management of green buildings. A cloud base platform proposed by (Wang, Qiu, Dong, & Wang, 2017) for the management of the building as presented in Figure-2. The main goal of the authors is to realize the building management operation using cloud computing and the Internet of Things. The cloud servers provide a lot of services such as data storage, hosting, and computing. Cloud-based platforms for green buildings management give various functionalities such as controlling, monitoring, data processing, and service customization. Furthermore, an architecture that is developed by the integration of smart automation, IoT, and cloud computing known as "Green Cloud" present in (Singh, Sinha, & Sharma, 2020). Another prototype base architecture proposes by (Dutta & Roy, 2017) for the green buildings that use surfacing technologies like cloud, fog, and the Internet of Things as shown in Figure-3. The demand for smart and green buildings increases at a rapid level, but the main stumbling block is its high cost. To overcome this issue and improve the standard of life in buildings, and architecture with the incorporation of IoT, cloud and fog based proposed by the researchers. This architecture automate the IoT device within building through user smartphone or computer with low cost and without compromising Quality of Services.

Figure 2. Cloud-based platform for green building management (Wang et al., 2017)

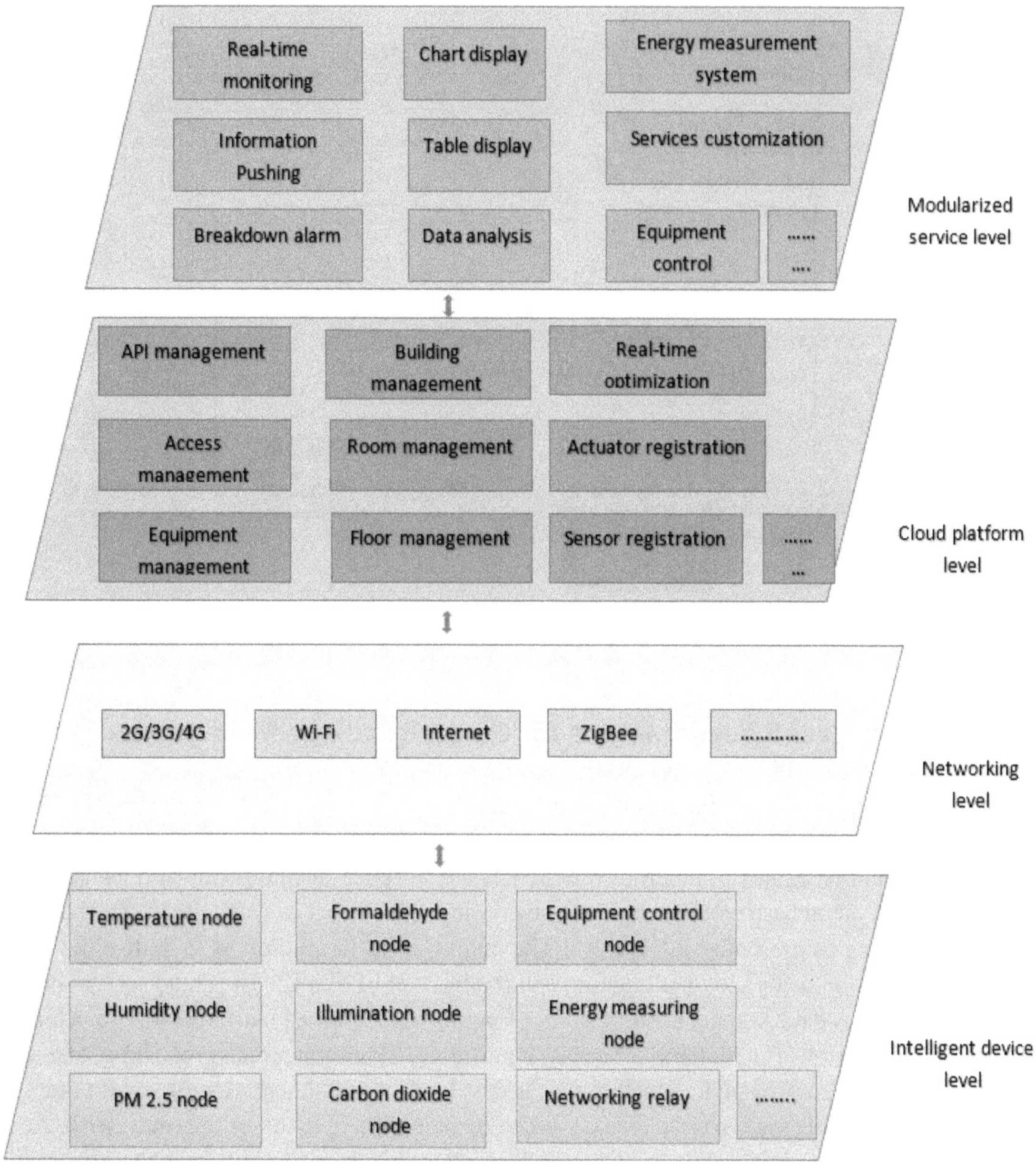

Energy Management in Green Buildings

Energy management in green buildings using the Internet of Things is an important research area because buildings consume 60% of global electricity. However, Buildings Management Systems (BMSs) are very expensive and the justification of the BMSs for small and medium-sized buildings is highly difficult. The IoT devices perform monitoring tasks in the buildings and collect a large amount of data from the different context of the buildings and then store this data into the processor of the BM System.

Figure 3. Typical diagram of fog based management architecture (Dutta & Roy, 2017)

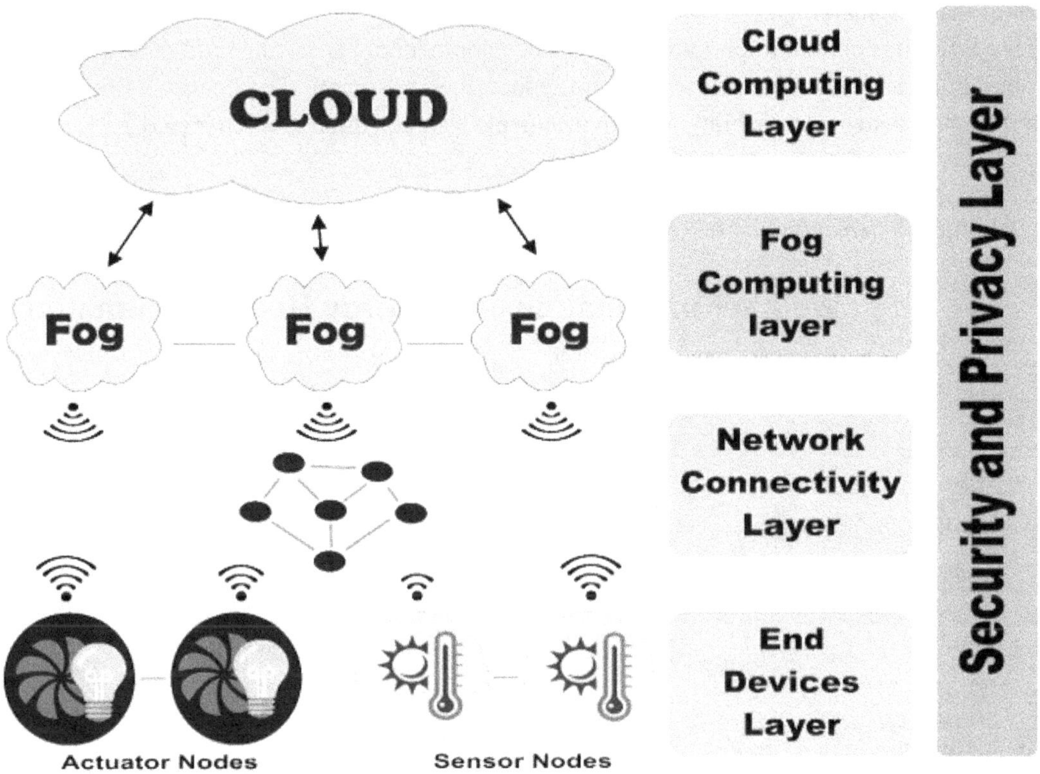

Numerous architectures and techniques are proposed by the researchers to manage the power in the green buildings. The artificial intelligence-based mechanisms used by the researchers to improve the performance of the BMSs. Furthermore, intelligence is integrated into the BMS to manage and monitor the energy consumption of the building in a cost-effective manner. The IoT based application and BMS used signal processing based techniques to separately manage the energy in the buildings. The machine learning and IoT based BMSs framework proposed by (Tushar et al., 2018) as present in Figure-4. This framework demonstrates how to extract the high-level occupancy information of the building by using low-cost IoT sensors and also analyzes the impact of human activities on the energy usages of a building. This information is used to design energy conservation measures to reduce energy consumption in the buildings. The typical diagram of the proposed framework is shown in Figure-4. Furthermore, the energy efficiency of the buildings is vital for global sustainability and the environment. LEED-gold–certificate with IoT technology used by the researchers to build an efficient and intelligent system to manage the energy in the building. The buildings with centralized and static controls may be known as "green" by design but these buildings are not energy efficient. In (Pan et al., 2015) the researchers use the "energy proportional computing" property of modern computers to propose an IoT base framework. This framework uses smart location-based automated and networked energy controls with cloud-computing and smartphone technologies to enable multiscale energy proportionality including user, building, and organizational-level energy proportionalities. The researchers also build a control system prototype and IoT network to analyze the effectiveness of the proposed framework in the real-world. The proposed framework improves the intelligence of building network, save energy, and has a great social impact in term of global sustainability.

Green Buildings play an important role in the development of harvesting renewable energy. Because green buildings use less carbon footprint and reduce the consumption of energy within the building which results in better sustainability and environment (Kanna, Ahmad, & Khan, n.d.).

Figure 4. Machine learning and IoT based platform for energy management (Tushar et al., 2018)

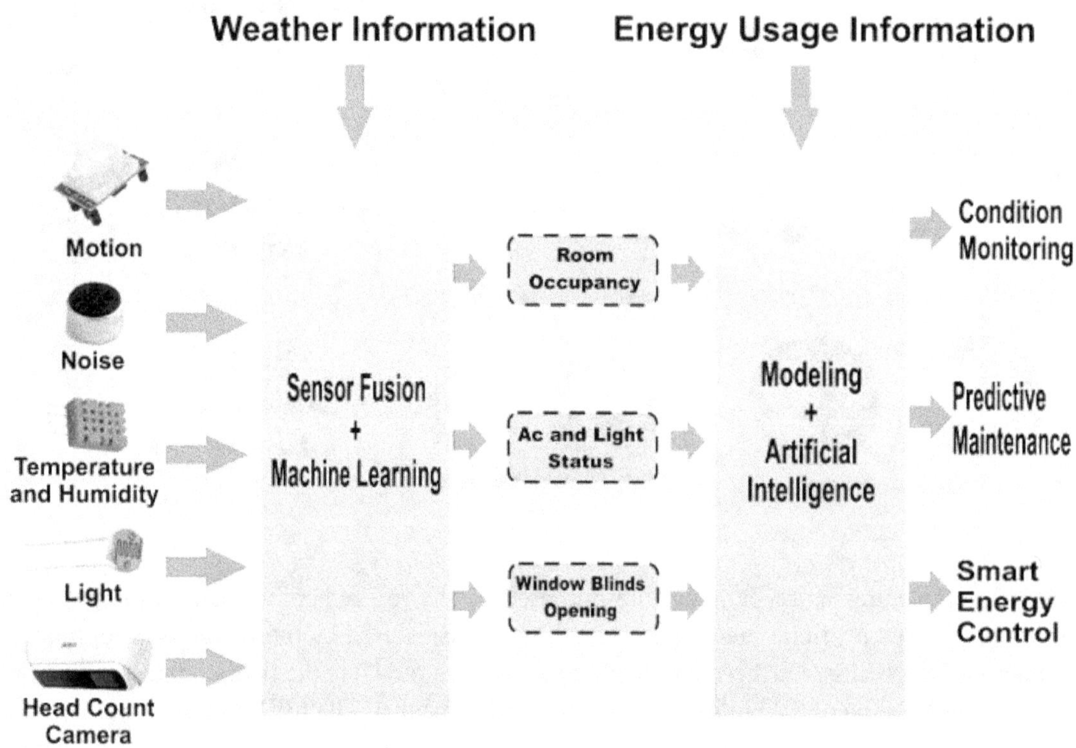

Energy Molding and Monitoring in Green Buildings

In this section, a brief overview of IoT based technologies for energy molding and monitoring in the green buildings is prescribed. Information and Communication Technologies (ICT) play a vital role to reduce the effect of greenhouse emissions on a sustainable environment. The advancement in sensing environmental and physical information, processing, and communication, enabled the monitoring of energy behavior in real-time in buildings. The IoT devices and protocols make easy access to get information for energy monitoring and molding in buildings. Consequently, the creation of Buildings Information Models (BIM) and digital repositories enabled the development of rich and complex energy models that can be used for simulation and prediction goals. However, better models are created by mixing these two information categories to detect inefficient or unwanted behaviors of the energy in the buildings. An Architecture that is the integration of IoT, BIM, meteorological services and Geographical Information System (GIS) proposed by the (Bottaccioli et al., 2017) to simulate and manage the energy behavior in the buildings. The architecture has the following properties: i) Visualize information about energy consumption in the building context and ii) measure the performance of the building by evaluating the energy modeling in real weather conditions.

IoT For Energy Monitoring

In this section, the IoT based energy monitoring technologies have described that help the researchers and industry to make efficient development of green buildings. Wireless Sensor Network (WSN) and Smart Meters are IoT based technologies that are used to monitor the energy consumptions in the buildings. Generally, these technologies allow realizing seamless, reliable, energy-efficient, and low-cost remote monitoring and control solutions in smart applications. The smart meters are advanced internet-connected devices that can give information for monitoring, controlling, and billing purposes (An, 2011). Furthermore, the information is also useful to handle systems (form the operator's point of view) to monitor, model, and prevent contingencies, unconventional, and faults behavior. Generally, the core functionalities of the smart meters are i) recording/data collection, ii) two-way communication, iii) load control (Bahmanyar et al., 2016), and programming capabilities. Consequently, the most common communication technologies are i) Power Line Communication (PLC) and ii) Radio Frequency (RF) (Zheng, Gao, & Lin, 2013). The most innovative uses of the IoT based devices are the Non-Intrusive Appliance Load Monitoring (NIALM), this is known as a disaggregation signal processing technique (Zoha, Gluhak, Imran, & Rajasegarar, 2012). The NIALM techniques are used to detect energy consumption of appliances form a signal point of measurement (i.e. smart meter). The WSN is known as a network of devices that are used to gather information form a monitored area (Rawat, Singh, Chaouchi, & Bonnin, 2014). The WSN works through wireless links. The collected data are sent to the concentrator that uses this data locally or send to the other network (e.g. Internet through a gateway). The WSN is divided into three different subsets (Fadel et al., 2015): Home Area Networks (HANs) Neighborhood Area Network (NANs) and Wide Area Network (WANs), these networks are used in smart and green buildings for the communication of IoT devices. For example, the HANs network establishes a communication path between home appliances, plug-in devices, and smart meters in the context of the building. These networks collect environmental information from the end-users where they live. Furthermore, the HANs based applications need low-cost communication protocols and technologies (e.g. Wi-Fi, Spirit, ZigBee, and Bluetooth) due to low bandwidth and power requirements. Differently, the NANs based network is established to collect data from smart meters in a neighborhood area. Usually, the NANs network is good for short-range communication typologies such as RF mesh technologies and Wi-Fi (H Ghayvat, Mukhopadhyay, & Gui, 2016; X. Han & Zhao, 2014; Kumar & Hancke, 2014; Lloret, Macías, Suárez, & Lacuesta, 2012; Rahman & Shah, 2016; Song, Han, Zhang, & Yang, 2012; Trinchero et al., 2011) and they collect data from smart meters and then transmit gather data to a data concentrator. Lastly, the WAN network is used to establish a communication path among data concentrator and data services provider. Generally, all these networks are based on LTE, mobile network (such as 2G or 3G standards), power lines communication network, and fiber. The deployment of WSN based technologies in the context of green buildings gives several benefits that include lower costs, reliability, scalability, flexibility, and accuracy in a wide range of IoT applications (Bottaccioli et al., 2017). The heterogeneous integration of

IoT devices in the buildings introduce a lack of interoperability among them. This is a serious problem in the context of IoT based energy monitoring devices. To overcome this problem, in (Patti, Acquaviva, & Macii, 2013) a middleware approach proposed by the authors that abstract the hardware functionalities into web services. This approach also allows us to use unified API (Application Programming Interface) for the transparent access to IoT resources.

Lighting Management

The efficient use of energy in green buildings plays a vital role in global sustainability. Different devices such as air conditioners, lights, sensors, etc. that are present in the context of green buildings consume a lot of energy due to centralized and static energy control on the devices. Among these energy consumers devices lights that are present in the area of green buildings consume a large amount of energy. Hence, the management of the lights is necessary for the buildings to reduce the consumption of energy by lights. In (Jeyasheeli & Selva, 2017) the authors present a LEDs based design that uses environmental sources such as sunlight and temperature to dynamically change the power supply among the lights. This design uses smart autonomous control and improves energy efficiency through the dynamical distribution of energy between the clusters of lights.

CONCLUSION AND FUTURE DIRECTIONS

This section gives a summarized form of research in the area of green building management using IoT. An extensive survey is present regarding green buildings management using IoT in this chapter. This chapter briefly describes IoT based architectures and frameworks for the management of green buildings. Cloud and Fog computing based architectures and platforms also present in this chapter. The lighting and energy management, monitoring, and modeling techniques for green buildings are also described. This chapter helps the authors and researchers to understand the concept of green buildings using IoT and fined new areas for research related to green building management. Several researchers present their ideas for the management of the individual context of building e.g. energy management, lighting management, etc. but there is a need for a complete management system or architecture that performs collective management of all the context of the building. Furthermore, there is a need for a dynamic management system instead of static or centralized. Since more, the world is moving toward Artificial Intelligence (AI) based systems in almost every field so, the AI-based systems are also required for the management of green buildings.

REFERENCES

Agarwal, V., Banerjee, N., Chakraborty, D., & Mittal, S. (2013). USense--A Smartphone Middleware for Community Sensing. In *2013 IEEE 14th International Conference on Mobile Data Management* (Vol. 1, pp. 56–65). IEEE.

Al-Turjman, F., Altrjman, C., Din, S., & Paul, A. (2019). Energy monitoring in IoT-based ad hoc networks: An overview. *Computers & Electrical Engineering*, *76*, 133–142. doi:10.1016/j.compeleceng.2019.03.013

An E.-A.-U. (2011). *A Discussion of Smart Meters And RF Exposure Issues*. Academic Press.

Angelopoulos, C. M., Evangelatos, O., Nikoletseas, S., Raptis, T. P., Rolim, J. D. P., & Veroutis, K. (2015). A user-enabled testbed architecture with mobile crowdsensing support for smart, green buildings. In *2015 IEEE International Conference on Communications (ICC)* (pp. 573–578). IEEE. 10.1109/ICC.2015.7248383

Armbrust, M., Fox, A., Griffith, R., Joseph, A. D., Katz, R., Konwinski, A., ... Stoica, I. (2010). A view of cloud computing. *Communications of the ACM*, *53*(4), 50–58. doi:10.1145/1721654.1721672

Bahmanyar, A., Jamali, S., Estebsari, A., Pons, E., Bompard, E., Patti, E., & Acquaviva, A. (2016). Emerging smart meters in electrical distribution systems: Opportunities and challenges. In *2016 24th Iranian Conference on Electrical Engineering (ICEE)* (pp. 1082–1087). IEEE.

Bottaccioli, L., Aliberti, A., Ugliotti, F., Patti, E., Osello, A., Macii, E., & Acquaviva, A. (2017). Building energy modelling and monitoring by integration of IoT devices and building information models. In *2017 IEEE 41st Annual Computer Software and Applications Conference (COMPSAC)* (Vol. 1, pp. 914–922). IEEE. 10.1109/COMPSAC.2017.75

Chen, M., Mao, S., & Liu, Y. (2014). Big data: A survey. *Mobile Networks and Applications*, *19*(2), 171–209. doi:10.100711036-013-0489-0

Chhabra, A., Vashishth, V., & Sharma, D. K. (2017). Seir: A stackelberg game based approach for energy-aware and incentivized routing in selfish opportunistic networks. In *2017 51st Annual Conference on Information Sciences and Systems (CISS)* (pp. 1–6). IEEE. 10.1109/CISS.2017.7926113

Chhabra, A., Vashishth, V., & Sharma, D. K. (2018). A fuzzy logic and game theory-based adaptive approach for securing opportunistic networks against blackhole attacks. *International Journal of Communication Systems*, *31*(4), e3487. doi:10.1002/dac.3487

Dutta, J., & Roy, S. (2017). IoT-fog-cloud based architecture for smart city: Prototype of a smart building. In *2017 7th International Conference on Cloud Computing, Data Science & Engineering-Confluence* (pp. 237–242). IEEE.

Evangelatos, O., Samarasinghe, K., & Rolim, J. (2013). Syndesi: A framework for creating personalized smart environments using wireless sensor networks. In 2013 IEEE international conference on distributed computing in sensor systems (pp. 325–330). IEEE. doi:10.1109/DCOSS.2013.35

Fadel, E., Gungor, V. C., Nassef, L., Akkari, N., Malik, M. G. A., Almasri, S., & Akyildiz, I. F. (2015). A survey on wireless sensor networks for the smart grid. *Computer Communications*, *71*, 22–33. doi:10.1016/j.comcom.2015.09.006

Fulk, C., Hobar, G., Olsen, K., El-Tawab, S., Rahman, F., & Ghazizadeh, P. (2019). Cloud-based Low-cost Energy Monitoring System through the Internet of Things. In *2019 IEEE International Conference on Pervasive Computing and Communications Workshops (PerCom Workshops)* (pp. 322–327). IEEE. 10.1109/PERCOMW.2019.8730823

Gaikwad, P. P., Gabhane, J. P., & Golait, S. S. (2015). A survey-based on the Smart Homes system using Internet-of-Things. In *2015 International Conference on Computation of Power, Energy, Information, and Communication (ICCPEIC)* (pp. 330–335). IEEE. 10.1109/ICCPEIC.2015.7259486

Galinina, O., Mikhaylov, K., Andreev, S., Turlikov, A., & Koucheryavy, Y. (2015). Smart home gateway system over Bluetooth low energy with wireless energy transfer capability. *EURASIP Journal on Wireless Communications and Networking*, *2015*(1), 178. doi:10.118613638-015-0393-3

Ghayvat, H., Liu, J., Babu, A., Alahi, E. E., Gui, X., & Mukhopadhyay, S. C. (2015). Internet of Things for smart homes and buildings: Opportunities and Challenges. *Journal of Telecommunications and the Digital Economy*, *3*(4), 33–47. doi:10.18080/jtde.v3n4.23

Ghayvat, H., Mukhopadhyay, S. C., & Gui, X. (2016). Issues and mitigation of interference, attenuation, and direction of arrival in IEEE 802.15. 4/ZigBee to wireless sensors and networks based smart building. *Measurement*, *86*, 209–226. doi:10.1016/j.measurement.2016.01.045

Ghosh, P., & Ghosh, S. (2020). IoT and Machine Learning in Green Smart Home Automation and Green Building Management. *Journal of Alternate Energy Sources and Technologies*, *10*(3), 8–36.

Gray, A. R. (2020, February 4). *Automated monitoring for changes in energy consumption patterns*. Google Patents.

Guoqiang, S., Yanming, C., Chao, Z., & Yanxu, Z. (2013). Design and implementation of a smart IoT gateway. In *2013 IEEE International Conference on Green Computing and Communications and IEEE Internet of Things and IEEE Cyber, Physical and Social Computing* (pp. 720–723). IEEE. 10.1109/GreenCom-iThings-CPSCom.2013.130

Han, L., & Wanhui, H. A. N. (2019, May 2). *Electric automobile energy monitoring and swapping network in remote monitoring of cloud computing network*. Google Patents.

Han, X., & Zhao, C. (2014). Distributing monitor system based on WIFI and GSM supporting SCPI. In *2014 13th International Symposium on Distributed Computing and Applications to Business, Engineering, and Science* (pp. 272–274). IEEE. 10.1109/DCABES.2014.59

Hu, W., Liu, J., Zhu, Y., Dong, J., Ma, X., Li, S., Zhang, J., Zhu, X., & Chen, W. (2019). Analysis of energy monitoring for a double-pulsed CO 2 integrated path differential absorption lidar at 1.57 μm. *Applied Optics*, *58*(3), 616–625. doi:10.1364/AO.58.000616 PMID:30694245

Jain, R., Goel, V., Rekhi, J. K., & Alzubi, J. A. (2020). IoT-Based Green Building: Towards an Energy-Efficient Future. In Green Building Management and Smart Automation (pp. 184–207). IGI Global.

Jeyasheeli, P. G., & Selva, J. V. J. (2017). An IOT design for smart lighting in green buildings based on environmental factors. In *2017 4th International Conference on Advanced Computing and Communication Systems (ICACCS)* (pp. 1–5). IEEE. 10.1109/ICACCS.2017.8014559

Kanna, P. R., Ahmad, S. S., & Khan, M. A. U. (n.d.). *A Study of Green Building-Renewable Energy Harvesting Technique*. Academic Press.

Khanna, A., Arora, S., Chhabra, A., Bhardwaj, K. K., & Sharma, D. K. (2019). IoT architecture for preventive energy conservation of smart buildings. In *Energy Conservation for IoT Devices* (pp. 179–208). Springer. doi:10.1007/978-981-13-7399-2_8

Kim, J. Y., Lee, H.-J., Son, J.-Y., & Park, J.-H. (2015). Smart home web of objects-based IoT management model and methods for home data mining. In *2015 17th Asia-Pacific Network Operations and Management Symposium (APNOMS)* (pp. 327–331). IEEE. 10.1109/APNOMS.2015.7275448

Kowalska-Pyzalska, A., & Byrka, K. (2019). Determinants of the willingness to energy monitoring by residential consumers: A case study in the city of Wroclaw in Poland. *Energies*, *12*(5), 907. doi:10.3390/en12050907

Kumar, A., & Hancke, G. P. (2014). Energy efficient environment monitoring system based on the IEEE 802.15. 4 standard for low cost requirements. *IEEE Sensors Journal*, *14*(8), 2557–2566. doi:10.1109/JSEN.2014.2313348

Lloret, J., Macías, E., Suárez, A., & Lacuesta, R. (2012). Ubiquitous monitoring of electrical household appliances. *Sensors (Basel)*, *12*(11), 15159–15191. doi:10.3390121115159 PMID:23202205

Mehrban, S., Nadeem, M. W., Hussain, M., Ahmed, M. M., Hakeem, O., Saqib, S., ... Khan, M. A. (2020). Towards Secure FinTech: A Survey, Taxonomy, and Open Research Challenges. *IEEE Access: Practical Innovations, Open Solutions*, *8*, 23391–23406. doi:10.1109/ACCESS.2020.2970430

Mell, P., & Grance, T. (2011). *The NIST definition of cloud computing*. Academic Press.

Middei, R., Bianchi, S., Petrucci, P. O., Ursini, F., Cappi, M., De Marco, B., ... Matt, G. (2019). High-energy monitoring of NGC 4593 II. Broad-band spectral analysis: Testing the two-corona model. *Monthly Notices of the Royal Astronomical Society*, *483*(4), 4695–4705. doi:10.1093/mnrasty3379

Nandugudi, A., Maiti, A., Ki, T., Bulut, F., Demirbas, M., Kosar, T., . . . Challen, G. (2013). Phonelab: A large programmable smartphone testbed. In *Proceedings of First International Workshop on Sensing and Big Data Mining* (pp. 1–6). Academic Press.

Nati, M., Gluhak, A., Abangar, H., & Headley, W. (2013). Smartcampus: A user-centric testbed for the internet of things experimentation. In *2013 16th International Symposium on Wireless Personal Multimedia Communications (WPMC)* (pp. 1–6). IEEE.

Pan, J., Jain, R., Paul, S., Vu, T., Saifullah, A., & Sha, M. (2015). An internet of things framework for smart energy in buildings: Designs, prototype, and experiments. *IEEE Internet of Things Journal*, *2*(6), 527–537. doi:10.1109/JIOT.2015.2413397

Patti, E., Acquaviva, A., & Macii, E. (2013). Enable sensor networks interoperability in smart public spaces through a service-oriented approach. In *5th IEEE International Workshop on Advances in Sensors and Interfaces IWASI* (pp. 2–7). IEEE. 10.1109/IWASI.2013.6576081

Rahman, R. A., & Shah, B. (2016). Security analysis of IoT protocols: A focus in CoAP. In *2016 3rd MEC international conference on big data and smart city (ICBDSC)* (pp. 1–7). IEEE. 10.1109/ICBDSC.2016.7460363

Rawat, P., Singh, K. D., Chaouchi, H., & Bonnin, J. M. (2014). Wireless sensor networks: A survey on recent developments and potential synergies. *The Journal of Supercomputing*, *68*(1), 1–48. doi:10.100711227-013-1021-9

Riahi, M., Papaioannou, T. G., Trummer, I., & Aberer, K. (2013). Utility-driven data acquisition in participatory sensing. In *Proceedings of the 16th International Conference on Extending Database Technology* (pp. 251–262). 10.1145/2452376.2452407

Samuel, S. S. I. (2016). A review of connectivity challenges in IoT-smart home. In *2016 3rd MEC International conference on big data and smart city (ICBDSC)* (pp. 1–4). IEEE. 10.1109/ICBDSC.2016.7460395

Singh, A., Sinha, U., & Sharma, D. K. (2020). Cloud-Based IoT Architecture in Green Buildings. In *Green Building Management and Smart Automation* (pp. 164–183). IGI Global. doi:10.4018/978-1-5225-9754-4.ch008

Song, Y., Han, B., Zhang, X., & Yang, D. (2012). Modeling and simulation of smart home scenarios based on the Internet of Things. In *2012 3rd IEEE international conference on network infrastructure and digital content* (pp. 596–600). IEEE. 10.1109/ICNIDC.2012.6418824

Trinchero, D., Stefanelli, R., Brunazzi, D., Casalegno, A., Durando, M., & Galardini, A. (2011). Integration of smart house sensors into a fully networked (web) environment. In *SENSORS, 2011 IEEE* (pp. 1624–1627). IEEE. doi:10.1109/ICSENS.2011.6127374

Tushar, W., Wijerathne, N., Li, W.-T., Yuen, C., Poor, H. V., Saha, T. K., & Wood, K. L. (2018). *Iot for green building management.* arXiv Preprint arXiv:1805.10635.

Wang, M., Qiu, S., Dong, H., & Wang, Y. (2017). Design an IoT-based building management cloud platform for green buildings. In *2017 Chinese Automation Congress (CAC)* (pp. 5663–5667). IEEE. 10.1109/CAC.2017.8243793

Yang, C.-T., Chen, S.-T., Liu, J.-C., Liu, R.-H., & Chang, C.-L. (2020). On construction of an energy monitoring service using big data technology for the smart campus. *Cluster Computing*, *23*(1), 265–288. doi:10.100710586-019-02921-5

Zheng, J., Gao, D. W., & Lin, L. (2013). Smart meters in the smart grid: An overview. In 2013 IEEE Green Technologies Conference (GreenTech) (pp. 57–64). IEEE.

Zoha, A., Gluhak, A., Imran, M. A., & Rajasegarar, S. (2012). Non-intrusive load monitoring approaches for disaggregated energy sensing: A survey. *Sensors (Basel)*, *12*(12), 16838–16866. doi:10.3390121216838 PMID:23223081

Chapter 8
Challenges and Issues of IoT Application in Heating Ventilating Air Conditioning Systems:
Energy Conservation Using IoT

Kapilan N.
Nitte Meenakshi Institute of Technology, Bangalore, India

Vidhya P.
SRSMN Government First Grade College, Barkur

ABSTRACT

The internet of things (IoT) is a system consisting of computing, mechanical, and electronic devices, which are having ability to transfer data in network without human interaction. The sensors used in IoT collect and transfer the data to the cloud, which is further processed using software to perform an action. The IoT is one of the fastest growing industries, and in recent years, it is most widely used in HVAC systems in residential and commercial applications to reduce the energy consumption as building consumes by approximately 40% of total energy. The IoT reduces the energy consumption of the building by optimizing the process variables of HVAC system components, increases life of system components, enhances the comfort of the occupants, and provides remote control of the system. However, there are challenges in data security and privacy, and also there is a lack of IoT platforms specifically oriented towards the proper processing, management, and analysis of such large and diverse data.

DOI: 10.4018/978-1-7998-6709-8.ch008

Figure 1. Schematic of the HVAC system

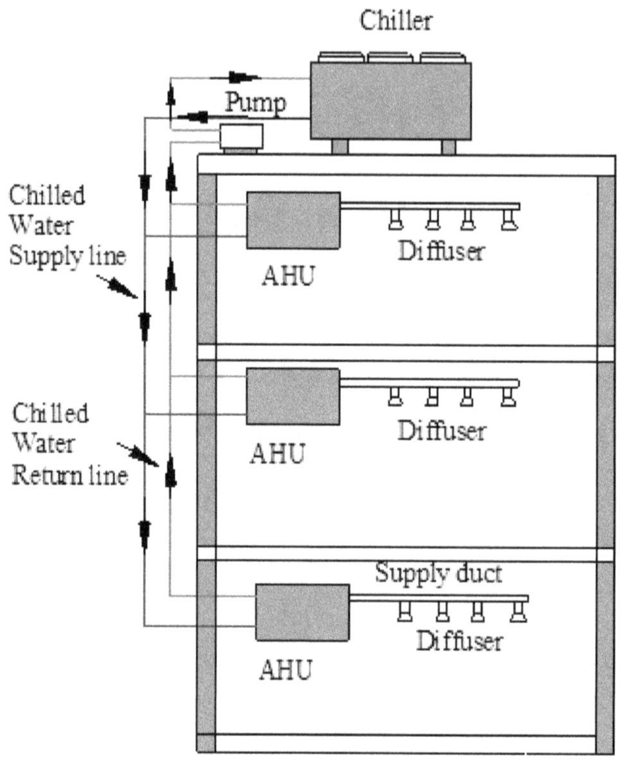

INTRODUCTION

The heating, ventilation, and air conditioning (HVAC) system is the mechanical air handling systems used in building to provide necessary indoor air quality by controlling temperature, humidity, purity, velocity etc. by conditioning the air to provide comfort indoor space for the occupants. The indoor air is replaced with outdoor air at regular intervals to dilute and remove contaminants and to remove odour. The infiltration or exfiltration may affect the quality of the indoor air. The important parts of the HVAC are chiller, air handling unit, air distribution system, air exhaust system etc. Figure 1 shows the schematic of the HVAC system. The chiller produces the cooling effect and it is transferred to the water. This chilled water is circulated to the air handling unit (AHU) using a pump. The room air is drawn to the AHU and is filtered to remove dusts, mold etc., cooled using cooling coil, humidified or dehumidified and finally circulated to the room. The humidifier and dehumidifier control the humidity of the air. In winter, heating coil is used to heat the air. Some percentage of room air is replaced with the outdoor air to control the carbon dioxide level. The air distribution system consists of fans, filters, dampers, ductwork etc. supplies the conditioned air to the room to maintain required indoor conditions. The return air system draws the room air and is send it to the AHU for conditioning. This return air is mixed with outdoor air and is re-conditioned using filter, cooling coil, humidifier / dehumidifier and re-circulated to the indoor space. Figure 2 shows the schematic of the AHU.

Figure 2. Schematic of the AHU system

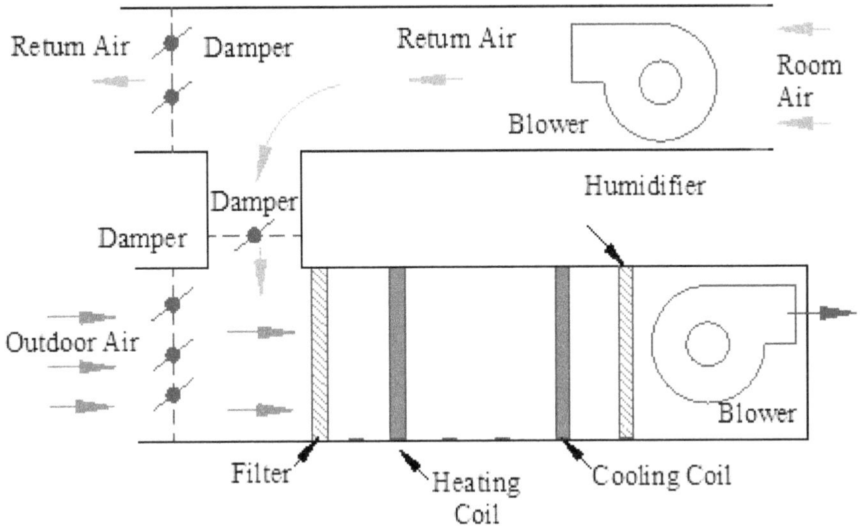

BACKGROUND

The figure 3 shows the energy consumption by different equipment used in commercial building. It shows that the major energy is consumed for space heating, ventilation, space cooling and lighting (Bawaneh et al., 2019). The HVAC system is available in different capacity and selection is based on application and building design. It consumes 40 to 50% of the total energy consumed by the building. The HVAC

Figure 3. Energy consumption in commercial building

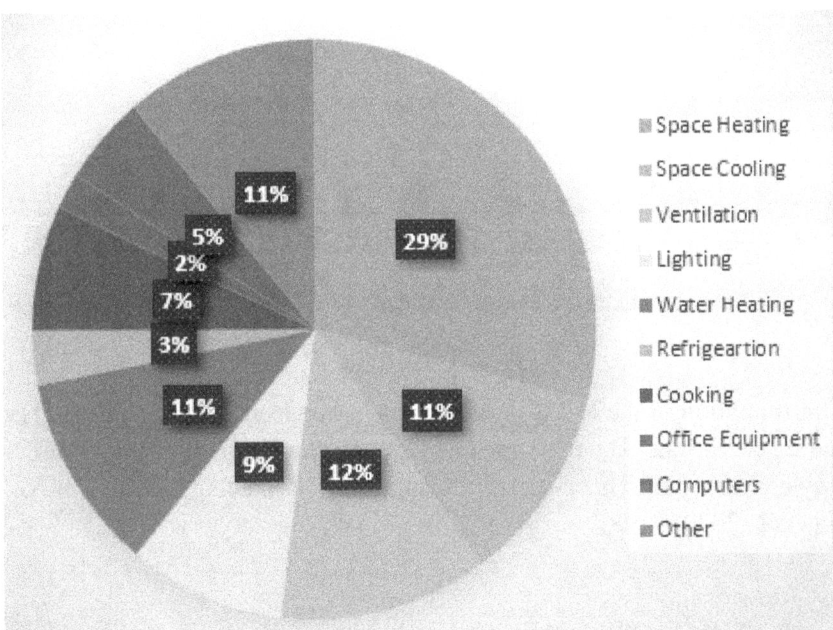

systems are most widely used in commercial building such as IT Park, hotels, resorts, office, malls, airports, railway station, educational institutions, hospitals, etc.

The climate condition, characteristics of building and occupancy rate and patterns significantly affects the HVAC energy demand. The HVAC systems consume major amount of the building energy in developed countries like, USA, Canada, European Union, Australia etc. The HVAC demand of the world is increasing exponentially, which may cause serious strains on electrical grids. The increase in HVAC demand is higher in developing nations such as India and China and countries with hot climate such as Middle East countries. The International Energy Agency (IEA) report estimated that the world energy demand from HVAC systems may triple by 2050. The commercial refrigeration systems consume about 40 to 60% of the energy consumed in the supermarkets. Hence the HVAC and refrigeration systems may cause strain on electrical grids and IEA suggest new efficiency standard, efficiency improvement etc. to reduce the energy demand. The increase in energy demand adds new power plants which increases the CO_2 emission. Hence it is necessary to introduce a new energy efficient technology to reduce the energy consumption and also to augment indoor comfort of the building occupants (International Energy Agency, 2020). Figure 4 shows the building energy consumption in developed and developing countries such as USA, EU and China. From this figure, we observe that the space cooling and space heating and ventilation consumes more than 50% of the building energy.

Figure 4. Building energy consumption pattern in different countries

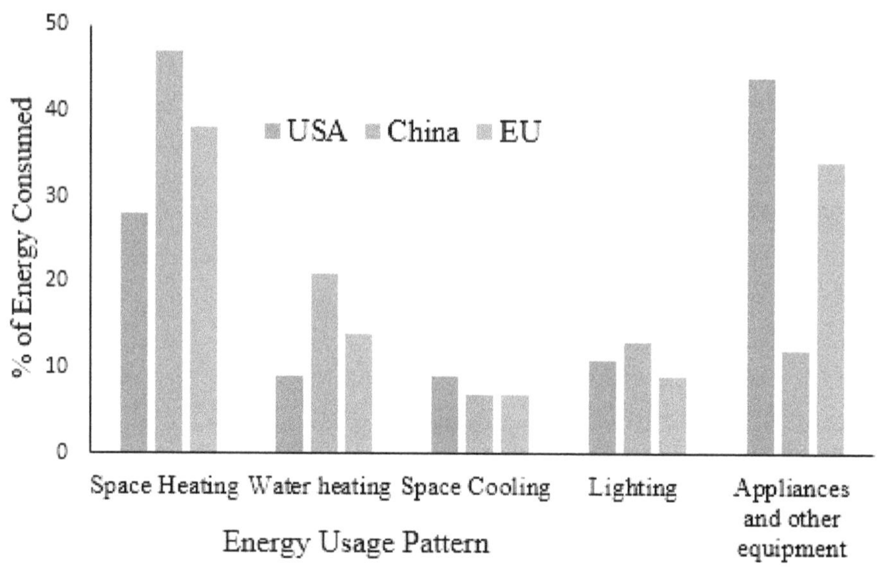

The energy consumption of the HVAC systems can be reduced by operating the HVAC systems at optimized operating conditions. The real time energy consumption monitoring of building will help to reduce the energy consumption. There is a significant potential in energy savings of commercial building (Khalilnejad et al., 2020).

ISSUES

Presently, the following issues are faced with the HVAC systems

1. The buildings consume 59% of global energy consumption and HVAC system consumes about 48% of the energy consumed by the building. A marginal improvement in HVAC system's operating efficiency will cause significant energy savings. To achieve energy efficient HVAC systems in buildings it is essential to enhance the designs of their various integrated mechanical and electrical components. More efficient HVAC systems will lead to a significant reduction in power consumption in buildings (Zhuang et al., 2020).
2. The optimization of the HVAC system performance parameters will help to reduce the energy consumption and hence there is a significant potential in energy savings (PSM Market Research, 2020).
3. The occupants' comfort in terms of temperature and humidity is important and hence it is necessary to operate different components of the HVAC system as per the comfort requirements of the occupants.
4. The poor indoor air quality (IAQ) results in fatigue among the occupants. The poor ventilation, lack of fresh outside air, bad odour, water leakage, contaminated air, etc. are the reasons for the poor IAQ.
5. It is necessary to monitor the performance of the HVAC components so that preventive maintenance can be carried out. It will assist in breakdown free operation of HVAC systems and to enhance life of the components of the HVAC system.

Hence in recent years, IoT system are integrated with HVAC system to overcome the issues and to increase the smartness of the HVAC systems and help the plant engineer or facility manager of the building.

The IoT systems are most widely used in various applications such as livestock monitoring in poultry and farming sector, healthcare applications, energy and power sector, retail sector, agriculture, smart city projects, automotive digital technology, wearable devices etc. An IoT system consists of sensors which collect the data from the equipment or devices and send it to the cloud for processing using software and to perform an action. The interrelated computing devices, mechanical and digital machines etc. are provided with unique identifiers in the IoT system and data is transferred to cloud network without any manual intervention. In cloud, software processes the data and performs an action such as sending an alert to the user or automatically turns on or off the device without the help of the user (Wikipedia, 2020).

It is projected that the installed base of IoT connected devices in the world to reach 75.44 billion by 2025. This increase will be fivefold in 10 years. Figure 5 shows the IoT connected devices installed worldwide from 2015 to 2025 (Statista, 2020).

IoT Key Features

The important key features of the IoT are as follows (LinkedIn, 2020)

1. The IoT system are becoming Intelligent due to AI algorithm and computing facilities which enhances the data collection capacity. The AI based algorithm which helps in analyzing data and to take immediate action.

Figure 5. IoT connected devices from 2015 to 2025 (in billions)

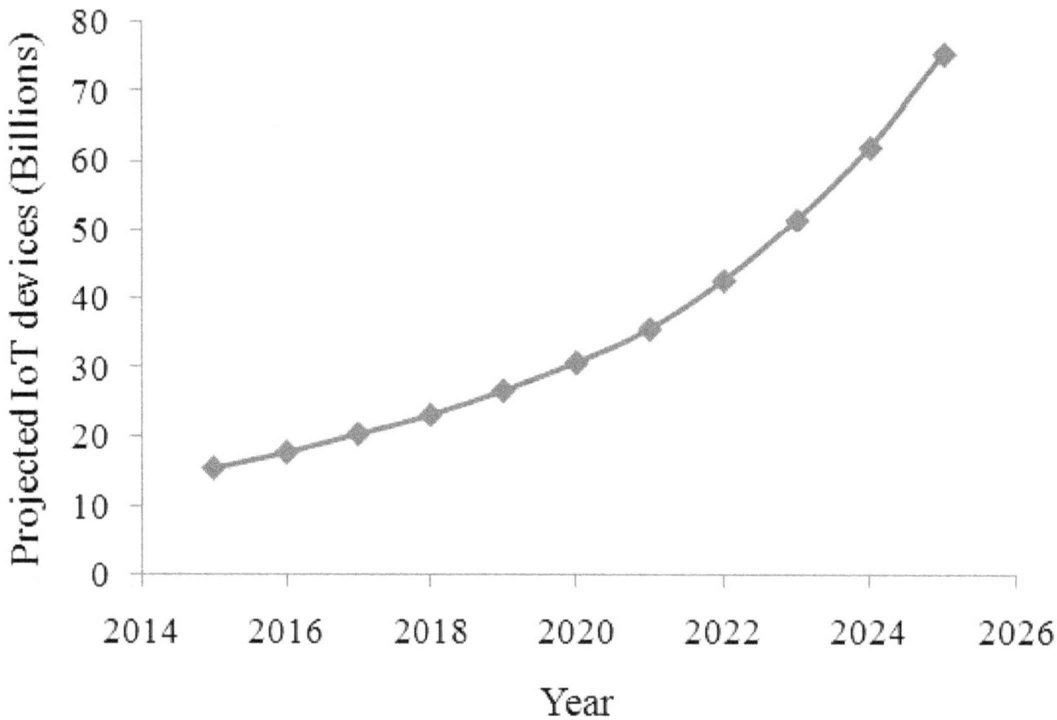

2. The IoT system has higher sensing capability due to improvement in sensor development. This helps the system to be active and increases its capability in real time integration.
3. The IoT connectivity helps in accessibility is getting on a network and compatibility produces the necessary data.
4. The IoT provides dynamic response due to AI algorithm and networking
5. The IoT is provided with sensors, actuators and processors which interact with each other and hence scalability is not problem.
6. The IoT security features help to reduce hacking of the IoT components.

Components of Internet of Things

The major components of IoT are as follows

1. The sensors and actuators are important parts of IoT. The sensor is a device which detect changes in the environment. These changes are converted into electrical impulses and is analyzed. The actuator introduces changes such as controlling of valves based on the signals received from the sensors.
2. The gateway is a hardware which executes the tasks using application software provided in it. It connects the sensors and the cloud. It receives the signal from the sensor and transfers it to the cloud.

3. The IoT cloud consists of number of cloud services and standards. It helps in connecting, controlling and managing various IoT components and applications.
4. The IoT Analytics is the data analysis software which analyze the data received from various IoT sensors.
5. User interface helps the devices to interact with each other and assist in data transmission to higher level system.

Working Principle of IOT

An IoT system consists of sensors which sense the data such as temperature, humidity, velocity, noise etc. of a room and transfer it to the cloud through proper connectivity. The data is processed by software in the cloud and action such as sending an alert to the user's mobile / computer or automatically adjusting the devices so that required conditions are maintained in the room. However, if the user input is required, then the alert is sent to the user and requesting to take corrective action based on the data analyzed. The action taken by the user is sent in the opposite direction, i.e. from cloud to the sensor and action is initiated. Also, the IoT system takes real time data and takes suitable action. For example, an air conditioning system should be automatically turn on by 6 pm for the home owner when he returned from office to the home. If that house owner struck in traffic and he may reach home by one hour late, then the IoT system can use the real time location of the person using google data and delay the turn on of the air-conditioning system by one hour without any external input. This saves lot of energy and does not need the action of the person who uses IoT system. The following steps to be considered while designing an IoT system.

1. The selection of IoT hardware and software based on the application and performance requirements. The required hardware and software are designed and integrated to perform the task. The capacity of the IoT system depends upon the data acquisition, connectivity and management.
2. The IoT systems can be remote controlled using mobiles / computers. This helps in remote monitoring and control of the process and device.
3. The networking protocols should be followed during Integrating various devices as it reduces or eliminates human to computer interaction.
4. The IoT hardware, connectivity and software should have a strong security features to avoid hacking. The IoT security protect the internet enabled devices that connect on wireless networks.

Figure 6 shows these components and the working principle of IoT system (Data Flair training, 2020). In recent years, smaller companies or manufacturers also started using IoT in their products to meet the demand of the customers because the use of IoT in mechanical systems offers predictive maintenance and better services and increase life-span of the mechanical components. It is also useful to the original equipment manufacturers to stay connected with clients and increase their satisfaction by continuously improving the quality of the products.

Types of IoT Connection Methods

The different types (All about circuits, 2020) of common IoT connection methods are

Figure 6. Major components and working of IoT system

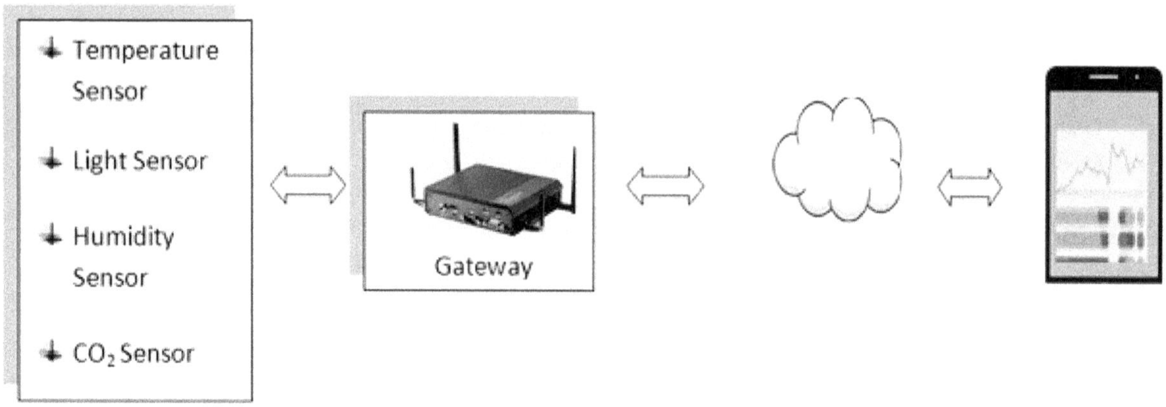

1. **Ethernet**: In IoT, ethernet is used to connect fixed and stationary devices of the IoT system. It can be used as a wired medium to connect computers and to provide the Internet connectivity.
2. **Wi-Fi**: It is a wireless protocol and is a replacement to ethernet. It is easy to fix and easy to use shortrange wireless connectivity. It supports narrowband and broadband IoT applications in a common platform.
3. **Low Power WAN:** It is a type of wireless network and is preferred when low power, long range data communications. It has data rates lower than 100Kbps and hence not comparable to broadband and preferred in data transmission from sensors used in IoT systems.
4. **Cellular:** It connects the sensors used in the IoT system to the internet using mobile networks, such as 2G, 3G or 4 G. It uses M2M simcards and suitable connectivity to meets the application of IoT. It is designed to meet the requirement of low power and long-range applications.
5. **Satellite:** It uses mobile satellite network and provides flexibility, speed and better connectivity.
6. **Bluetooth:** Bluetooth technology can be used in IoT applications. However, it cannot be used in every IoT system. The programming is easy and is compatible with other devices.

IoT Platforms for Building IoT Projects

The different types of IoT platforms are used in building IoT projects. An open source IoT platform which provides different services is Kaa. It is easy and fast and hence used by the developers to build complex IoT projects. The integration of Arduino, Raspberry and other platforms with different Internet services can be done with Temboo platform which uses choreos. The other famous platforms are Ubidots and myDevices. The myDevices platform has a rich set of features due to good data visualization. The open IoT platform, Thing Speak, uses Matlab support and has excellent IoT analytics and data visualization. The Matlab support helps to get real-time data visualization and data analysis and processing also can be added. The Things.iot platform provide simple IoT solution development and also assist in connecting devices remotely. The Xively IoT platform plays a key role to manage the communication, security, and firmware update. The set of tools offered by the Google Cloud IoT platform helps to manage connected devices and the date at the edge level. The bidirectional communication between connected devices can be done with Microsoft Azure IoT platform which uses IoT standard protocols. The IoT platform developed by IBM also used in building IoT solutions (Dzone, 2020).

Working Principle of IoT in HVAC

The various components of IoT used in HVAC systems are as follows

1. Different types of actuators to control HVAC modules
2. Wireless sensor network
3. Gateway which connects the local network with the cloud
4. Embedded devices to remotely manage the HVAC systems

The properly designed IoT HVAC system permits data collection, filtering, sharing to the cloud, data analysis and action with the help of the components.

The building management system is most widely used in commercial building and it can be integrated with IoT to develop a powerful system in building applications. The weather information such as temperature, humidity, wind speed, etc. are important in building management system. These data can be collected and can be used with the building data collected with different types of sensors for planning and operation of devices used in building. It is necessary to add various types of wireless sensors and other intelligence components for real time monitoring of temperature, humidity, light, motion, noise and occupancy rate. Also, it is possible to monitor and collect data related to occupancy levels in the building at different timings and days, water usage pattern of the occupants, lighting requirement of occupants etc. These data are collected in centralized control room and can be analyzed to understand how users are using the building services and also for planning purposes. Generally, these data are analyzed by machine learning and AI software. The action is initiated by the software to control system so that HVAC system and other systems are turned ON or OFF. Sometime other functions are performed based on the signal or data given by the sensor. The temperature, airflow quantity and speed, humidity, closing or opening of vents are varied to meet the comfort of the occupants. The analyzed data is communicated to the user or can be accessed using computers, tablets or centralized control panels. This helps in carry-

Figure 7. IoT based building management system

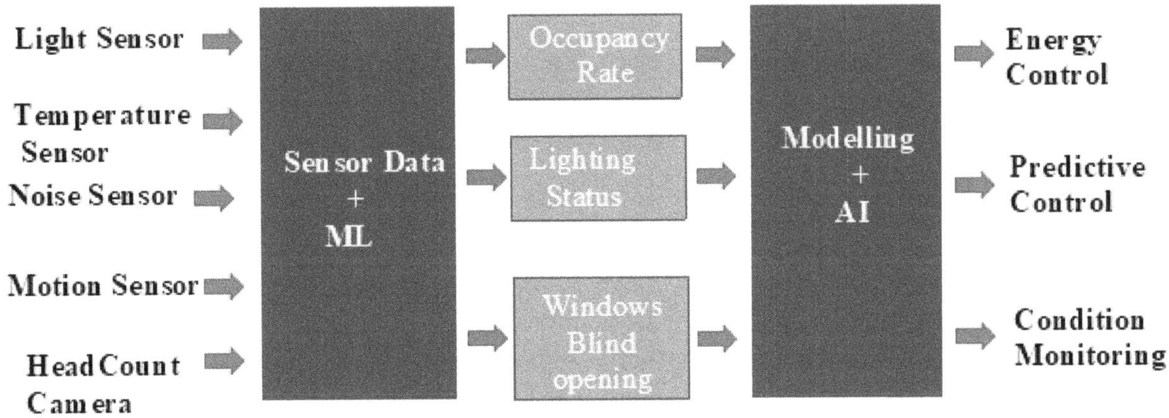

Figure 8. IoT system in HVAC environment

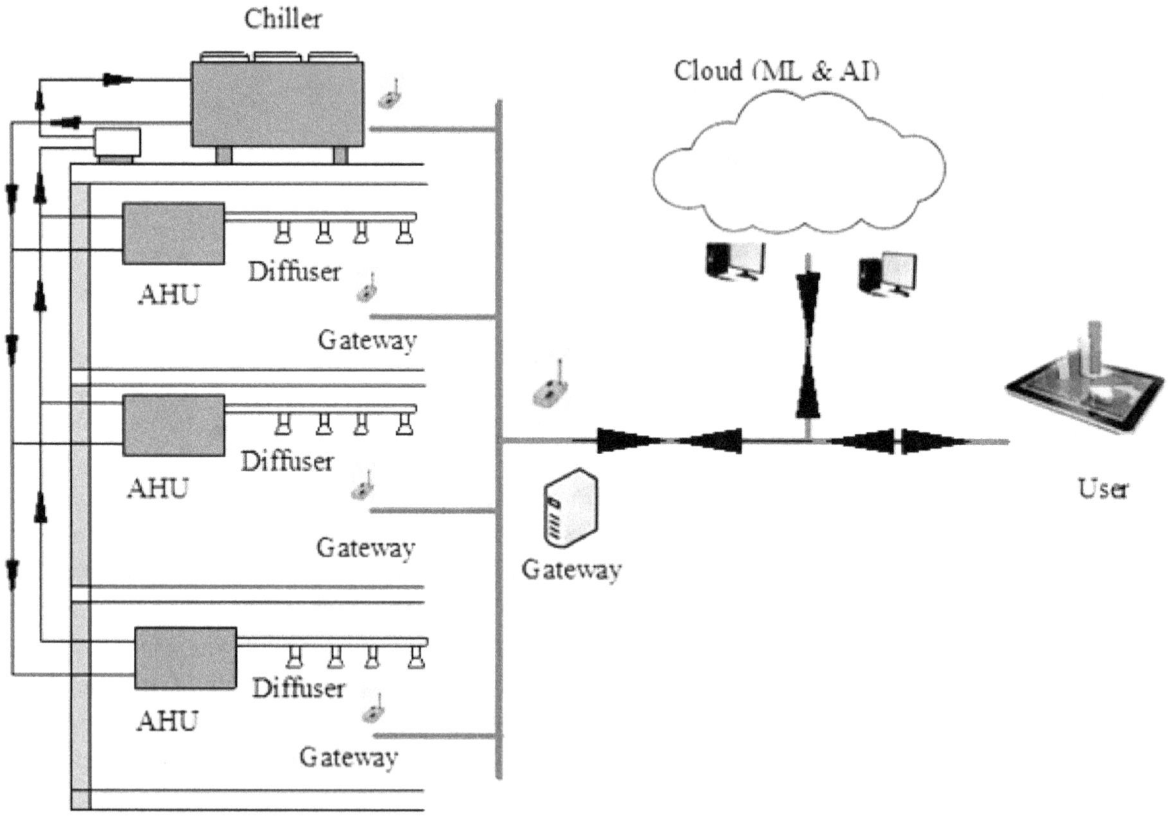

ing out preventive maintenance to avoid breakdown. The energy consumption can be reduced by using smart energy control which optimizes the system variables and reduces energy consumption. Figure 7 shows the IoT based building management system.

The HVAC systems are used in buildings to provide comfort to the occupants. The HVAC system consumes 40% of the total energy consumed in the building. Hence it is necessary to increase the performance of the HVAC systems so that the energy consumed by these devices can be reduced. The introduction of IoT in HVAC systems enhances the performance of the HVAC systems and also enhances the occupants' comfort. It is necessary to add various IoT components such as different types of wireless sensors, smart vents control in each room and other intelligence components for real time monitoring and controlling of various HVAC system devices. The sensor sends data to the control system so that HVAC system can be controlled to get good performance. The HVAC systems such as chiller, air handling units etc. are operated at optimized conditions to get better temperature, airflow quantity, air speed, humidity and purity for the comfort of the occupants. Also, it is possible to monitor and collect data related to motors, compressors, fans, pumps, heater, etc. used in HVAC systems. These data are collected in centralized control room and can be analyzed using machine learning and AI software to optimize the operating variables of different devices used in HVAC system. These analyzed data is sent to the user or facility manager. Figure 8 shows the IoT system in HVAC environment

The real time data analysis facilitates HVAC components to adjust themselves depends upon occupancy level, temperature, weather, timing of the day etc. Also, the sensors continuously monitor the performance of the system in HVAC real time and do real-time diagnostics for preventative maintenance. If any problem is noticed by the sensor, then alert is given to plant in-charge so that the repair can be taken up immediately to avoid any major breakdown. In recent years, IoT systems are embedded with intelligence so that fan, pump, and temperature set point are configured automatically to get best energy performance.

The real time data provided by the IoT system controls the HVAC system such that industry standard comfort temperature and humidity are maintained and also pollutants level, dust, bad odors, mold and bacteria are reduced which improve the employee health. Similarly, in industries such as food processing units and pharmaceutical companies, comfort working condition improves the production and health of the workers

Figure 9. Performance optimization using IoT

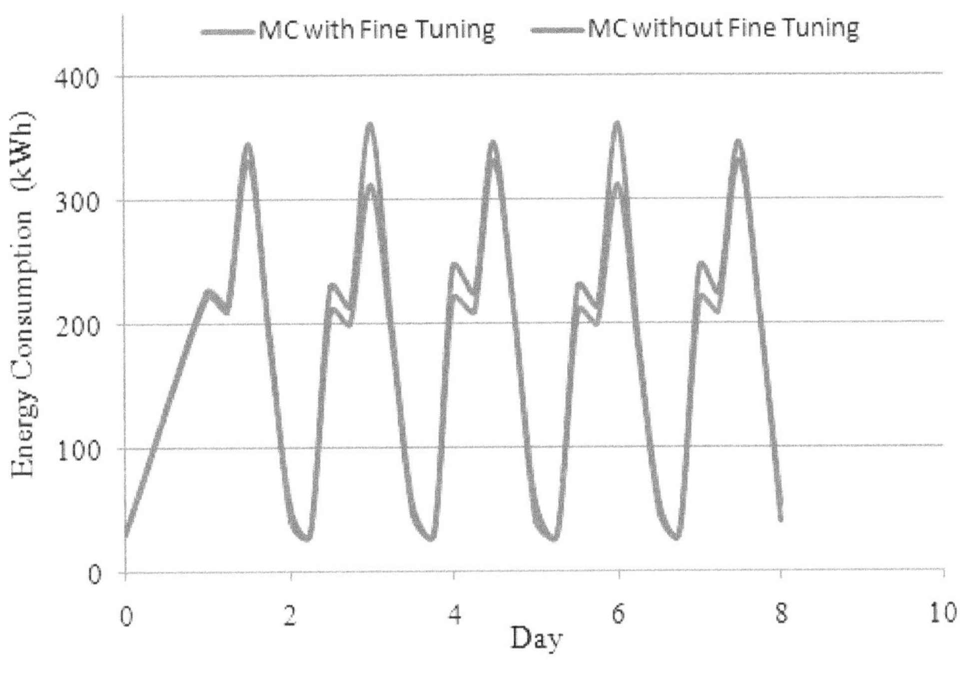

Figure 9 shows the performance optimization of HVAC system using IoT machine learning and AI algorithm. The measured energy consumption with fine tuning is compared with measured energy consumption without fine tuning. The fine tuning of measured energy consumption reduces the energy consumption significantly due to use of machine learning and AI algorithm which optimizes the performance of the system components. Also, the energy consumption with fine tuning is lower than the base line value.

Figure 10 shows the real time data of HVAC systems 1 and 2 recorded using IoT systems and these data were compared. The capacity of these HVAC plants is same; however, the energy consumptions are different. The HVAC 1 uses IoT based system and another system does not have it. From the figure we understand that the plant 1 performance is better than plant 2. Hence the energy consumption of

Figure 10. Comparison of performance of different HVAC plants

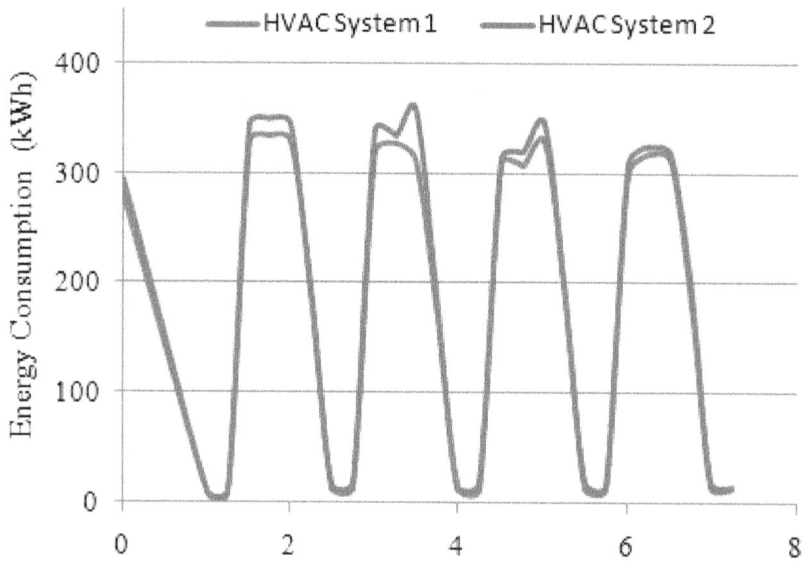

same capacity HVAC plants may vary and hence optimization is required for each plant to get higher performance and also to reduce breakdown.

The radiative sky cooling is a passive process that can be harnessed to sub cool refrigerants in air conditioning and refrigeration systems, thereby increasing the cooling capacity and improving the efficiency of the base cooling system (Goldstein et al., 2018). Generally, most locations can provide net night sky radiation above 40 W/m^2, while the nocturnal coolers have the potential for reducing space temperatures by 2 to 4°C and can yield 14 to 48% savings in the energy demand of a building (Nwaigwe et al., 2010). The ventilating cooling has been receiving attention in recent years because of the energy saving potential for all types of buildings. The study carried out on night ventilation for supermarkets with

Figure 11. Night Time Ventilation using IoT

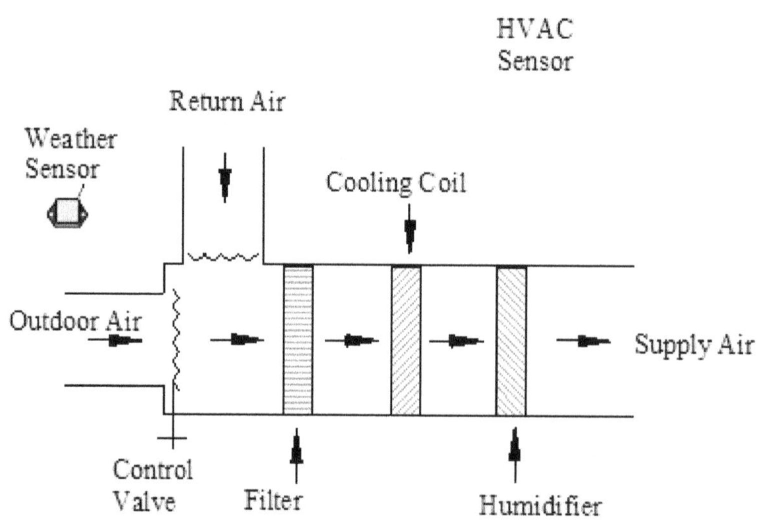

high cooling demand indicate that effective night ventilation can reduce the duration of active cooling during trading times and achieve 17% reduction in cooling annual energy use and 3.3% in total annual energy use (Mylona et al., 2018). The night sky radiative cooling technology using thermal panels and night time ventilation shows significant energy savings. The phase change materials also can be used with this system (Bourdakis et al., 2015). Figure 11 shows the night time ventilation system which uses IoT system to sense the weather condition using suitable sensor and use the low temperature outdoor air for cooling in night and automatically reduces the cooling load.

The solar powered cooling systems are getting popularity due to advancement in solar collector technologies (Hwang et al., 2008). The solar-assisted HVAC system that can operate for 24 h needs energy storage systems (Review of solar assisted HVAC systems, 2019). The thermal energy storage systems are used in HVAC systems to reduce the peak load demand and also to reduce the operating cost as the cost of electricity in peak load is higher than the off-peak loads. The cooling energy produced during non-peak load can be stored and can be utilized during peak load condition. The phase change materials are used in thermal energy storage system due to its capacity to store large amount of heat (Products, 2020). The real time weather condition may be monitored by suitable sensors and the data collected by the sensor can be analyzed by suitable software so that the solar thermal system can be integrated with the HVAC system. Figure 12 shows the solar powered HVAC system. The utilization of the solar energy reduces the power required to operate the HVAC system and also it can be used during peak cooling load demand so that the operating cost can be reduced. Hence there are enough scopes in integrating renewable energy sources into HVAC system using IoT systems.

Figure 12. Integration of solar thermal system with HVAC unit using IoT

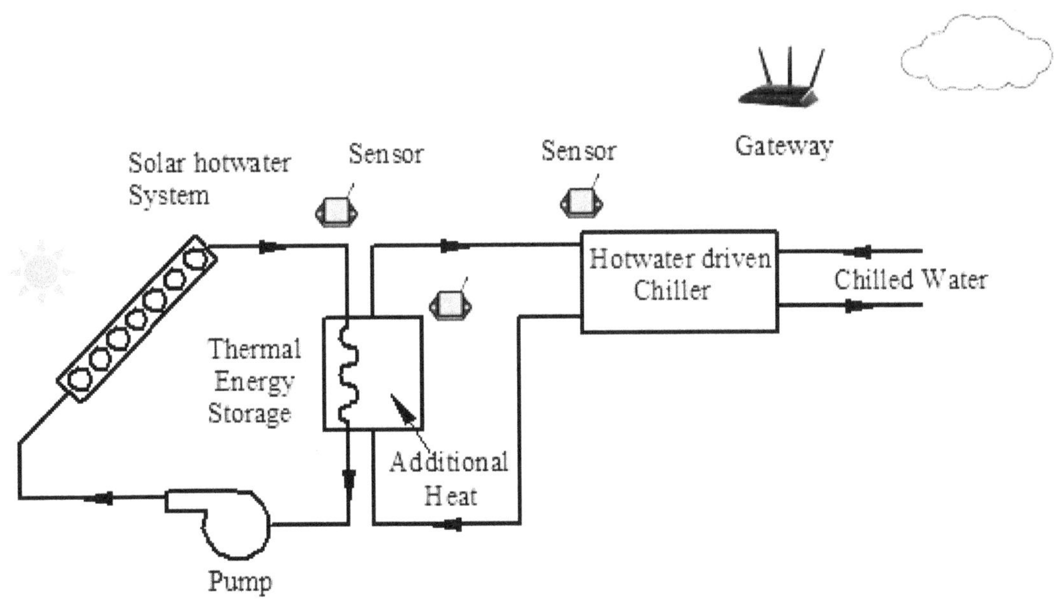

The recent HVAC systems come with their own smart sensors and hence intelligence of the systems increases. However, the older HVAC systems needs integration of sensors to create the preferred con-

Figure 13. IoT system in HVAC equipment monitoring and diagnosis.

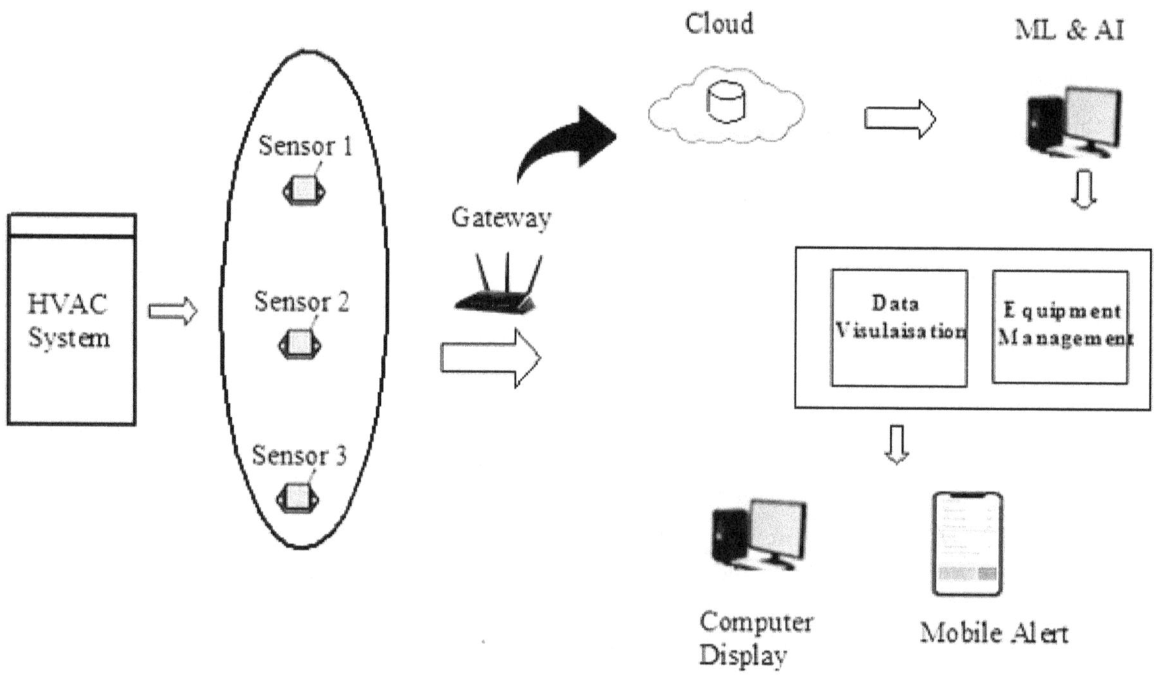

trol or management and it helps in addressing issues faced by the older traditional HVAC systems. The addition of IoT smart sensors turn on or off the heater or compressor when there is a need of heating or cooling of the room respectively. The IoT monitoring and automation system determine the state of the room and make necessary arrangements automatically to control temperature, humidity, indoor air quality, air flowrate etc., based on the occupancy level and need.

It is reported that the HVAC manufacturers such as Daikin, Johnson Controls, Honeywell etc. have introduced IoT in their HVAC products to increase the system energy efficiency. The inbuilt IoT systems in the HVAC system uses AI technologies estimate the occupants' preferences and automatically control HVAC system considering the outdoor weather data.

IoT in HVAC Equipment Monitoring and Diagnostics

The facility manager can monitor the conditions of the various electrical and mechanical devices used in the HVAC systems using IoT devices. The collection and analyzing of the data help the facility manger to understand the condition of the equipment and take preventive maintenance. For example, when in operation, the compressor produce specific sounds or vibration and it can be sensed by vibration sensors to determine the state of the compressor. This compressor data can be compared and analyzed with other compressor by machine learning and AI system to find the fault and alert is sent to the facility manager to take-up routine maintenance or service checkups. Also, this data can be shared with the compressor manufacturers for their analysis and immediate corrective steps. Figure 13 shows how the IoT is used in HVAC equipment monitoring and diagnosis. The smart features provided in the IoT systems sends reminders for oil or filter replacements, routine maintenance, service checkups etc.

Another benefit of using this system is to find out equipment failures, excess runtime, valve leakage, stuck dampers etc. Also, it is possible to compare and analyze past and present trends and issues. This type of approach will increase the life of existing equipment, maintenance and repair cost and reduces the downtime. Also, these data can be used by the original equipment manufacturers to improve the design aspects of their products and corrective steps to improve service productivity and future repair or replacement concerns. The IoT also provides visualization of the real time data, analysis and corrective steps taken

Figure 14. Predictive maintenance analysis using IoT system

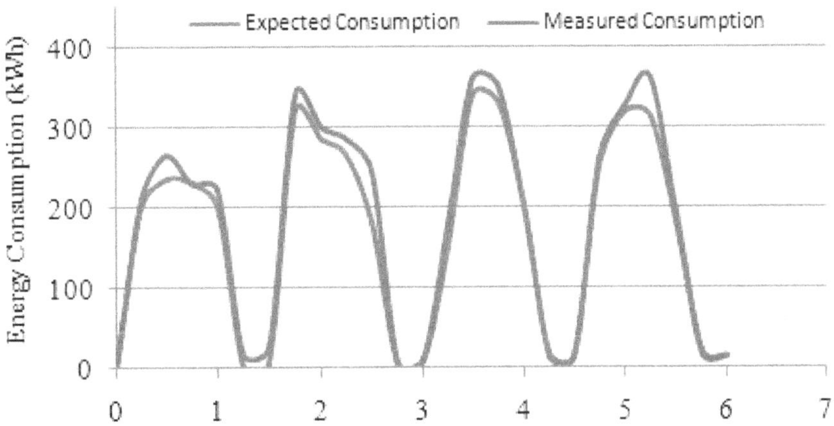

Figure 14 shows the measured consumption, expected consumption, upper and lower bounds. When the expected energy consumption of HVAC system is more than the upper bound an action is taken by the IoT to send an alert to the user about energy consumption and emergency action is followed to set right the issue. The data relevant to similar type of problem occurred previously is analyzed and emergency action plan is communicated by the IoT system to the facility manager or plant in-charge.

HVAC Big Data Analysis

The bigdata refers to high volume, high velocity and large variety of information assets that require new forms of processing to enable enhanced decision making, insight discovery and process optimization. The IoT sensors used in the HVAC systems send lot of data continuously to the cloud and these data can be analyzed for energy consumption, systems efficiency and peak demand reduction. The big data management is used to analyze the HVAC data and HVAC system is driven big data analytics which includes predictive maintenance, complexity mapping and visualization. Also, it eliminates the manual energy audits which take lot of time. The comprehensive analysis of bigdata provided by the various sensors such as facility operating conditions and the outdoor environment to provide detailed monitoring of indicators that an issue has occurred at the facility. The analysis of the data includes optimized environment agents, measuring system performance against industry standards, achieving sustainability goals, benchmarking plug loads and lighting, and optimizing financial performance.

About 30 operating parameters of a chiller can be monitored in HVAC system. If these data are recorded every 15 min then about 1 million data are recorded in a year from a single chiller. For 600 HVAC systems with higher number of chillers, the data collected will be over 21 million records every year. For 5,000 similar buildings, over 105 billion records will be captured annually, translating into over 4.2 terabytes of data storage. The sensors provided in the IoT system collect weather condition, facility data, performance and operational data of the HVAC systems etc., in every min or sec. If the number of variables collected is more than large volume of data are generated and can be analyzed. However, we need to identify the most important variables' data for analysis to avoid misleading analytics. The bigdata analytics starts with big data management and then the data analytics. The big data and analytics will help the building owner to understand the building and equipment performance and also to take up preventive maintenance to increase equipment life, minimize disruption and operating costs. It also enables benchmarking of a building's HVAC system performance against industry standards or benchmarks. It also brings much additional value for HVAC systems manufacturers and service providers. The manufacturer of the HVAC system can improve their design based on how their equipment performance.

The real time monitoring and analyzing the energy and other activities helps to minimize the energy and water consumptions using big data analytics. This process also helps the facility or building manager is able to identify information on human activity, weather condition, the micro-climate within the building and related energy wastage. Indeed, to do so, algorithms need to be designed that can accurately extract the inter correlation between the load consumption, human occupancy and movement activity, energy wastage, renewable energy generation, weather condition, and hence design effective model from real-time large-scale datasets in order to perform predictive maintenance.

In recent years, solar energy systems are integrated with HVAC systems to reduce the conventional energy use. The solar PV or solar thermal system can be integrated with the HVAC systems. The weather conditions can be sensed using IoT system and the solar heat which can be absorbed by the solar thermal system or power generated by the solar PV system can be predicted. The IoT system uses big data analytics to predict how much hot water or electrical power will be needed for operating the HVAC system. The available renewable energy resources are used optimally and HVAC systems are operated by using machine learning techniques to reduce electricity consumption of the building. If the temperature in night is low, then the systems uses fresh low temperature air for cooling and the load on the HVAC system is reduced.

Case Studies / Success Stories

The IoT system was developed and integrated with Building information modeling (BIM) to carry out the predictive maintenance and to enhance the effectiveness of the facility maintenance management (FMM). A data-driven predictive maintenance was planned for the FMM by integrating BIM and IoT. The IoT system consists of two layers such as information and application layers. The information layer consists of collection of sensor data and integration of this data with BIM models, and IoT networking. The application layer consists of 4 modules such as condition monitoring, assessment, prediction and maintenance planning modules. These modules help to carry out the predictive maintenance. The performance parameters such as chilled water pressure at evaporator and condenser, flow rates of water at evaporator and condenser and temperatures at various locations such as evaporator inlet and outlet and condenser inlet and outlet temperature were measured using IoT sensors and these data were visualized using BIM models. The trends of these data trends were visualized using BIM model. The abnormal signals or data provided by the sensor indicates the malfunctioning of HVAC component and these data were recorded for predicting future condition of the equipment. The artificial neural network, SVM and ML algorithms predicts the condition of the HVAC components. The information layer constantly supplies the machine condition data to the application layer to predict the condition of the HVAC components using ML algorithms. The data analysis provides information about future condition of the HVAC components and provides data for the facility management maintenance planning (Cheng et al., 2020).

A novel IoT based system was developed to improve the energy efficiency of the HVAC system used in office. The real time data related to the indoor office environment was collected. This data was analyzed to implement the IoT system to enhance the efficiency of the HVAC system. The office staff usually set lower set-temperature and forgets to switch off the HVAC system when they leave the meeting hall or office. A IoT based CO_2 sensor system was installed in the office and meeting hall. The CO_2 sensor senses the CO_2 level and determines the occupancy. When the CO_2 level reduces or remains the same, then the IoT system automatically switches off the HVAC system in the unoccupied area of the office such as meeting room, individual office room etc. Also, it detects the occupancy level and adjust the set-temperature value according to the preferred range of the occupants. This operating strategy significantly reduces the energy consumption and provided thermal comfort to the occupants (Li et al., 2020).

Recent Advances In Application of IoT

In recent years, IoT has made lot of changes in which a global network of systems and devices capable of interacting with each other were developed (Lee, 2019). The IoT Energy Platform helps in the management of IoT energy data and provides robust and versatile application. This platform was tested successfully in real use in three building with hundreds of sensors (Cheng et al., 2020). The model based predictive control technique provides significant amount of energy savings in existing buildings, under normal occupancy conditions and suggested that this model can be used in commercial application which uses self-powered wireless sensors (Terroso-Saenz et al., 2019). An intelligent antifragile platform learns and performs suitable action using IoT devices and Big-data technologies for effective control of ventilation systems (Lachhab et al., 2018). The building information data can be processed and corrective steps can be initiated (Gao & Pishdad-Bozorgi, 2019). An IOT platform was developed to integrate BIM and IoT devices to analyze the energy usage using IFC models and through open messaging standards (Dave et al., 2018). An IoT with machine learning system was used to predict the thermal comfort of individual and the study was carried out for three-week. The HVAC control system used fuzzy logic to calculate a set point and the study results shows improvement in thermal comfort of the occupants (Aguilera et al., 2019). An algorithm was developed by a Singapore university to reduce energy consumption of HVAC systems used in commercial building using suitable software, hardware, networking and their integration. The decentralized control increases the scalability and smartness of the system. The results show that by combining decentralized control with IoT decreases the energy consumption and increases the energy savings by 20% (Png et al., 2019).

The energy consumption of the building can be reduced using sophisticated energy management systems (EMS) and building automation systems (BAS). The BAS helps to identify the occupants' preferences and automatically manages HVAC system. The building EMS is controlled by EMS controls based on atmospheric condition. The energy consumption of the HVAC system was reduced by the BAS without affecting the comfort levels of majority occupants (Carreira et al., 2018). A commercial Wi-Fi enabled

IoT device was used for occupancy detection and counting and it is reported that this device achieves good occupancy detection and crowd counting (Zou et al., 2018).

The energy conservation of the building is affected by the selected model order and parameter estimation technique as the energy optimization depends upon calculation power and data availability (J et al., 2017). The model based predictive control can be used in the existing building with normal occupancy level to reduce energy consumption with self-powered wireless sensors and IoT system (Silva & Ruano, 2018). An ARM-compliant IoT security framework was developed and tested in smart buildings and the results shows reasonable improvement in energy savings, services discovery and authorization (Hernández-Ramos et al., 2015).

The energy conservation of the residential HVAC systems can be improved with the utilization of thermostat data, energy use data and history-based model (Cetin & Kallus, 2016). The remotely programmable and controllable thermostats can be used in residential HVAC systems with suitable algorithm to manage power demand of a group of HVAC units. These thermostats do not need any additional electrical infrastructure (Adhikari et al., 2018). An active controller based smart heating system which uses the combination of internal set points and external signal from local control entity minimize the size, cost energy consumption and reliability of smart microgrid (Hakimi, 2017).

A real-time certification system uses IoT technologies to measure real-time data of indoor comfort, energy efficiency of electricity, water, and gas usage. This system provides information on a house's current status and collects the real-time data using IoT technologies. This result shows the difference between static and dynamic data (Matsui, 2017). The home energy management system which uses the stored data was effective in reducing electricity consumption while maintaining indoor comfort according to each resident's short-term indoor comfort preference. Also, it improves indoor comfort and provides user with random information (Matsui, 2018).

An IoT based ambient assisted living system was developed for elderly monitoring and it focuses on cloud based centralized data management repository, city wide data capturing layer and risk analysis and prediction module (Almeida et al., 2019). The LEED certifications use static values such as energy efficiency of electricity, water and gas usage. However, the real time dynamic data provided by IoT technologies has the potential to provide information on a home's current status and can be used for online certification (Matsui, 2017). The use of process history-based model in the residential HVAC systems improves annual energy saving of residential buildings (Cetin & Kallus, 2016). The energy consumption models may predict annual energy consumption data; however, none of this model meets ASHRAE criteria for a calibrated model when looking at hourly interval data. The error is the model may be due to appliance and lighting energy overestimates (Glasgo et al., 2017).

Advantages of IoT

The advantages of using IoT in HVAC systems are as follows

1. Higher energy efficiency and management of HVAC systems
   The IoT system optimize the operating conditions and controls the equipments' which helps in energy wastage and reduces the energy consumption. This reduction in energy loss improves the efficiency of the HVAC system. The integration of IoT with HVAC systems helps in real time
2. Better maintenance, service and lower downtime.

The IoT helps in real time monitoring of the equipment and devices for the break down and take action on service-related activities. This helps in regular maintenance of HVAC system and reduces the breakdown and improves the safety and life of the equipment.

3. Smartness of the HVAC system
Smart system learns the occupant's preferences and adjusts these automatically based on the occupancy and other programmed conditions.
4. Process control and optimization
The monitoring and process controlling of the various devices of the HVAC system can be done with the IoT system. Also, the IoT helps in optimization of various process parameters.
5. Easy problem detection
The IoT helps in real time monitoring and diagnostisation of the HVAC system components. Hence it is easy to identify the defect and to take corrective steps immediately.
6. Remote control of the system
The IoT helps in remote control of the HVAC system using control valves and switches. Hence the manual interventions are reduced significantly and machine intelligent level is increased with AI technology.
7. Data can be stored in cloud and can be accessed easily
The real time data collected from the HVAC system is stored in the cloud and hence the plant manager or maintenance staff can access the data easily.
8. Reducing operational costs while improving comfort
The real time data is analyzed and suitable action is taken immediately based on optimization. This controls the process parameters and HVAC system is operated at the optimized condition. This results in lower operating cost. The IoT system detect number of occupants and their location. The process variables are adjusted according to the comfort of the occupants.
9. Remote diagnostics
The remote diagnostics involves collecting data and analyze it to find out any operational issues. This helps in understanding the operational status of the HVAC system and take quick action based on current conditions.

Challenges Faced in Integrating IoT with HVAC Systems

The challenges faced in integrating IoT with HVAC systems are as follows. The research work may be carried out in this direction to solve the industrial problem.

1) Security flaws: The sensors used in the IoT system can be accessed by the hackers through internet and can create major security issues.
2) Privacy issues: The consumers and user data are fed into the IoT system and is available in the cloud. If proper security system is not implemented then the hackers can steal the data.
3) Incompatibility between various components: The incompatibility between various components, IoT platforms, data models and information models may affect the working of the IoT system.
4) Quality of connectivity: The IoT connectivity refers to connection between various sensors, gateways, platform, application and other components. The poor network connection and bandwidth affects the IoT system.

5) Maintenance related issues: The preventive maintenance reduces emergency repairs and reduces the revenues of maintenance contractors. Hence there may be lack of cooperation and support from maintenance department staff.
6) Coordination among stake holders: The cooperation between different stake holders such as device provider, platform providers, application providers and operator is crucial for the successful implementation of the IoT system. Any delay and non-cooperation from any one stakeholder may affects the IoT integration with HVAC system.
7) Scalability: The scalability of IoT for larger HVAC system is s challenge. The control algorithms used for savings energy and their implementation on low cost resource constrained hardware

CONCLUSION

The IoT has amalgamated hardware and software to the internet to create a smarter world. The increase in global temperature increases with increase in usage of HVAC systems. The HVAC systems consume more than 40% of the total energy consumed by the building. In recent years, the growth of HVAC systems is significant and hence it is necessary to develop energy efficiency systems and suitable control mechanisms. The IoT based systems are increasingly available and installed in buildings, enabling new, data-driven methodologies for the operation of smarter, more sustainable building systems. The connectivity of different devices and machine learning techniques are being applied to HVAC systems leading to promising results. The IoT based systems has significant potential to save the energy consumed by the HVAC systems. Also, it gives real time performance and helps to carryout preventive maintenance. The amount of energy saved by IoT based HVAC systems depends upon number of variables. It is estimated that 24 billion IoT devices would be installed by the year 2020, generating revenue of over 300 billion. The IoT has been growing at a significant pace and offers a multitude of opportunities for government and business. The number of job openings for IoT professionals in recent years too high. Also, it is necessary to explore the true potential of this technology for HVAC systems and other service sectors. However, there are few challenges in the implementation of IoT based systems for HVAC systems which will be solved in near future. The target audience of this book chapter are researchers, faculty members, students and practicing engineers will be benefitted from this book chapter.

ACKNOWLEDGMENT

This work did not receive any funding. Authors thank Mr.Nagaraja, Director, M/s Bluevolt Services, Bangalore, India for his suggestions.

REFERENCES

Adhikari, R., Pipattanasomporn, M., & Rahman, S. (2018). An algorithm for optimal management of aggregated HVAC power demand using smart thermostats. *Applied Energy, 217*, 166–177. doi:10.1016/j.apenergy.2018.02.085

Aguilera, J. J., Kazanci, O. B., & Toftum, J. (2019). Thermal adaptation in occupant-driven HVAC control. *Journal of Building Engineering, 25*, 100846. doi:10.1016/j.jobe.2019.100846

All about circuits. (2020). https://www.allaboutcircuits.com/industry-articles/utilizing-the-different-types-of-common-iot-connection-methods/

Almeida, A., Mulero, R., Rametta, P., Urošević, V., Andrić, M., & Patrono, L. (2019). A critical analysis of an IoT—Aware AAL system for elderly monitoring. *Future Generation Computer Systems, 97*, 598–619. doi:10.1016/j.future.2019.03.019

Bawaneh, Nezami, Rasheduzzaman, & Deken. (2019). Energy Consumption Analysis and Characterization of Healthcare Facilities in the United States. *Energies, 12*(9), 3775.

Bourdakis, E., Olesen, B. W., & Grossule, F. (2015). *Night time cooling by ventilation or night sky radiation combined with in-room radiant cooling panels including phase change materials*. Paper presented at 36th AIVC Conference, Madrid, Spain.

Carreira, P., Costa, A. A., Mansur, V., & Arsénio, A. (2018). Can HVAC really learn from users? A simulation-based study on the effectiveness of voting for comfort and energy use optimization. *Sustainable Cities and Society, 41*, 275–285. doi:10.1016/j.scs.2018.05.043

Cetin, K. S., & Kallus, C. (2016). Data-Driven Methodology for Energy and Peak Load Reduction of Residential HVAC Systems. *Procedia Engineering, 145*, 852–859. doi:10.1016/j.proeng.2016.04.205

Cetin, K. S., & Kallus, C. (2016). Data-Driven Methodology for Energy and Peak Load Reduction of Residential HVAC Systems. *Procedia Engineering, 145*, 852–859. doi:10.1016/j.proeng.2016.04.205

Cheng, J. C. P., Chen, W., Chen, K., & Wang, Q. (2020). Data-driven predictive maintenance planning framework for MEP components based on BIM and IoT using machine learning algorithms. *Automation in Construction, 112*, 103087. doi:10.1016/j.autcon.2020.103087

Data Flair training. (2020). https://data-flair.training/blogs/how-iot-works/

Dave, B., Buda, A., Nurminen, A., & Främling, K. (2018). A framework for integrating BIM and IoT through open standards. *Automation in Construction, 95*, 35–45. doi:10.1016/j.autcon.2018.07.022

Dzone. (2020). https://dzone.com/articles/12-iot-platforms-for-building-iot-projects

Electronics and ICT. (2020). https://electronicsandict.com/iot-in-hvac-systems/

Gao, X., & Pishdad-Bozorgi, P. (2019). BIM-enabled facilities operation and maintenance: A review. *Advanced Engineering Informatics, 39*, 227–247. doi:10.1016/j.aei.2019.01.005

Glasgo, B., Hendrickson, C., & Azevedo, I. L. (2017). Assessing the value of information in residential building simulation: Comparing simulated and actual building loads at the circuit level. *Applied Energy*, *203*, 348–363. doi:10.1016/j.apenergy.2017.05.164

Goldstein, E. A., Nasuta, D., Li, S., Martin, C., & Raman, A. (2018). Free Subcooling with the Sky: Improving the efficiency of air conditioning systems. *International Refrigeration and Air Conditioning Conference*. Paper 1913, https://docs.lib.purdue.edu/iracc/1913

Hakimi, S. M. (2017). A novel intelligent control of HVAC system in smart microgrid. *Journal of Electrical Systems and Information Technology*, *4*(2), 299–309. doi:10.1016/j.jesit.2017.01.005

Hernández-Ramos, J. L., Moreno, M. V., Bernabé, J. B., Carrillo, D. G., & Skarmeta, A. F. (2015). SAFIR: Secure access framework for IoT-enabled services on smart buildings. *Journal of Computer and System Sciences*, *81*(8), 1452–1463. doi:10.1016/j.jcss.2014.12.021

Hwang, Y., Radermacher, R., Al Alili, A., & Kubo, I. (2008). Review of Solar Cooling Technologies. *HVAC & R Research*, *14*(3), 507–528. doi:10.1080/10789669.2008.10391022

International Energy Agency. (2020). https://www.iea.org/reports/the-future-of-cooling

J., V., G., V. H., D., S., & L., H. (2017). Model selection for continuous commissioning of HVAC-systems in office buildings: A review. *Renewable and Sustainable Energy Reviews, 76*, 673–686.

Khalilnejad, A., French, R. H., & Abramson, A. R. (2020). Data-driven evaluation of HVAC operation and savings in commercial buildings. *Applied Energy*, *278*, 115505. doi:10.1016/j.apenergy.2020.115505

Lachhab, F., Bakhouya, M., Ouladsine, R., & Essaaidi, M. (2018). Towards an Intelligent Approach for Ventilation Systems Control using IoT and Big Data Technologies. *Procedia Computer Science*, *130*, 926–931. doi:10.1016/j.procs.2018.04.091

Lee, I. (2019). The Internet of Things for enterprises: An ecosystem, architecture, and IoT service business model. *Internet of Things*, *7*, 100078. doi:10.1016/j.iot.2019.100078

Li, W., Koo, C., Hong, T., Oh, J., Cha, S. H., & Wang, S. (2020). A novel operation approach for the energy efficiency improvement of the HVAC system in office spaces through real-time big data analytics. *Renewable & Sustainable Energy Reviews*, 127.

LinkedIn. (2020). *IoT characteristics*. https://www.linkedin.com/pulse/internet-things-iot-characteristics-kavyashree-g-c

Matsui, K. (2017). Proposal and implementation of real-time certification system for smart home using IoT technology. *Energy Procedia*, *142*, 2027–2034. doi:10.1016/j.egypro.2017.12.406

Matsui, K. (2018). An information provision system to promote energy conservation and maintain indoor comfort in smart homes using sensed data by IoT sensors. *Future Generation Computer Systems*, *82*, 388–394. doi:10.1016/j.future.2017.10.043

Mylona, Z., Kolokotroni, M., & Tassou, S. A. (2018). Coupling night ventilative and active cooling to reduce energy use in supermarkets with high refrigeration loads. *Energy and Building*, *171*, 26–39. doi:10.1016/j.enbuild.2018.04.021

Nwaigwe, K. N., Anthony, O. C., Ogueke, N., & Anyanwu, E. E. (2010). Review of nocturnal cooling systems. *International Journal of Energy for a Clean Environment*, *11*(1-4), 117–143. doi:10.1615/InterJEnerCleanEnv.2011003225

Png, E., Srinivasan, S., Bekiroglu, K., Chaoyang, J., Su, R., & Poolla, K. (2019). An internet of things upgrades for smart and scalable heating, ventilation and air-conditioning control in commercial buildings. *Applied Energy*, *239*, 408–424. doi:10.1016/j.apenergy.2019.01.229

ProductsP. C. M. (2020). https://www.pcmproducts.net/files/cool_thermal_energy_storage.pdf

PSM Market Research. (2020). https://www.psmarketresearch.com/market-analysis/hvac-market

Review of solar assisted HVAC systems; Its performance analysis using CO_2 as a refrigerant. (2019). *Energy Sources, Part A: Recovery, Utilization and Environmental Effects*, *41*(24), 2957-2974.

Silva, S., & Ruano, A. (2018). The IMBPC HVAC system: Wireless Sensors and IoT Platform. *IFAC-PapersOnLine*, *51*(10), 1–8. doi:10.1016/j.ifacol.2018.06.227

Statista. (2020). *IoT Number of connected devices worldwide*. dhttps://www.statista.com/statistics/471264/iot-number-of-connected-devices-worldwide/

Terroso-Saenz, F., González-Vidal, A., Ramallo-González, A. P., & Skarmeta, A. F. (2019). An open IoT platform for the management and analysis of energy data. *Future Generation Computer Systems*, *92*, 1066–1079. doi:10.1016/j.future.2017.08.046

Wikipedia. (2020). *Internet of Things*. https://en.wikipedia.org/wiki/Internet_of_things

Zhuang, L., Chen, X., & Guan, X. (2020). A decentralized method for energy conservation of an HVAC system. *Building Simulation*, *13*(1), 155–170. doi:10.100712273-019-0569-6

Zou, H., Zhou, Y., Yang, J., & Spanos, C. J. (2018). Device-free occupancy detection and crowd counting in smart buildings with Wi-Fi-enabled IoT. *Energy and Building*, *174*, 309–322. doi:10.1016/j.enbuild.2018.06.040

Chapter 9
MPPT-Based Solar Powered Water Pumping With RMS:
Augmentation of IoE Technology

Sushri Mukherjee
Indian Institute of Technology, New Delhi, India

Sumana Chattaraj
Indian Institute of Technology, Rajasthan, India

Dharmbir Prasad
Asansol Engineering College, Asansol, India

Rudra Pratap Singh
Asansol Engineering College, Asansol, India

Md Irfan Khan
Supreme & Co. Pvt. Ltd., India

ABSTRACT

In the present era of technology drift, the Internet of Things (IoT) is a promising technology that shows the considerable transition from enterprises to businesses. In this context, renewable energy has played critical roles in developing the agricultural sector with its functionality to generate power over distant areas and different terrain. There has been a transformation in the way we consume electricity and the way it is operated by the utilities. The smart monitoring features are an efficient way to utilize energy and minimize losses. These losses might be the transmission and distribution losses or losses by the inefficient household wiring or appliances. These losses can be minimized if they are monitored well with relevant data and analytics. This book chapter presents an application of the Internet of Energy (IoE) technology in the renewable sector. The study was carried applying Maximum Power Point Tracking (MPPT) functions and corresponding Remote Monitoring System (RMS) functions. The undertaken site

DOI: 10.4018/978-1-7998-6709-8.ch009

for Solar Water Pump (SWP) installation for irrigation is Bihar Sharif (Nalanda, Bihar). The proposed IoE based technology served maximum power output with high efficiency during operating hours. Here, the farmers may do automatic start/shut, remote start/shut from SMS/web with a health check, and self-diagnostic mechanisms for easy maintenance. Thus, the SWP system emerged as a cost-effective and environment-friendly irrigation solution for remote farmers throughout the year. The solar potential assessment of the site and circuital development were done using PVsyst 7.0 and KiCad software tools, respectively. The successful working of the developed product has been observed at the site.

INTRODUCTION

Renewable energy has driven the resource gathering process for power generation into more dynamical supply chain operation, since the sources of the renewable energy are diversified throughout the earth in the form of solar, wind, biogas, geothermal and few more. These resources build the confidence to generate and distribute the energy more efficiently with its deep penetration into the micro-industries such as farming, cattle rearing and the types of industries among the rural population directly affecting the livelihood. No matter how tremendous the energy consumption pattern is among the urban population but the basic necessities are fulfilled by the micro-industries, which provide us food and shelter. On the other hand, millions of people living in remote areas are not having sufficient access to the electricity due to unavailability of central grid. Furthermore, the expansion of power grid to such places with low power demand and having dispersed population is not economically feasible. Hence, for such places solar energy is emerged as one of the most inexhaustible source of energy with boundless potential. Solar powered pump is a cost effective solution to the farmers by which they can manage harvesting upfront cost and even becoming an optional revenue stream for them (as presented in Table 1). Most of the farmers are dependent on conventional sources based pumping system like electric and diesel/kerosene water pumps for irrigation in India. Due to its continuous increase, approximately 8-12% of total greenhouse gases (*i.e.*, 45 million tonnes of CO_2) emitted to the atmosphere (Rathore et al., 2018).

Table 1. Competitive features of SWP system

Parameters	Features
Cost	• No fuel cost compared with diesel price • Savings on high cost of providing last mile electricity connections • Savings on current diesel and agricultural power subsidy
Power	• Peak load management between urban and rural areas • Zero distribution losses due to decentralized power generation
Operations and Maintenance	• Long operating life of pumps • Highly reliable, durable/negligible maintenance over time • Easy to operate and maintain
Environmental	• Reduction in greenhouse gas emissions • Reduction in pollution due to diesel combustion
Societal	• Generation of local employment in villages • Prevention of mass migration of villagers to urban areas • Enabling economic development of rural farmer

IMPORTANCE OF IRRIGATION WATER MANAGEMENT

The agriculture sector has major contribution in livelihood of households and to the development of state economy in Bihar (Bahinipati & Viswanathan, 2019). The water resources management to meet supply and demand of water for irrigation and drinking are amongst the most critical policy issues. Many farmers from remote eastern India have inadequate access to electricity and there is no easy access to diesel fuel too for cultivation of crops. There appears, one possible way to farm during dry months is to rent pumps that run on diesel or kerosene, but that option shrinks their profit after the high costs of the rental and fuel. This makes it difficult to avail groundwater required to grow crops beyond the monsoon season. Most of the neighbouring river basins are facing water abundance (*i.e.,* flood) and scarcity (*i.e.,* drought) as the common issues (Reddy et al., 2020).

As per the Economic Survey (2019), the flood irrigation is the most common method in India for farming. In this mode of irrigation, water is allowed to flow in the field and seep into the soil, which results wastage of water since excess water flows off the surface or seeps into the soil without being completely utilized by the plants. This circumstances compelling many farmers to travel to cities from rural India each year to pursue physically demanding migrant workers. Thus, separating families at a time when crop prices are higher because supply is poor (Agriculture Department, Government of Bihar). The proposed SWP utilize the sun's energy to convert into electrical energy for the pump to operate and enable us to move water from its remote source to where we need it, without access to electricity. Hence, in India solar water pumping systems market exhibited substantial growth over the past few years. Rapid population increase and growth in food demand have forced the farmers to adopt intensive irrigation techniques and thus, non-uniform extraction of water. This results into depletion and deterioration of groundwater head and quality, respectively (Patnaik & Biswal, 2020).

At present, over exploitation of groundwater has become one of the major issues in various regions of the world. However, the groundwater is replenishable, but its overuse leading towards more mining than the recharging level of the soil (Adimalla et al., 2020). To overcome it, by curtailing water consumption limit or drift irrigation or participatory irrigation management (Ghosh et al., 2019).

Installation Sites' Overview

In India, the state of Bihar provides an examplory case of SWP application for irrigation purpose, which explore options of greener means of water pupming around under served regions of electrcity. The proposed study on SWP system has been carried out, where agriculture is the backbone of the economy and it is changing fast. This state is shining with a high growth rate for the first time in the post-independence era. The system was installed at Bihar Sharif (Nalanda, Bihar) (co-ordinates - latitude: 25.21°N and longitude: 85.51°E), which is well known region for extensive production of vegetables. Despite the rising footprint of grid-electricity, this study finds a significant application for the area with no access or unreliable access to the national grid. The proposed approched has the potential to be scaled-up across the entire nation.

Case Study Highlights of SWP System

Recently, the solar water pump was emerged as a leapfrogging technology for the rural India. In this context, the SWP system (rated capacity: 5 HP) was installed for the Public Health Engineering Department, Bihar. The entire control system of the pumps was enclosed in the IP62 cabinet, which can be installed in house or with the solar panels. The control cabinet was installed in the pump house. The solar structures were dual axis manual tracking type, that can regularly tilted towards the direction of the sun. The daily tracking frequency were 3 to 5 times, a day and it resulted in 15% additional output compared to seasonal tracking. This seasonal tilting adjustment depends on the local wind condition. Special care was taken to keep the solar panels at horizontal position in case of heavy storm and wind conditions. The farmers were highly satisfied because of the water availability throughout the irrigation period.

PAPER ORGANIZATION

The rest portion of this chapter is presented as follows: in Section 2, operating principle of the proposed model is explained. Then after, background study on recent IoT based revolutions has been reviewed in Section 3. While, in Section 4, a mathematical modelling is briefed. Furthermore, in Section 5, architecture of the proposed SWP system is presented. Section 6 is having the simulation of PV pumping system and Section 7 highlights integration of IoE concepts to the SWP solution. However, Section 8 presents results' assessment and finally, summary of the current study is given in Section 9.

Figure 1. Solar water pumping arrangement at the site

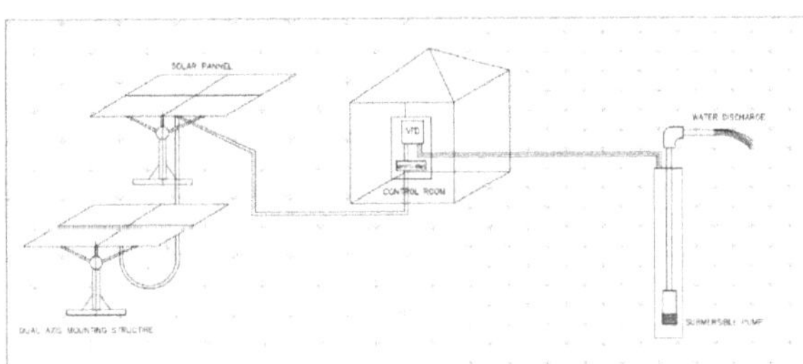

OPERATING PRINCIPLE OF SWP MODEL

The SWP system discharge was maximum during summer which eased the dependency on arrival of monsoon. The MPPT tracking was iteration-based tracking for the maximum power which can be extracted from the solar panels and can be used to drive the motor. The motor is controlled by VFD, which receives signal from the MPPT to increase the frequency (*i.e.,* speed of the motor). While the motor speed is increased it drags power from the solar panels. The speed increases gradually as the motor power consumption is lesser than the power available from the solar panels. The MPPT checks the voltage and current of the solar panels as given in Figure 1. The control signal is given to the VFD from MPPT. These devices measuring the voltage and current are of high value and are protected by opto-coupler.

BACKGROUND STUDY ON RECENT IOT BASED REVOLUTIONS

Recently, most of the physical objects have been upgraded by electronic devices over the internet to create local intelligence and make communication with cyberspace. Intelligent decisions are made based on the information retrieved from raw data through sensors, machines and other devices, which are discussed in succeeding sections:

Advancement in IoT Technology

The IoT is a network paradigm in which physical, digital and virtual objects are equipped with identification, detection, networking and processing functions to communicate with each other and with other objects (termed as "things") and services. A brief literature survey on IoT technology and its applications are presented as below:

Hersent et al., (2011) have studied a heterogonous CPS structure for an IoT based smart grid, where security and privacy of data acquisition, storage, and retrieving are crucial for IoT as well as CPS security. It cause serious challenge for IoT integration to it, on node-to-node interoperability due to different hardware and software usage on the same platform and meagerness in commonly embraced security architecture. Thenafter, in order to ease in application and trobleshooting Hajiheidari et al., (2019) have categorised IDSs in the IoT among following parts - (centralized, distributed, hybrid), (anomaly based, signature based, specification based and hybrid), (denial of service attack, Sybil attack, replay attack, selective forwarding attack, wormhole attack, black hole attack, sinkhole attack, jamming attack, false data attack). In the same year, another group of authors have discussed the role of BSD in MIoT. They have conceptualized IoT and BSD focusing on mobility, highlighted relationship among BSD and MIoT and at the end covered the machine learning approaches in BSD analytics (Zafar et al., 2019).

Recently, Savaglio et al., (2020) have presented ABC to face multifaceted requirements and development issues e.g., large scale deployments, cyber physicality, heterogeneity, distributed smartness, interoperability, self-management etc. This ABC allows modeling both SO/IoT system with multiple design choices before the deployment phase and programming (technical, syntactical and semantic) interoperability, autonomicity and distributed intelligence. In continuation with this development, Diène et al., (2020) have presented an insight into the data storage, service response time and overall management approaches for IoT. As it has been seen that edge computing is gradually emerging as one of bridges between cloud computing centers and mobile terminals. And, for this, Ning et al., (2020) have investigated generalize framework for its platform considering business needs, longterm maintenance solutions. Liu and Zhang (2020) have tried to enhance the security of IoT communication in block chains by encrypting information using ECC asymmetric algorithm without compromising ease of use. They have applied data compression reconstruction to accelerate data storage. Its deployment challenges has been discussed by Alamri et al., (2020).

Xu et al., (2020) have analyzed multi-heterogeneity sensor networks from micro details and macro trends. In order minimzed data transmission times, they have appied centralized wireless routing method to find the optimal path from all the sensing nodes to sink. There after, Geraldi et al., (2020) have investigated application of SPL engineering and implemetaion of VM model for IoT based on identified gaps related to context management complying constraints, contextual applications, energy-aware self-adaptive systems and consumption variants, lack of multi-perspective process, modeling of functionalities, in embedded open-source software ecosystems, context-aware systems and VM in cross-organization business requirements and process. Again in 2020, for RAP detection in embedded IoT devices by Agyemang et al., (2020) they have implemented information theoretic alogorithm through a lightweight client server application that runs on the WiFi gateway and also there was an agent application daemon that runs on the IoT device to boost layer of defense.

Application of IoT in Industry

As an indispensable technology in the future development, the IoT links people's lives in all aspects. The adoption of emerging technological trends and applications of the IoT in the industrial systems is leading towards the development of IIoT. The vision of Industry 4.0 to achieve high operational efficiency, increased productivity and better management of industrial assets and processes through product customization, intelligent monitoring applications for production floor shops and machine health, and predictive and preventive maintenance of industrial equipment can be provided by adopting IIoT and BDA processes as core components for value creation like bar-code technology, ERP, smart grid *etc*. Here, few studies have reviewd to highlight multi-dimensional growth in IIoT sector:

There are a number of challenges associated with SDN-enabled smart grids, such as single-point-of-failure due to a centralized controller, compromised network switches, grid devices such as RTUs, relays or SCADA slaves, and compromised SDN controllers or SDN controller applications, protecting hypervisors enabling virtual network slices or virtual SDNs. In this regard, Parra et al., (2019) have evaluated DPI approaches as a security tool for smart grids. They have proposed a conceptual SDN-based security monitoring framework based on SDN, NBA, Deep learning models and DPI attack corroboration, as well as a conceptual forensic-driven security monitoring framework where digital forensics and investigation capabilities are integrated to inform security monitoring. In the year 2019, authors have applied BDA method to provide generic frameworks for data engineering that can lead to development of intelligent IIoT systems (ur Rehman et al., 2019).

Thenafter, Khan et al., (2020) have evaluated mainly three different areas including IIoT architectures and frameworks, communication protocols and data management schemes. They have addressed major issues like collaborations between heterogeneous IIoT systems, public safety, coexistence of wireless technologies and protocols in IIoT, enabling decentralization on the Edge etc. In 2020, Liu et al., (2020) have presented barcode technology with ERP system to establish an industrial automatic bar code acquisition system based on IoT. Their solution provide product information quickly with unified statistics and analysis in the background system, which provides convenience for enterprise management.

(Huang, 2020, pp. 421-428) has proposed IoT data fusion method, an IoT heterogeneous data fusion method (namely wavelet transformation) based on intelligent optimization algorithm to get rid of problems like - low fusion accuracy, slow speed etc. Hence, he has checked reliability and stability of wireless network remote monitoring system while collecting data from multiple nodes. Duan et al., (2019) have explored peak load shifting, guiding EV charging options by electricity price to change the demand response of each node, thereby regulating the power quality for stable operation of the power system. Then, Feng and Hu (2020) have monitored and analysed the electric vehicle behaviour based on Hadoop network data mining model to ascertain reliability of the IoT system quickly.

Renewable Energy Harvesting using IoE Technology

The IoE is a subset of the IoT which covers all aspects of electrical energy systems and provides secure connectivity and interoperability between power grid and Internet. It consist a cloud-based and hybrid wireless mesh communication framework for bi-level, nested, distributed optimization of networked clusters of microgrids. It has appeared as an apt technology in the power sector by integrating various communication technologies and sensors. Zhihong and Li have found internet as a new driving force for clean energy across the world (Zhihong et al., 2018; Li et al., 2020).

In continuation with it, blockchain could play a pivotal role in deployment of smart grid (Humayun et al., 2020) and peer-to-peer (P2P) power trading (Miglani et al., 2020). This bidirectional P2P power transactions empowered passive consumers to be prosumers (Mahmud et al., 2020). Then after, integration of 5G internet facilities makes such communication faster with low time complexity (Lin et al., 2020). And at the same time, IoE inherits security vulnerabilities from its constituents that should be taken care for reliable operation of power system (Rose et al., 2020; Zeadally et al., 2020).

Conversely, a smart grid framework manage and control itself mainly by IoT devices and other underpinning components, where, collecting information from sensors, such as PMUs, have the capability to issue commands to actuators. This establishes the fact that IoT based renewable energy systems upgrades many network-based activities at main levels of power generation and transmitting side unanimously, which could be further served drinking and irrigation water to the rural people using solar energy.

Ground water is a major source for irrigation and drinking water supply in India. In remote areas where access to electric supply is not practical, the use of diesel pumps is very common. However, the system suffers with number of problems, such as fuel supply, technical know how and of course the operating cost. The water pumping systems are frequently used for irrigation, livestock watering and in households' water supplies and in many cases are cost effective. These systems are operated with ac or dc, directly coupled or battery operated, using various pumps, running under different climatic conditions. The PV systems are becoming increasingly popular, especially in remote areas where power from a grid is not available or is too costly to be installed. And, also due to the low quality of grid electricity supply, pumpsets are oversized to avoid burnout due to voltage and frequency fluctuations and to achieve enough water flow during the short periods of electricity availability due to heavy subsidies. Recent progress in this fields has been presented as below:

In 1993, Chowdhury et al., (1993) have taken a comprehensive look at PV-powered water pumping systems from conceptualization to design and implementation in the state of Wyoming, USA. They suggested that PV-powered water pumping could be a cost-effective alternative to distribution line extension or other conventional means of water pumping. Al Suleimani and Rao (2000) have investigated the pump's output at different wind speeds and monthly water output against average wind velocity and compares the results with the design output values provided by the system manufacturer. This study was conducted on pilot project namely MWR Heelat Ar Rakah Camp (Oman). The suggested assessing and analysing the role of wind power amongst other renewable energy sources, in the abstraction and desalination of groundwater supplies.

In the year 2010, Fernández-Ramos et al., (2010) have tried to overcome problems associated with PVPS based on SFCs and implementation of MPPT. They have applied PLC to increase the overall performance of the system. Corrêa et al., (2012) have developed a prototype stand-alone solar pumping model to reduce initial cost and its payback period. Their study suggest power savings up to 8% compared with the typical constant Volts/Hertz reference using a maximum power point tracking algorithm.

In the year 2013, Benbelkacem et al., (2013) have proposed an AR based support system to improve performance of the photovoltaic pumping system, reduction of human error in maintenance tasks and reduction of its cost. In order to locate the repairer in the working environment and then to overlay the virtual objects, a tracking method was suggested.

In continuation with it, Ebaid et al., (2013) have presented a design of a standalone photovoltaic system used to power continuously a submersible water pump located at the Disi aquifer. They suggested system sizing factors to operate it over the 25 years operational time-period are as follows - number and type of solar panels required to capture the available solar energy, the capacity and number of batteries, inverter rating, cable sizing, charge controller numbers and rating to ensure the maximum reliability of the system. In this work, the operation of a PV system having - PV generator, buck converter and DC motorpump was tested. The buck converter controls the solar array operating point in order to track MPP by using a non linear control strategy based on feedback linearization. This technique is shown to be able to obtain a linear control of the PV system and can dramatically increase its dynamical performance (Andoulssi et al., 2013).

A study on clean water availabilty based on PVRO systems in remotely located communities has beeen conducted by Elasaad et al. (2015). They have evaluated problems (like - technical, cost and operational problems) associated with small PVRO systems with production on the order of 1 m^3/day. In this research works, Chen et al., (2016) have proposed a superstructure model to determine the optimal power allocation while satisfying the power demands on the island with minimum power losses during transfer and storage. They have presented HPS design problem as a MILP for the proposed superstructure of the remote K Island on the other side of Taiwan Strait. Here, PHS is used in the hybrid HPS to buffer the impact of intermittent behavior of wind energy.

Then after, Closas and Edwin (2017) have studied the impacts, opportunities and limitations of SGPI technology within the WEF nexus. They suggested - state policies with financial subsidies lowering the cost of solar technology can mask the real costs of this technology for farmers (Khan et al., 2020). Moreover, the focus on cost and technology-efficiency comparisons (diesel pumps vs. solar powered pumps) has distracted until now the discussion on the impact of state intervention in market regulation and financial incentives. A conundrum here arises as, on the one hand, subsidies can lead to market distortions but on the other they are required if this technology is to be made accessible to all. Additionally, without positive incentives such as reduced trade barriers, the financial and economic feasibility of this technology could be compromised. The recommendation here is that subsidies need to be targeted and tied to design regulations, water requirements and accompanied with the monitoring of maximum pump and well extraction capacities.

In the year 2018, Playán et al. (2018) have assessed TM and RC systems for irrigation in Spain. Their studies clarify future evolution and support policy updates on this technology. In this study, (Gupta, 2019, pp. 598-609) has adopted difference-in-differences approach to understand informal impact of solar water pumps on the use of water, electricity and diesel as well as the gross cropped area under fruits and vegetables alongwith profits of farmers across many districts Rajasthan such as Jaipur, Sikar, Sri Ganganagar, Bikaner, Jaisalmer and Chittorgarh.

In continuation with it, an IoT-based prototype model for smart farm irrigation was developed by Al-Ali et al., (2020) for regions around the world, which face water scarcity and power shortage. This renewable energy monitoring system makes it easily accessible through the web via a computer, tab or mobile phone devices. In this paper, authors have carried out dispatch strategy based performance analysis to provide power supply using the hybrid renewable energy configurations (eg., Lead Acid and Lithium-Ion batteries with Photovoltaic / Hydro / Battery and Photovoltaic / Wind / Hydro / Diesel Generator / Battery) for a remote rural area in the state of Karnataka (India). This study is reflecting, that a stand-alone system may be a feasible option for unserved regions of electricity (Ramesh and Saini, 2020). In this paper (Garrido-Rubio et al., 2020), authors have introduces an operative application of a RS-SWB to obtain an RS-IWA for irrigation management in arid or semi-arid environments.

MATHEMATICAL MODELLING

The electricity generation behavior of a PV cell is often described by a diode model. In this model, the PV cell (generator) is represented by an electric current generator which is equivalent to a current source parallel to a diode. The VI equation of the solar generator is given by (1) (Mohammedi et al., 2014; Mokeddem et al., 2011),

$$I_p = I_{ph} - I_s\left[e^{\frac{V_p}{V_T}} - 1\right] \tag{1}$$

Here, V_T is can be expressed as in (2) (Mohammedi et al., 2014; Mokeddem et al., 2011),

$$V_T = \frac{nK_BT}{q} \tag{2}$$

The equivalent mechanical output power of the pump is calculated as (3) (Mohammedi et al., 2014; Mokeddem et al., 2011),

$$P_{hyd} = mgh_t = \eta_t P_{mot} = \eta_t \omega T_L \tag{3}$$

The load torque of the pump is now given by (4) (Mohammedi et al., 2014; Mokeddem et al., 2011),

$$T_L = \frac{mgh_t}{\eta_t \omega} = \frac{V_d gh_t}{\eta_t} \tag{4}$$

The maximum theoretical amount of energy stored may be given by (5) (Morabito and Hendrick, 2019),

$$E = \frac{\rho VgH}{3600} \tag{5}$$

The amount of energy consumed in storing energy in the whole upper reservoir may be formulated as in (6) (Morabito and Hendrick, 2019),

$$E_p = \frac{E}{\eta_p} \tag{6}$$

Figure 2. MPPT photovoltaic radiation characteristics

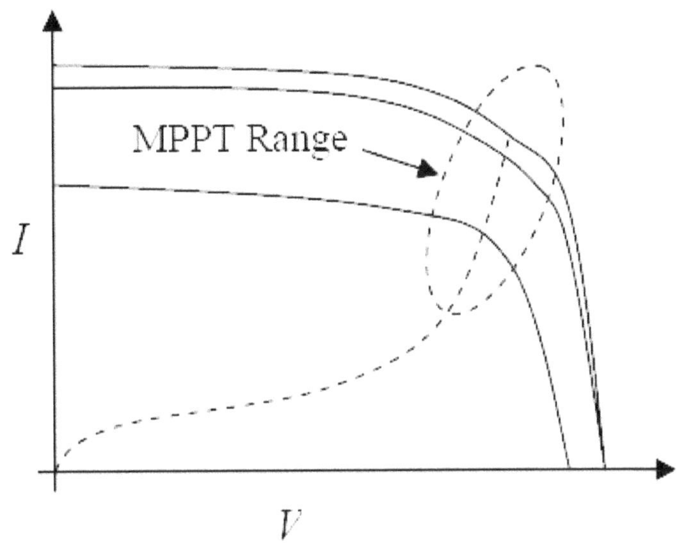

In case of unavailablity of exploitable head and its varition considerably during a whole cycle, then it must able recharge the upper reservoir by (7) (Morabito and Hendrick, 2019),

$$H_{P,\max} = H_{th} + H_l \tag{7}$$

The pump specific speed may be given by (8) (Morabito and Hendrick, 2019),

$$N_{sp} = \frac{N_p\sqrt{Q_p}}{\left(H_p\right)^{0.75}} \tag{8}$$

The power generated by the pump installed may be given by (9) (Morabito and Hendrick, 2019),

$$P_p = \frac{\rho g H_p Q_p}{\eta_p} \tag{9}$$

Thenafter, solar panel and pump efficiency as applicable may be formulated by (10)-(11) (Mohammedi et al., 2014; Mokeddem et al., 2011),

$$\eta_{pv} = \frac{P_{pv}}{E_s S N_s N_p} \tag{10}$$

$$\eta_{pump} = \frac{\rho g Q_v h}{3600 P_{pv}} \tag{11}$$

Figure 3. Architecture layout of SWP System

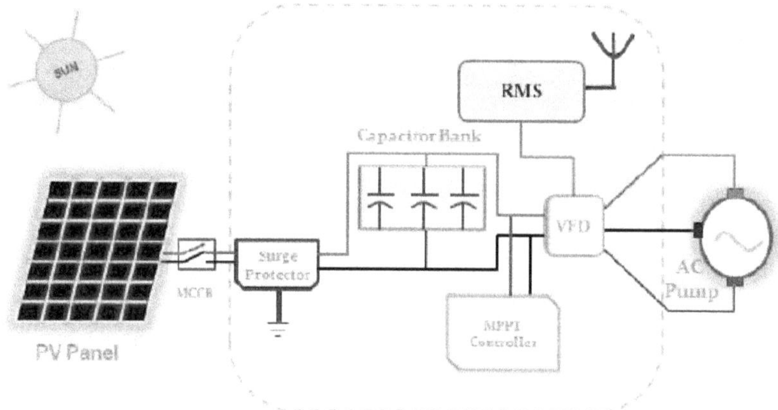

Then, the total efficiency of the PV pumping system may be formulated as in (12) (Mohammedi et al., 2014; Mokeddem et al., 2011),

$$\eta_t = \eta_{pv} . \eta_{pump} \tag{12}$$

The maximum power point characteristics has been shown in Figure 2 and corresponding mathematical formulation may be expressed by (13) (Vijayakumari, 2020; Mohanty et al., 2014),

$$P_{mpp} = V_{mpp} I_{mpp} \tag{13}$$

Furthermore, the maximum output power from a PV panel in terms of FF may be given by (14) (Vijayakumari, 2020; Mohanty et al., 2014),

$$P_{mpp} = V_{oc} I_{sc} \times FF \tag{14}$$

This constant value of FF of the PV panel under varying weather condition may be presented as in (15) (Vijayakumari, 2020; Mohanty et al., 2014),

$$\text{FF} = \frac{\left(\frac{V_{oc}}{nkT / q}\right)}{\left(\frac{V_{oc}}{nkT / q}\right) + 4.7} \tag{15}$$

The incremental voltage increases the power and its performace evaluation of the MPPT, which could be formulated as in (16)-(17) (Vijayakumari, 2020; Mohanty et al., 2014),

$$V_r = V_i + dV \text{ if } \frac{dP}{dV} > 0 \tag{16}$$

Figure 4. Schematic of SWP Controller

$$V_r = V_i - dV \text{ if } \frac{dP}{dV} < 0 \quad (17)$$

ARCHITECTURE OF THE PROPOSED SWP SYSTEM

The performance evaluation of the undertaken case has been made to ascertain viability of the IoT technology. Hence, the IoE emerged as solution, where the scattered devices can be operated with the power of internet and transmit data to the web based platform. This collection of devices needs not to be communicated with same protocol but can be extracted and gathered from different inter-communicating

Figure 5. Optimization flowchart of standalone SWP system

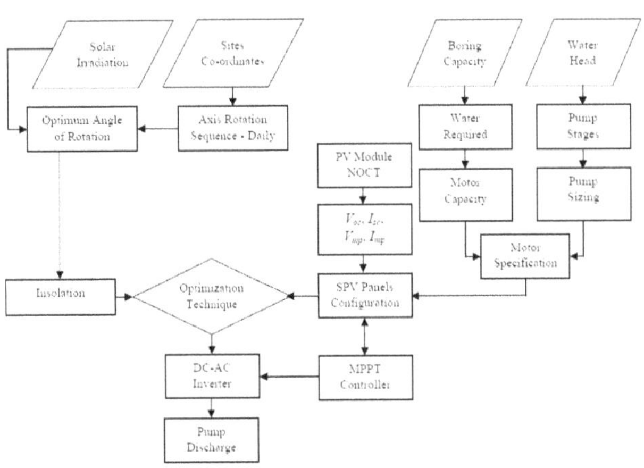

Table 2. Configuration of PV pump controller

Parameters	Descriptions
Dimensions	• 400mm × 370mm × 650mm (MS powder coated enclosure)
Environment	• Operating temperature: 0-70°C • Ingress protection: IP52 • Operating humidity: 10-90% RH, non-condensing • Operates on existing 3 phase EB supply
Linear current boosting control	• Inbuilt MPPT tracking for maximum power harvesting • Dynamic PI based control feedback for variable speed ranges • Field configurable performance settings for site-specific needs
Variable frequency drive	• Output: 3 ph, 200-240V AC/ 380-500V AC • Operational capacity: 1-10 HP • Allowable voltage input fluctuation: + 10%, -15%
Power supply	• 100-900V DC from solar panel • 3 ph, 415V AC • Automatic/programmable switchover between EB and solar panel
Network interfaces	• MODBUS over RS485/RS232 • Serial port debugging interface GSM/GPRS two way communication
Remote monitoring & control features	• Real-time pump status, frequency, discharge, energies, power monitoring and uplink to web server at predefined intervals • SMS/web-based monitoring and control for controlling pump operation, frequencies and troubleshooting • SMS based on/off mechanism for local farmers/operators web-based application for central office monitoring and control
Central data features	• Scalable solution with several concurrent monitoring units • User defined granularity on data access and visibility • Modular software model: admin, user, visualization, monitoring *etc* • Data dashboard, analysis, trending, alarms, SMS • Cloud based data model with redundancies • Central web server for remote multiple client interfaces

Figure 6. Flowchart for MPPT operations

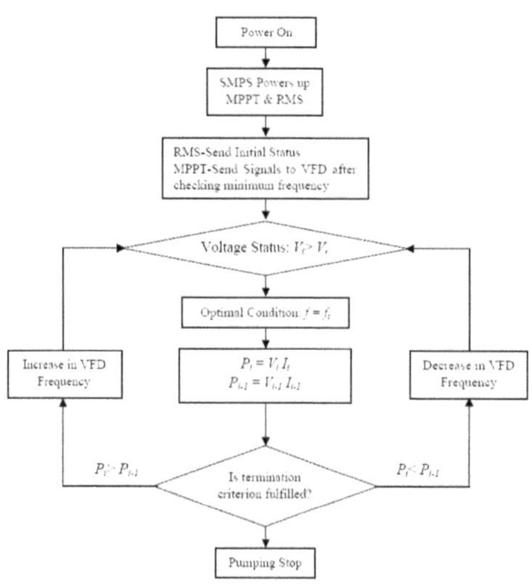

Figure 7. Flowchart for RMS operations

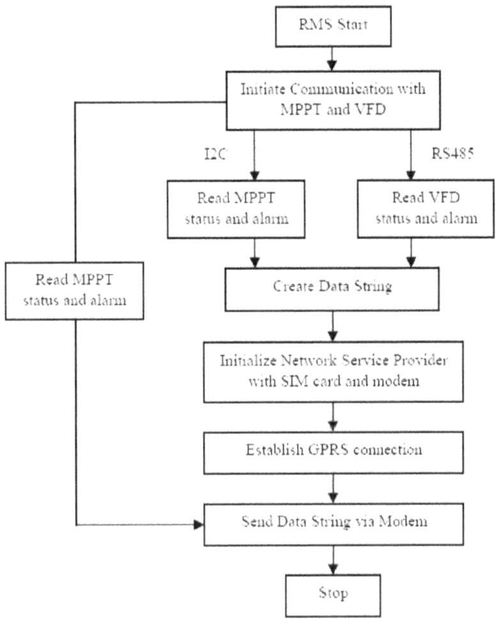

devices. The key factor is internet and once the device is connected data will be in cloud. This architecture keeps the device more robust with efficient processors and handy control logic. The system comprises following major components *e.g.,* solar panel, structure (dual axis manual tracking), VFD, RMS, MPPT, capacitor, submersible pump with AC motor, protection accessories *etc* (as shown in Figure 3). The cloud computing makes the IoT devices powerful with computation and analytics at the server level, data provided by the internet connected field devices. Major components have described as below:

Solar Pump Controller

The solar water pump is an off grid solution to harvest water with renewable energy and minimum RoI period. These solutions are especially for the location which are far away from grid or the grid supply is intermittent and the farmers need to switch to diesel for power generation. The solar water pumps installed at remote locations and are difficult to reach (refer to Figure 4). The effectiveness of the solar pumps can be estimated if these are monitored well with all the critical parameter. Solar water pump made of electrical and electronics component which may not perform as designed or can malfunction. The data are also important to troubleshoot the system and its various parameters are shown in Table 2:

The flowchart depicts the optimization techniques to achieve maximum water discharge from the SWP. The efficacy of the solar water pump converges to the total discharge throughout the day. The right combination of pump and pipes sizing to mitigate the stack head losses is critical when the pump thrust at power less than 50% of the capacity available (as shown in Figure 5). The solar panels voltage is minimum to enhance the motor speed and minimum current to achieve the torque to overcome the head losses. Here, the component linear current booster is used to boost the starting current surge required during morning time.

Table 3. Technical aspects of transmitter

Parameters	Description
Trans receiver Input	4-20 mA, 0-10V DC, 0-24V DC, Digital (On/Off)
Protocol	MODBUS-RTU and HART
Serial port	RS485/RS422/RS232
MODBUS slave	Multiple MODBUS slave (32 max without repeater)
Baud rate	300 - 1,15,200 Bauds
Digital input (Integral)	2 No 24VDCDI (On/Off)
Log interval	30 sec to 60 min
No. of alarm	Multiple Alarm Web configured
Type of alarm	Email / SMS
Communication (GSM/GPRS)	Quad band GPRS Class 12
GPRS failure	Facility to save historical data into SD Card
Power supply	9-24 V DC
Power consumption	4W
Temperature	55°C
Humidity	10- 95% (Non condensing)

The MPPT controlled SWP system keeps on tracking the max available power from the PV panels. Then allow the motor to operate under such conditions. The core objective of the MPPT is to keep the flow of current in control. The increase in current leads to drop in voltage which is subjected to maximum under certain fixed insolation (as prsented in Figure 6). The change in radiation leads to deviation of the maximum power point of the voltage (V_{mpp}) and corresponding current (I_{mpp}). The process is dynamical control needs of function instantaneously to respond by controlling the frequency of VFD with respect to the available power.

Temperature Transmitter

Temperature being an important aspect which plays critical role for any system failure, prior information of such data plays crucial role to provide predictive maintenance before any failure could happen. These data is also critical to overview system efficiency which gets detoriated and the time of use of the device. Cool down period could be also be established for any continuously working devices. Temperature of the solar panels also is an important criterion for the generation of solar power; hence the temperature measurement of the solar panels is important. These temperature transmitters can directly produce signal to the RMS and the can add it to the strings of data being sent.

RMS System

The RMS system plays vital roles by delivering the key data which might be the reason of the system failure. Sometimes these solar water pumps are interconnected with grid and can run on solar and grid as well. The operator may opt to run on grid in spite of available solar energy from the solar cells just

Table 4. Pumping system descriptions

Parameters	Description
Well characteristics (Diameter)	15 cm
Static level depth	-40 m
Pump level	-50 m
Storage tank (Volume)	5.0 m^3
Feeding by bottom (Bottom level)	0.0 m
Hydraulic circuit (Piping length)	60 m
Number of elbows	1
Water needs (Yearly average)	137.56 m^3/day
Pump technology	Centrifugal multistage
Integrated inverter	MPPT type (100-600 V)
Operating conditions water head wise (m)	Minimum: 28.0
	Nominal: 50.0
	Maximum: 75.0
Corresponding maximum flow rate (l/min)	Minimum head: 370.0
	Nominal head: 270.0
	Maximum head: 130.0
Required power (W)	Minimum: 3529
	Nominal: 3900
	Maximum: 3707

for comfort. These are important for the end user if the solar power is being used effectively (refer to Figure 7).

Water and Process Flow Transmitter

Solar power generated is directly related to the flow of water in solar pumps hence to measure the efficiency of the solar pump. The flow of water measurement is important. The flowmeter also plays important parameter for any distribution line to estimate the capacity of the system and its usage. The flow meters are of mostly pulse type which can provide pulse signal to the RMS directly and some of the flowmeter have communication protocol if the flow-meters have inbuilt smart technology it can store data for certain time (given in Table 3). These flow-meters can provide flow rate, total fluid displaced per day and many other criteria.

Features on the Dashboard for Quick Response: It also has feature to create and tag alarm online over web application as followings:

- Historical reports of the consumption over time.
- Estimations of when it's time to refill.
- Alarms when certain thresholds are reached or if the content levels drop too fast (e.g., theft condition).

Table 5. Description for capacity sizing of deep-well to a storage facility

Parameters	Description
Specific drawdown	-10.00 m/m³/h
Lower dynamic level	-45 m
Diameter	1.8 m
Height (full level)	2.1 m
Pipes DN65 (D_{int})	69 mm

Table 6. PV array characteristics of the proposed system

Parameters	Description
PV module	Si-Poly
Model	Poly 250 Wp, 60 cells
No. of PV modules (in series)	24
No. of PV modules (in parallel)	2 strings
Unit normal power	250 Wp
Array global power (at STC)	6.00 kWp
Array global power (at operating conditions)	5.37 kWp (50°C)
Array operating characteristics (50°C)	V_{mpp}: 324 V I_{mpp}: 17 A
Total area	Module area: 38.6 m^2 Cell area: 35.0 m^2
Tracking plane (two-axis)	Minimum tilt: 0° Maximum tilt: 80°
Rotation limitations	Minimum azimuth: -120° Maximum azimuth: 120°

Table 7. Features of power conditioning unit (MPPT-AC inverter)

Parameters	Description
Minimum MPP voltage	100 V
Maximum MPP voltage	600 V
Maximum array voltage	600 V
Maximum input current	40.0 A
Nominal power	8000 W
Power threshold	80 W
Maximum efficiency	96.0%
EURO efficiency	94.0%

Table 8. Simulation outcome for SWP systems' loss throughout the year

Month	E_{ef} (kWh/m²)	E_{UE} (kWh)	E_{PO} (kWh)	E_{UE} (kWh)	h_t (m)	W_P (m³)	W_u (m³)	W_m (m³)
Jan	173.1	919	0.000	0.000	0.000	0.000	2.500	3098
Feb	192.8	986	1.600	0.000	40.57	0.057	0.057	3360
Mar	233.7	1157	4.845	0.000	42.56	0.768	0.768	4339
Apr	240.2	1159	0.000	0.000	0.000	0.000	0.000	4500
May	240.3	1162	1.605	0.000	41.24	0.124	0.124	4650
Jun	194.8	964	3.238	0.000	43.13	0.625	0.625	4499
Jul	169.0	850	9.708	0.000	42.98	1.787	1.787	4648
Aug	177.0	884	3.260	0.000	44.56	0.912	0.912	4649
Sep	177 4	885	8.065	0.000	42.32	1.161	1.161	4499
Oct	184.2	931	4.827	0.000	41.79	0.537	0.537	4649
Nov	190.9	972	3.277	0.000	42.34	0.469	0.469	3900
Dec	176.7	930	0.000	0.000	0.00	0.000	0.000	3410

Figure 8. Characteristics of normalized productions (per installed kWp)

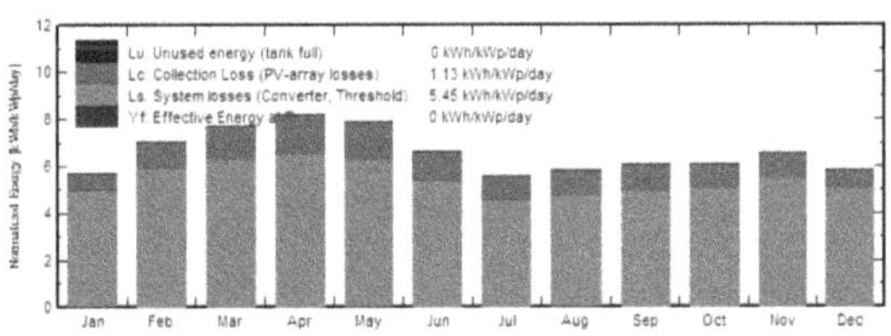

Figure 9. Daily electricity (input/output) characteristics

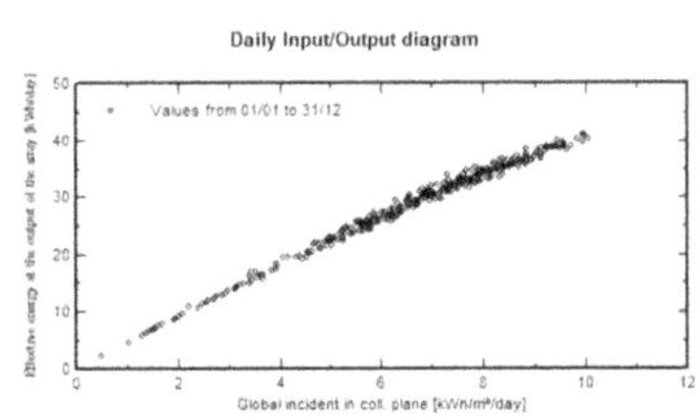

Figure 10. Characteristics of daily array temperature and effective irradiance variation

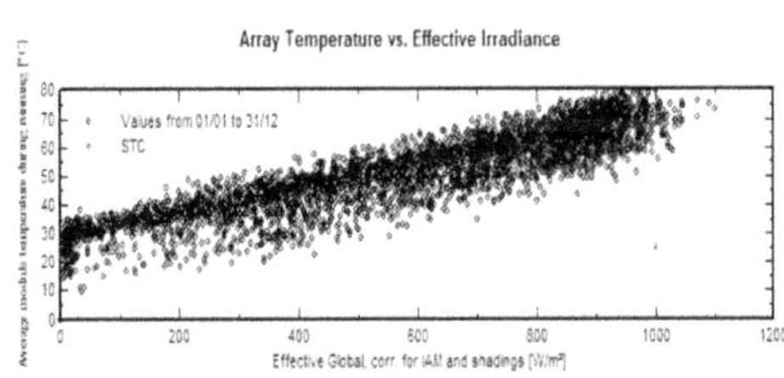

Figure 11. Sequential power loss characteristics over the year

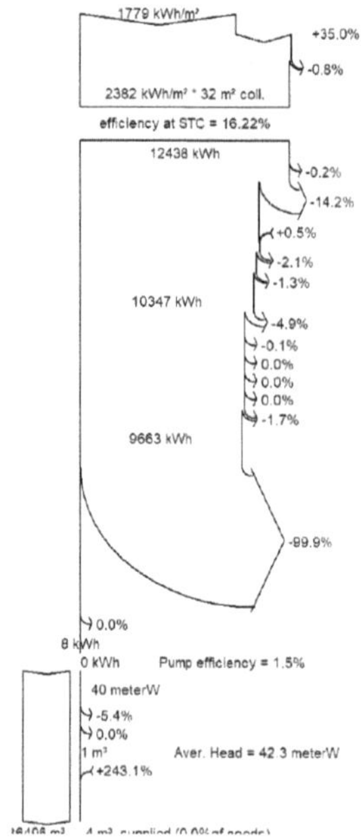

- Alarms can be set on several thresholds and sent via e-mail, RSS-feeds, SMS.

Applications:

- Access management: Alarm for opening of the door.
- Power metering: Power generation / consumption
- Rectifier/Battery management: Monitor the health of solar batteries.
- Generator management: Start and stop of DG from web monitoring of important parameters.
- Fuel management: To monitor fuel level in diesel tank.
- kVA level solar power plant (On/Off Grid)

Field Device

Now-a-days, use of renewable energy for various energy driven application is the most common. These sources of energy are available throughout geographically and can be utilised for multiple usages as and

Figure 12. Circuit schematic of MPPT-RMS interface

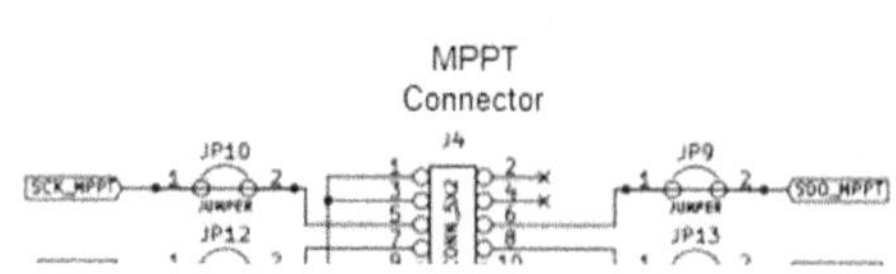

Figure 13. Circuit schematic of RMS for SWP system

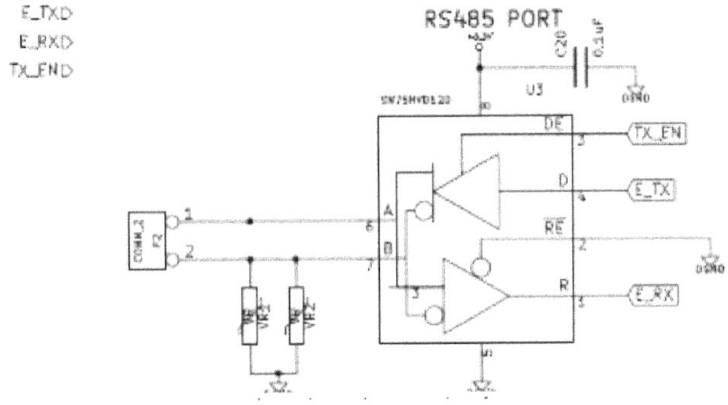

when required. The operation, control and RMS have wide range of use which plays crucial roles in the various application of IoT. The device is installed at remote location to monitor the variable data of the system along with the critical parameter. These data are stored locally on the device and broadcasted to the server at programmed time delay.

SIMULATION OF PV PUMPING SYSTEM

This paper presents solar water pumping solution for the remote places. Initially, the proposed set up has been examined using the PVSyst software tools and corresponding outcome have been briefed as below (refer to Table 4-8) and graphical representation (refer to Figure 8-11):

INTEGRATION OF IOE CONCEPTS TO SWP SOLUTION

The IoT field devices are prone to hacking and security threat, these concerns were identified to avoid any kind of data theft or device mishandling. Two layers of security concern with the RMS devices:

Figure 14. Circuit schematic for VFD data readings

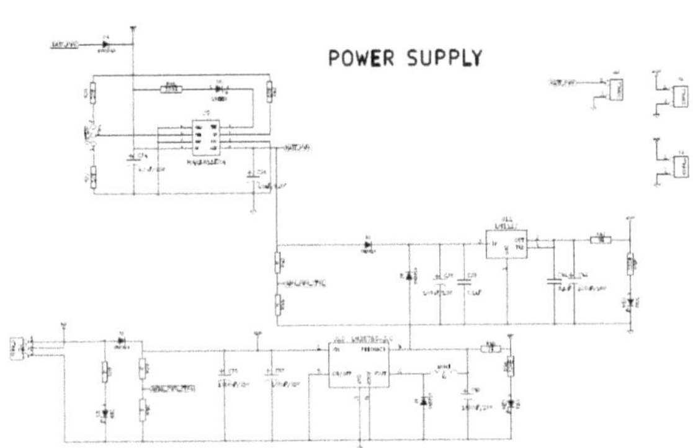

Figure 15. Water discharge variation graph

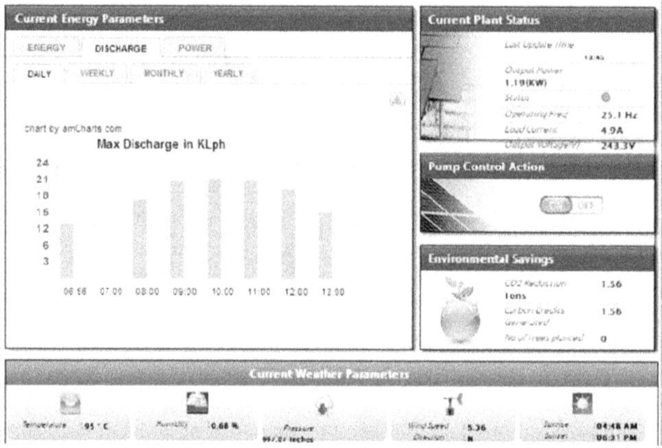

Field Device Level

These RMS devices where installed at the various sites and are used as communication gateway between field controller, such as inverter, meters, MPPT controller, flow meter and other monitoring devices. In such a scenario communication should be safe enough to communicate from the field devices and as well the remote data transmission should also be protected. The communications established between the devices are through various protocols such as I2C, MODBUS, HART. The physical communication medium should be tampered proof, hence the cable used where special purpose signalling cables with shielded ground and the connectors where also lock type which can be removed only if the device is opened. There where various mechanism used which was if the device cover is removed it will raise an alarm to the server and will cease to function if the token respond is not received against the alarm. These token response time where programmable depending on the site condition and signal (refer to Figure 12). Token response *I* generated by the admin who shall confirm if it was intentional maintenance work or system tampering. Mutual authentication between RMS, field devices, gateway and backend through PKI.

The RS485 communication is used to communicate with VFD reading the parameter from the VFD via MODBUS protocol. Where RMS is the master device and VFD is slave. Schematic shows the communication link between the RS485/232 and the Microchip 32bit processor. DB9 connector is used as an interface to connect with the external device. MPPT is connected with the RMS and protocol used is I2C.

RMS Cloud Servers

The communication between the RMS and Server where data encrypted and the encryption keys can be changed over time. To ensure if the hacker has gained any access to the data, the data retrieved will be encrypted and will be of no use. Data encryption in all data exchange to ensure privacy and integrity. Security updated FOA (Firmware On Air) with renewable device keys. Offset maximum levels of diesel through use of renewable and energy storage (possible demand response/shift) while maintaining reliability (as shown in Figure 13). Rural electrification systems may have no fossil back-up and instead focus on achieving the best service for the available resources. Reduce costs and carbon while providing

Figure 16. Power variation throughout the day

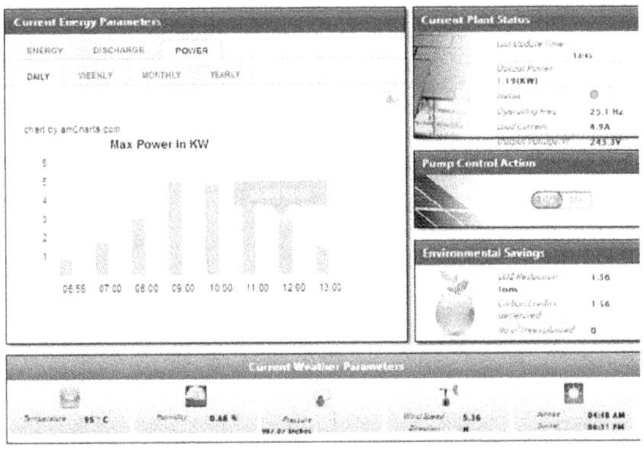

reliability. Dispatch resources to maintain frequency and voltage of system while maintaining adequate spinning reserves for load and generation fluctuations. Major benefits are as follows:

- Electrification
- Cost and carbon reduction
- Community empowerment

Table 9. Low voltage VFD discharge

Frequency	Duration of discharge
6 Hz	43.5 sec
9 Hz	23.8 sec
12 Hz	20 sec
15 Hz	16.1 sec
18 Hz	13.3 sec
21 Hz	11.3 sec
24 Hz	10.1 sec
27 Hz	9 sec
30 Hz	8 sec
33 Hz	7.2 sec
36 Hz	6.5 sec
39 Hz	6.2 sec
42 Hz	6 sec
45 Hz	-

Power Supply Unit

The power supply circuit has 12V DC input which can be sourced from any 12VDC standard battery or SMPS which can tap power directly from the high voltage DC from solar panel connections used to drive the pump (as shown in Figure 14). The Output 3.3V DC is used by the processor and 5 VDC is used by the MPPT board and few other components on the PCB. Provision for battery charging circuit is also provided in order to provide data redundancy and not solely depending on the power from the main circuit. The battery backed RMS provides data of the system in no power condition monitoring the critical parameter and also sending the details of the system parameters.

RESULT DISCUSSIONS

Most important parameter for the solar water pump is the water discharge which is directly proportional to the return on investment brings up the overall economic benefits by of solar water pump. The discharge can be observed by reading from the flowmeter. Optional method to calculate the flow rate is by the reading the frequency and formulating it with the flow rate. The graph shows the total discharge per hour for the particular day and is nil if there is no discharge (as shown in Figure 15).

Table 10. DAC output started with frequency (for working board)

DAC	U7		U5				
	PIN 2&3	*PIN 6&7*	*PIN 3&2*	*VIA & CC*	*PIN 1&3*	*PIN 3 wrt GND*	*PIN 2 wrt GND*
100	0.018	0.124	0.004	1	1.084	2.066	2.071
150	0.029	0.178	0.004	1.441	0.728	2.09	2.099
200	0.036	0.236	0.005	1.985	0.361	2.116	2.13
250	0.046	0.298	0.003	2.163	0.271	2.141	2.252
300	0.057	0.342	0.002	3.021	0.625	2.176	2.17
350	0.064	0.409	0.002	3.215	1.03	2.195	2.192
400	0.072	0.457	0.001	3.678	1.378	2.21	2.219
450	0.083	0.514	0.001	3.971	1.92	2.236	2.242
500	0.091	0.566	0.001	4.521	2.316	2.261	2.266
550	0.1	0.633	0.001	5.09	2.671	2.285	2.29
600	0.108	0.678	0.001	5.375	3.079	2.31	2.315
650	0.117	0.721	0.001	5.998	3.216	2.33	2.34
700	0.126	0.783	0.001	6.546	4.021	2.359	2.365
750	0.135	0.849	0.001	6.93	4.052	2.384	2.388
800	0.145	0.903	0.001	7.38	4.971	2.408	2.414
850	0.151	0.957	0.001	7.78	5.351	2.434	2.438
900	0.16	1	0.001	8.22	5.968	2.457	2.465
**902 DAC started with frequency 41.5 Hz (Programmed frequency is 45 Hz)							

Table 11. Summary of the proposed SWP solution

Application	Irrigation, sanitation and household
Requirement	More than 1 Lac LPD @ 40 m head average
Peak generation	>1,50,000 Ltrs per day
Monitoring	• Cloud computed remote monitoring and control. • Web based user interface.
Project design	• The solar water pump was designed with MPPT controller for maximized power extraction, inbuilt remote monitoring and control.
	• Capacitor based resiliency to support radiation fluctuation.
	• Dual axis manual tracker daily and seasonal.
	• SMS and web based on-off mechanism.
	• Data analysis report - voltage, current, power, frequency, discharge and radiation.
Benefits	• MPPT based controlled extracting maximum power from the solar panels
	• Capacitor bank minimized motor trip to efficiently control the power delivery
	• Remote monitoring lowered the maintenance time and failure rate of the pump.
	• Remote data provided efficient troubleshooting and minimized field visit.
	• Solar panel generation and forecasting the drop in generation due to weather and dusty atmosphere condition
	• Power generation maximized by minimizing the shutdown time of numbers of installation with alerts and reports.
Challenges and Learning	• Solar water pump leveraged the power utilization compared to intermittent grid supply or diesel generator pumps. • The system advancement needed proper training for the end user and incorporation of remote monitoring minimized the intervention of unskilled user at remote location. • The solar DC power supply was 500 kW with 700V DC which is fatal hence special conduit was provided to protect the cables from rodent and human reach.

Power bar graphs average the output power for one hour to provide graphical data with the power out from the pump for every hour (as shown in Figure 16). These data are averaged from the number of data strings received from the RMS device in hour. The right side panel on the screen shows the live status of the pump (given in Table 9-10).

CONCLUSION

As we know, life without water is beyond our imagination and pumping water is the most common way to meet those basic needs. The PV water pump is emerged as a viable solution to the rural India, where availability of the grid power is uncertain and dependence on the crude oils based irrigation system is uneconomical. These factors motivated us to develop IoE technology based SWP system. In this context, the MPPT based SWP system incorporating RMS has been developed and further, installed at Bihar Sharif (Nalanda, Bihar) site in India. Its installation capacity was 5HP having four in numbers. The primary objective of installation is to meet the local irrigation needs; however, it has also served their sanitation and household requirements to the some extents. The proposed web-based solution is apt to monitor and control significant SWP system parameters like voltage, current, power, frequency, discharge and radiation etc. In order to attain maximum power on daily and seasonal basis, the MPPT based controlled extracting with dual axis manual tracker. The effectiveness of irrigation management using proposed SWP system has been summarised in Table 11. However, in near future, same solution could be expanded for energy management of conventional and renewable energy fields.

REFERENCES

Adimalla, N., Dhakate, R., Kasarla, A., & Taloor, A. K. (2020). Appraisal of groundwater quality for drinking and irrigation purposes in Central Telangana, India. *Groundwater for Sustainable Development, 10*, 100334. https://state.bihar.gov.in/krishi/CitizenHome.html

Agyemang, J. O., Kponyo, J. J., Klogo, G. S., & Boateng, J. O. (2020). Lightweight Rogue Access Point Detection Algorithm for WiFi-Enabled Internet of Things (IoT) Devices. Internet of Things, 100200.

Al-Ali, A.R., Al Nabulsi, A., Mukhopadhyay, S., Awal, M.S., Fernandes, S. & Ailabouni, K. (2020). IoT-solar energy powered smart farm irrigation system. *Journal of Electronic Science and Technology*, 100017.

Al Suleimani, Z., & Rao, N. R. (2000). Wind-powered electric water-pumping system installed in a remote location. *Applied Energy*, *65*(1-4), 339–347. doi:10.1016/S0306-2619(99)00099-9

Alamri, M. (2019). Blockchain for Internet of Things (IoT) Research Issues Challenges & Future Directions: A Review. *Int. J. Comput. Sci. Netw. Secur*, *19*, 244–258.

Andoulssi, R., Draou, A., Jerbi, H., Alghonamy, A., & Khiari, B. (2013). Non linear control of a photovoltaic water pumping System. *Energy Procedia*, *42*, 328–336. doi:10.1016/j.egypro.2013.11.033

Bahinipati, C. S., & Viswanathan, P. K. (2019). Incentivizing resource efficient technologies in India: Evidence from diffusion of micro-irrigation in the dark zone regions of Gujarat. *Land Use Policy*, *86*, 253–260. doi:10.1016/j.landusepol.2019.04.017

Benbelkacem, S., Belhocine, M., Bellarbi, A., Zenati-Henda, N., & Tadjine, M. (2013). Augmented reality for photovoltaic pumping systems maintenance tasks. *Renewable Energy*, *55*, 428–437. doi:10.1016/j.renene.2012.12.043

Chapter 7: Agriculture and Food Management. (n.d.). *Economic Survey 2019-20*. https://www.indiabudget.gov.in/economicsurvey/eco_sur_vol(iiallchapter).php

Chen, C. L., Chen, H. C., & Lee, J. Y. (2016). Application of a generic superstructure-based formulation to the design of wind-pumped-storage hybrid systems on remote islands. *Energy Conversion and Management*, *111*, 339–351.

Chowdhury, B. H., Ula, S., & Stokes, K. (1993). Photovoltaic-powered water pumping-design, and implementation: Case studies in Wyoming. *IEEE Transactions on Energy Conversion*, *8*(4), 646–652.

Closas, A., & Rap, E. (2017). Solar-based groundwater pumping for irrigation: Sustainability, policies, and limitations. *Energy Policy*, *104*, 33–37.

Corrêa, T. P., Seleme, S. I. Jr, & Silva, S. R. (2012). Efficiency optimization in stand-alone photovoltaic pumping system. *Renewable Energy*, *41*, 220–226.

Diène, B., Rodrigues, J. J., Diallo, O., Ndoye, E. H. M., & Korotaev, V. V. (2020). Data management techniques for Internet of Things. *Mechanical Systems and Signal Processing*, *138*, 106564.

Duan, B., Xin, K., & Zhong, Y. (2019). Optimal Dispatching of Electric Vehicles Based on Smart Contract and Internet of Things. *IEEE Access: Practical Innovations, Open Solutions*, *8*, 9630–9639.

Ebaid, M. S., Qandil, H., & Hammad, M. (2013). A unified approach for designing a photovoltaic solar system for the underground water pumping well-34 at Disi aquifer. *Energy Conversion and Management*, *75*, 780–795.

Elasaad, H., Bilton, A., Kelley, L., Duayhe, O., & Dubowsky, S. (2015). Field evaluation of a community scale solar powered water purification technology: A case study of a remote Mexican community application. *Desalination*, *375*, 71–80.

Feng, X., & Hu, J. (2020). Research on the identification and management of vehicle behaviour based on Internet of things technology. *Computer Communications*.

Fernández-Ramos, J., Narvarte-Fernández, L. & Poza-Saura, F. (2010). Improvement of photovoltaic pumping systems based on standard frequency converters by means of programmable logic controllers. *Solar Energy, 84*(1), 101-109.

Garrido-Rubio, J., González-Piqueras, J., Campos, I., Osann, A., González-Gómez, L., & Calera, A. (2020). Remote sensing–based soil water balance for irrigation water accounting at plot and water user association management scale. *Agricultural Water Management*, *238*, 106236.

Geraldi, R. T., Reinehr, S., & Malucelli, A. (2020). Software product line applied to the internet of things: A systematic literature review. *Information and Software Technology*, 106293.

Ghosh, S., Kolady, D. E., Das, U., Gorain, S., Srivastava, S. K., & Mondal, B. (2019). Spatio-temporal variations in effects of participatory irrigation management (PIM) reform in India: A panel data analysis. *Agricultural Water Management*, *222*, 48–61.

Gupta, E. (2019). The impact of solar water pumps on energy-water-food nexus: Evidence from Rajasthan, India. *Energy Policy*, *129*, 598–609.

Hajiheidari, S., Wakil, K., Badri, M., & Navimipour, N. J. (2019). Intrusion detection systems in the Internet of things: A comprehensive investigation. *Computer Networks*, *160*, 165–191.

Hersent, O., Boswarthick, D., & Elloumi, O. (2011). The internet of things: Key applications and protocols. John Wiley & Sons.

Huang, X. (2020). Intelligent remote monitoring and manufacturing system of production line based on industrial Internet of Things. *Computer Communications*, *150*, 421–428.

Humayun, M., Jhanjhi, N. Z., Hamid, B., & Ahmed, G. (2020). Emerging Smart Logistics and Transportation Using IoT and Blockchain. *IEEE Internet of Things Magazine*, *3*(2), 58–62.

Khan, A. (2020). N. Z. Jhanjhi, Mamoona Humayun, and Muneer Ahmad. "The Role of IoT in Digital Governance. In *Employing Recent Technologies for Improved Digital Governance* (pp. 128–150). IGI Global.

Khan, W. Z., Rehman, M. H., Zangoti, H. M., Afzal, M. K., Armi, N., & Salah, K. (2020). Industrial internet of things: Recent advances, enabling technologies and open challenges. *Computers & Electrical Engineering*, *81*, 106522.

Li, L., Zheng, Y., Zheng, S., & Ke, H. (2020). The new smart city programme: Evaluating the effect of the internet of energy on air quality in China. *The Science of the Total Environment*, *714*, 136380.

Lin, L., Guan, X., Hu, B., Li, J., Wang, N., & Sun, D. (2020). Deep reinforcement learning and LSTM for optimal renewable energy accommodation in 5G internet of energy with bad data tolerant. *Computer Communications*.

Liu, K., Bi, Y., & Liu, D. (2020). Internet of Things based acquisition system of industrial intelligent bar code for smart city applications. *Computer Communications*, *150*, 325–333.

Liu, Y., & Zhang, S. (2020). Information security and storage of Internet of Things based on block chains. *Future Generation Computer Systems*, *106*, 296–303.

Mahmud, K., Khan, B., Ravishankar, J., Ahmadi, A., & Siano, P. (2020). An internet of energy framework with distributed energy resources, prosumers and small-scale virtual power plants: An overview. *Renewable & Sustainable Energy Reviews*, *127*, 109840.

Miglani, A., Kumar, N., Chamola, V., & Zeadally, S. (2020). Blockchain for Internet of Energy management: Review, solutions, and challenges. *Computer Communications*, *151*, 395–418.

Mohammedi, A., Mezzai, N., Rekioua, D., & Rekioua, T. (2014). Impact of shadow on the performances of a domestic photovoltaic pumping system incorporating an MPPT control: A case study in Bejaia, North Algeria. *Energy Conversion and Management*, *84*, 20–29.

Mohanty, P., Bhuvaneswari, G., Balasubramanian, R., & Dhaliwal, N. K. (2014). MATLAB based modeling to study the performance of different MPPT techniques used for solar PV system under various operating conditions. *Renewable & Sustainable Energy Reviews*, *38*, 581–593.

Mokeddem, A., Midoun, A., Kadri, D., Hiadsi, S., & Raja, I. A. (2011). Performance of a directly-coupled PV water pumping system. *Energy Conversion and Management*, *52*(10), 3089–3095.

Morabito, A., & Hendrick, P. (2019). Pump as turbine applied to micro energy storage and smart water grids: A case study. *Applied Energy*, *241*, 567–579.

Ning, H., Li, Y., Shi, F., & Yang, L. T. (2020). Heterogeneous edge computing open platforms and tools for internet of things. *Future Generation Computer Systems*, *106*, 67–76.

Parra, G. D. L. T., Rad, P., & Choo, K. K. R. (2019). Implementation of deep packet inspection in smart grids and industrial Internet of Things: Challenges and opportunities. *Journal of Network and Computer Applications*, *135*, 32–46.

Patnaik, S., & Biswal, B. (2020). Importance of Nutrient Loading and Irrigation in Gross Primary Productivity Trends in India. *Journal of Hydrology (Amsterdam)*, 125047.

Playán, E., Salvador, R., Bonet, L., Camacho, E., Intrigliolo, D. S., Moreno, M. A., Rodríguez-Díaz, J. A., Tarjuelo, J. M., Madurga, C., Zazo, T., & Sánchez-de-Ribera, A. (2018). Assessing telemetry and remote control systems for water users associations in Spain. *Agricultural Water Management*, *202*, 89–98.

Ramesh, M., & Saini, R. P. (2020). Dispatch strategies based performance analysis of a hybrid renewable energy system for a remote rural area in India. *Journal of Cleaner Production*, 120697.

Rathore, P.K.S., Das, S.S. & Chauhan, D.S. (2018). Perspectives of solar photovoltaic water pumping for irrigation in India. *Energy Strategy Reviews, 22*, 385-395.

Reddy, V. R., Rout, S. K., Shalsi, S., Pavelic, P., & Ross, A. (2020). Managing underground transfer of floods for irrigation: A case study from the Ramganga basin, India. *Journal of Hydrology (Amsterdam)*, *583*, 124518.

Rose, T., Kifayat, K., Abbas, S., & Asim, M. (2020). A hybrid anomaly-based intrusion detection system to improve time complexity in the Internet of Energy environment. *Journal of Parallel and Distributed Computing*.

Savaglio, C., Ganzha, M., Paprzycki, M., Bădică, C., Ivanović, M., & Fortino, G. (2020). Agent-based Internet of Things: State-of-the-art and research challenges. *Future Generation Computer Systems*, *102*, 1038–1053.

Sidhu, B. S., Kandlikar, M., & Ramankutty, N. (2020). Power tariffs for groundwater irrigation in India: A comparative analysis of the environmental, equity, and economic tradeoffs. *World Development*, *128*, 104836.

ur Rehman, M.H., Yaqoob, I., Salah, K., Imran, M., Jayaraman, P.P. & Perera, C. (2019). The role of big data analytics in industrial Internet of Things. *Future Generation Computer Systems, 99*, 247-259.

Vijayakumari, A. (2020). A non-iterative MPPT of PV array with online measured short circuit and open circuit quantities. *Journal of King Saud University-Engineering Sciences*.

Xu, J., Tao, F., Liu, Y., Hu, C., Xu, Y., Keivanimehr, F., & Nabipour, N. (2020). Data transmission method for sensor devices in internet of things based on multivariate analysis. *Measurement*, 107536.

Zafar, S., Hussain, R., Hussain, F., & Jangsher, S. (2019). Interplay between Big Spectrum Data and Mobile Internet of Things: Current solutions and future challenges. *Computer Networks*, *163*, 106879.

Zeadally, S., Shaikh, F. K., Talpur, A., & Sheng, Q. Z. (2020). Design architectures for energy harvesting in the Internet of Things. *Renewable & Sustainable Energy Reviews*, *128*, 109901.

Zhihong, J., Jian, H., Wenzhou, L., Zhe, C., Ning, L., Siyuan, W., Xiao, Z., & Chang, L. (2018). Energy internet-a new driving force for sustainable urban development. *Energy Procedia*, *152*, 1206–1211.

APPENDIX

Glossary of Abbreviations

ABC: Agent based computing
AR: Augmented reality
BDA: Big data analytics
BSD: Big spectrum data
CPS: Common part sublayer
DPI: Deep packet inspection
ECC: Ellipse curve cryptography
ERP: Enterprise resource planning
EV: Electric vehicle
FF: Fill factor
HPS: Hybrid power system
IDS: Intrusion detection system
IIoT: Industrial IoT
IoE: Internet of energy
IoT: Internet of things
MILP: Mixed integer linear program
MIoT: Mobile IoT
MPP: Maximum power point
MPPT: Maximum power point tracking
MWR: Ministry of water resources
NBA: Network behavior analysis
PHS: Pumped hydroelectric storage
PLC: Programmable logic controller
PMU: Phasor measurement unit
PVPS: Photovoltaic pumping systems
PVRO: Photovoltaic powered reverse osmosis
P2P: Peer to peer
RAP: Rogue access point
RC: Remote control
RMS: Remote monitoring system
RoI: Rate of invenstment
RS-IWA: Remote sensing based irrigation water accounting
RS-SWB: Remote sensing based soil water balance
RTU: Remote terminal unit
SDN: Software defined networking
SFC: Standard frequency converters

SGPI: Solar based groundwater pumping for irrigation
SO: Smart object
SPL: Software product line
SWP: Solar water pump
SCADA: Supervisory control and data acquisition
TM: Telemetry control
VFD: Variable frequency drives
VM: Variability management
WEF: Water energy food

List of Symbols

E Energy storage capacity (kWh)
E_{ef} Effective per unit area energy realted to IAM and shadings (kWh.m^{-2})
E_P Energy storage in pumped mode (kWh)
E_{PO} Pump operating energy (kWh)
E_{UE} Unused energy (tank full) (kWh)
g Acceleration due to gravity (m.s^2)
h_t Total water head (m)
H Available water head (m)
H_l Loss in head (m)
H_P Pump storage head (m)
$H_{P,\max}$ Maximum pump storage head (m)
H_{th} Theoretical head (m)
I_P Current of PV unit (A)
I_{ph} Photocurrent dependent of solar irradinace (A)
I_s Reverse saturation current of PV cell (A)
K_B Boltzman's constant (=1.38×$^{-23}$)J/K
m Mass flow rate (s^{-1})
n Idealistic facvtor for PN junction
N_p Pumping speed (rpm)
P_{pv} Operating electric power of the system (W)
q Electron charge
Q_p Flow rate (m^{3}s^{-1})
S Area of one module (m^2)
T Absolute temperature (K)
V Available water volume (m^3)
V_d Effective volumetric displacement of the pump (rad^{-1})
V_p Voltage of PV cell (V)
V_T Thermodynamic potential (V)
W_p Water volume pumped (m^3)
W_u Water drawn by the user (m^3)
W_m Missing water (m^3)

ω Angular rotaional speed of the pump (rad.s-$^{1)}$

ρ Water density (kg.m-$^{3)}$

η_p Pumping effeincy (%)

η_t Overall efficiency of the pump (%)

Chapter 10
Green Energy in Data Centers Using Internet of Things

Vasaki Ponnusamy
Universiti Tunku Abdul Rahman, Malaysia

Bobby Sharma
Assam Don Bosco University, India

Gan Ming Lee
https://orcid.org/0000-0002-0993-1130
Universiti Tunku Abdul Rahman, Malaysia

ABSTRACT

Green energy infrastructure with the internet technologies relies on five important domains: green machine to machine (M2M), green cloud computing (CC), green data center (DC), green ICT, and green cellular. The ever-increasing demand for cloud computing and heavy dependence on cloud for storage, processing, and applications results in the need for more data centers with high capacity. Power management using wireless sensor networks (WSN) can be a potential solution as there has been a lot of works suing WSN for power management for green buildings, green home, and green farming. The same design can be applied to data centers with modifications to cater for data centers. Since WSN is part of IoT, various IoT-related solutions can be proposed for green data center solutions. A hybrid model that consists of virtualization, cooling systems, and IoT shows energy efficient data center designs. There have been various efforts as such, and this research will present green energy designs and mainly IoT-related initiatives for green-aware data centers.

INTRODUCTION

The increase in need of data processing has led to a demand in obtainability for cheaper, faster, efficient and larger data management system. This has further introduced a significant problem in terms of energy consumption and energy usage as mentioned in (Barroso. L.A, 2005). Since the trend of Internet

DOI: 10.4018/978-1-7998-6709-8.ch010

Infrastructure is moving towards services-based computing, data center plays a key role in this new computing architecture. It is used widely in variety of services independent encapsulating any vendor, product and technology including web hosting, application services, electronic markets, outsourced storage, online-monitoring system, mashups, digital nervous system and other network services. Fostering renewable energy production, developing better renewable technologies and improving the environmental performance of conventional electricity benefits heightened through improving the long-term strategy of data center performance. Environmental Protection Agency (EPA) shows that the data center and database management system in the US consume power doubled between 2000 and 2006 and may multiply again within the next decade. According to EPA, it is projected that energy consumption by data centers for the year 2008 is more than 80 billion kWh. (US Department, 2013). Report to the Congress, EPA's Energy Star Program estimates that increase efficiency in data centers could yield energy savings as the burgeoning efforts to make data centers greener is the energy and the word 'green' adds multiple meaning when applied to the data centers. (Koomey J, 2011).

Many works have been done to produce efficient result in reducing the energy usage for the purpose of data center and data management for example by Cisco System Inc., Google Inc., and many more (Barroso.L.A,2005). Some of the approaches were to design highly efficient systems at the peak performance point with maximum utilization of resources. But very few studies have addressed the energy problem comprehensively (US Department, 2013).

PROBLEM STATEMENT

In today's technological world, it is proven that there is a significant relationship between data center and investment in capital overlay and outgoing costs. Table 1 shows the detailed breakdown on the distribution of the cost from data center (Greenberg et al., 2008).

The design of computing system is being optimized based on the execution time and operates at low consumption due to entrapment of resources and fragmentation of data. No concern was made for the issue of energy consumption by the system. Uncontrolled usage of energy in data centers has negative effects on the reliability, density, scalability of information processing and the environment. Need and awareness are raised across nation through multiple solution and disciplines to optimize and save energy use in data management system (Barroso. L.A, 2005). System or even its components like CPUs, memory and disk are hardly distributing energy efficiently. It trades power for performance (US Department, 2013). Power management has become a critical issue due to large energy usage and all over the world government has started imposing the taxes on the carbon emission, ultimately aiming to reduce the emission and help make computing into green computing (Bansal et al., 2011).

Although research was conducted in designing power aware applications, an effort for implementing has not been taken place. Large expenses and allocations need to be augmented when data center reaches maximum provisioned power (Bansal et al., 2011). In the very near future, energy efficiency is expected to be one of the key procuring arguments in the society. Research should focus in building power-aware database management systems (DBMS) or GreenDB.

Analysis shows that giant companies are also working towards energy efficiency using green database system. Huge volume of energy usage goes to the data center. The global business market intelligence is a billion dollar market and still growing to see double-digit growth rates (US Department, 2013). With the growth of the internet data center (IDC), data systems are also running its management workload

Table 1. Guide On The Distribution Of Data Center Costs.

Amortized Cost	Component	Sub-component
~ 45%	Servers	CPU, memory, storage systems.
~ 25%	Infrastructure	Power distribution and cooling.
~15%	Power draw	Electrical utility costs.
~15%	Network	Links, transit, equipment.

a. Source of data: A. Greenberg et al, "The cost of a cloud: research problems in data center networks", ACM SIGCOMM Computer Communication Review, 39(1), pp. 68, 2008

such as search engines, storage of multimedia like video and audio, streaming, web analytics, and cloud computing. Dataset are growing rapidly and workload is also increasing. Together with it, the energy usage has increased. Companies like Google are working towards reducing the energy usage excessively in data centers (US Department, 2013). Data center has become the major immobilizer in the topic of energy wastage.

Clients particularly are searching for high scalability setup, lower unit price and simplicity of management. As to maintenance costs, the second largest bill for large database centers comes from the power company. Statistic shows that in US alone, data centers consumed about 60 billion kilowatt-hours of electricity in 2006 which equaling to 1.5% of electricity consumption in US (US Department, 2013). Demand for a solution is in the high urge and is being researched by many giant companies.

Figure 1. Critical Analysis Flow Model

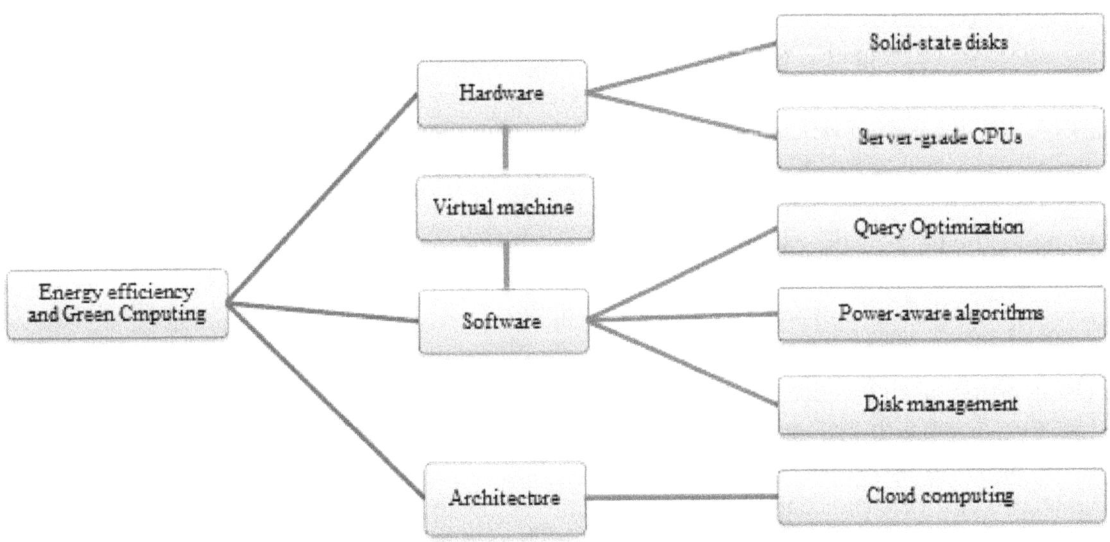

Many companies, organizations and individuals have proposed methodology to resolve this problem based on the categories shown in Figure 1. This research paper analyzes similar papers which has solution but needs minor integration to propose new innovation with better and efficient energy saving. Analysis focuses on power consumption patterns and identification of power-saving opportunity in GreenDB. Critical analysis done based on the following observation.

- Power-aware database.
- Data management through software and hardware.
- Data management through disk power management.
- Green computing architecture
- Renewable energy.

THE USE OF IOT IN GREENING DATA CENTER

Every day millions of data have been generated all around. Streamlining and distributing data from various data source remains a challenge. The entire economy is based on collective data. According to recent survey the world will have 24 billion Internet of Things devices online. It will generate a huge number of data. The IoT devices collect data continuously, process them and send these to respective data centres. As a result of this, it becomes a burden for the entire system. It not only consumes resources but at the same time it also concern with storage. Data centre may have to change their architecture from conventional way to a different architecture to cop with data from IoT devices. Even people may think of changing the budget with respect to storage, bandwidth and management. It should be proactive instead of traditional reactive approach. Data centre must specifically develop aiming to keep data comes from IoT devices. At the same time, it must concern with data security and privacy. Unlike previous data centre, with the IoT, there needs significant improvement in terms of the number of connections and the speed of access. To cop with data scalability, even it requires cost effective, multiple location based data centre. With exponential growth of Internet and use of ioT device must concern with energy consumption. It impacts a lot in environment. Hence Green Policy has come up with few policies including public awareness, energy efficient process etc. By looking the current scenario, it becomes mandatory for the people to follow green policy.

Most of the enterprises handle a huge amount of data every sec which finally consumes lots of energy. Indirectly it effects the environment too. It now becomes a big challenges for all organizations to have environmental friendly on demand data centers. The global server power demand has crossed 30 giga watts, that leads the enterprises to think of an alternative data centre like green data centre with the concept of renewable energy.

Use of artificial intelligence in the field of data centers has several positive impact in terms of resource saving. Even using artificial intelligence across business operation may reduce the effectof green house emission by 4% by 2030.

Objective of green data centre is to use optimum energy to give positive impact to the environment [14]. While storing data through Green data centre, it shouldn't compromise with data modification, deletion, addition etc. It is the repository used for data storage, management and dissemination. Green data centers are influenced by many factors including

1. Use of renewable energy instead of non-renewable energy
2. Environmental friendly use of data centre
3. Optimum resource use to manage maximum data

Figure 2. Green Data Center

4. Adaptive and on demand use of server for saving unnecessary use of energy
5. Use of virtual data centre etc.

Latest Efforts in Green Data Center

Jin et, al, (2016) mentioned about the concern over high energy consumption as well as high amount carbon emission by the data centre due to massive amount of data management on daily basis. As an alternative, researchers are coming up with green data centre. According to author, green data centre must concentrate on energy efficiency, resource management, thermal control and maintenance of green matrices. Liu et a., (2009) mentioned about power consumption issues of data centers. Concept of virtual machines is widely used by data center due to its seminal features like reliability, flexibility, integrity etc. Green cloud may be concept to reduce power consumption by traditional data centers.

Sharma et al., (2017) mentioned about Massachusetts Green High Performance Computing Center (MGHPCC) which adopts green design to achieve good results related to energy efficient operation of data centers.

Mujawar et al., (2020) mentioned that data centre comprises of Servers, storage devices, networking devices, power distribution system, cooling system etc. For every operation on data processing, it consumes lots of energy by the different components of data centers. A study done by International Data Corporation showed that in 2011, 1.8 trillion gigabytes (GB) of digital information were processed while in 2012, it increases to 2.8 ZB and likewise it will be 40 ZB by the year 2020. Avgerinou et al., (2017) mentioned about fastest emission of CO2 through data centers. It is mentioned that data centers have huge impact on Earth's energy as they consume lots of energy while they emits lots of global greenhouse gas. E-waste is also considered as another byproduct of data centre refresh process. Objective of green data centre is to use minimum energy to give maximum output.. At the same time, it must have positive impact to the environment.

Green data centers are designed in such a way that it ensures cost reduction and lower the organizational carbon footprint with respect to the execution of various devices including centralized humidification systems, high-efficiency HVAC, servers, routers, and LED lighting, Networking devices etc.

Due to environmental changes and use of optimum energy, green data centers are becoming popular amongst the stuff. Even the capability of providing on demand data from green data centre with less resources, makes the whole system very popular. Rapid growth of enterprises along with their massive data generates a high demand of green data centre. A market report of green data center coverage is given in Table 2.

Table 2. Green Data Center Market Report Coverage (Ankita and Solani, 2019)

Report Coverage	Details		
Base Year:	2018	Market Size in 2018:	6.5 Billion (USD)
Historical Data for:	2014 to 2018	Forecast Period:	2019 to 2026
Forecast Period 2019 to 2026 CAGR:	23.5%	2026 Value Projection:	35 Billion (USD)
Pages:	250	Tables, Charts & Figures:	254
Geographies covered (18):	U.S., Canada, UK, Germany, France, Spain, Benelux, China, India, Japan, Singapore, Australia, Indonesia, Brazil, Mexico, Argentina, GCC, South Africa		
Segments covered:	Component, End-Use		
Companies covered (20):	Airedale Air Conditioning, Dell Inc., Eaton Corporation, EcoDataCenter, Fujitsu Ltd., General Electric, HCL Technologies Limited, Hitachi, Ltd., Huawei Technologies Co. Ltd., IBM Corporation, Microsoft Corporation, PCCW Solutions Limited, Schneider Electric SE, Siemens AG, Tech Mahindra Limited, Telefonaktiebolaget LM Ericsson, The 3M Company, The Hewlett-Packard Company, Verne Global, Vertiv Co.		
Growth Drivers:	· Rapid shift towards renewable power sources in the U.S. and Europe · Transition to cloud-based solutions in the U.S. · Increase in organizational data traffic and big data analytics in the UK and Germany · Government initiatives to lower PUE in the UK and Singapore · Growth of mega data centers and trend of collocation in China, India, and Mexico · Increase in construction of data centers and rise in SME demand in South-East Asian countries · Decline in costs of semiconductor components in Asia Pacific · Rising electricity prices across Europe · Growing focus of players to establish data centers in South America due to abundant resources and high cost benefits · Rising urbanization in the MEA		
Pitfalls & Challenges:	· High cost of equipment over traditional technologies · Security issues in green storage technologies · Incompatibility of green solutions with existing data centers		

Generally green data centre require the following,

- Cloud data service
- Internet services
- Server services
- Security features etc.
- Sustainability

While designing green data centers, It should be committed to

- Green
- Powered by REBA (Renewable Energy Buyers Alliance) which works for customers, suppliers, policy makers etc.
- Future of Internet Power
- Cost Efficient
- Sustainability
- Leading the Way customer needs.

In many countries, data centers are shifted to Green data centers. In that situation, these must be certified. As for example in USA, green data centers should be certified through proper authority. The Leadership in Energy and Environmental Design (LEED), developed by the U.S. Green Building Council, certifies the green data centers as silver, gold or platinum certification. While designating, it depends on criteria of green data centre like environmental friendly, availability of resources etc. In US data centers are also certified by U.S. Environmental Protection Agency and the U.S. Department of Energy under the National Data Center Energy Efficiency Information Program by Energy Star.

While formulating the data centers, it must emphasizes on the facts like Low powered but efficient server, reusing of energy like heat energy, manage humidification, manage cooling effect etc.

Another example of powerful green data centre may be Green Data Centre in Ferrera Erbognone. The HPC5 was revealed on February 2020. It is considered as world's most powerful green data centre with powerful industrial super computer. The centre has been certified for its energy efficiency, environmental friendly and reliability for data storage and access.Even the PUE(power usage effective) of this centre is 1.171 against the average PUE of the world i.e. 1.67

Google is constantly trying to reduce the energy consumption since 2014. Till then it is capable of reducing energy upto 50%. Though the entire process involves a big investment yet for the safe of our energy and resource, people must work on it. In case, people can follow the guidelines provided by data centre infrastructure management (DCIM). Even DCIM can provide services on data management, modeling, analytics et. which finally results in lower cost, reduced operation etc.

Even in India, the concept of green data centers is followed through IGBC (Indian Green Building Council). At the same time, IGBC is also taking steps to formulate the "green data centre rating systems" via which it extend help for enabling and construction of green data centres in the country. The concept of "Digital India" is totally based on reliable data centres. So, with huge numbers of data in data centre must be transformed to green data centre for the environmental safety of the country.

It is mentioned about The green data centre practice guide which provides the practice for data centre wit respect to Hong Kong. It basically aims for describing energy efficiency and environmental friendly practices. It emphasizes mainly on heat absorption and cooling system, power system adjustment, IT equipment deployment Energy efficiency, control over telecom and networking sector, green disposal etc.

Day by day, use of data and reliability on data is increasing. People are running behind big and big data centre. It directly effects the environment due huge amount of energy dissipation. And it finds that it is a never ending process.

Hence to avoid the worst situation, adoption of green data centre becomes an essential for all the stake holders of data. Moreover it has been proved that green data centres are found to be environmental friendly, secured and energy efficient data centres.

Green Data Center Using IoT Initiatives

Zhou 2019, proposed a Green Edge architecture (Figure 3) towards a sustainable energy goal which consists of device to device communication and energy harvesting components. This architecture consists of a set of wireless IoT devices that can be easily detected by the nearest base stations and all IoT devices are equipped with heterogenous IoT applications. An edge datacenter is deployed to extend resources to the IoT devices, and this edge datacenter is connected to cellular base station via gateway. Each IoT devices can communicate with the base station using the established cellular link and can also establish device

Figure 3. Green Edge Architecture (Zhou 2019)

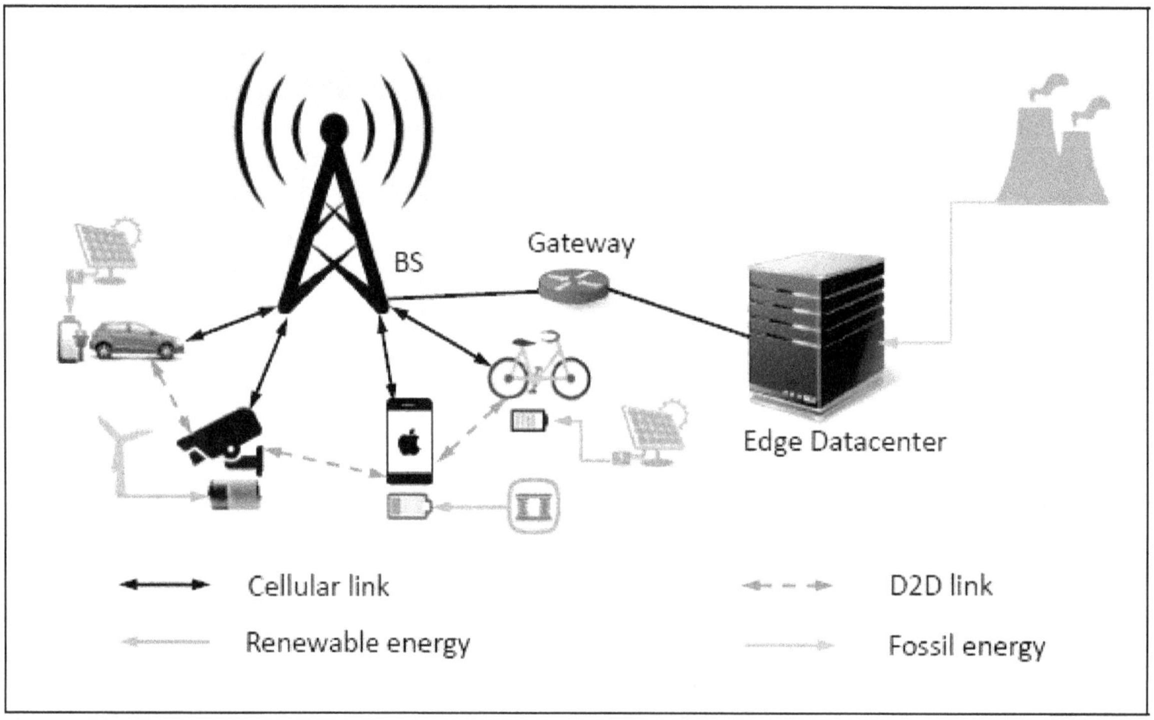

to device link with other IoT devices within the proximity. With this setup there can be three modes of how a particular task can be executed: i. local mobile, ii. device to device offload and iii. edge offload.

Han et al., (2015) designed a cloud-assisted mobile ad-hoc network (Figure 4) that can support multimedia and data services in which the network is formed with cloud of interconnected servers and mobile ad-hoc network that consists of mobile devices, sensors and computers. This mobile cloud has advantage of supporting rich applications that is the main aspect of multimedia services. Energy efficiency has been studied in this work because data searching is the main aspect of cloud based mobile ad-hoc network for multimedia services. An overlay of network as service delivery has been focused with the focus on the operation of network transactions.

Shard et al. (2015) proposed a system (Figure 5) that can trace the energy usage in the data center which provides an indication when there is unnecessary usage of energy. The proposed system consists of input devices like temperature sensor, humidity sensor and power consumption sensor, the processing unit that consists of Arduino and ZigBee components whereas the output consists of LED and alarms. The proposed system is an automated energy monitoring system at a university environment and the results show a minimal energy consumption.

Yang et al. 2018 introduced an AI enable green cloud (Figure 6) framework that consists of application, middleware, virtualization and resource layer with the focus on scheduling control engine and intelligent refrigerating engine which is suitable for dynamic resource consumption and good for energy optimization in data centers. The framework developed shows a low power consumption in the cloud platform and an efficient energy consumption data center operation.

Figure 4. Architecture of Cloud-assisted Mobile Ad Hoc Networks in 5G (Han et al., 2015)

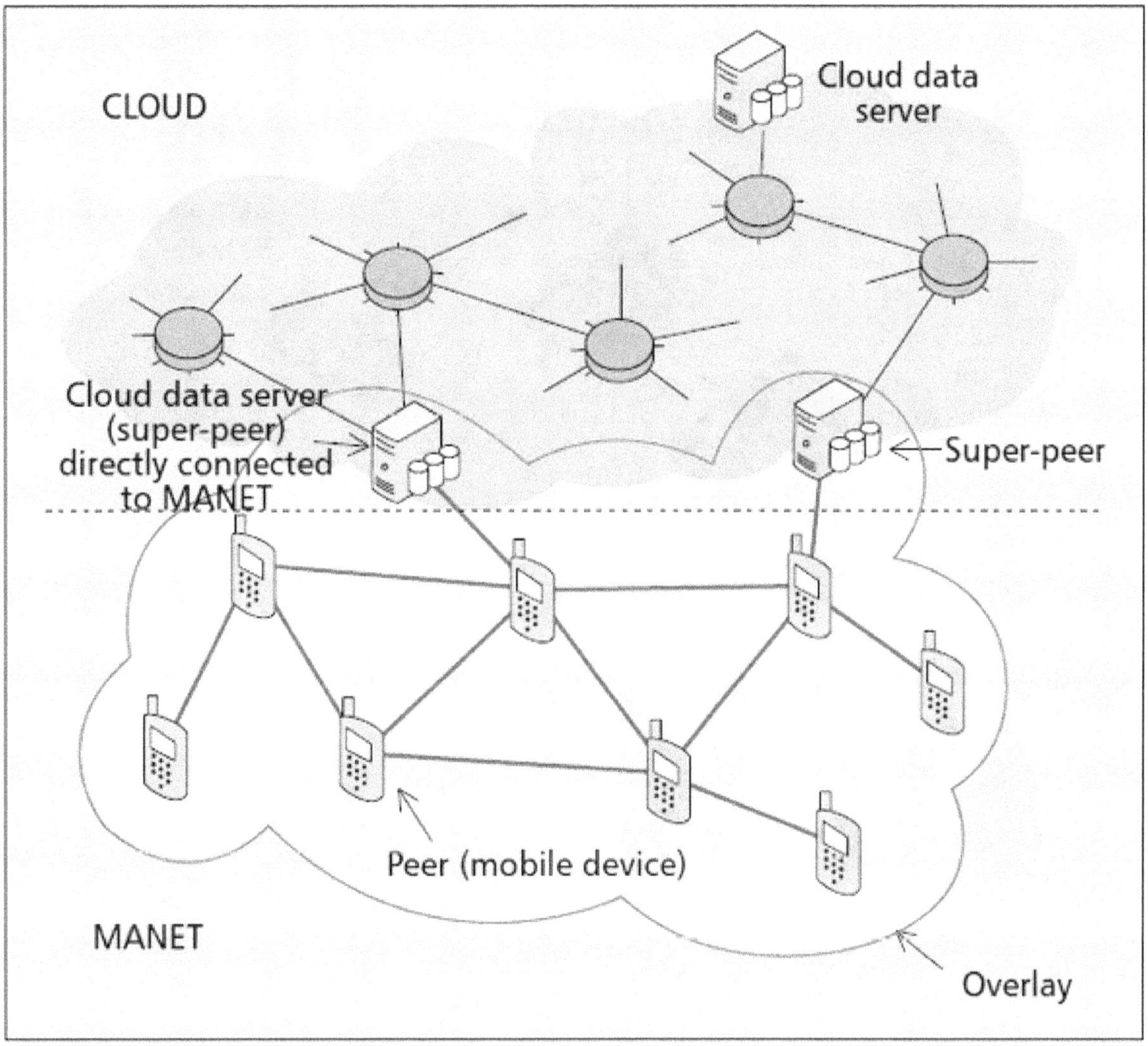

POWER-AWARE DATABASE

Power-aware computing research at the application level in the aspect of hardware and operating system level found to be synergistic for energy proficiency in the core systems. Xu. Z., 2011 describes on creating a power-aware database management (P-DBMS) and initial ideas on the design as such systems, with the focus on a power-aware query optimization module in the DBMS. This particular paper has proposed and created a power model to accurately measure the energy costs of query executions and plan selection. It is in contrast with the heuristic-based adaptive solutions that rely on manual timing and extensive empirical evaluation. It is meaningless without considering the database performance. It is either to make the query run faster by optimizing the query execution or let the system run in lower power state while sacrificing the performance to some extent. They proposed that the throughput is improved while maintaining the power line. It indirectly reduces the power cost while not sacrificing the

Figure 5. The hierarchy of devices in large-scale deployment in Shard et al. (2015)

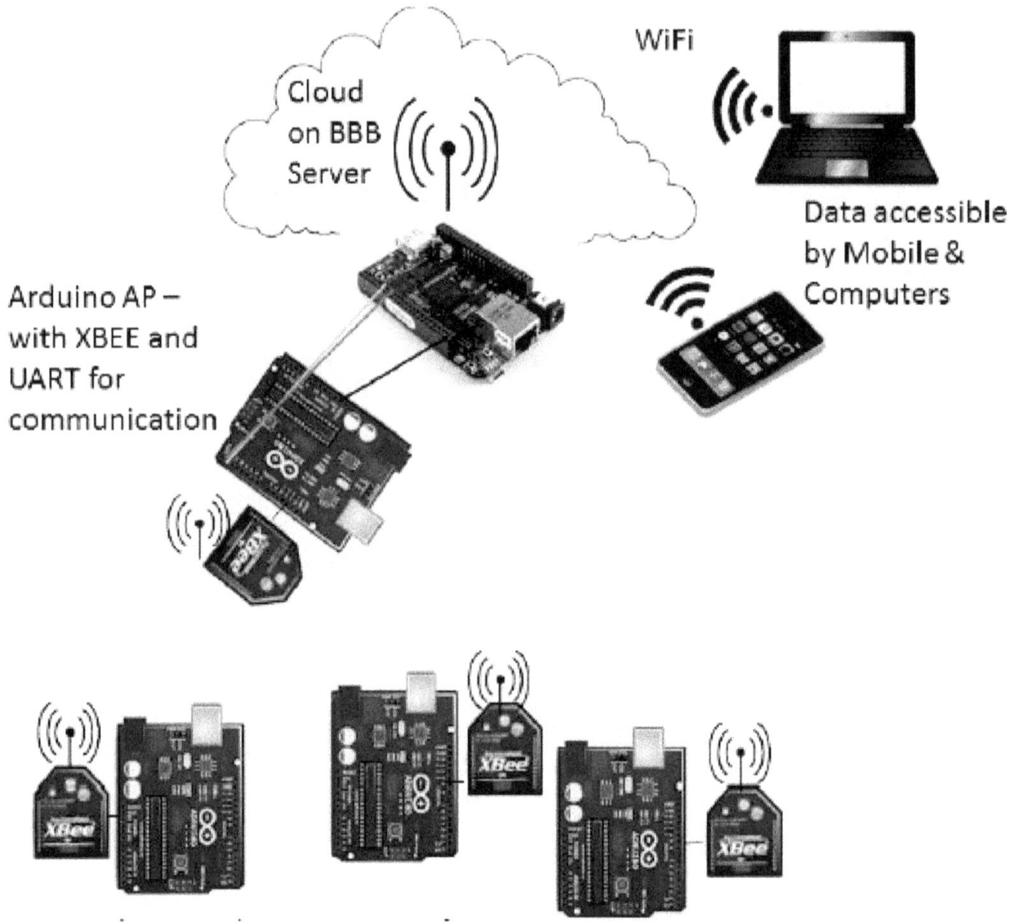

Figure 6. The architecture of ai-enable green cloud. Yang et al. 2018

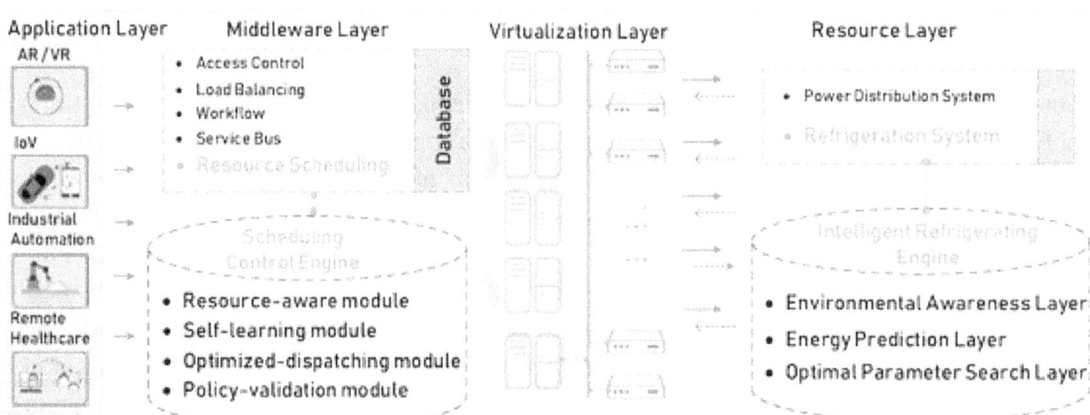

throughput performance. Implementation has been taken place through initial design in algorithm based software named PostgreSQL rather than power-saving through hardware and operating system levels modification. The proposed solution resolves the power aware of the database as seen in Table 3. Only setback would be database management system needs to be revisited or redesigned and initial efforts be invested. The plan evaluation model has a stumbling block where the system is stable and foreseeable but the database is barely stable and predictable. Thus, modeling error occurs and estimation per operation cost cannot be 100% correct.

Table 3. CRitical Overview On PostgreSQL System

Attribute	PostgreSQL system	
	Description	*Analysis*
Hardware	Hardware Optimisation	x
	Power Management	x
Software	Software Optimisation	/
	Query management and optimisation	/
Architecture	Architecture Manuveuring	x
	Distributed System	x

a. Evaluation based on the result of proposed research paper by Author

DATA MANAGEMENT THROUGH SOFTWARE AND HARDWARE

Synergy between database management through algorithm and hardware architecture are promising in energy efficiency computing. Harizopoulus, S. et al., 2009 recommends that adjusting existing system-wide configuration and design algorithm for energy proficiency is not the same as designing algorithm based on the throughput and performance. Deploying an energy-efficient data management platform is by choosing hardware components with good performance-per-watt characteristics by choosing high-performance components (server-grade CPUs, solid-state disks etc.). Installing an energy-efficient system was also done through software-based approaches for reducing wasted energy in data management systems. This paper proposes to power down the unused hardware components, revisit and redesign data structures and algorithm from a wider perspective. The 'push-to' design components offers more control over power performance trade-offs for example usage of multicore chips which can be activated at any time. The increase of hardware heterogeneity offers technology refresh cycle like the virtual machine. The disadvantage here is that some workloads may be able to use these additional resources while others will underutilize them and create a waste of power. Table 4 shows that hardware configuration and application-agnostic power management is only part of the solution proposed in this paper.

Data Management Through Disk Power Management

Reducing energy consumption through hardware depends on many aspects. Earlier research shows that the result from monitoring data center is that the average idle period for a server disk is very small compared to the time taken in spinning up and down. This significantly limits the conservation of energy. Zhu, Q.

Table 4. CRitical Overview On Stavros Review

Attribute	Synergistic system	
	Description	*Analysis*
Hardware	Hardware Optimisation	/
	Power Management	x
Software	Software Optimisation	/
	Query management and optimisation	/
Architecture	Architecture Manuveuring	x
	Distributed System	x

et al., 2004 proposed reducing energy consumption of disk storage using power-aware cache management with algorithm which provides more opportunities for causal disk power management structures in energy efficiency. So far only a few studies (D. Colarelli and D. Grunwald, 2002 ; Gurumurthi et al., 2003 ; Gurumurthi et al. 2003 ; Carrera et al., 2003) addressed the problem in energy efficiency for the storage systems at the data centers. Based on these studies, multi-speed disks or multiple rotational speed disks shown promising results by using synthetics workloads. Therefore, power-aware cache management algorithm was designed whereby it is aware of the base disk energy ingestion and power management plan for read and writes accesses.

It is an energy-optimal cache replacement algorithm which uses the dynamic programming in polynomial time leading to 16% less energy and lead to 50% better average response time. But this study has its limitation such as it is only designed for multiple disks and focuses on high-end storage systems. Single disk may not work with this algorithm. Also it has low effect on prefetching as well. It does not consider the storage cache energy consumption and focuses on smaller part of the overall management as shown in Table 5.

Table 5. CRitical Overview On Power-Aware Cache Management Algorithm

Attribute	Synergistic system	
	Description	*Analysis*
Hardware	Hardware Optimisation	x
	Power Management	/
Software	Software Optimisation	/
	Query management and optimisation	x
Architecture	Architecture Manuveuring	x
	Distributed System	x

a. Evaluation based on the result of proposed research paper by Author

GREEN COMPUTING ARCHITECTURE

Virtual machine (VM) has been globally used in the data center environment due to its important feature and architecture. Liu. L. et al. 2009 uses the latest technology trend which is the cloud computing as a guaranteed performance and designed to reduce the data center energy consumption. It is effective and enables the widespread online-monitoring, live VM migrating and placement optimization. It utilizes the live migration feature of VM Technology such as XEN (Bansal et al.,2011) to implement the architecture for example Amazon Elastic Compute Cloud (Zu. Z. 2011) provides a virtualized computing environment hosted with different kind of Linux-based service. Evaluation results show that 27% energy can be saved when applying Green computing architecture. VM Migration allows lesser usage of computer. This architecture assures real-time performance requirement as well as saves the consumption of energy by Internet Data Center (IDC). Green computing architecture with the integration of VM technology is an effective method in reducing server power consumption.

The problem may arise in the angle of automatic scheduling decision on dynamic migration or consolidating the VMS among physical servers to meet the workload requirements meanwhile saving energy, especially for performance sensitive. The aspect of performance trade-off between the operating system and applications or middleware has not been tested which can be a flaw. But based on Table 6, this technology alone may not be sufficient to provide efficiency.

Table 6. CRitical Overview On Green Computing and VM Technology

Attribute	Green Computing and VM Technology System	
	Description	*Analysis*
Hardware	Hardware Optimisation	/
	Power Management	/
Software	Software Optimisation	/
	Query management and optimisation	x
Architecture	Architecture Manuveuring	/
	Distributed System	x

a. Evaluation based on the result of proposed research paper by Author

RENEWABLE ENERGY

Exploiting green energy sources is another opportunity but it is variable and intermittent. So far, there has been less work on integrating renewable energy with database system. Chen, C., et al., (2012) uses the hardware optimization to save energy and retaining the execution part of the hardware. They proposed through integration of renewable energy which is the solar to power the execution of database management system. This research bridges the gap between renewable energy and database management system with ReinDB (Renewable Energy Integrated Database). The purpose would be to minimize the usage on carbon intensive energy consumption on a database server with both renewable energy and

Table 7. CRitical Overview On Integration of Renewable Energy

Attribute	Integration of Renewable Energy	
	Description	*Analysis*
Hardware	Hardware Optimisation	/
	Power Management	x
Software	Software Optimisation	/
	Query management and optimisation	x
Architecture	Architecture Manuveuring	/
	Distributed System	x

a. Evaluation based on the result of proposed research paper by Author

carbon intensive energy supplies. This works on Supply Driven Execution (SDE) on a database server with two mechanisms.

a) Workload scheduling to meet workload deadlines
b) Dynamic configurations on hardware setting
c) Brown-aware workload shifting
d) Battery-aware workload shifting

Although it increases the usage of renewable energy (solar usage) but there are still some disadvantages to take into consideration. Studies show that the mismatch between workloads and green energy supplies challenges in the utilization of energy. Divergence in the context of data center from brown energy and green causes excessive usage of energy (Goiri et al,. 2011; Goiri et al,. 2012 ; Krioukov et al 2012 ; Liu et al. 2011). Clients are required to change the hardware to new system which may need initial investment that may cost high. Also the latency time (slack time) is also another issue whereby the clients have to tolerate a large slack time to complete a task with the green techniques. In the economical aspect, the cost involves in manufacturing solar panel and battery provision. Table 7, shows that, there are many other aspect to be incorporated.

ANALYSIS AND PROPOSED GREEN SOLUTIONS

The concept of green database and energy efficiency is emerging high and gaining more popularity and everyone wants to be on top of it and they have started researches on the concept of energy saving in the DBMSs. Companies like IBM, Oracle that develops their own DBMSs want to decrease the energy consumption. All researches are focusing on individual area only. Integration of success methods from each research can result in wholesome solution.

Evaluation and analysis done to multiple proposal and project paper resulting in integration of the best modules for many different sources tosses excellent solution. Comparisons shown in Table 8-12 define the strength and weakness based on the critical analysis done.

Table 8. Physical Usage

Project/ Attribute	*Hardware Modification*	*Software Modification*	*Solar Integration*
Zichen Xu	No	No	No
Cheng Chen	Yes	Yes	Yes
Liang Liu	No	Yes	No
Stavros Harizopoulos	Yes	Yes	No
Qingbo Zhu	No	Yes	No

Table 9. Power Coverage

Project/ Attribute	*Power State*	*Power Cost*	*Power efficiency*
Zichen Xu	Low	Reduced	30%
Cheng Chen	Very low	Reduced	28-53%
Liang Liu	Low	Reduced	27%
Stavros Harizopoulos	Low	Reduced	45%
Qingbo Zhu	Low	Reduced	16-35%

VIRTUALIZATION AND GREEN CLOUD

Virtualization allows software and hardware virtualization by providing virtual environment. This reduces the need for overwhelming use of servers. More processing can be done with lesser energy consump-

Table 10. Query and Throughput

Project/Attributes	Query Execution Speed	Query Optimizer	Throughput time
Zichen Xu	Faster	Increased	Maintain
Cheng Chen	No change	Scheduled Query	peak performance
Liang Liu	No change	Optimized	Guaranteed real time
Stavros Harizopoulos	No change	Optimized	Peak performance
Qingbo Zhu	Faster	Optimized	Limited

Table 11. Other Aspects

Project/Attributes	Scalability	Reliability	Slack time
Zichen Xu	Inefficient	Inefficient	Retained
Cheng Chen	Efficient	Efficient	Retained
Liang Liu	Efficient	Efficient	Retained
Stavros Harizopoulos	Efficient	Efficient	Retained
Qingbo Zhu	Efficient	Inefficient	Retained

Table 12. Application and Remark

Project/Attributes	Application level	Remark/ New invention
Zichen Xu	Very little	Cost evaluation model and power evaluation model
Cheng Chen	Very little	ReinDB – brown and green energy usage with additional equipment to be installed
Liang Liu	Increasing	VM, Heuristic search algorithm and performance sensitive
Stavros Harizopoulos	Low	Algorithm based and redesign the system component
Qingbo Zhu et al.	Low	Works only for multiple disk and focuses on high-end storage system.

tion. Every single component of IT can be virtualized from the servers, routers, switches, input-output, applications, operating systems and desktop. Virtual machine allows more than one operating system to be operated in one computer. Each operating system runs its own virtual machine and resources like I/O, storage and processor can be assigned to each virtual machine. It is to automatically make the scheduling decision on dynamically migrating or consolidating Virtual Machines among physical servers to meet the workload requirements while saving energy, especially for performance sensitive. The heuristic algorithm is used in virtual machine migration to achieve the optimization. States of server power is adjusted through the dynamic voltage or frequency scaling.

RENEWABLE ENERGY SOURCES

Datacenters can be equipped with renewable energy sources such as solar powered energy and hydroelectric energy supply. Solar powered energy can be good source of energy for data centers located at locations that has wide availability of solar energy. The solar panel nodes can be charged during the day with the sunlight and stored in backup sources to be supplied during the absence of sunlight. Although installation of such facility is comparatively costly, but the outcome can be very promising as these are renewable energy sources. Moreover energy generation through hydroelectric source is another promising factor. This is applicable in areas where there plenty of water sources for hydro energy generation. A good example would be the Facebook datacenter located in Sweden which is supplied with hydroelectric power without interruption due to the abundant resources of hydroelectric power. In conclusion, the choice of renewable energy is very much dependent on the geographical areas where the datacenters are located (1). More research work can be carried on vibration energy generation from the vibration caused by the servers and the other high end equipment present in the datacenter. This requires extensive research on the strength of vibration generated and amount of energy can be drawn from that.

DISK POWER MODEL & MANAGEMENT

Managing the multi disks speed to increase the amount of energy saved with data center workloads by identifying the active, idle and standby mode of the disk. During the active mode, read and write mode is actively processing in the platters spin and the head. Likewise, during the idle, no disk is required whereby energy can be saved by idling the rotation and spinning. Mathematically, a lower rotational speed mode consumes lesser energy compare to higher rotations. Reduction in Input/ Output Operations per Second (I/Ops) directly leads to reduction in the energy consumption which is shown in (1). An average latency is the time taken by the disk that is accessed to rotate to the position under read/ write head. Meanwhile the average seek time would be the time (ms) taken for the hard drive's read/write head to locate itself over the read/ write track.

I/Ops Calculation:

1 / ((σεεκ τιμε / 1000) + (λατενχψ / 1000) = Ι/Οπσ (1)

Ενεργψ Χονσυμπτιον= Ποωερ (Ωαττ) ξ Ι/Οπσ

POWER-AWARE DATABASE MANAGEMENT SYSTEM

It monitors the query optimizer, buffer manager, storage manager and many other database query optimization modules.

Cost can be reduced by improving the query optimization time whereby the processing time can be improved and obtained through the setup of the server which is shown in (2).

Cost estimation, C:

Χ = (Θυερψ Οπτιμιζατιον ξ τιμε) ξ Προχεσσινγ Τιμε (2)

DATACENTER COOLING SYSTEM

Cooling system in the datacenter is very important aspect as it guarantees proper operation of datacenter equipment such as servers and storage devices. A suitable cooling method to maintain the thermal condition of the CPU's can provide efficient energy usage. The Facebook datacenter in northern Sweden provides an energy efficient cooling system as compared to other datacenters. Besides that, the arctic climate in the northern region allows the Facebook datacenter to be in a colder temperature condition requiring less cooling needs. The river along Lulea provides a lot of resources for hydroelectricity supply for the entire datacenter without the need for generators. The datacenter depends on the climate chilling system rather than depending on air conditioning system. An intelligent heat, ventilation and air conditioning (HVAC) system can be designed with proper cooling system (Parisio et al.; Mousavi et al.) Datacenters with central air systems, direct liquid cooling and humidification can allow for proper cooling system.

POWER MANAGEMENT USING WIRELESS SENSOR NETWORK

One way to make datacenters energy efficient is by providing good network infrastructure. The servers, routers, switches and other equipment linked with an efficient networking facility can decide the energy consumption of the datacenter. Congested network linkage allows unnecessary communication among

Figure 7. WSN Data Center Monitoring System

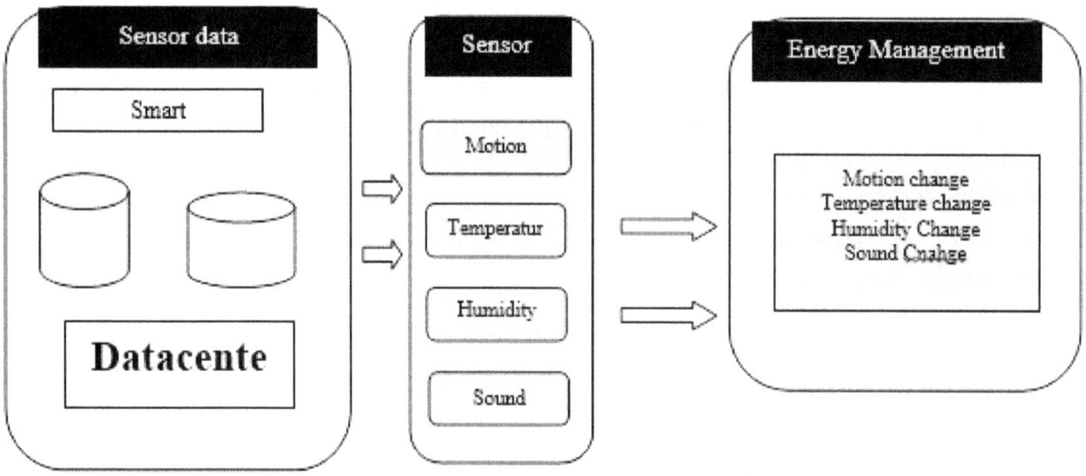

devices and allows for additional energy consumption. One potential networking infrastructure that could facilitate with proper energy management is through employing wireless sensor networks (WSN). WSN provides intelligent services by deploying various types of sensors to monitor the environment in the datacenter to make it a smart datacenter. The smart datacenter is intended to be designed to create a system that performs energy management in terms of energy consumption and optimizing power consumed by datacenter devices. The framework of the system is shown in Figure 7. The system is categorized into fault detection and energy management. Fault detection stage should identify malfunctioned datacenter equipment and resources that is unnecessarily consuming power and wasting energy. The energy management stage should be able to manage energy consumption, optimize power consumption and perform automated notification. The fault detection system consists of data acquisition from various sensors and information processing using fuzzy logic engine. The energy management system performs event detection and action mechanism based on changes in the sensor data and motion detection. In this system, datacenter environment parameters such as temperature, humidity, light and sound are detected for deciding whether electrical appliances and lights to be switched on or off for energy management in buildings. Besides that, motion sensors are used for detecting movement so that lights can be switched on and off based on occupancy. Motion sensors can also be used to detect heat emitted from the equipment so that notification alarm is triggered to take necessary action. The temperature and humidity sensors can be used to decide whether the datacenter cooling system to be turned on or off. This system can be created using some machine intelligent capabilities to make the system self-autonomic. Moreover WSN is a cost effective solution as it does not require extensive IT infrastructure and data communication.

HYBRID MODEL

A hybrid model (Figure 8) can take advantage of a combination of energy efficient solutions including virtualization by cloud, WSN monitoring system and renewable energy system (Gholamhosseinian, and Khalifeh, 2012). Machine learning systems such as Case Based Reasoning (CBS) or decision trees can be integrated together with the virtualization concepts so that lesser servers are needed for data storage and processing (Pawlish et al., 2014). These intelligent systems allow servers to make automatic decisions and require lesser tasks. Moreover intelligent agents such as pattern recognition, fuzzy logic and neural network can be integrated with WSN so that energy usage at datacenter can be monitored intelligently. A proper selection hardware that requires less cooling system is also an added advantage. For example the Open Computer hardware has been designed in a simplified way for energy efficiency. It is just a piece of metal sheet with the motherboard that allows for proper air flow. A casing free system can allow for faster cooling process in datacenters. In conclusion the hybrid model should provide the following advantage:

1. Reducing energy consumption compared to the traditional datacenter model
2. More utilization of virtualization by employing green cloud architecture by allowing lesser usage of storage, machines and processing without compromising the number of applications can be operated
3. Utilizing natural resources for renewable energy for datacenter. Since natural resources are ample depending on the geographical location, a proper energy harvesting method could solve energy usage at datacenters

Figure 8. Datacenter Hybrid Model

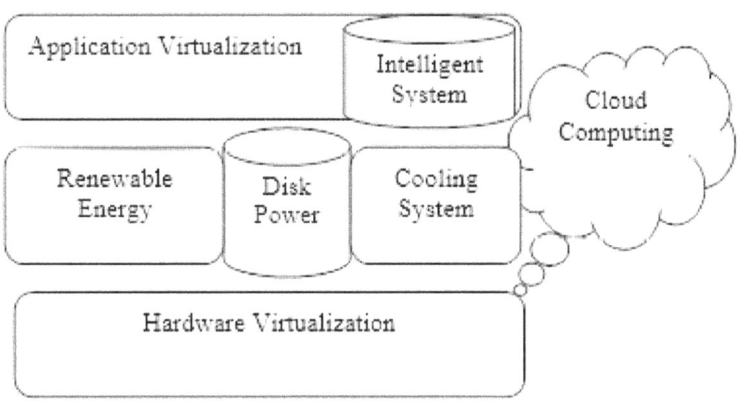

FUTURE WORK

Looking at the hybrid model proposed, a complete framework and simulation of the model will prove the feasibility of the model. Taking renewable energy into consideration, future work in constructing data centers should focus on natural resources available. More emphasis should be given to locations near the northern poles to reduce the need for cooling system. Focus on areas that have plenty of sunlight to harvest energy using solar system or locating data centers near hydroelectric power generators. Virtualization should be highest importance by deploying systems using cloud computing infrastructure. Provide an integrated system operation with software, hardware, controls and sensors to be coupled together with the grid electricity. An automation system using self-healing energy management can be considered in this case. The system stores huge amount of energy in the storage via renewable energy and constantly monitor the level of energy at the grids and allowing power to be supplied during energy crisis.

CONCLUSION

The green energy at data centers will help to reduce the serious energy consumption issues at data centers to reduce the critics from authorities. Total energy consumption of data centers in the US for example can be reduced from 2% to even lower. The suggested solutions may not only reduce the greenhouse effect but offer a more cost effective data center setup. In this paper, several approaches of reducing energy consumption at data centers have been proposed. The proposals are validated by adopting some of the solutions addressed in the literature providing reasonable metrics. Ultimately the aim and objectives are to minimize the wastage of energy and to retain the efficiency. It is foreseen that implementation of technology collaborated with the green energy will ultimately enhance the technology to be ecofriendly.

REFERENCES

Amazon Elastic Compute Cloud (EC2). (n.d.). Retrieved July, 2008, from https://www.amazon.com/cc2/

Avgerinou, M., Bertoldi, P., & Castellazzi, L. (2017). Trends in data centre energy consumption under the european code of conduct for data centre energy efficiency. *Energies*, *10*(10), 1470. doi:10.3390/en10101470

Avgerinou, M., Bertoldi, P., & Castellazzi, L. (2017). under the European Code of Conduct for Data Centre Energy Efficiency. *Energies, MDPI*, *10*(10), 1470. doi:10.3390/en10101470

Bansal, K. (2011). A critical review on concept of green database. *Global Journal of Business Management and Information Technology*, *1*, 113–118.

Barroso, L. A. (2005). The price of performance. *ACM Queue, 7*, 48-53.

Carrera, E. V. (2003). *Conserving disk energy in network servers*. ICS.

Chen, C. (2012). *Green databases through integration of renewable energy*. Nanyang Technological University & Sun Yat-sen University.

Colarelli, D., & Grunwald, D. (2002). *Massive arrays of idle disks for storage archives*. Academic Press.

Gholamhosseinian, A., & Khalifeh, A. (2012). *Cloud computing and sustainability: Energy efficiency aspects*. Academic Press.

Goiri, I. n. (2011). Greenslot: scheduling energy consumption in green datacenters. *International Conference for High Performance Computing, Networking, Storage and Analysis, SC '11*, 20:1–20:11. 10.1145/2063384.2063411

Goiri, I. n. (2012). Greenhadoop: leveraging green energy in data-processing frameworks. *7th ACM european conference on Computer Systems, EuroSys '12*, 57–70. 10.1145/2168836.2168843

Greenberg, A., Hamilton, J., Maltz, D. A., & Patel, P. (2008). The cost of a cloud: Research problems in data center networks. *Computer Communication Review*, *39*(1), 68–73. doi:10.1145/1496091.1496103

Gurumurthi, S. (2003a). Interplay of energy and performance for disk arrays running transaction processing workloads. *ISPASS*, 123–132.

Gurumurthi, S. (2003b). DRPM: Dynamic speed control for power management in server class disks. *ISCA*, 169–179.

Han, N. D., Chung, Y., & Jo, M. (2015). Green data centers for cloud-assisted mobile ad hoc networks in 5G. *IEEE Network*, *29*(2), 70–76. doi:10.1109/MNET.2015.7064906

Harizopoulos, S. (2009). *Energy efficiency: the new holy grail of data management systems research.* HP Labs & UCLA.

Jin, X., Zhang, F., Vasilakos, A. V., & Liu, Z. (2016). *Green data centers: A survey, perspectives, and future directions.* arXiv preprint arXiv:1608.00687.

Koomey, J. (2011). *Growth in data center electricity use 2005 to 2010.* Analytical Press.

Krioukov, A. (2011). Integrating renewable energy using data analytics systems: Challenges and opportunities. Bulletin of the IEEE Computer Society Technical Committee, 1–9.

Liu, L., Wang, H., Liu, X., Jin, X., He, W. B., Wang, Q. B., & Chen, Y. (2009, June). GreenCloud: a new architecture for green data center. In *Proceedings of the 6th international conference industry session on Autonomic computing and communications industry session* (pp. 29-38). 10.1145/1555312.1555319

Liu, Q., Ma, Y., Alhussein, M., Zhang, Y., & Peng, L. (2016). Green data center with IoT sensing and cloud-assisted smart temperature control system. *Computer Networks*, *101*, 104–112. doi:10.1016/j.comnet.2015.11.024

Liu, Z. (2011). Greening geographical load balancing. ACM SIGMETRICS joint international conference on Measurement and modeling of computer systems, SIGMETRICS'11, 233–244. doi:10.1145/1993744.1993767

Manual, X. U. (n.d.). *Citrix: Zen Server.* Retrieved June, 3, 2013, from http://bits.xensource.com/Xen/docs/user.pdf

Mohammadi, E., Karimi, M., & Heikalabad, S. R. (2011). A novel virtual machine placement in cloud computing. *Australian Journal of Basic and Applied Sciences*, *5*(10), 1549–1555.

Pawlish, M., Varde, A. S., Robila, S. A., & Ranganathan, A. (2014). A call for energy efficiency in data centers. *SIGMOD Record*, *43*(1), 45–51. doi:10.1145/2627692.2627703

Powers, C. (2011). *Leadership in Green IT. U.S. Department of ENERGY, The National Renewable Energy Laboratory*. NREL.

Ranganathan, P., Rivoire, S., & Moore, J. (2009). Models and metrics for energy-efficient computing. *Advances in Computers*, *75*, 159–233. doi:10.1016/S0065-2458(08)00803-6

Sharad, S., Sivakumar, P. B., & Anantha Narayanan, V. (2015, July). A novel IoT-based energy management system for large scale data centers. In *Proceedings of the 2015 ACM Sixth International Conference on Future Energy Systems* (pp. 313-318). ACM.

Sharma, P., Pegus, I. I. P. II, Irwin, D., Shenoy, P., Goodhue, J., & Culbert, J. (2017). Design and operational analysis of a green data center. *IEEE Internet Computing*, *21*(4), 16–24. doi:10.1109/MIC.2017.2911421

US Department of Energy. (2013). *Pacific Northwest National Laboratory*. Retrieved June, 28, 2013, from http://eioc.pnnl.gov/ research/sitawareness.stm

US Environment Protection Agency. (2013). *Report to Congress on Server and Data Center Efficiency: Public Law 109-431*. Energy Star. http://www.energystar.gov/ia/ partners/prod_development/downloads/EPA_Datacenter_Report_Congress_Final1.pdf

Xia, M., Li, T., Zhang, Y., & De Silva, C. W. (2016). Closed-loop design evolution of engineering system using condition monitoring through internet of things and cloud computing. *Computer Networks*, *101*, 5–18. doi:10.1016/j.comnet.2015.12.016

Xu, Z. (2011). *Building a power-aware database management system*. University of South Florida.

Yang, J., Xiao, W., Jiang, C., Hossain, M. S., Muhammad, G., & Amin, S. U. (2018). AI-powered green cloud and data center. *IEEE Access: Practical Innovations, Open Solutions*, *7*, 4195–4203. doi:10.1109/ACCESS.2018.2888976

Zhou, Z. (2019, October). GreenEdge: Greening Edge Datacenters with Energy-Harvesting IoT Devices. In *2019 IEEE 27th International Conference on Network Protocols (ICNP)* (pp. 1-6). IEEE. http://www.datacenterknowledge.com/inside-facebooks-lulea-data-center/ http://www.theguardian.com/technology/2014/oct/22/cloud-storage-data-centre-sweden-arctic-hydro66

Zhu, Q. (2004). *Reducing energy consumption of disk storage using power-aware cache management*. Department of Computer Science, University of Illinois at Urbana Champaign.

Chapter 11
Green Energy Harvesting and Energy-Efficient Routing Protocol in Internet of Underwater Things

Muhammad Imran Majid
Institute of Business Management, Karachi, Pakistan

Osama Mahfooz
Institute of Business Management, Karachi, Pakistan

Wajahat Ahmed
Institute of Business Management, Karachi, Pakistan

ABSTRACT

The ability to sustain communication between nodes as per Industry 4.0 standard within the waterways is still not mature enough. It is prudent to make use of natural resources that are abundantly found in the form of strong wind (wind energy) and underwater tidal waves generation (tidal energy). Climate change and environmental sustainability are now becoming more important for a sustainable smart world. This is discussed in detail along with the decentralized mode of crop management. In this chapter, it is shown that energy efficiency can lead to the selection of different system models for the internet of underwater things (IoUT) where systems are sustainable and can tap previously unexplored resources. This resource is useful as a reference to further explore IoUT and green energy harvesting techniques.

INTRODUCTION: GREEN ICT FEATURES AND IOT NETWORKS

Information and Communication Technology refers to the technology that is responsible for all types of communication between one transmitter to one receiver or from one transmitter to many receivers. It also refers to the vast amount of data processing that the whole communication infrastructure entails

DOI: 10.4018/978-1-7998-6709-8.ch011

upon the end user. Green ICT refers to the convergence of several technologies towards energy efficiency whereby less energy is wasted and used more efficiently in hardware; however, it is strictly limited not only to hardware but software design and implementation is also considered. Big data technologies has been discussed with respect to green communication as in (Wu et al., 2016) which emphasizes the importance of IoT and green energy future.

Features of Green ICT

In 2009, UN Secretary General launched a new group A Global Green New Deal consisting of business professionals and knowledge experts who consulted on energy and climate change challenges, identifying key issues to help combat climate change. As per the Green Economy post (2009), UN's five sectors key to sustainable recovery and the global green new deal, March 2009:

1. Improving the energy efficiency of old and new buildings
2. Increasing viability of renewable energy including wind, solar, geothermal, and biomass
3. Development of sustainable transport including hybrid vehicles as well as high speed rail and bus rapid transit systems
4. Preserving the planet's ecology including freshwater sources, forests, soils and coral reefs
5. Developing sustainable agriculture including organic production.

Points 1 to 3 are relevant to energy usage in which mainly green energy is of relevance. Point 4 is the closest one gets to green ecology based on underwater life and features while point 5 relates to the oldest settlement-based project that humans have adopted which is farming and crop management.

Although not directly evident green new deal involves delivery installation and use of ICT products. For buildings, the concept of Smart Grids to regulate renewable energy usage within and without consumer and commercial entities is of interest. For evaluating wind solar geothermal and biomass the infrastructure to regulate the battery power requires processing of storage and charging which requires heavy data usage and hence development of ICT services, applications and content. The concept of smart cities helps to understand smart transportation which aims to make use of hybrid electric as well as fossil fuel. For large scale infrastructure projects, the traffic management algorithm is intense and requires not only managing traffic for one type but multiple types. It also includes various traffic routing strategies, vehicle to vehicle technologies as well as vehicle to base station infrastructure technology.

IoT Networks

IoT or the Internet of Things is an internetwork of sensors or 'things' connected via gateways and controllers to a LAN (local area network). IoT can be considered as a sub class of M2M (machine to machine which include transportation, energy, industrial, sales and payment, security, and healthcare can be regarded as pioneers before IoT systems. IoT networks would include an array of sensors, controllers, and routers. It is imperative to note that autonomous sensors are used to study the environment under sea, in the air and in remote areas where the presence of humans is poised for danger. IoT networks would eventually become more acceptable and noteworthy.

A key element to sustain industry 4.0 standard over the ocean water ways is still not mature enough. It is prudent to make use of natural resources which are abundantly found in the form of strong wind (wind energy) underwater tidal waves generation (tidal energy). Climate change and environmental sustainability is now becoming more and more important for sustainable smart world. Decentralized mode of crop yield is thus preferred. We aim to explore how energy efficiency can lead to selection of different models where systems are sustainable and can tap previously unexplored resources.

Internet of Underwater Things

The Internet of Things Technology (IoT) is developing so rapidly. One of IoT implementation is an Internet of Underwater Things (IoUT) which is used for underwater monitoring (Berlian et al., 2017). The Internet of Underwater Things (IoUT) is a unique category of Internet of Things (IoT) that encompasses practical knowledge for the development of smart cities by interconnecting many applications under water (Kao et al., 2017).

According to (Kao et al., 2017) following are some applications of IoUT

1. Environment Monitoring (water quality, pollution monitoring, oil and gas pipeline monitoring)
2. Underwater Exploration (lost treasure discovery and natural resource discovery)
3. Disaster Protection (flood and earthquake detection)
4. Military (submarine and mine detection)
5. Others (Sports analysis and Navigation assistance)

Open challenges for IoUT are (1) delay, (2) reliability, (3) bandwidth, (4) energy, and (5) mobility. Routing protocols will have an important role in controlling ways to forward data towards the adequate number of" things"(Liou et al., 2018). As per commitment to UN sustainable development goals Energy efficient protocols can be used to overcome high energy consumption problem and the global energy crisis, especially one that is bound to envelope after the episode of Covid-19.

Survey of IoUT Routing Protocols

Protocols For IoUT

Following are the protocols used in IoUT according to (Liou et al., 2018)

Table-Driven

Table-driven routing protocol is derived to reduce delay and minimize the latency introduced as a result of the route discoveries just before sending data packets from a source sensor node to a destination sink node. To minimize the real-time latency, the table-driven routing protocol maintains records of destinations and their routes by periodically distributing routing information/tables all the way through the communication network. In this way, updated and checked routing information from every node to every other node can be ensured. When there are data ready to send out, the sensor nodes can search the routing tables to forward data along appropriate paths and directions hereby ensuring the legacy and reliability to the table driven approach.

On-Demand

On-demand routing protocol finds a journey for a packet. Specifically, a route discovery process is initiated by a sensor node only when it needs to transmit data to a destination node. In the beginning, the route discovery process would overcome the network with a route request packet. Then, if the destination node receives the request packet, the destination node will send back the matching route reply packet. Accordingly, a sensor node can find out a route to its destination and then transmit data packets through this route.

Geographic

Compared to regional or on ground mapped routing, it somehow appears more probable to find a solution for an Internet of Underwater Things. Table driven and "on-demand routing" since geographic routing protocols can reduce to a very large extent, the overhead associated with relaying a message uniformly to all recipients. This is done by exploiting the geographic location information of the sensor nodes together with the information on where the knowledge to that node is originated or basically the source nodes. Secondly these are the nodes which have geographical proximity referred to as neighbor nodes. Finally, the third type of nodes are those nodes where the data or knowledge will be finally heading for the destination nodes. A point or intersection can decide on its next data exchanging action or hop what action needs to ascertain the location of its neighboring node(s) and of the destination node(s) and forward data packets in the right (or a reasonable) direction.

Opportunistic

Opportunistic routing makes relentless and reckless misuse of the broadcasting nature or one transmitter to many receiver's nature of wireless communications. When a sensor node sends out data packets, all its neighboring nodes receive the packets; then each of the receivers decides whether to forward the packets according to a probability model or a predetermined condition. This is based on certain criteria. These could be the condition that the measured value is distance between the receiver and the sink. If the condition is fulfilled (e.g., the receiver is closer to the sink than the sender), the receiver would decide to re-broadcast the packets. In this way, because the receivers will determine whether to forward packets by themselves when a sender has packets to transmit, it can send out the packets directly.

Multicast-Tree-Based

Multicast-tree-based routing is a state-of-the-art replacement technique for routing in turbid and unpredictable underwater environments. Distinct from previous routing protocols, this class of routing protocol would build a tree topology for route optimization. To attain a precise goal (e.g., delay minimization or energy minimization), these protocols are used in constructing the corresponding specific tree (e.g., the shortest path tree or minimum spanning tree). Note that instead of flooding packets throughout the network, multicast tree-based routing protocols are more likely to result in flooding packets on the specific tree topology. This is so as to achieve the specific goal.

Figure 1. IoUT routing protocols (Liou et al., 2018)

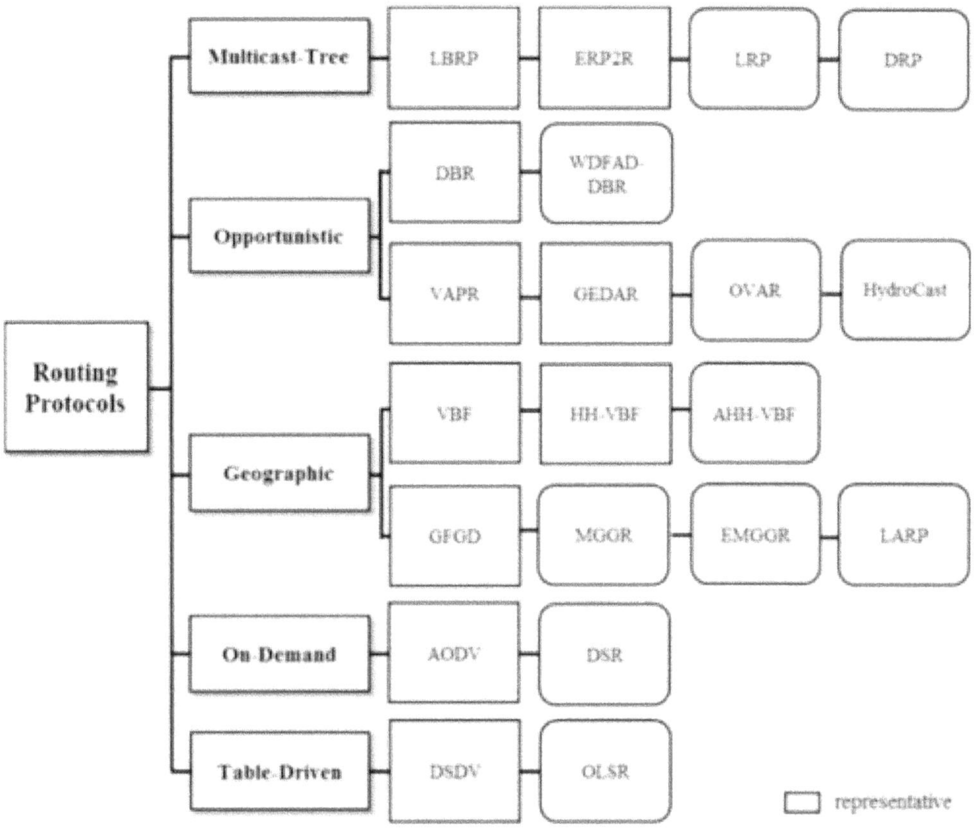

As shown in Figure 1, a summary of routing protocols is shown where each class of protocols is as defined under the section: Survey of IoUT Routing Protocols. These include Table-Driven, On-Demand, Geographic, Opportunistic, Multicast-Tree-Based protocols and have been previously elaborated.

Table 1 demonstrates the summary of the challenges faced for the second-tier protocols as specified in Figure 2. Here opportunistic types are the ones where mobility is the dominant characteristic. Multicast tier dominates in delay, energy and bandwidth. This and further information is provided in Table 2.

Energy Efficient Protocols

According to Table 1 following are energy efficient routing protocols. These are followed by a brief synopsis and description:

LRP Link quality-aware queue-based spectral clustering routing protocol (Faheem et al., 2017)

It is a layer based distributed routing protocol that addresses routing table complexity, node failures, and routing over- heads is proposed. The proposed protocol uses the idea of per-contact routing to carry the data packets

Table 1. IoUT protocols and challenges relationship (Liou et al., 2018)

Protocol	Delay	Reliability	Bandwidth	Energy	Mobility
DSDV	√	√	X	X	X
OLSR	√	√	X	X	X
AODV	X	√	Δ	Δ	√
DSR	X	Δ	Δ	Δ	√
VBF	√	Δ	Δ	Δ	√
HH-VBF	√	Δ	Δ	Δ	√
GDGD	X	Δ	√	√	X
MGGR	X	Δ	√	Δ	Δ
EMGGR	X	Δ	√	√	Δ
AHH-VBF	√	Δ	Δ	Δ	√
LARP	X	√	Δ	Δ	Δ
DBR	√	Δ	X	X	√
VAPR	X	√	X	X	√
GEDAR	Δ	√	X	X	√
WDFAD-DBR	√	Δ	X	Δ	√
OVAR	√	√	X	X	√
HYDRO-CAST	√	√	X	Δ	√
ERP2R	√	X	√	√	X
LBRP	√	X	√	√	X
DRP	√	Δ	√	√	X
LRP	√	Δ	Δ	√	X

DRP: An Energy-Efficient Routing Protocol for Underwater Sensor Networks (Chao et al., 2017)

DRP is a distance-vector routing protocol. The route selection process in DRP not only considers the transmission distance between a sender and the receiver but also the residual power for each node along the path. The DRP works properly with ocean currents as far as the distance between nodes is obtained through an existing localization scheme

Energy-Efficient Routing Protocol based on Physical distance and Residual energy (Sahana et al., 2017)

In this protocol the physical distance is taken into consideration and the residual energy that is required to balance the load of the sensor nodes. This parameter improves the delay and also meets the energy requirement. Due to this reason the lifetime of the network also increases.

Load Balanced Routing Protocol (Ze et al., 2017)

LBRP is a hybrid approach of LBMRP Load balanced multicast routing protocol and Segmented data reliable transfer protocol SDRT. It adopts the concept of network stratification, transferring multimedia data packets block by block and layer by layer.

Energy Harvesting and Wireless Power Transfer

Energy harvesting is scavenging ambient energy from environment. Here we propose real life options and discuss the different paradigms of a reconfigurable energy harvester design.

We also envisage that scalable options to energy harvesting will acquire harvesting only ambient RF power. Since RF devices are expected to exist across testbed we discuss various types of wireless power transfer types such as microwave transmission, inductive and resonance coupling.

A relatively new class of network layer protocols described as low power wide area (LPWA) networks are coming to the fore as they consume less power but cover a wide range at cost of smaller data transmission rate. We explore LORA, NB-IoT, SigFoX in terms of scalable deployment of IoT utilizing far field wireless power transfer techniques

We aim to highlight the notion of sensors in the situation where the channel can be under water such that we explore underwater wireless power transfer.

Green Architectures

In this section we aim to describe green architecture exploiting link layer protocols which are energy efficient. Here we perform a review to identify features of protocols which are used to enhance resource allocation. Specifically, architecture for spectrum efficiency (in terms of full duplex, mm Wave for green RF-WPT) will be evaluated.

IoUT ARCHITECTURE WITH ENERGY EFFICIENT ROUTING PROTOCOL: AN ANALYSIS

The UWSN architecture includes numerous sensor hubs that are joined to the base of the sea and are remotely associated with each other. Information is imparted among the sensor arrange from the base to top, surface station through multi-hop ways. The submerged gateways are the predetermined hubs outfitted

Table 2. IoUT protocols advantages and disadvantages (Liou et al., 2018)

Category	Advantages	Disadvantages
Table-Driven	1.Suitable for static networks 2.Low latency for data packets On-Demand	1.Spends much time on periodically exchanging control messages 2.Error recovery can be extremely slow 3.High signaling overhead 4.High energy consumption
On-Demand	1.Suitable for dynamic networks 2.No need to periodically exchange control messages	1.Long latency for route discovery and thus for data transmission 2.Much overhead for route discovery
Geographic	1.Forwards data packets in the right (or a reasonable) direction 2.Low signaling overhead	1.Needs geographic information 2.Packets are easily forwarded into communication holes
Opportunistic	1.Each sensor node decides whether to re-broadcast packets by itself 2.Suitable for dynamic environments	1.High probability of packet collision 2.Low efficiency in dense networks
Multicast-Tree-Based	1.Low propagation delay 2.Low energy consumption 3.Can achieve the specific goal	1.Extra control signaling overhead 2.Low stability for mobile nodes

with vertical and even transceivers. The first node is utilized to send instructions and arrange information to the sensor hubs and get the gathered information from it. The subsequent one is utilized to relay the checked information to the ocean surface station. Other than shallow water, vertical correspondence is normally required for a long-run in profound water to accomplish information conveyance toward the surface station. The acoustic correspondence is utilized to play out numerous equal interchanges to assemble information from sensor hubs. Where radio correspondence typically settled with a satellite is utilized to hand-off accumulated information to the coastal sink.

Figure 2. Internet of underwater things model (Qiu et al., 2019)

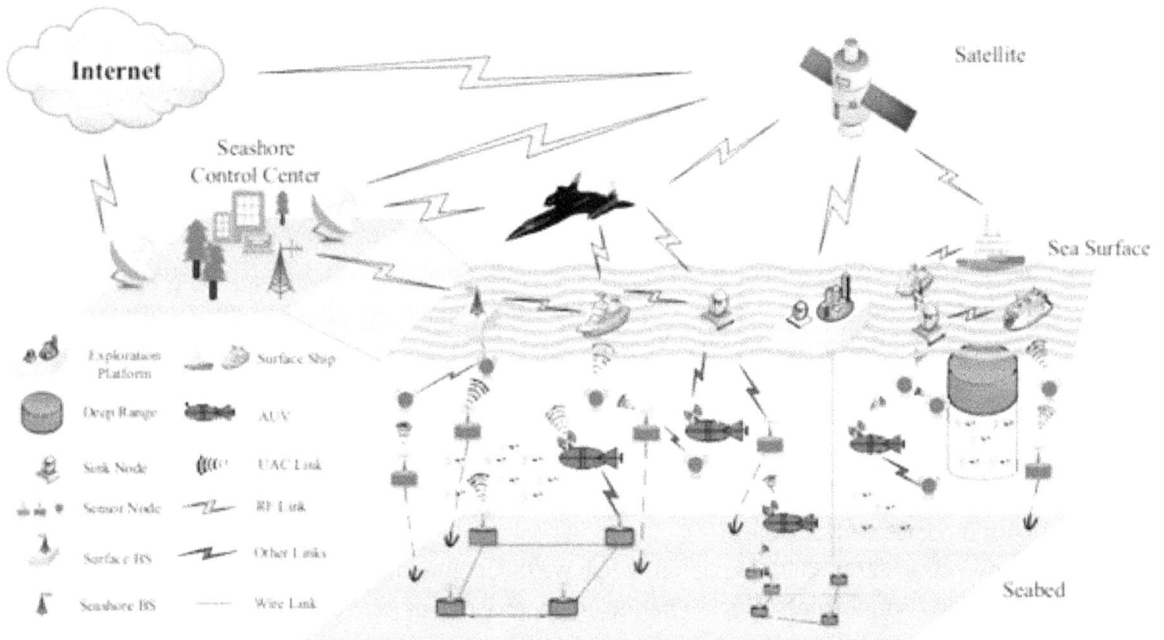

Differently from (Rice & Diego, 2001) where only the at the bottom of the ocean sensor nodes are considered and the Autonomous Underwater Vehicles (AUV) are used to transfer data from the bottom to the surface.

In (Tommaso Melodia et al., 2013) Sensor nodes are installed in various depth to sense the given data. Floating buoys are attached to surface station or the ocean bottom to keep itself floating at the specific depth. The floating buoy changes its depth accordingly along with the sensor node with it, by variable wire length that shifts it to the bottom. Floating buoys has a vulnerable aspect of security as it is easily detected on the sea surface (Jouhari et al., 2019).

The underwater acoustic signal propagation depends on many environmental factors like salinity, temperature, pressure, and depth of transceivers. Other concerns are absorption and transmission losses. Movement of transceivers and changes in environment cause scattering and Doppler effect. Underwater noises also have significant impact on the acoustic signal like man-generated and natural noise creating elements are important to take care of. Unmanned Underwater Vehicle (UUV) or Autonomous Underwater Vehicle (AUV) are used from wide-range UWSN to cover areas that are out of range. Also, terrestrial GPS does not work accurately for sensor nodes anchored to the ocean bottom. For this purpose, acoustic channel should be considered.

For the above, the proposed network assumes that data plane and control plane of the communication interface has limitless energy source. In real life, the sources are limited especially for the buoy and sensors underwater which need extra power to communicate and to keep their position and attitude. It is therefore important that methods to explore power adoption are considered. Since the rough environment of sea tend to discourage laying of wires. Further as the goal is to protect the natural habitat of the water life. But the wires only entangle or make the habitat not worth living. It is hence appropriate to explore wireless modes of power transfer to nodes. The concept is explored in detail in the following two sections.

Wireless Power Transfer & Green Architectures: A Review

Wireless power transfer has always been so fascinating since tesla implemented it using microwaves. The idea to transfer power or electricity without wires has been capturing interest of scientists and engineers for decades. Nikola tesla influential work in 1893 and 1917 on wireless power transfer motivated scientist and engineers all around the world. It was nearly after some year's that brown demonstrated a process comprising of microwave beams to power a helicopter (Taofeek Orekan & Zhang, 2019). Then later in 1971, Don Otto followed the work of brown and developed a trolley which was based on induction principle of wireless power transfer (Taofeek Orekan & Zhang, 2019). In 1973, National lab of Los Alamos demonstrated world first ever RFID system. Nikola tesla aka father of electricity has always been the motivation behind wireless power transfer.

In early days, only a considerable amount of work has been done on wireless power transfer until it has been revolutionized by resonant induction technique in 2008. A wireless charging product PRIMOVE developed by Sony and Bombardier was based on inductive energy wireless power transfer technique (Taofeek Orekan & Zhang, 2019). Later in years between 2010-2017 several such products have been developed with moderate efficiencies.

Wireless power transfer is challenging but, when we talk about underwater wireless power transfer, it becomes even more complex due to the dynamic characteristics of ocean itself. In underwater wireless power transfer, the transmitting side of WPT is in continuous motion and thus, the time-varying coefficient and transient characteristics in underwater WPT becomes more challenging as compared to WPT in air. Moreover, further challenges such as coil structures, power capacities and leakage inductances of both transmitting and receiving sides has to be addressed too. One viable solution was mentioned in (Taofeek Orekan & Zhang, 2019) in which a tunable capacitance using variable capacitor, a technique impedance control was implemented for dynamic WPT in air but, it is very difficult to implement in underwater WPT but, research is being modernized in underwater WPT and it is only a matter of time.

Why We Need Underwater Wireless Power Transfer?

The basic and most important question is that why we need it? As fascinating it sounds, there must be a need and reason to implement such complex technique. To explore this question, we must first know how underwater sensor network works. So, we get into its basic components and devices that are a major player in underwater sensor network mystery box.

Underwater sensor networks are utilized for data collection routines and collaborative monitoring tasks (Awan et al., 2019). It comprises of sensor nodes, that is none other than embedded systems that contains several units such as sensing unit for sensing the underwater parameters such as temperature, pressure, toxin level etc. Processing unit to process the data using a digital signal processor or else, communication unit for communication with other nodes and radio stations. Finally, power unit for powering the battery required for all these tasks. The main problem is the limited capacity of battery, that may run out depending on the life of the battery. If a sensor node dies out due to the battery dried out, it could possibly compromise the reliability of the network.

Autonomous underwater vehicles (AUV) primary tasks are exploration of sea, military purposes, search operations, recovery protocols etc. These AUV's relies on batteries and have limited battery capacity. Therefore, these batteries either must be charged or replaced. Traditional methods for charging the AUV is to put it out of the water and load it onto the dock where a cable connection is provided;

a copper wire which corrodes due to salty water. Furthermore, for the removal of AUV from the very underwater, a large vessel is required in addition to labor which increases the cost. Moreover, a particular type of AUV can only be charged with a specific type of charger which adds on to the costs. In past times, docking stations were proposed which were build underwater for charging of AUV'S but, did not make up to the expectations because of several issues such as high cost, susceptibility (Taofeek Orekan & Zhang, 2019). Moreover, these docking stations are a buildup for dirt and other toxin contaminants and as time passed, these could interfere in proper functioning of the system. An alternative solution is to replace the battery, but it's not suitable to dive out underwater and check for nodes batteries and their maintenance as well. For reliable, effective and ongoing completion of tasks, an automated system has to be designed and the possible solution has to be happened underwater and it has to be underwater wireless power transfer so that remote operations can be performed easily without manual collaboration.

Thus, based on the application and architecture of the underwater sensor network, additional units can be added such as energy harvesting module, or transmitter and receiver coils for underwater wireless power transfer.

Figure 3 shows an underwater sensor node architecture. As it can be seen, that out of the basic units, it also contains an additional unit energy harvester for harvesting the energy from ambient environment.

Figure 3. Modular approach to sensor node architecture

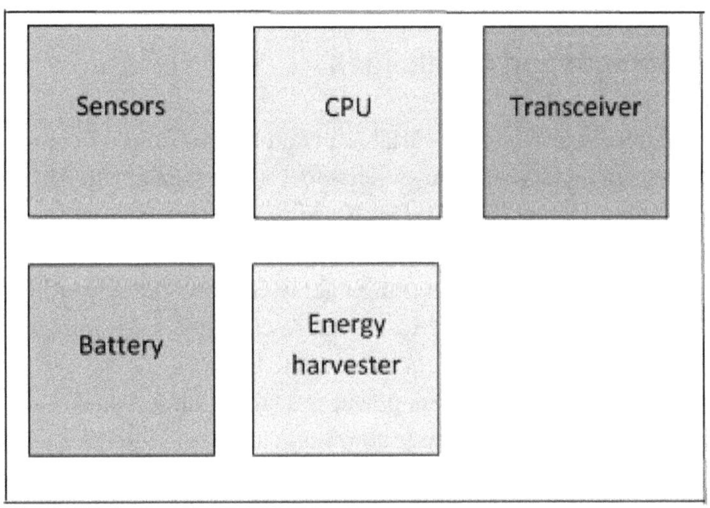

Now that we know that why we need underwater wireless power transfer, it's time to get into the depth that what challenges underwater sensor networks are currently facing and how underwater wireless power transfer has been done in the past and how it can be improved in near future.

Wireless Power Transfer Vs Energy Harvesting

Energy harvesting is basically harvesting energy from the environment. These are also called renewable energy or green energy or sustainable energy. Wireless power transfer is quite different from energy harvesting as it does not involve any ambient energy or green energy. WPT technologies are wire free

Table 3. Comparison between wireless power transfer and energy harvesting

Wireless power transfer	Energy harvesting
In Radio-frequency, the signal can transfer both information and power at the same time.	Only transfer power.
WPT does not depend on environmental sources and hence, more controllable (Jayakody, Thompson, Chatzinotas, & Durrani, 2017)	Totally depends on environmental sources and very unpredictable in nature and hence, less controllable
Is suitable for both small-scale and large-scale systems	Suitable for large scale systems because of the heavy mechanisms of the energy harvester device (Jayakody et al., 2017)

and its basically transmission of electricity without wire. Transmission and harvesting are totally different as transmission means from once place to another and harvesting means utilizing surrounding energy through the use of energy harvester. Table 3 shows a short comparison between WPT and energy harvesting to grasp a better picture.

Underwater Environment and Challenges

Underwater environment characteristics are variable in nature. Its communication channel, localization (one of trending researches in underwater sensor networks) discussed in (Su, Ullah, Liu, & Choi, 2020), environmental factors, routing protocols, MAC (Media access control) are totally different and complex as compared to sensor networks in air. The current challenges in underwater sensor networks are somewhat broad such as bandwidth limitation, high propagation delay, 3D topologies, routing protocols, energy and resources utilization and power constraints. Most of these challenges are somewhat out of the scope of this chapter but it's important to highlight them to motivate the need for energy harvesting or utilization of green environment underwater and wireless power transfer in underwater.

To further motivate the need for underwater communication and alternatively, underwater wireless power transfer, some of their applications are aquatic surveillance, oceanographic data collection and compilation, sea pollution discovery, monitoring habitats and exploiting the sea environment (Awan et al., 2019). Unlike, normal sensor nodes, underwater sensor nodes are not static and constantly flowing with the water when water currents hits them and thus sensing data only gets sense when localization comes into place (Su et al., 2020). Energy utilization protocols does not work efficiently in underwater sensor networks (Awan et al., 2019) due to the fact, that the nodes become mobile because of the variable nature of water. Therefore, energy harvesting and underwater wireless power transfer becomes so efficient in reliable underwater communication and monitoring.

Before getting into the depths of underwater wireless power transfer, let's have a look at some of wireless power transfer techniques.

Wireless Power Transfer Techniques

Wireless power transmission techniques can be technically grouped into radioactive based charging, and no-radioactive based charging (Taofeek Orekan & Zhang, 2019). Radio-active based charging is utilized for far-field transmission which includes, microwave transmission, millimeter wave transmission, optical and acoustic based techniques. Whereas, in non-radioactive based charging is utilized for near field transmissions which rely on the principle of electromagnetic field coupling such as magnetic resonance, capacitive and inductive based coupling techniques (Taofeek Orekan & Zhang, 2019). Wireless power transfer techniques has been widely discussed by (Zhang et al., 2019) in which technical challenges in WPT has been elaborated.

Microwave Power Transmission

As shown in Figure 4, Microwave transmission is a far-field wireless power technique which involves wireless power transmitting through free space. It consists of two sections; transmitting section and the receiving section. In transmission block, there is microwave power source which generates microwaves and it can be controlled by electronic circuits but, microwave source must have perfect directional axis or alternatively laser beam. In microwave transmission, usually magnetron is used as microwave source. The next block is the wave guide circulator which is connected through the wave guide adaptor and it basically protects the source from the reflected power. Then there is tuner, which balances the impedance between source and transmitting antenna. Then, based on the direction of propagation signal, the attenuated signals (due to attenuation) gets separated by direction coupler. In the receiving part, the receiving antenna converts microwave power into DC. The matching and impedance circuit sets the output impedance.

Figure 4. Microwave wireless power transfer (Taofeek Orekan & Zhang, 2019)

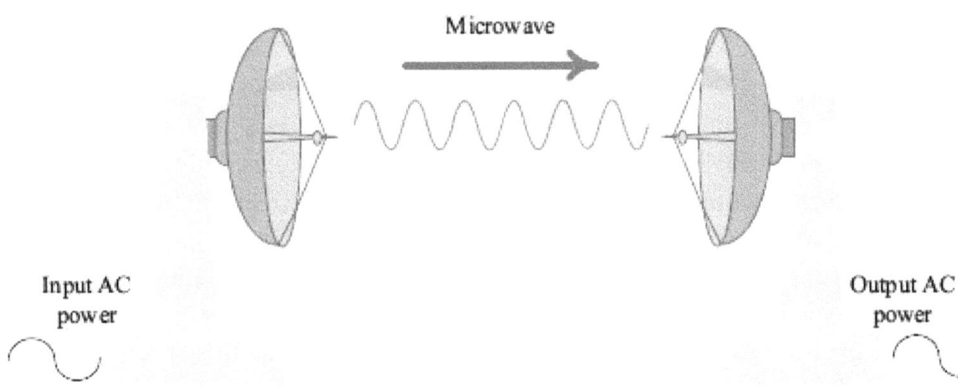

Acoustic Power Transfer

Acoustic transfer technique has been recently proposed as illustrated in (Taofeek Orekan & Zhang, 2019). This technique makes use of piezoelectric transducers for sending and receiving ultrasonic waves through

Figure 5. Acoustic wireless power transfer (Taofeek Orekan & Zhang, 2019)

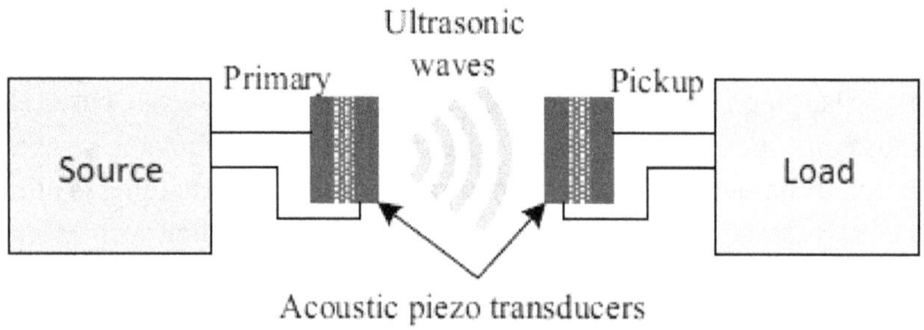

metal as shown in Figure 5. This wireless power transfer technique is a feasible approach for achieving suitable efficiencies (Taofeek Orekan & Zhang, 2019). Furthermore, due to shorter wavelengths acoustic power transfer can achieve higher beam directive (Taofeek Orekan & Zhang, 2019).

However, acoustic power transfer is facing some challenges just like other techniques. Some of them are considerable spatial fluctuations in the acoustic intensity due to the wave interference phenomena and its efficiency drops significantly when its receiver is not align centered on the acoustic beam. Particularly, spatial resonance basically limits the receiver deployment and therefore, limits its application.

Optical Power Transfer

This technique of wireless power transfer is based on laser sources. It is similar to microwave transmission but, somewhat less efficient due to lower energy conversion efficiency and higher attenuation. However, unlike microwave transmission, optical power transfer does not require the large-aperture antenna for focusing the microwave beam. It can focus in small area as well as depicted in Figure 6.

Optical power transfer also offers others advantages over other far-field WPT techniques such as large coverage and long transmission range (Taofeek Orekan & Zhang, 2019).

Inductive Coupling

Inductive coupling is near-field wireless power transmission technique as in Figure 7. It involves two coils; transmitter coil and receiving coil. If we connect AC source to transmitting coil and pass an electric current to it and then place receiver coil in its field, the electric current will transfer to receiver coil. In inductive coupling transmission, the energy transferred is due to mutual induction. This technique works for short distances well, if we place second coil far apart from first coil, it will get out of its field and power seizes. When the coils axes are same, energy transfer is efficient, when misalignment occurs between two coils axis, energy transfer can be affected because magnetic flux shaping.

Design Challenges

The design of the underwater wireless power transfer system must be reliable that can withstand underwater dynamics such as force of waves, fluctuation of temperature, environmental parameters, corrosive salts etc.

Figure 6. Optical wireless power transfer (Taofeek Orekan & Zhang, 2019)

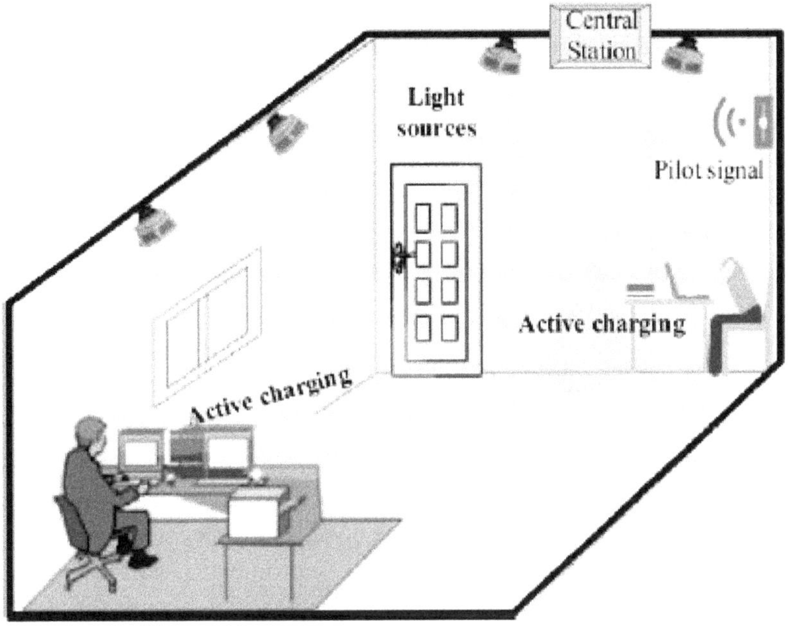

Other design challenges such as the transmission of electromagnetic signals due to the fact of seawater conductance because of its inherent properties. Moreover, coherent salts increase the conductivity further and eddy currents can be formed due to alternative paths. Eddy currents are not good, they can diminish the strength and productivity of underwater wireless power transfer (Taofeek Orekan & Zhang, 2019).

Another challenge is the design of the coil itself. Too large coils will be a cumbersome task to be integrated into existing devices and systems and too small coils size will have throughput issue. So, the size of the coils must be optimal (Taofeek Orekan & Zhang, 2019).

Figure 7. Inductive wireless power transfer (Taofeek Orekan & Zhang, 2019)

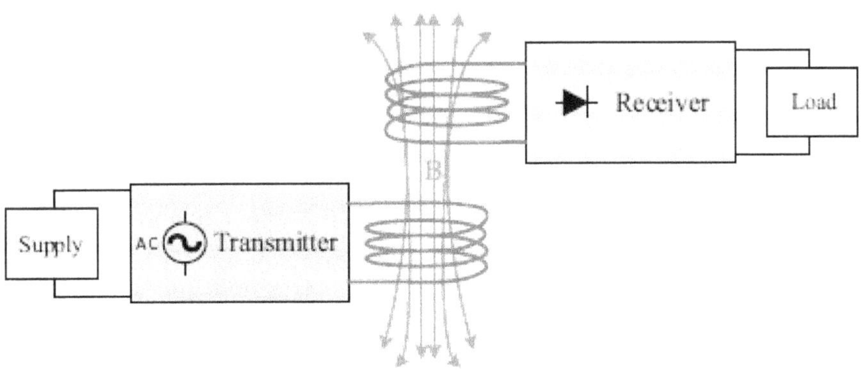

Green Energy Harvester Source for IoUT Architecture

Energy harvesting is scavenging an ambient energy from the environment. The idea is to get the energy from the surroundings to charge the batteries, even powering the nodes directly from an ambient energy source. The ambient energy can be of different types such as solar energy, wind energy, thermoelectric energy and tidal energy. It depends on the application of WSN as to which energy is used. As for IoUT, tidal energy, wave energy and mechanical energy are most suitable. Although, RF can penetrate through the water as well.

Sustainable energy or green energy is crucial element for internet of underwater things because these devices are energy-hungry and can die out any time. These energy constraint devices have to be operated constantly for reliable data collection, analysis and collaborative monitoring. It is very essential to use sustainable or green energy resources for these crucial tasks.

The energy harvesting device is installed in the node which collects ambient energy from the surroundings. Energy harvesting device collaborates with energy storage element for the storage of harvested energy. Power management is used for the control of energy harvesting device and processing and control is used to generate fixed voltage supply suitable for the node. The reservoir or energy storage element can be a battery but, now a day, super capacitors are used as reservoir because of their super capacitance and high charging and discharging rates which makes them suitable to be used with energy harvesting device.

Energy Sources and Devices

In the following sections we will explore the suitability of energy sources separately from solar and wave energy as well as tidal energy. Then we approach systems which take these energy sources in synergy as well as microbial and tidal. Finally, we chose another hybrid variation of energy source.

Solar Energy

Solar energy has been most popular for very long. With the advantage of high power output with moderate efficiency, this green source is reliable and long-term as long as the sun is there. PV cells being an alternative to traditional photodetectors has revolutionized the modern era of green energy and so, it has been researched deeply in underwater wireless communications (Sun et al., 2020). There are special kind of solar cells that can be deployed underwater for harvesting underwater sunlight and convert into usable electricity. Traditional underwater solar cells have been using silicon material which has a narrow band-gap and thus, the efficiency is low in underwater as studied in (Rohr et al., 2020). Detailed balanced calculations are performed in (Rohr et al., 2020) which suggests to make use of semiconductors materials which has a wide band-gap and then underwater solar cells can operate with 65% efficiency.

Wave Energy

Wave energy has not achieved much attention as compared to other renewable energy sources. But, for underwater energy harvesting wave energy comes into play and it is only a matter of time before it becomes an attraction for scientist and engineers. To withstand the dynamic and harsh environment underwater, hydrodynamics assessment of wave energy device performance is crucially important (Taofeek Orekan & Zhang, 2019).

Figure 8. Underwater solar cell (Rohr et al., 2020)

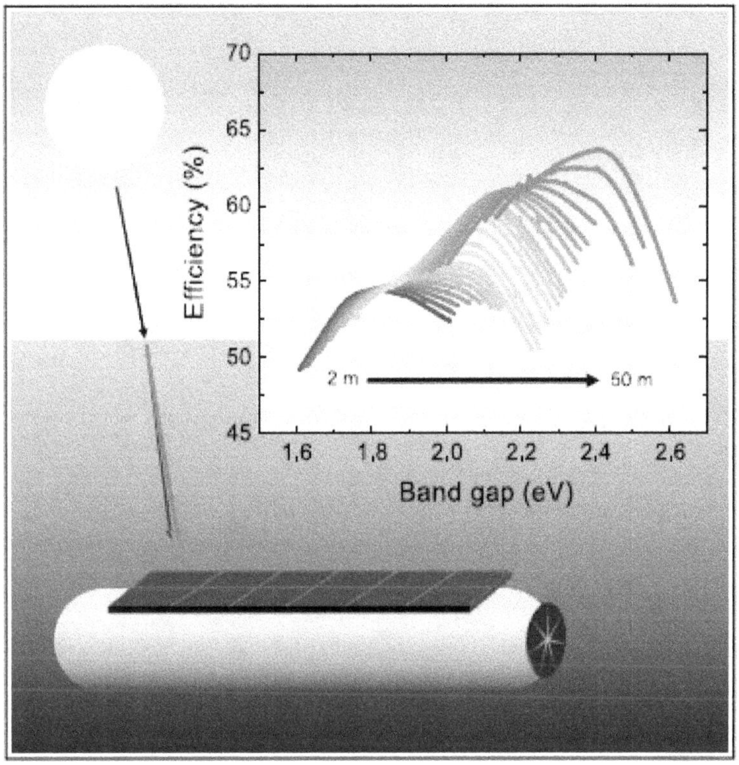

Wave energy converter (WEC) is a device that extracts energy from motion of the waves however, past research has not been much reliable in terms of the wave converter efficiency (Taofeek Orekan & Zhang, 2019). Past researches assume sine-wave pattern for incident waves as studied in (Taofeek Orekan & Zhang, 2019), whereas in wave energy converters, it does not have a single sine-wave pattern only, but other patterns as well. To improve the efficiency of a wave-energy converter complex non-linear equations of the system has yet to be considered (Taofeek Orekan & Zhang, 2019). In extreme weather conditions, the load on the device is 100 times its normal load (Taofeek Orekan & Zhang, 2019) and therefore efficiency of the system must be high in order to compensate.

(Taofeek Orekan & Zhang, 2019) studied the recent and effective design for wave energy converter which can be seen in Figure 9, it's a single point absorber based WEC. Like regular WEC, it also consists of heaving buoy which oscillates at the ocean surface and is connected to cylindrical shape based spar mounted on the seabed. A power takes off (PTO) mechanism in this model is different than previous PTO models such as hydraulic based PTO which is currently most popular, a PTO converts motion of waves into electricity. Unlike, other PTO, the mechanism in this model differs in a way that it extracts energy from the motion of buoy through a direct drive linear generator which in turns does not involve moving parts for regular maintenance and hence, a good design.

Figure 9. Single point absorber based WEC (Taofeek Orekan & Zhang, 2019)

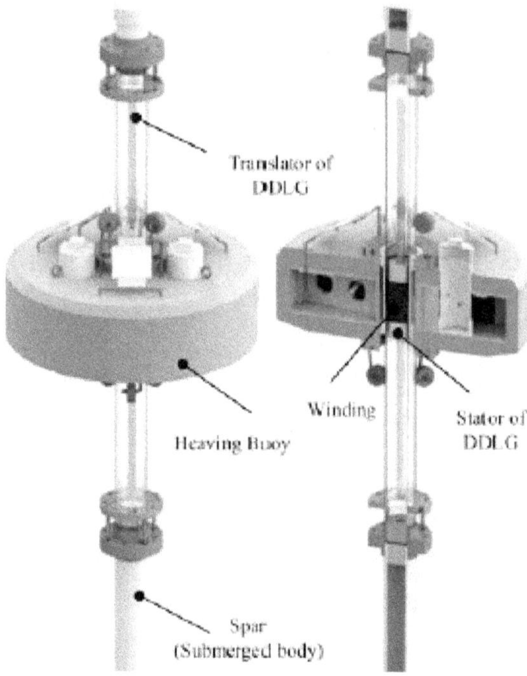

Multi-Source Energy Harvester

Not many energy harvesting techniques can fulfill the energy requirement underwater. Solar cells can only work in daylight. Wave energy converters relies on the motion and shape of waves. Tidal power can be damageable to habitats underwater but, according to (Sai Srujana, Neha, Mathews, & Harigovindan, 2015) two green sources can be combined and forms multi-source energy harvester. According to (Sai Srujana et al., 2015) Microbial fuel cell and piezoelectric could eliminate energy crisis underwater. For this, we must first understand these two harvesting techniques.

Microbial Fuel Cell

This microbial fuel cell is an electrochemical based system which utilizes metabolic reactions in bacteria found in nature particularly underwater to produce electricity (Sai Srujana et al., 2015). At the anode of this device, there is the bacteria house growing on to the anode surface, and due to metabolic activities going on, it basically degrades the substrate and transfers the electrons to the anode. With the help of an external circuit, these electron reaches the cathode chamber to where dissolved oxygen at the cathode is reduced to water. In (Sai Srujana et al., 2015) author proposed a different microbial fuel cell in which, the anode is to be dipped in the silt which contains bacteria and cathode is dipped in water so that it can use the dissolved oxygen to accept electrons from the anode. Basic structure of MFC is as shown in the Figure 10.

Figure 10. Two-channel MFC basic structure (Sai Srujana et al., 2015)

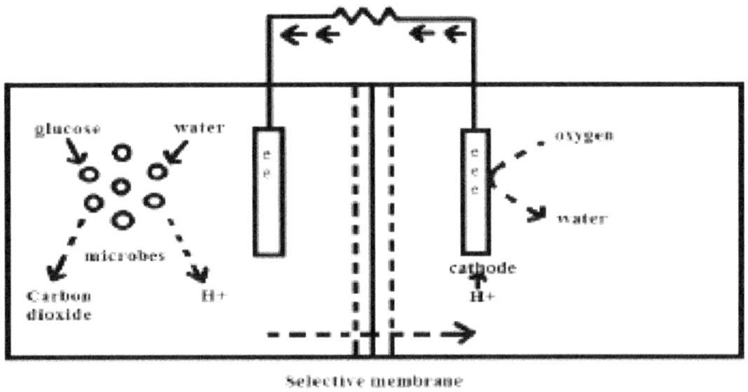

Piezoelectric Harvesters

Acoustic resonators are basically used for amplification of sound in musical instruments. Static resonant waves are formed when an acoustic resonator operating at their resonant frequencies is excited by an incident wave (Sai Srujana et al., 2015). This is the basic principle behind the mechanism of acoustic-based piezoelectric harvesters. When an unclamped piezoelectric beam is placed in a resonator, pressure difference within the beam is created due to the spatially varying acoustic resonant pressure. This pressure difference controls the piezoelectric beam vibration at its resonant frequency and thus resulting in generating electricity (Sai Srujana et al., 2015).

Figure 11 shows (Sai Srujana et al., 2015) architecture for multi-source energy harvesting system

Underwater Kites

One source of energy harvesting is tidal power although, it has not been implemented at a larger scale, researchers have been using creative ways to make use of this type of energy. One fact about tidal power is that it is more effective at high-pace and fast tidal streams and if we look at ocean waves, they have low-pace and slow tidal currents which are certainly not effective to harness enough tidal energy. If we take an example of horizontal axis turbines which primary purpose is to extract ocean energy from the movement of water, if it sits stationary on seabed, it requires 2.5m/s ocean current to harness energy effectively or we can say cost-effectively.

Figure 11. Hybrid harvesting system (Sai Srujana et al., 2015)

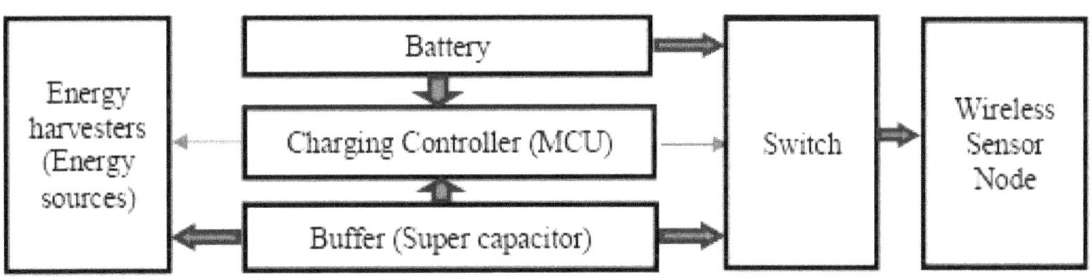

Figure 12. Eight shape-like pattern

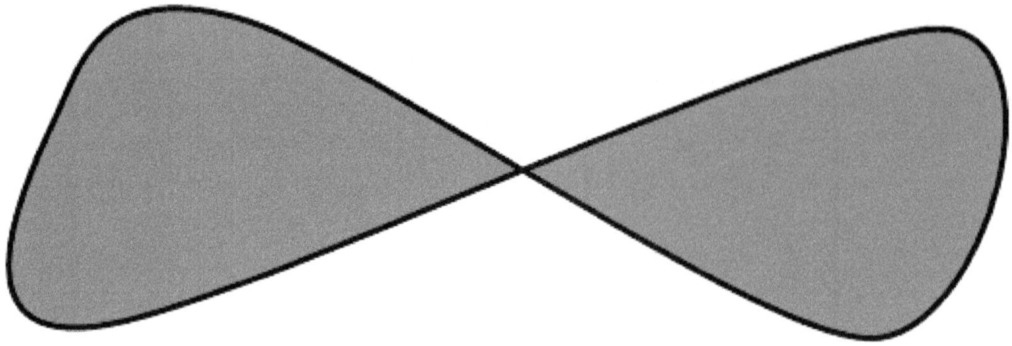

This technology of underwater kites works on the principle of hydrodynamic lift, an upward force is necessary for the lift and in this case, ocean currents provides it for the underwater kites. It has an underneath turbine for generation of electricity and wings which are tethered to the seabed using a cable. As the kite flies something called figure 8 pattern shown in Figure 12, the ocean water flows along over its wing span and underwater kite sweeps through the ocean current which in turn produces kinetic energy. There is an onboard generator and turbine converts this kinetic energy into usable electricity and this electricity is transmitted through tethered to cables.

Underwater Gliders

Underwater gliders (UG) are one particular type of autonomous underwater vehicle (AUV) used for ocean sampling and monitoring purposes. UG shapes like an airplane but, with only one wing as shown in Figure 13. Underwater gliders are basically known for long traveling range (Wu, Yu, Yuan, & Tan, 2019). They are also known for high-energy efficiency and secret lies in their buoyancy-driven mechanism which propels by the synthetic effect of hydrodynamic lift force and buoyancy. Most of the energy is consumed when buoyancy is switched between surfacing and diving (Wu et al., 2019). Whatsoever, like other AUV's, they are also battery dependent and have a life of about 3 to 6 months so, author (Cao et al., 2019) presented s smart float design for underwater gliders in which according to the author has an additional advantage of long endurance capability and battery capacity of up to 3 years. Author (Cao et al., 2019) achieves this by combining underwater gliders abilities with float profilers capabilities. Float profilers are known for drifting with ocean currents but, are not suitable for trajectory tracking. Their primary task is to capture environmental data related to ocean currents, thermocline and halocline(Cao et al., 2019). Author (Cao et al., 2019) understands the mechanics of both float profilers and underwater gliders and explained that float profilers only drift with the ocean flow and thus, there is no need to change the altitude so, author (Cao et al., 2019) designed in it such a way that the platform center of buoyancy equals center of gravity and therefore, float remains vertical and stable in ocean water. On the opposite, underwater gliders needs to change its altitude which is clearly based on the situation of surfacing and gliding, so it needs to glide forward during operation however, the author (Cao et al., 2019) ensures underwater glider to be bottom heavy, its central buoyancy and central gravity are designed in that way that it is not aligned to its central longitudinal axis.

Figure 13. Underwater glider internal mechanism (Cao et al., 2019)

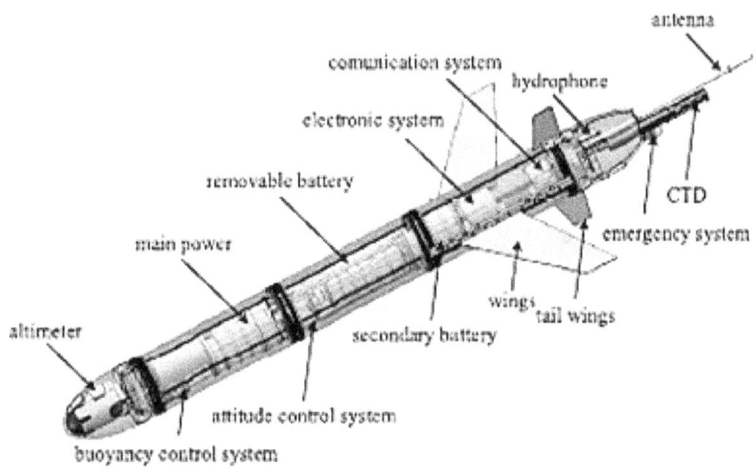

According to author (Wu et al., 2019), the speed and maneuverability of underwater glider is much lower because of the fact that it's heavy and controlling of the trajectory in 3 dimensions is cumbersome task. So, the author (Wu et al., 2019) developed a dolphin based underwater glider which has some attributes of a dolphin such as flippers and flukes (Wu et al., 2019) which helps merging propulsion modes and thus, combining two robots into one. Author (Wu et al., 2019) gliding robot dolphin switches between gliding and swimming. These two propulsion modes can enhance each other control mode as

Figure 14. Mechanical structure of gliding dolphin robot (Wu et al., 2019)

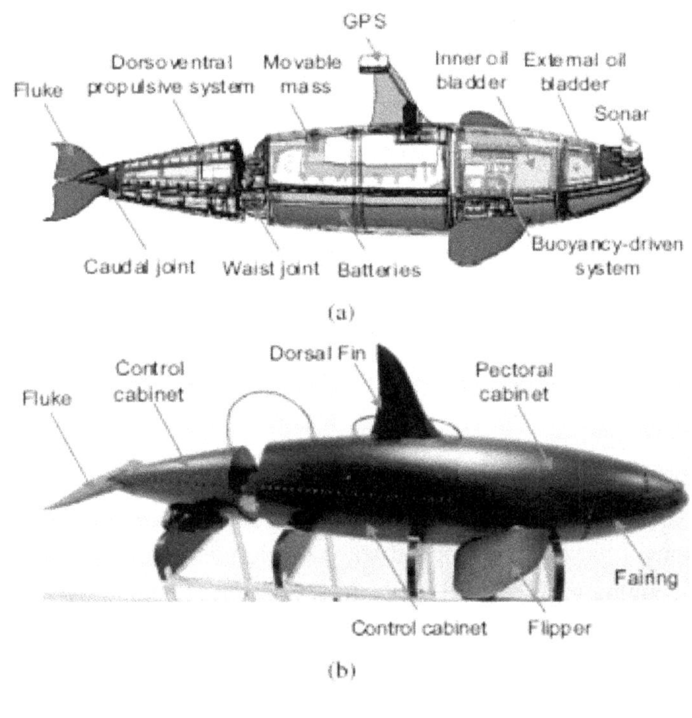

Figure 15. Mass distribution of UG (Side view) (Cook & Zhang, 2019)

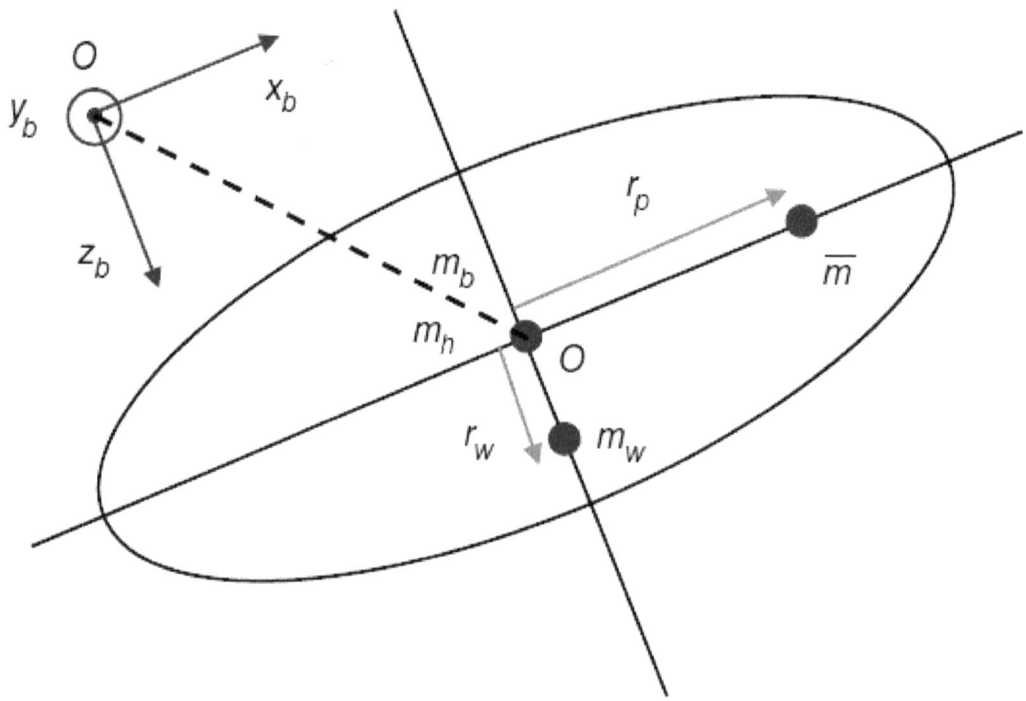

well (Wu et al., 2019). It's basically a glider dolphin having hybrid motions and controllable fins that can enhance or manipulate the gliding maneuvers. Figure 14 shows the gliding dolphin robot design.

One of the difference between surface normal vehicles and underwater vehicles is that in the former motion is limited to two dimensions x-y plane whereas in the latter motion is three dimensional x-y-z plane and expressing altitude in three dimension as pitch, yaw and roll (Cook & Zhang, 2019). Author (Cook & Zhang, 2019) modelled such 3 dimensional underwater glider system as rigid-body system considering that UG has three inclination systems for motion including the control system, mass distribution system and buoyancy system. The buoyancy system basically changes the net buoyancy of UG by transferring fluid in and out of an internal reservoir. The mass distribution system basically uses a kind of linear and rotary actuator mechanism to push and spin a mass, in this case a (battery pack) to alter the alignment from center of the mass of the UG. The control system basically includes a rudder maneuver and stabilization. Figure 15 shows mass distribution of UG. The whole point of this modelling is to simulate so that problems can be identified specially in controllability of UG, novel designing and increasing efficiency of UG.

Future Research Directions

Ocean is just another world. are all familiar with the extending range and the depth of oceans so, the AUV's, docking stations, tidal barrages, ocean wave converters can be anywhere and in the coming decades, the most challenging problem will be the transmission of energy over long distances in underwater such as transmitting power between wave energy converter and AUV or even between multiple

AUV through wireless power transfer will be a challenging task. This is due to the fact that attenuation increases rapidly when the distance between Tx and Rx coils increases.

CONCLUSION

In this book chapter, we introduced the idea of energy harvesting using IoUT and wireless power transfer as applied in the context of the UN SDG driven Green ICT systems and established by the International Telecommunication Union (ITU). A review of various measures of wireless power transfer techniques follows including Design challenges, temperature fluctuation and environmental concerns as well as EM performance variations underwater along with the design of the coil itself. The concept of energy harvesting from solar and wave source separately first followed by combination is carried out in tandem. Further hybrid harvesting and microbial fuel cell techniques are also introduced. The material provided in this book chapter is unique and novel and forms the basis of evaluating feasibility for green energy harvesting with reference to Internet of Underwater Things (IoUT). This area of expertise is expected to be emerging and have enough potential to be considered for studies in the futuristic mobile wireless 6G standardization as well as to sustain Industry 4.0 standard.

REFERENCES

Awan, Shah, & Iqbal, Gillani, Ahmad, & Nam. (2019). Underwater Wireless Sensor Networks: A Review of Recent Issues and Challenges. *Wireless Communications and Mobile Computing*.

Berlian, M. H., Sahputra, T. E. R., Ardi, B. J. W., Dzatmika, L. W., Besari, A. R. A., Sudibyo, R. W., & Sukaridhoto, S. (2017). Design and implementation of smart environment monitoring and analytics in real-time system framework based on internet of underwater things and big data. *Proceedings - 2016 International Electronics Symposium, IES 2016*, 403–408. 10.1109/ELECSYM.2016.7861040

Cao, L., & Li, Z. Yao, & Lian. (2019). Smartfloat: A Multimodal Underwater Vehicle Combining Float and Glider Capabilities. *IEEE Access, 7*, 77825–38. doi:10.1109/ACCESS.2019.2922171

Chao, C. M., Jiang, C. H., & Li, W. C. (2017). DRP: An energy-efficient routing protocol for underwater sensor networks. *International Journal of Communication Systems*, *30*(15), 1–10. doi:10.1002/dac.3303

Cook & Zhang. (2019). Dynamics Modeling of AUVs. *Mobile Robots*, 269–89. doi:10.1002/9781119534839.ch11

Faheem, M., Tuna, G., & Gungor, V. C. (2017). LRP: Link quality-aware queue-based spectral clustering routing protocol for underwater acoustic sensor networks. *International Journal of Communication Systems*, *30*(12), 1–17. doi:10.1002/dac.3257

Jayakody, T. Chatzinotas, & Durrani. (2017). Wireless Information and Power Transfer: A New Paradigm for Green Communications. Springer.

Jouhari, M., Ibrahimi, K., Tembine, H., & Ben-Othman, J. (2019). Underwater Wireless Sensor Networks: A Survey on Enabling Technologies, Localization Protocols, and Internet of Underwater Things. *IEEE Access: Practical Innovations, Open Solutions*, *7*, 96879–96899. doi:10.1109/ACCESS.2019.2928876

Kao, C. C., Lin, Y. S., De Wu, G., & Huang, C. J. (2017). A comprehensive study on the internet of underwater things: Applications, challenges, and channel models. *Sensors (Switzerland)*, *17*(7), 1375–1378. doi:10.339017071477 PMID:28640220

Liou, E. C., Kao, C. C., Chang, C. H., Lin, Y. S., & Huang, C. J. (2018). Internet of underwater things: Challenges and routing protocols. *Proceedings of 4th IEEE International Conference on Applied System Innovation 2018, ICASI 2018*, 1171–1174. 10.1109/ICASI.2018.8394494

Melodia, T., Kulhandjian, H., Kuo, L.-C., & Demirors, E. (2013). *Advances in Underwater Acoustic Networking*. Wiley-IEEE Press. doi:10.1002/9781118511305.ch23

Orekan, T., & Zhang, P. (2019). *Underwater Wireless Power Transfer* (1st ed.). Springer. doi:10.1007/978-3-030-02562-5

Qiu, T., Zhao, Z., Zhang, T., Chen, C., & Chen, C. L. P. (2019). Underwater Internet of Things in Smart Ocean: System Architecture and Open Issues. *IEEE Transactions on Industrial Informatics, PP*, (c), 1–1. doi:10.1109/TII.2019.2946618

Rice, J. A., & Diego, S. (2001). *ADAPTIVE Shallow Water Acoustic Networks.* Academic Press.

Rohr, Lipton, & Kong, Stephen, Taylor, & Ro. (2020). Efficiency Limits of Underwater Solar Cells. *Joule*, *4*, 1–10.

Sahana, S., Singh, K., Das, S., & Kumar, R. (2017). Energy efficient shortest path routing protocol in underwater sensor networks. *Proceeding - IEEE International Conference on Computing, Communication and Automation, ICCCA 2016*, 546–550. 10.1109/CCAA.2016.7813780

Srujana, Sai, & Neha, Mathews, & Harigovindan. (2015). Multi-Source Energy Harvesting System for Underwater Wireless Sensor Networks. *Procedia Computer Science*, *46*, 1041–1048.

Su, Ullah, Liu, & Choi. (2020). A Review of Underwater Localization Techniques, Algorithms, and Challenges. *Journal of Sensors*.

Sun, Ooi, Kang, Kong, Alkhazragi, Guo, & Ouhssain, Weng, Jones, & Ng. (2020). A Review on Practical Considerations and Solutions in Underwater Wireless Optical Communication. *Journal of Lightwave Technology*, *38*(2), 421–431.

Wu, J., Guo, S., Li, J., & Zeng, D. (2016). Big Data Meet Green Challenges: Greening Big Data. *IEEE Systems Journal*, *10*(3), 873–887. doi:10.1109/JSYST.2016.2550538

Wu, Yu. Yuan, & Tan. (2019). Towards a Gliding Robotic Dolphin: Design, Modeling, and Experiments. *IEEE/ASME Transactions on Mechatronics, 24*(1), 260–70. doi:10.1109/TMECH.2019.2891290

Ze, L., Ruichun, T., Ye, T., & Xin, S. (2017). Design of multicast protocol for underwater multimedia data transmission. *2016 3rd International Conference on Systems and Informatics, ICSAI 2016, Icsai*, 710–715. 10.1109/ICSAI.2016.7811045

Zhang, Z., Pang, H., Georgiadis, A., & Cecati, C. (2019). Wireless Power Transfer - An Overview. *IEEE Transactions on Industrial Electronics*, *66*(2), 1044–1058. doi:10.1109/TIE.2018.2835378

KEY TERMS AND DEFINITIONS

AUV: Autonomous Underwater Vehicle.

Green ICT: Technology for energy saving in information, communication, and technology.

IoUT: Internet of Underwater Things.

WPT: Wireless Power Transfer is the ability to transfer power remotely.:

Chapter 12
Incentive-Based Scheduling for Green Computational Grid

Low Tang Jung
Universiti Teknologi PETRONAS, Malaysia

Ahmed Abba Haruna
Skyline University, Nigeria

ABSTRACT

In the computing grid environment, jobs scheduling is fundamentally the process of allocating computing jobs with choices relevant to the available resources. As the scale of grid computing system grows in size over time, exponential increase in energy consumption is foreseen. As such, large data centers (DC) are embarking on green computing initiatives to address the IT operations impact on the environment. The main component within a computing system consuming the most electricity and generating the most heat is the microprocessor. The heat generated by these high-performance microprocessors is emitting high CO2 footprint. Therefore, jobs scheduling with thermal considerations (thermal-aware) to the microprocessors is important in DC grid operations. An approach for jobs scheduling is proposed in this chapter for reducing electricity usage (green computing) in DC grid. This approach is the outcome of the R&D works based on the DC grid environment in Universiti Teknologi PETRONAS, Malaysia.

INTRODUCTION

Increasing demand for computing resources in recent years has led to the development of grid computing (GC) (Abbas, 2004; Berman, Fox, & Hey, 2003). Computing grid is basically taking the advantage of the idle time of thousands or millions of computers throughout the world to perform computing tasks. In general, grid can be seen as a large-scale dynamic collection of heterogeneous independent computers which are geographically distributed and interconnected by networks for processing massive tasks.

Grid computing originated in the early 1990's, when Ian Foster (Foster & Kesselman, 1999) was promoting a program to elevate shared computing to some global level. Just like the Internet which is a tool for mass communications, grids make computer resources and storage spaces globally available for

DOI: 10.4018/978-1-7998-6709-8.ch012

users. In essence, grid computing is all about getting computer systems operating together to perform computing tasks. The grid has the ability to share heterogeneous distributed assets from programs, data, storage and servers. There are three major factors that have drawn much attention from grid researchers and developers. The first and the prime concern is resource sharing i.e. resource sharing is perhaps the root of grid computing philosophy. Grid is putting computer systems into a prime position in ensuring all grid users are benefiting from the efficiencies of sharing the computing resources. More importantly, grid gives users extra computing power to compute tasks that cannot be computed using just one or few computers. The second factor behind grid is safe access which is directly related to the first factor. It is logically understood that when there are numerous users sharing the computing resources, safe access to those resources is a must with regards to data/information privacy. The third factor is the efficient utilization of grid resources. This includes a mechanism to allocate jobs effectively in real-time among the many users. Consequently, efficient utilization of resources can lessen the jobs' waiting (or completion) time.

In grid environment, scheduling is fundamentally the process of allocating computing jobs with choices relevant to the resources available over the multiple administrative domains. Scheduling may include searching over multiple administrative domains to utilize a single machine or scheduling just one job to make use of multiple resources in a single site or across multiple sites. Alternatively, scheduling can be seen as a process that maps and handles the execution of interdependent tasks on distributed resources. Scheduling comprises two important elements. First, scheduling is a decision making function i.e. to determine a job schedule. Second, scheduling is a collection of principles, models and methods. The issue of mapping tasks on distributed resources goes to a class of problem referred to as NP-complete problems (Galstyan, Czajkowski, & Lerman, 2005). For such problems, no known algorithms can create an optimal solution within polynomial time. Thus, scheduling is a dynamic allocation of resources over time in order to perform a collection of tasks raised in a variety of scenarios.

As large scale distributed computing system tends to grow in size over time, adding more and more computing nodes and storage resources resulting in exponential increase in energy consumption (Orgeric, Lefevre, & Gelas, 2010). Consequently, many large corporate information technology (IT) departments have embarked on green computing initiatives to address their IT operations impact on the environment (Curry, Guyon, Sheridan, & Donnellan, 2012). Particularly in reducing the carbon emission footprint. Green computing and green technology refer to the study and practice of environmentally sustainable computing for information technology (IT). Green computing includes the implementation of best practices, such as energy efficiency central processing units (CPUs), peripherals and servers. In addition, green technology aims to reduce resource consumption and reduce the disposal of electronic waste. Green computing is important for all classes of systems, ranging from small handheld systems to large-scale data centers (Mittal, 2014).

In all computer systems, it is the electrical energy that provides power to the hardware components to perform computation where the dissipated rate of energy is calculated as power (in watts or joules per second) consumed. The ever increasing energy consumption in large-scale distributed computing systems brings along higher operation costs (e.g. electricity bills) and caused negative environmental impacts such as CO_2 emissions. Unless steps are taken to save energy and go green, the global data centers' (DC) share of carbon emission is estimated to rise from 307 million tons in 2007 to 358 million tons in 2020 (Cook, 2011).

Electricity consumption for cooling purpose in DC is known to be the most expensive operational cost. Inefficient cooling may lead to high temperature (hot spot) and this in turn leads to more frequent hardware failures. It is common in most of the DCs nowadays in housing up to tens of thousands of high performance computers consuming tens of thousands megawatts of electricity annually. Koomey (Koomey, 2011) stated that in DC the electricity is spent mainly on cooling and computing. Hence, these two factors are becoming the prime focal points for energy savings.

For good cooling in DCs it is a norm to have all equipment installed and arranged in a specific layout to avoid potential heat accumulation. It is known that temperature mis-management is the major cause of equipment failure in DCs (Technica, 2008; Yoshida, 2008) Electricity consumed in DC is eventually converted to heat in proportion to the allocated computing loads. Effective cooling mechanism is needed to ensure that the servers in DC are operating at the vendor specified temperature thresholds for normal and long term reliable performance. The main component within a computing system that consumes the most electricity and generates the most heat is very often the microprocessor unit (Kong, Chung, & Skadron, 2012). The high heat generated by the high-performance-oriented microprocessor architecture is beyond the limit of free air cooling.

As the component density of current electronic circuits increases dramatically computers are prone to overheating issue. Furthermore, high temperature has adverse effect in computer reliability and in increasing the timing errors (Wu, Yang, Yuh, & Chang, 2005). Based on the Arrhenius time-to-fail model (Hale, 1986), at every increase of 10°C, system failure rate may be doubled. Consequently, resource management with thermal considerations (thermal-aware capability) is of importance in DC operations. Therefore, an approach to effectively reduce electricity usage (green computing) would be deemed necessary to provide efficient cooling environment for the grid computing facilities (Koomey, 2011). In this perspective, an approach for jobs scheduling is proposed in this chapter for reducing electricity usage (green computing) in DC grid. This approach is the outcome of the R&D works based on the DC grid environment in Universiti Teknologi PETRONAS (UTP), Malaysia.

BACKGROUND

Gas District Cooling Data Center (GDC-DC)

As electricity consumption for cooling DC is known to be the most expensive operational cost, Okitsu *et al.* (Okitsu, Naono, Khamis, Haruna, & Zakaria, 2013) proposed a Gas District Cooling Data Center (GDC-DC) integrated model with the aim to efficiently manage the chilled water supply-demand gap. The GDC plant supplies UTP campus with electricity and chilled water produced using natural gas. Since 2003, the plant has been supplying electricity and chilled water to campus buildings that house the "Campus Grid" – a network of thousands of personal computers in UTP. The effective reuse of wasted energy in the proposed GDC-DC model to reduce the cooling electricity consumption (by 40%) as well as CO_2 emission is one of the promising economic way to provide cooling for the DC. The conceptual GDC-DC model is shown in Figure 1.

The GDC plant generates and supplies electricity (ELE) and chilled water (CHW) to the DC facilities. The GDC plant consists of gas turbine generators (GTGs) to generate electricity (ELE) supply to run the electric chiller (EC) to produce chilled water supply (EC CHW supply). The CHW is produced to fulfill the CHW demand in DC. The ELE supply is also used to provide electricity for the GDC EC unit power demand. Take note that GTG produces waste heat. This wasted heat is absorbed by the heat recovery steam generators (HRSG) to generate steam. Output of HRSG is fed into steam absorption chillers (SACs) to generate chilled water (CHW). The SAC CHW supply is also used to fulfill the CHW

Figure 1. GDC-DC Conceptual Model

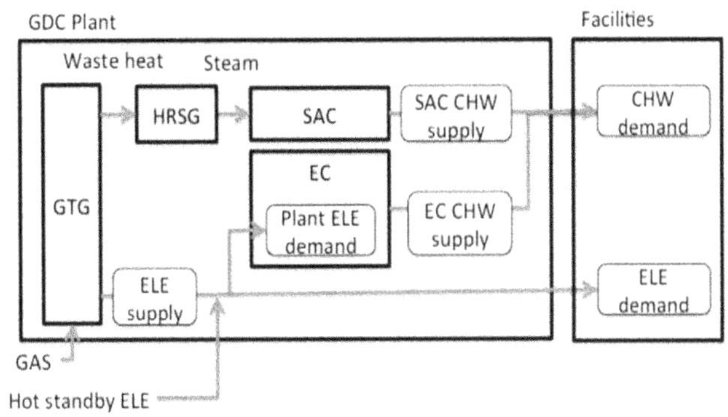

demand for cooling the DC facilities. GDC is therefore designed (with redundancy) to provide CHW either by SAC or by EC. However, it should be mentioned here that there is an external (hot redundancy) standby electricity generator source to provide CHW and ELE supply and ELE demand in case of emergency situation.

The GDC plant is in fact recycling the waste heat from GTG to produce SAC CHW with the aim to reduce CO_2 emission. However, the present GDC CO_2 emission tends to be higher than expected due to the chilled water supply-demand gap issue (Okitsu et al., 2013). This is because, the CHW supply generated by recycling the waste heat from the gas turbine cannot satisfy the CHW demand in DC. This leads to the fact that additional electricity for EC has to be generated to make up for the CHW demand. This CHW supply-demand gap resulted in the increase of CO_2 emission.

The GDC-DC model proposed by Okitsu *et al.* (Okitsu et al., 2013) is yet to totally eliminate the supply-demand gap. Therefore, it triggers an idea on how to exploit this model by integrating proper job scheduling to reduce or perhaps to eliminate the CHW supply-demand gap. According to Okitsu *et al* (Okitsu et al., 2013), to realize a full-scale GDC-DC model, 1MW chillers electricity should be controlled in the UTP campus. For a fundamental research purpose, we are looking into how a job scheduler can exploit the GDC-DC model specifically on a small-scale DCs. This research is targeting the experimental DC platform in the High Performance Computing Center (HPCC) of UTP.

Masato *et al.* (Maruyama, Takahashi, Yajima, Takeuchi, & Yamashita, 2013) proposed a method for eradicating hot spots thus to reduce the cooling energy consumption in a Telecom Equipment Room (TER) in Japan. Various heat controlling methods were implemented in their research to eradicate hot spots. However, applying their methods in tropical countries (like Malaysia) could be very challenging. That is, in addition to reduce cooling energy consumption in DC it is necessary to find ways to avoid hot spots or thermal stress circumstances (thus thermal violations) when scheduling heavy jobs to the DC servers.

Scheduling and Deadlines

Job scheduling in computational grids can be classified into two types: controlled grid and market-like grid (Xiao, Zhu, Ni, & Xu, 2008). The key difference between these two lies in who makes the scheduling decisions. In controlled grid, the grid system decides when to execute which job and on which resource(s). In the market-like grid, such decisions are made by each resource provider or the resource consumer, and all the individual participants utilize some market instruments such as price to achieve the grid system wide objectives.

A considerable amount of work has been devoted to tackle the problem of scheduling for grid computing in a controlled grid (non-market like grid environment). Majority of the previous research works, based on the literature found in (Aggarwal, Du, & Robert, 2010; Behera, Khuntia, & Nayak, 2012; Haruna et al., 2014; Hwang, Choi, & Kim, 2010; Kamalam & Bhaskaran, 2011; Li & Shen, 2010), only focused on optimization with respect to system-centric or application-centric performance metrics. Examples of such metrics include system throughput, turnaround time of job, system resource utilization, job tardiness and job response time. It can be noticed that in most of the controlled resource scheduling methods (Behera et al., 2012; Hwang et al., 2010; Kamalam & Bhaskaran, 2011), (Aggarwal et al., 2010; Li & Shen, 2010) there are various concerns relating to energy saving are still not taken into account or properly addressed.

These concerns can be classified into two categories, namely concerns from grid users and those of grid providers. From the perspective of grid users, they are very much concerning on meeting the respective jobs' completion deadlines and how they shall be compensated when a job deadline is missed. Grid users always anticipate a successfully completion of their submitted jobs without missing deadlines, and at the lowest possible cost (Anitha & Rajkumar, 2006; Xu & Yang, 2015). Users tend to lose interest in the grid system and may finally leave the grid system if submitted jobs frequently missing the deadlines, or the compensation is not up to expectation (no fairness issue). The major concerns from the perspective of grid providers are ways for saving the electricity cost in cooling their resources and how to minimize jobs missing deadlines.

Green Scheduling Motivation

Green computing has rapidly becoming a focus point over the past few years (Wang, 2008). Green computing is about improving or maintaining computing performance while reducing energy consumption and CO_2 emission footprint (Harmon & Auseklis, 2009; Wang, 2008). Recent studies showed that the increase in server power density has led to a concomitant increase in DC heat density. Servers require approximately 1 to 1.5 watts of cooling for each watt of power used (Schmidt & Shaukatullah, 2002). The ratio of cooling power to server power requirement will continue to increase as DC server density increases over time. Such high density DCs typically use two or three times more the overall amount of electrical power used for the IT equipment and mostly used for cooling purposes (Dietrich, Schmidt, Hogan, & Allen, 2007).

Transitioning to green computing may involve a number of strategies, such as integrating new approaches to reduce power consumption for cooling, computing jobs by using energy-efficient hardware, etc. Various new methods for virtualization, software, and power and workload management can be found in (Dietrich et al., 2007). These approaches are proposed to optimize the DCs operation efficiency by lowering DC running costs and lessening the negative impact of heavy computing to the environment. These factors are among the basic motivations for many researchers to look into the green scheduling algorithms in grid environment. Essentially when optimizing the efficiency of DC schedulers, at resource providers' perspective they are more interested in saving the electricity cost for cooling purposes; and at the resource users' perspective they would be happy to see an incentive scheme (incurring compensation) is put in place when their jobs missed deadlines. These lead to some interesting research questions such as:

1. How to integrate GDC-DC model into job scheduling to support green computing grid environment?
2. How to formulate a cooling-efficient job scheduling mechanism that is applicable in a small-scale DC that could not accommodate a full-scale GDC-DC model?
3. How to devise a green incentive based scheduler that can reduce the cooling electricity consumption cost in DC and can compensate the resource users when their submitted jobs missed the deadlines?

Green Scheduling Objectives and Scopes

As mentioned earlier, in grid or DC job scheduling, at resource provider's level the main concern is about saving electricity cost for cooling the DC, and at the resource user's level, they are expecting that some kind of incentives can be given when their job deadline(s) is(are) missed. In addition, the scheduling should provide fair benefits between both the resource provider and the resource user. In an effort to optimize the benefits to both sides the objective of incentive based scheduling can be stated as follows:

1. To integrate GDC-DC model into job scheduling to provide a green heterogeneous grid environment.
2. To propose a new cooling-efficient job scheduling mechanism for a small-scale DC that is not able to accommodate a full scale GDC-DC model.
3. To propose green incentive based scheduling algorithms that can save cooling electricity consumption cost in DC and also able to compensate grid users when their submitted jobs missed the job completion deadline(s).

This research is to be of fundamental basis and thus the scopes have been carefully laid down as the followings:

1. Computing are done in a distributed locations to be within the same time zone. In fact the test site is to be in Universiti Teknologi PETRONAS (UTP). The test site has a small scale DC with server rooms located in different parts of the campus i.e. geographically distributed. Jobs are scheduled to the DC servers (at different rooms) for jobs execution.
2. It is assumed that air-conditions can be programmed/controlled. Therefore, at each hour, if the temperature sensed is greater than the threshold temperature, then the scheduler's cooling (On-AC) feature will be triggered. Meaning the room cooling is not enough to cool the resources for jobs execution at that particular hour, so air condition is to be turned on. Conversely, if the temperature level is less than the threshold, then there is no need to use cooling (Off-AC), because the cooling capacity is enough to cool the resources throughout the job execution at that particular hour.

METHODOLOGY

Research Process Flow

The phases involved in this research are shown in Figure 2. A brief description for the overall requirements or outcomes in each of the phase is given in this sub-section. This sub-section is aimed to allow the reader to gain an overview of the R&D work.

1. State-of-the-Art Review: Several factors are to be considered for job scheduling. For example, scheduling policies, job workflow, job priority, job queue management policies, time and space constraints and job deadlines. Therefore, as a first step in achieving the research objectives, state-of-the-art literature review is critical to draw attention to the current mechanisms and practices based on Deadline and Slack Time Based Grid Scheduling and DC Thermal-Aware Scheduling Methods.
2. Identification of Key Features of the Problem: The key problems are the cooling energy consumption in DC in relation to the GDC-DC model, and meeting the jobs' deadlines. There are two major concerns in job scheduling from the perspective of grid users. These are, meeting the jobs' respective deadlines and incurring of compensation when the job deadline(s) is (are) missed. These two concerns are of significance since grid users generally hope to successfully complete their submitted jobs without missing deadlines and at the lowest possible cost. If jobs frequently missing deadlines, or the compensation is low (no fairness), then users may tend to lose interest in the grid system and finally leave the grid system. From the perspective of grid providers, there are two major concerns as well, i.e. saving the cooling electricity cost and minimizing the missing of submitted job deadlines.
3. Integrating GDC-DC into Job Scheduling: A scheduling algorithm shall be designed and developed for a heterogeneous grid environment in mind. The full-scale GDC-DC model and its architecture from (Okitsu et al., 2013) shall be used to test its performance and efficiency in job scheduling in a small scale DC. This is for estimating the number of resources needed for achieving the GDC-DC model in the UTP campus grid-based DC.
4. Grid Scheduling Algorithm Development: The grid scheduling algorithms shall be designed and developed for a network with geographically distributed computing sites. The focus is on the dynamics of the model that are relevant to the scheduling and its scheduling efficiency. In view of the UTP distributed DC layout, this research is expecting a large number of users (UTP students and faculty members) are participating in the grid environment and each user is associated with a particular site. In the proposed grid scheduling, it is supposed that each user submits a job and that each job would be scheduled to a DC server room with minimum temperature among the available server rooms at its respective time of execution. The schedulers shall avoid high thermal stress circumstances such as large hotspots to reduce the cooling energy consumption in DC. Also, the schedulers shall minimize the missing of jobs' deadline and compensate the user once their jobs missed deadlines. This is to encourage the users for staying in the grid for their jobs processing.

5. Designing and Development of Green Scheduling Algorithms: The development and the designing of green incentive scheduling algorithms is to be based on deadline and slack time for a green computational grid environment. In addition to that, a cooling efficient job scheduling algorithm shall be developed for a small-scale DC (that can be realized at DC with small-scale computational resources) that could not accommodate a full-scale GDC-DC model. Take note that a full-scale GDC-DC model proposed in (Okitsu et al., 2013) can only be realized per the specification at controlling 1MW of CHW electricity demand.
6. Testing of Developed Algorithms: The proposed solution shall be integrated with experimental grid using the DC facilities in UTP. The performance of the grid scheduling algorithms would be

Figure 2. Research phases overview and the research process flow

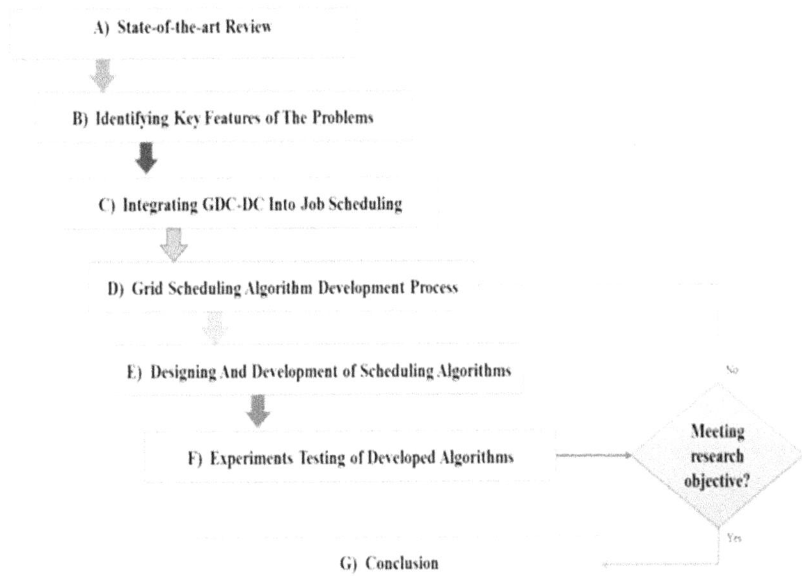

evaluated per the defined research objectives. If the research objectives are not met, then research process flow of D, E and F shall be repeated until the research objectives are met.

7. Conclusion: Concludes all the research findings with remarks on future directions to extend the outcomes of this research.

It should be mentioned here that this chapter will focus only on Phase D for sharing the R&D of this project with the readers. Phase E and Phase F are planned to be the sequels to this chapter that hopefully shall be published in the near future.

Grid Scheduling Algorithm Development

Phase D in Figure 2 is expanded into more defined steps as shown in Figure 3. Each step is discussed in its relevant sub-section below.

Benchmark Traces Files

Benchmark traces files will be required in evaluating the performance of the algorithms for this research work. This research work is to benchmark against the benchmark trace files available from Grid Workload Archive ("Grid Workloads Archive,"). Grid Workload Archive offers (anonym) workload traces from grid infrastructure to scientists as well as to professionals as follows:

1. DAS-2 traces: a wide-area distributed system consisting of 400 CPUs located at five Dutch Universities. DAS-2 is a research test bed, with the workload composed of a large variety of applications. DAS-2 was provided by the Advanced School for Computing and Imaging, the owner of the DAS-2 system (Bal).

Figure 3. Grid Scheduling Algorithm Development Flow

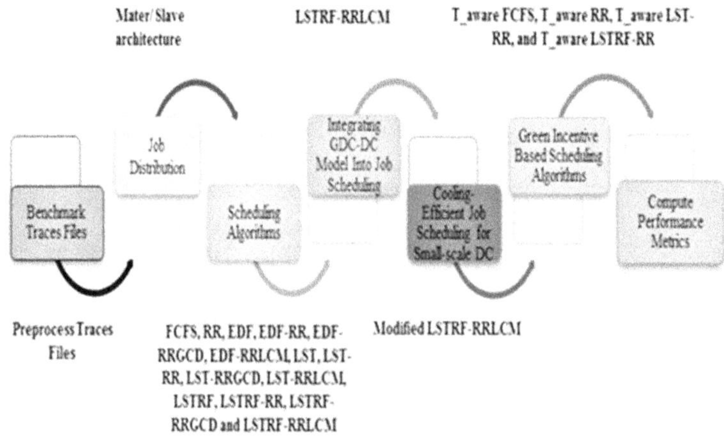

2. Grid'5000 traces: Grid'5000 is an experimental grid platform composed of 9 geographically distributed sites in France. Each site comprises one or more groupings inside the Grid'5000. It was made available from the Grid'5000 team (Dr. Franck Cappello and Dr. Olivier Richard), the proprietors from the Grid'5000 system by the OAR team. The traces collected from Grid'5000 include programs in the regions of physics, biomed, math, chemistry, climate, astronomy, language, existence, finance, etc. Additionally, the Grid'5000 traces include experimental programs for parallel and distributed systems research (Cappello).
3. NorduGrid traces: NorduGrid is really a production grid for academic scientists in Denmark, Estonia, Finland, Norway and Sweden. Since 2002, NorduGrid has been in continuous operation and development. Also, since 2003 industrial, scientific or private organizations with curiosity about grid computing happen to be asked to link their compute energy towards the NorduGrid as collaborators. The traces were made available by the NorduGrid team (Dr. Balasz konya), the owners of the NorduGrid system. The traces collected from NorduGrid include programs in the regions of CAS, chemistry, graphics, biomed, and HEP (Konya).
4. AuverGrid traces: AuverGrid is a production grid platform composed of 5 groupings situated geographically within the Auvergne region, in France. Groupings are comprised of dual 3GHz Pentium-IV Xeons nodes running Scientific Linux, LPC ("Laboratoire Physique Corpusculaire") or (Laboratory of Corpuscular Physics) of Universite Blaise-Pascal, Clermont-Ferrand, France. The Auvergrid project is really a regional grid area of the EGEE project (Enabling Grids for E-science in Europe). It is used mostly for biomedical andenergy physics researches. Auvergrid was provided by the owner (Dr. Emmanuel Medernach) of the Auvergrid system (Medernach).
5. SHARCNET traces: SHARCNET is structured like a "cluster of clusters'' across south western, central and northern Ontario (Canada), made to satisfy the computational needs of scientists in a diverse of research areas and also to facilitate the introduction of leading-edge tools for top performance computing. This trace consists of up to and including year's price of accounting records in the SHARCNET groupings installed at a number of schools in Ontario. SHARCNET was provided by John Morton and Clayton Chrusch, who also helped with background information and interpretation in high performance computing (Morton).

6. LCG traces: This log consists of 11 times of activity from multiple nodes that comprise the LCG (Large Hadron Collider Computing Grid). Customers submit serial or parallel jobs to resource brokers. The resource brokers find appropriate assets for undertaking the computation, and send jobs for execution around the different computational resources. LCG was provided by the eScience group of HEP, at Imperial College London, and made publicly available by Hui Li through the parallel workloads archive (London).

In all these trace files there were no job deadline parameters. The deadline parameters are therefore generated using the Monte Carlo distribution methods (Rubinstein & Kroese, 2016). Monte Carlo methods are a set of computational algorithms that rely on repeated random sampling to compute intended results. This method is most commonly used for computing an exact result with a deterministic algorithm.

Job Distribution

An extensive analysis was required to evaluate the efficiency of the various scheduling algorithms in a heterogeneous environment (based on the DC facilities in UTP). A master-slave architecture in Figure 4 was thus designed for evaluating the proposed scheduling algorithms.

Figure 4. Master/Slave Architecture

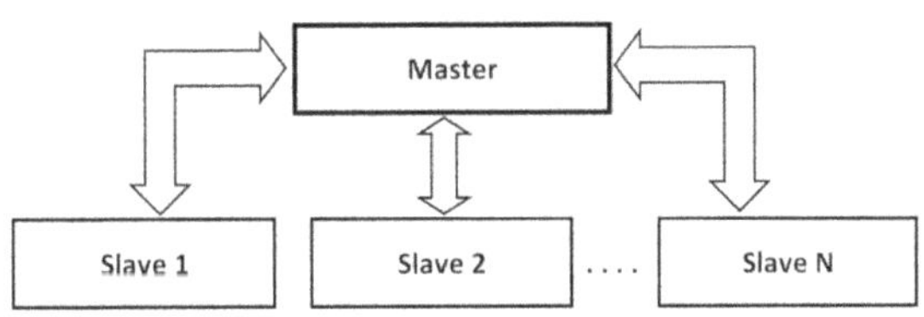

This involves the use of HPCC (High Performance Computing Center) cluster in UTP. The master takes jobs as its input and distributes the processes/jobs to the processors using adopted allocation strategy for parallel computation from (Baker & Carpenter, 2000). In this case, all workload traces are used as input. Master takes the inputs and divides them (as equally as possible) by the total number of processors available in the DC at the moment of job scheduling. The divided number of jobs are distributed to each slave where the scheduling algorithm would be used to schedule the jobs computation. Each of the slaves will receive jobs described by a unique job process ID, the job arrival time, job burst time and the job completion deadline. After each slave received the jobs, then the scheduler will schedule and execute the jobs respectively depending on the scheduler priority.

Scheduling Algorithms

These algorithms defined the way how tasks are assigned to available resources in DC. The primary requirement associated with the scheduling policy was taken into consideration for practical applicability. In this research, fourteen different versions of scheduling algorithms i.e. FCFS, RR, EDF, EDF-RR, EDF-RRGCD, EDF-RRLCM, LST, LST-RR, LSTRRGCD, LST-RRLCM, LSTRF, LSTRF-RR, LSTRF-RRGCD, and LSTRF-RRLCM were implemented in the designed architecture for their performance evaluation. For the heterogeneous grid environment in this research, EDF, LST, RR and LST-RRGCD were used as baseline approaches in evaluating the performance of the other scheduling algorithms. The generic flow of the scheduling algorithm based on the master-slave architecture is given in Figure 5.

As shown in Figure 5, the master takes jobs as the input and distributes the jobs based on the generic equation (1) to allocate the jobs among the available slaves:

$$x_i = \frac{n}{n_{si}} \tag{1}$$

where,

n is the number of job i and
n_{si} is the number of slaves for job i

The Slaves take the input from master, where each job is described by its process ID, arrival time, burst time, deadline and then execute the job at CPU level for its given burst time (i.e. the demand) based on the respective scheduling policies (Pre-emptive or Non pre-emptive). This process will continue until the jobs pool (ready queue) is empty. The value of turnaround time, waiting time and tardiness for each job are then computed and return to the Master.

Note: It should be mentioned here that the main concept of the scheduling algorithms and the performance metrics for evaluating the scheduling algorithms will be published in the sequels to this chapter.

Integrating GDC-DC Model into Job Scheduling

According to Okitsu *et al.* (Okitsu et al., 2013), there is a need to develop a scheduling algorithm in a true heterogeneous grid environment to test the energy efficiency of the GDC-DC model. Their GDC-DC model required job scheduler to control 1MW of electricity in UTP campus environment in order to achieve the CHW supply-demand gap for the full-scale GDC-DC. Based on the conceptual model in Figure 1, accordingly jobs that need heavy computation resources were scheduled to be processed at night time; small jobs that need small computation resources were executed during day time. The GDC plant supplies the UTP campus with electricity and chilled water using natural gas. The plant has been supplying electricity and chilled water to campus buildings that house the "Campus Grid" – a network of thousands of personal computers. The UTP GDC plant is fitted with Steam Absorption Chillers (SAC) and a SAC scheduling controller (introduced by Hitachi Japan in 2013. This controller allows the integration between SAC scheduling and the job scheduler, so as to control the chilled water demand in UTP campus) (Okitsu et al., 2013).

The server rooms are rooms housing the DC facilities, such as DC servers and cluster computers. These DC servers and cluster computers are geographically distributed throughout the campus and controlled from a central DC job scheduler. Jobs submitted to the central scheduler in the main DC server can be channeled to any suitable servers or clusters. The electricity is used to operate DC servers or clusters. The chilled water is used to remove heat (cooling down) in the server rooms. A series of experiments

Figure 5. Generic Scheduling Algorithm Flow

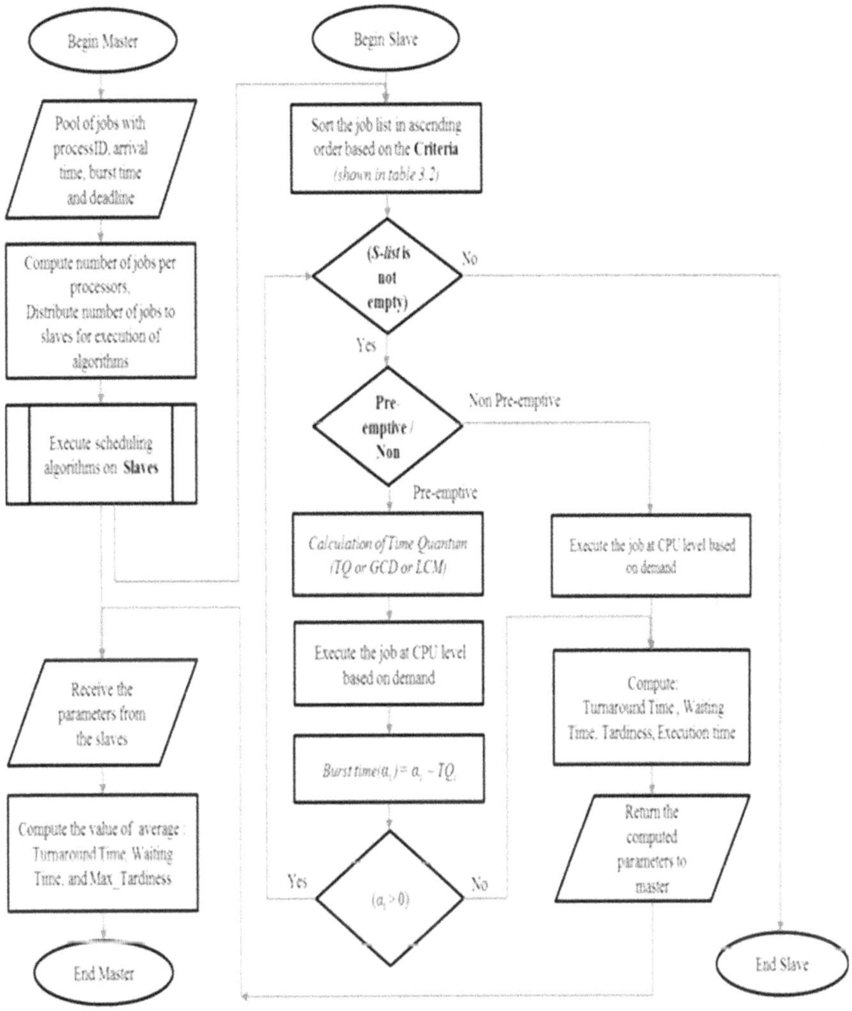

were conducted in this research to primarily estimate the number of resources needed for achieving the full-scale GDC-DC model and also to verify the GDC-DC effectiveness by integrating the GDC-DC model into job scheduling in a heterogeneous grid environment.

LSTRF-RRLCM scheduling algorithm was adopted in conducting these experiments. The scheduling algorithm is an extension of a prior version that has been shown to perform with competitive average turnaround time, waiting time and maximum tardiness in heterogeneous grid environments. Heavy jobs (whereby millions of jobs using the aforementioned benchmark trace files) were scheduled to the servers using the LSTRF-RRLCM scheduling algorithm such that the computational job schedulers can control the electricity demand in the DC in UTP campus.

Cooling-Efficient Job Scheduling for Small Scale DC

This chapter proposes a cooling efficient job scheduling algorithm for a small-scale DC that could not accommodate a full-scale GDC-DC model. The proposed scheduler algorithm takes advantage of the GDC-DC model (that has a central scheduler attached) to reduce the cooling energy consumption in a scaled down DC.

LSTRF-RRLCM is been modified (as a control strategy) to become the cooling efficient job scheduling algorithm. A control strategy is used such that the job scheduler is capable to shift heavy jobs to be executed at night time and light jobs to be scheduled during the day time. This is based on the fact that electricity consumption depends on the temperature difference between indoor and outdoor temperature of the DC server rooms. This concept can be found in (Maruyama et al., 2013). It is therefore sensible that heavy jobs are not executed during day time when the temperature difference between in-door and out-door would be high (saving electricity cost). It is worth mentioning here that the proposed method allows the increase of air-con temperature setting by 1°C (Maruyama et al., 2013) i.e. from 20°C to 21°C, in other to increase the saving rate of the cooling energy consumption in DC. The proposed cooling-efficient job scheduling algorithm is shown in Figure 6.

This algorithm provides an alternative solution for a small-scale DC that could not accommodate the full-scale GDC-DC model. However, it should be highlighted here that the proposed scheduling strategy may leads to more jobs missing deadlines. Imagine shifting heavy jobs to the night time when the outcomes of these jobs could be crucial during the day time! Therefore, if deadline of job (heavy) is to be at day time or very closely, then the non-green scheduler (LSTRF-RRLCM) should be used in such situations to satisfy the users' requirement in meeting job completion deadline. The details and the strategy adopted in this algorithm shall be published in the sequel of this chapter. Do note that this algorithm is the refinement of the generic flow shown in Figure 5.

Green Incentive Based Scheduling Algorithm

Four (4) algorithms namely, LST-RR, LSTRF-RR, FCFS and RR (as baseline approaches) were adopted and integrated with three green elements. These are, (i) Thermal-awareness based on forecasted temperature profile, (ii) Air-con (AC) controlling based on temperature threshold, (iii) Allows the setting of AC temperature be increased by 1°C or 1.5°C (Xu & Yang, 2015) e.g. from 19°C to 20.5°C.

Thermal-awareness profiling was integrated with the 4 selected scheduling algorithms based on an hourly forecasted temperature data from eight (8) DC server rooms in UTP campus. Data are collected from 00 hour to 23 hour. Taking load balancing into consideration, for LST-RR, LSTRF-RR, and RR scheduling algorithms the pre-emption of a job for execution is based on system defined variable (Time Quantum) in a cyclic manner (Round Robin approach). The time quantum of each job was sets to 60 minutes (1 hour). Meaning that, a job cannot be executed for more than its allocated time of 1 hour. This is to create fairness between the submitted jobs as well as to reduce computational loads in each respective server room when executing jobs.

When a job is ready for execution. The algorithms compare the temperature profile with the job arrival time, and map the job to the respective server room (among the DC's server rooms) that has the minimum temperature at that particular hour. After the job was mapped for execution to the respective server rooms, at each hour the scheduler will check the room temperature level. If it is greater than the threshold temperature, then cooling (On-AC) will be triggered, because the room cooling is not enough

Figure 6. Proposed Cooling-efficient Job Scheduling

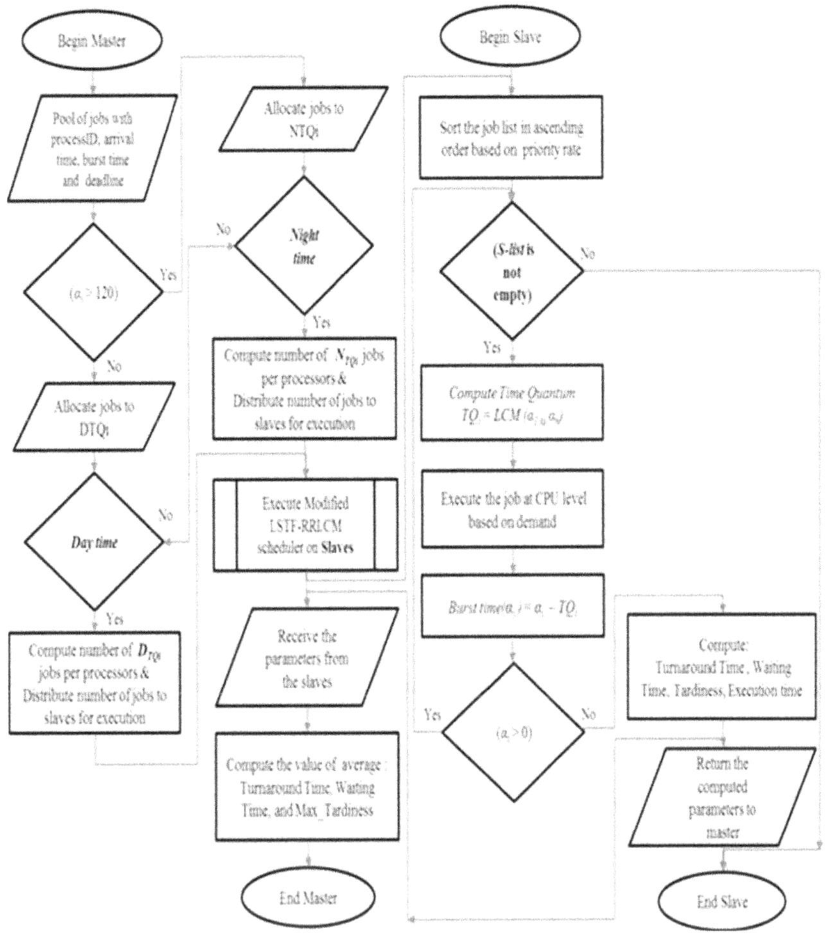

to cool the room for the job execution at that particular hour. However, if temperature level is less than the threshold temperature, then no need to use cooling (Off-AC), because the room cooling is enough to cool the room throughout the job execution at that particular hour.

The minimum temperature and threshold constraint were applied in these algorithms. This is to avoid a situation whereby the temperature of a server room is too high and workload is allocated to servers in that room. The server will be slowed down and may overheats and thus demands extra electricity for additional cooling needs. As such, the proposed green scheduling algorithms allows the increase of AC temperature settings by 1.5°C (Maruyama et al., 2013). With this strategy, the proposed method can avoid high thermal stress as well as thermal violations to minimize hardware failure rate.

Unfortunately, the proposed green scheduling algorithms may lead to more missing deadlines. At this juncture, the proposed green scheduling algorithms only addresses the concerns of the resources providers i.e. saving cooling electricity consumption in DC. The absence of compensation for missed deadlines may dissuade users from submitting jobs to the grid. To solve this, a compensation technique was adapted from (Anitha & Rajkumar, 2006), in relation to this green scheduling prospective to create a balance between resources providers and resource users and the grid environment at large.

The proposed incentive (compensation) policy is hereby explained. At each hour a job missed its deadline, the resource user will be compensated 100% cost of what he has paid per hour of each job for the resource usage in DC server rooms. This because if jobs frequently missing their deadlines, or the compensation is low, then the users tend to lose interest in the grid system and may finally leave the grid system. The 100% compensation will encourage the resource users to stay in the green grid environment as well as encouraging the resource providers to enhance their schedulers to avoid deadlines misses, so they can reduce the amount of penalties paid to resource users and able to enjoy better benefits of the green scheduler. This policy brings about the balance between the resource providers and the resource users i.e. at providers' level, saving the cooling consumption cost and at the users' level, incurring an incentive when a job missed deadline.

Accordingly, based on the proposed green incentive based scheduling algorithms, the total fine incurred due to missing deadline will be deducted from the cost of electricity saved by the proposed green schedulers and the cost of resource usage paid by the user. While in the existing method (Anitha & Rajkumar, 2006; Rubinstein & Kroese, 2016), the total fine incurred in missing deadline will be deducted only from the cost of resource usage paid by the user. The general descriptions of the proposed green incentive based scheduling algorithms is given as follow based on the First-Come-First-Serve (FCFS) scheduling:

1. Thermal-aware FCFS (T_aware FCFS): Schedules jobs to DC server room with minimum temperature at a particular hour between 00hr to 23hr. Each job are described by its process ID, arrival time, burst time and deadline.
2. When a job arrived, the scheduler checks the arrival time of the job with that of the hourly temperature profile (00hr to 23hr) of all the DC server rooms at that particular hour. The scheduler will choose the server room with the lowest temperature in that hour. The scheduler then allocates the jobs to the servers in the chosen server room. Once the jobs arrived to the respective server room, based on this algorithm the ready queue is preserved as a first come first served (FIFO) queue, dispatching jobs is from the head of the ready queue for execution by the processor.
3. Once a job has a processor, the scheduler checks the room temperature level at each hour. If it is greater than the threshold temperature, then invoke the cooling (On-AC). However, if temperature level less than the threshold level, then no need to use cooling (Off-AC). The job keeps running until it finished execution. Once the job finished executing it will be terminated and the next job is dispatched from the ready queue. This process will continue until the ready queue is empty. The scheduler will now check for jobs that missed deadline at each hour. If there are jobs missing their deadlines, the scheduler will now compute the fine/compound cost incurred respectively. That is 100% cost of what the resource user has paid per hour of each job for the resource usage in the DC server rooms. The computed fine/compound cost would be compensated to the respective users. The computed fine can also be used as a quality (QoS) indicator to the resource provider.

This policy is also integrated into other proposed schemes namely, Thermal-aware Round-Robin scheduling (T_aware RR), Thermal-aware Least Slack Time Round-Robin scheduling (T_aware LST-RR), and Thermal-aware Least Slack Time Rate First Round-Robin (T_aware LSTRF-RR).

Compute Performances Metrics

The critical performance metrics for evaluating the green incentive based scheduling algorithms in this research are discussed in the followings.

1. Total electricity consumption in a server (T_{PCEi}) – This refers to the computation of the estimated electricity consumption needed for a server while multiple jobs are being executed in that particular server.

$$T_{PCEi} = n_i \times PC_{Ei} \tag{2}$$

where, n_i is the portion of jobs allocated to the server for job i and PC_{Ei} is taken as the nominal power of 100W needed in a scrvcr

2. Total electricity consumption in a server at threshold temperature (T_{PCETTi}) – This is the computation of the estimated air conditioning electricity consumption needed for cooling a server while it is operating within the threshold temperature for job i.

$$T_{PCETTi} = \frac{T_{PCEi}}{COP_i} \tag{3}$$

where, COP_i is assigned a value of 3 as the nominal value for the cooling efficiency of a chiller

3. Total electricity consumption in a server after an increase of air-con temperature setting (T_{AECi}) – This is to compute the estimated air conditioning electricity consumption needed for cooling a server when the threshold temperature for job i is increased by 1.5°C.

$$T_{AECi} = \left(\frac{T_{di}}{1.5}\right)\left(AEC_i \times \frac{T_{PCETTi}}{PC_{Ei}}\right) \tag{4}$$

where, T_{di} is the difference between in-door and out-door temperature when job i is being executed in the server. AEC_i is adopted from the work of [27] and is defined with a value of 10. PC_{Ei} is taken as the nominal power of 100W needed in a server.

4. Total electricity consumption for job i completed in a server (T_{ECi}) – this is the summation of TPCEi and TAECi.

$$T_{ECi} = \Sigma(T_{PCEi}, T_{AECi}) \tag{5}$$

5. Power Usage Effectiveness (PUE): This refers to total power usage effectiveness by the job schedulers when scheduling job *i* for execution in a server.

$$PUE_i = \frac{T_{PCETTI} + T_{AECi} + T_{PCEi}}{T_{PCEi}} \tag{6}$$

6. Total electricity saved (T_{ECs}) – This is the difference between the total cooling electricity consumed by the non-thermal aware job schedulers and the total electricity consumed by the thermal aware jobs schedulers. It shows the total cooling electricity saved.

$$T_{ECs} = T_{ECntj} - T_{Ectj} \tag{7}$$

where, T_{ECntj} is total electricity consumed with non-thermal aware scheduler and T_{ECtj} is total electricity consumed with thermal aware scheduler

7. Deadline missing rate (D_{MR}) – This is the ratio showing the rate of job missing the deadline and therefore can be used for calculating compensation to the resource user (or as fine incurred to the resource provider). Take note that job missing deadline is when the job completion time is longer than the job deadline predefined upon job submission to the scheduler.

$$D_{MR} = \frac{N_{jmd}}{N_{jc} + N_{jmd}} \tag{8}$$

where, N_{jmd} is number of jobs missing deadlines and N_{jc} is number of jobs completed without missing deadline

8. Total fine incurred (T_{FI}) – This is computed as the total fine to be paid by the resource provider as compensation to the resource user as a token of incentive. The cost of resource usage is defined as RM0.005 * the burst time of the job submitted. Note that RM0.005 is the cost of electricity charged by the power supplier in Malaysia.

$$T_{FI} = T_{DT} \times C_{RU} \tag{9}$$

where, T_{DT} is the positive difference of time between the time of job completion and the job's predefined deadline i.e. the amount of time missed. C_{RU} is the cost of resource usage.

The critical processing steps of the incentive based scheduling algorithm is shown in Figure 7. Note that the variable T_{di} is the temperature difference between the in-door and out-door temperature of the server room. If there is no difference (T_{di} is not true) then it means there is no need to turn on the air-

Figure 7. Green Incentive Based Scheduling Algorithm Flow

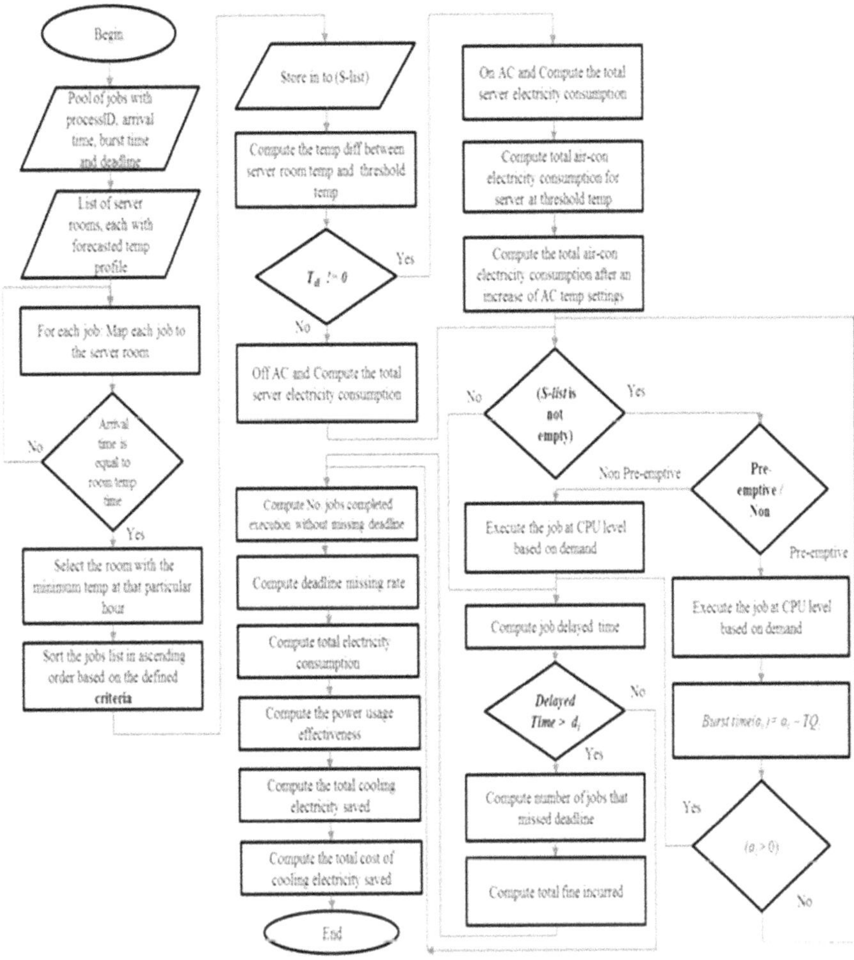

conditioner (Off AC) therefore to save the electricity for cooling purpose. The details of this algorithm shall be presented in the sequel to this chapter.

CONCLUSION

This chapter describes a R&D work on an incentive based scheduling algorithms based on the GDC-DC model for the small scale DC in the UTP campus. The methodology adopted to develop the algorithms is the main focus of this chapter. The jobs scheduling approach proposed in this chapter is aimed to reduce the electricity usage with compensation for the user jobs that missed the job completion deadline as the main green computing factor in DC grid. The pseudocode details and the performance metrics for the algorithms have not been exhaustively explained in this chapter. It is in fact the intention of the authors to have the in-depth explanation of the algorithms and the relevant performance metrics to be the sequels to this chapter.

Low Tang Jung Biography

Dr. Tang Jung, Low obtained his bachelor's degree in Computer Technology from Teesside University (UK 1989), MSc IT from National University Malaysia (2001), and PhD IT from Universiti Teknologi PETRONAS (Malaysia 2012). Dr. Low has been in the academic line for more 25 years involving with teaching various engineering and ICT courses for university students (UGs and PGs) and for ICT professionals. He is currently an Assoc. Prof. in UTP Computer and Information Sciences Department. Dr. Low is the panel member of Malaysia Quality Agency since 2012. His research interest includes wireless sensor network, embedded systems, and IoT. Some of his current R&D projects include blockchain consensus model, mobile cybersecurity, and IoT applications.

Ahmed Abba Haruna Biography

Dr. Ahmed Abba Haruna, FHEA, FIMC, CMC acquired his Ph.D. and M.Sc. in Information Technology from Universiti Teknologi PETRONAS (UTP), Malaysia. He obtained his Bachelor Degree in Information Technology from Infrastructure University Kuala Lumpur, Malaysia. He has recently acquired his Post Graduate Certificate in Higher Education (PGCHE), at Middlesex University. Moreover, he has achieved the status of Fellow of the Higher Education Academy, UK, Fellow of Instiitute of Management and Consultant, and a Certified Management Consultant. He has about eight years of invaluable experience in lecturing in Mauritius and in Malaysia. Currently he serves as a Senior Lecturer in Computer Science Department in Skyline University Nigeria and as a Governing Council Member in the same university. His research interest includes Green & Sustainable Computing, Scheduling, Resource Allocation, Machine Learning, and Higher Education Teaching and Learning Support.

REFERENCES

Abbas, A. (2004). *Grid Computing: A Practical Guide to Tech. and App.* Firewall Media.

Aggarwal, A., Du, P., & Robert, D. K. (2010). Grid Scheduling Optimization Based on Resource Characteristics. *Journal of Computer Information Systems*, *6*(14), 4609–4616.

Anitha, R., & Rajkumar, N. (2006). *Compensation Based Grid Scheduling*. Paper presented at the 2006 Annual IEEE India Conference. 10.1109/INDCON.2006.302816

Baker, M., & Carpenter, B. (2000). *Mpj: A proposed java message passing api and environment for high performance computing*. Paper presented at the International Parallel and Distributed Processing Symposium. Bal, D. H. DAS-2 team. Retrieved from http://gwa.ewi.tudelft.nl/pmwiki/reports/gwa-t-1/trace_analysis_report.html

Behera, H., Khuntia, S., & Nayak, S. (2012). An improved least-laxity-first scheduling algorithm for real-time tasks. *International Journal of Engineering Science and Technology*, *4*(04).

Berman, F., Fox, G., & Hey, T. J. M. G. I. R. (2003). *Overview of the book: grid computing–making the global infrastructure a reality*. Retrieved from http://www.grid5000.org/

Cook, G. (2011). *How dirty is your data: A look at the energy choices that power cloud computing*. Greenpeace.

Curry, E., Guyon, B., Sheridan, C., & Donnellan, B. (2012). Developing a sustainable it capability: Lessons from intel's journey. *MIS Quarterly Executive*, *11*(2), 61–74.

Dietrich, J., Schmidt, R., Hogan, M., & Allen, G. (2007). The Green Data Center. IBM Global Services, 1-20.

Foster, I., & Kesselman, C. (1999). *The globus toolkit*. Academic Press.

Galstyan, A., Czajkowski, K., & Lerman, K. (2005). Resource allocation in the grid with learning agents. *Journal of Grid Computing*, *3*(1-2), 91–100. doi:10.100710723-005-9003-7

Grid Workloads Archive. (n.d.). Retrieved from http://gwa.ewi.tudelft.nl/pmwiki/

Hale, P. (1986). Acceleration and time to fail. *Quality and Reliability Engineering International*, *2*(4), 259–262. doi:10.1002/qre.4680020409

Harmon, R. R., & Auseklis, N. (2009). *Sustainable IT services: Assessing the impact of green computing practices*. Paper presented at the PICMET'09-2009 Portland International Conference on Management of Engineering & Technology. 10.1109/PICMET.2009.5261969

Haruna, A. A., Zakaria, N. B., Jung, L. T., Pal, A. J., Naono, K., & Okitsu, J. (2014). *Performance comparison of least slack time based heuristics for job scheduling on computational grid*. Paper presented at the 2014 International Conference on Computer and Information Sciences (ICCOINS). 10.1109/ICCOINS.2014.6868841

Hwang, M., Choi, D., & Kim, P. (2010). *Least slack time rate first: New scheduling algorithm for multi-processor environment*. Paper presented at the 2010 International Conference on Complex, Intelligent and Software Intensive Systems. 10.1109/CISIS.2010.20

Kamalam, G., & Bhaskaran, V. M. (2011). An efficient hybrid job scheduling algorithm for computational grids. *International Journal of Computers and Applications*.

Kong, J., Chung, S. W., & Skadron, K. (2012). Recent thermal management techniques for microprocessors. *ACM Computing Surveys*, *44*(3), 1–42. doi:10.1145/2187671.2187675

Konya, D. B. (n.d.). *The NorduGrid team*. Retrieved from http://www.nordugrid.org/

Koomey, J. (2011). *Growth in data center electricity use 2005 to 2010*. Academic Press.

Li, B., & Shen, B. (2010). *Slack-based advance reservation for Grid jobs*. Paper presented at the 2010 3rd International Conference on Advanced Computer Theory and Engineering (ICACTE). Retrieved from http://lcg.web.cern.ch/LCG/

Maruyama, M., Takahashi, A., Yajima, H., Takeuchi, A., & Yamashita, N. (2013). Reducing electric power consumption for air conditioning by improving temperature distribution in telecom equipment rooms. *NTT Technical Review*, *11*, 10.

Mittal, S. (2014). *Power management techniques for data centers: A survey.* Retrieved from https://www.sharcnet.ca/my/front/

Okitsu, J., Naono, K., Khamis, M. F. I., Haruna, A. A., & Zakaria, N. (2013). *Towards an architecture for integrated Gas District Cooling with Data Center control to reduce CO2 emission.* Paper presented at the 2013 International Green Computing Conference Proceedings. 10.1109/IGCC.2013.6604468

Orgerie, A.-C., Lefevre, L., & Gelas, J.-P. (2010). *Demystifying energy consumption in grids and clouds.* Paper presented at the International Conference on Green Computing. 10.1109/GREENCOMP.2010.5598295

Rubinstein, R. Y., & Kroese, D. P. (2016). *Simulation and the Monte Carlo method* (Vol. 10). John Wiley & Sons. doi:10.1002/9781118631980

Schmidt, R. R., & Shaukatullah, H. (2002). *Computer and telecommunications equipment room cooling: a review of literature.* Paper presented at the ITherm 2002. Eighth Intersociety Conference on Thermal and Thermomechanical Phenomena in Electronic Systems (Cat. No. 02CH37258). 10.1109/ITHERM.2002.1012531

Technica, A. (2008). NVIDIA denies rumors of faulty chips, mass GPU failures. Academic Press.

Wang, D. (2008). *Meeting green computing challenges.* Paper presented at the 2008 10th Electronics Packaging Technology Conference. 10.1109/EPTC.2008.4763421

Wu, Y.-W., Yang, C.-L., Yuh, P.-H., & Chang, Y.-W. (2005). Joint exploration of architectural and physical design spaces with thermal consideration. *Proceedings of the 2005 international symposium on Low power electronics and design.* 10.1145/1077603.1077636

Xiao, L., Zhu, Y., Ni, L. M., & Xu, Z. (2008). Incentive-based scheduling for market-like computational grids. *IEEE Transactions on Parallel and Distributed Systems, 19*(7), 903–913. doi:10.1109/TPDS.2007.70791

Xu, H., & Yang, B. (2015). An incentive-based heuristic job scheduling algorithm for utility grids. *Future Generation Computer Systems, 49*, 1–7. doi:10.1016/j.future.2015.02.002

Yoshida, J. (2008). *The truth about last year's xbox 360 recall.* EE Times.

ADDITIONAL READING

Ahmad, A. H., Low, T., Nordin, Z., & Okitsu, J. (2015). Cooling-efficient job scheduling in a heterogeneous grid environment. *Procedia Computer Science. ScienceDirect, 69*, 104–115.

Chaudhry, M. T., Ling, T. C., Manzoor, A., Hussain, S. A., & Kim, J. (2015). Thermal-aware scheduling in green data centers. *ACM Journals. ACM Computing Surveys, 47*(3), 39. doi:10.1145/2678278

Haruna, A., Low, T., & Zakaria, N. (2015). Design and development of hybrid integrated thermal aware job scheduling on computational grid environment. In Proceedings of *IEEE International Symposium on Mathematical Sciences and Computing Research* (pp.13-17). 10.1109/ISMSC.2015.7594020

Jung, L. T., & Haruna, A. A. (2018). Green data center by incentive-based job scheduling approach. In *Proceedings of IEEE Conference on Open Systems* (pp. 13-18). 10.1109/ICOS.2018.8632811

Mhedheb, Y., Jrad, F., Tao, J., Zhao, J., Kołodziej, J., & Streit, A. (2013). *Load and thermal-aware VM scheduling on the cloud* (Vol. 8285). Lecture Notes in Computer Science. Springer. doi:10.1007/978-3-319-03859-9_8

Sun, H., & Stolf, P., & S., & Pierson, J. M. (2017). Spatio-temporal thermal-aware scheduling for homogeneous high-performance computing datacenters. *Future Generation Computer Systems. ScienceDirect, 71*, 157–170.

Xu, H., & Yang, B. (2015). An incentive-based heuristic job scheduling algorithm for utility grids. *Future Generation Computer Systems. ScienceDirect, 49*, 1–7.

KEY TERMS AND DEFINITIONS

Computing Grid: It is a technique of combining via networking the computing power of multiple computers which are geographically distributed but are directly interacting to each others to accomplish computationally intensive jobs.

Data Center: DC is a physical building that houses a physical network of computing and storage resources that enable the delivery of shared applications and data.

Green Computing: The deployment of computing devices and the related resources to give a minimal impact to the environment thus aims to attain economic viability and the ways the computing devices are used e.g. producing small CO_2 footprint in computing jobs.

Green IT: It is interchangeable with Green Computing and is concerning the use of IT resources to develop environmentally sustainable production practices e.g. improving recycling procedures.

Job Scheduling: It is the process of allocating computing resources to many different jobs/tasks by a job scheduler based on certain scheduling algorithm(s) to ensure that all jobs/tasks are carried out fairly and on time.

Scheduling Algorithm: A set of rules and policies defined to allocate jobs/tasks to the suitable computing resources such as CPU, memory, bandwidth, etc to attain an optimal system performance and system resources utilization.

Thermal-Aware Scheduling: It is a job scheduling technique with thermal management element to reduce thermal hotspots in computing devices thus aims to distribute the temperature uniformly among the computing resources.

Chapter 13
Self-Healing and Green Energy in Wireless Sensor Network: A Survey

Vasaki Ponnusamy
Universiti Tunku Abdul Rahman, Malaysia

N. Z. Jhanjhi
https://orcid.org/0000-0001-8116-4733
Taylor's University, Malaysia

Beh Zi Xuan
Universiti Tunku Abdul Rahman, Malaysia

ABSTRACT

Green energy can be classified into many domains such as green aware applications, green aware protocols, green energy harvesting, and many more. The main focus of this chapter is to analyze green energy in terms of energy aware routing protocols and mainly focusing on wireless sensor networks. So, in this domain, greening the routing protocols with energy efficient algorithms is investigated. Energy efficient routing algorithms can be designed by having self-healing concepts whereby routing algorithms can heal themselves when a routing hole is found or nodes can heal themselves when energy depletion occurs. Having that concept in mind, this chapter investigates both the green energy protocols and the self-healing protocols in the literature.

INTRODUCTION

Wireless sensor networks are created for monitoring situation such as temperature, sound, movement, location, to create "smart environments". (Youssef et al. 2002 & Noor Zaman et al. 2010-2011). Sensor network has its distinct features from traditional network by having limited energy resources (Noor Zaman et al. 2012). Most of the routing in wireless sensor is done through lowest energy route this eventually

DOI: 10.4018/978-1-7998-6709-8.ch013

exhausts energy at these nodes impacts the lifetime of the entire network. (Sha et al. 2005 & Noor Zaman et al. 2012). Energy efficient routing becomes the prime importance when considering extending the lifetime of sensor network. (Gui & Mohapatra 2003 & Noor Zaman et al. 2011). When sensor nodes exhaust their energy indefinitely, the nodes will eventually create a routing hole (Zaman N et al. 2014). Routing hole refers to the areas where data cannot be routed since there are no active nodes available (Zaman N et al. 2015). Therefore a self-healing mechanism is needed to detect this issue. So the technique of fighting routing holes in sensor network is denoted as self-healing methods. Therefore the two prominent features in wireless sensor network that need attention in research are energy-efficiency and self-healing. This paper targets to carry out an evaluation of several energy efficient and self-healing routing protocols in wireless sensor networks. This paper differs from other studies, as it performs a study on bio-inspired as well as non-bio inspired mechanisms for energy efficient routing protocols that have been submitted by others given the advantages and disadvantages of each proposed protocol. The paper is organized as follows: section 1 will look at some of the literature reviews based on the terms such as self-healing and energy efficiency. The literature starts by looking at what is meant by autonomic computing in which self-healing is one of the aspects of it. Further literature looks at what is meant by energy efficient routing in wireless sensor network. Section 2 consists of surveys on various energy efficient self-healing routing protocols discussing their protocol approach and results. Section 3 further gives a summary of each protocol giving a brief summary on its advantages and disadvantages. A complete comparison using various metrics is given at section 4. Section 5 concludes the paper with recommendation and future research issues.

Autonomic Behavior in Wireless Sensor Networks

Autonomic system is considered as a system that is capable of self-management services. Self-management activities can be categorized into self-healing, self-optimization, self-configuration, self-protection, self-awareness, self-knowledge, self-maintenance and self-service (Lanthaler,2000). According to Lanthaler (2000), there are four important tasks that an autonomic sensor node should perform for self healing purpose. The functions are: i) to monitor and to collect data from the system, ii) to analyze whether something is going wrong in the system, iii) to plan or to create desired changes in order to execute the changes and iv) to perform the desired actions (Figure 1).

Self-Healing Behavior in Wireless Sensor Networks

Self-healing in the context of wireless sensor network is the capability of sensor network to recover from any types of behavior such as malicious attack, disruption or failure. In terms of failure, there exist two natures of failure: the failure of node and the failure of system. Failure in sensor network is due to many reasons such as failure in software and failure in hardware; unavailability of communication link, medium access contention and failure in routing. This failure in WSN is considered one of the main challenges and yet opens to many areas of research and discussions. Routing failure occurs due to routing holes, node failure, energy depletion, natural disruption in sensor area and many more. Self-healing in the context of routing failure is the capability of sensor nodes to monitor, detect, analyze, plan and execute events in order to recover or heal from the phenomena (Lanthaler,2000). Moreover, self-healing by means of energy efficient routing is one of the main research focuses. There are various research projects going on that trying to tackle the self-healing problem in the context of software systems. The model proposed

Figure 1. Autonomic Computing Architecture

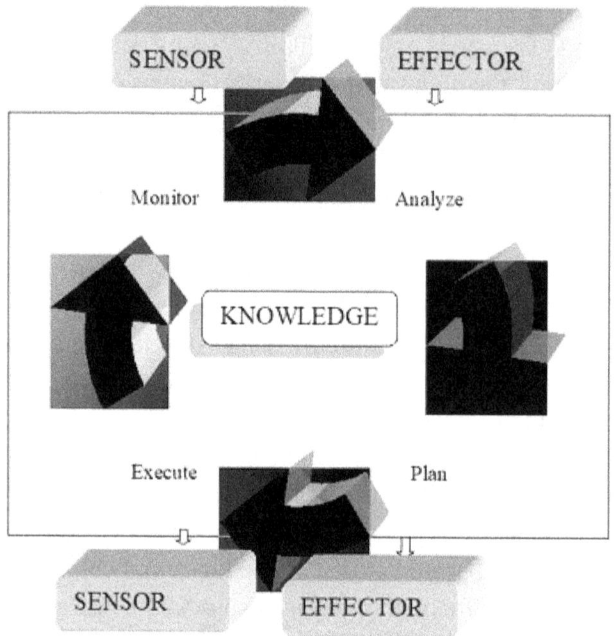

here can be used as guideline to perform self- healing system in terms of software, hardware, systems or network. In conclusion, significant features of self-healing schemes are fault model, system response, system completeness and design context. Self-healing systems research is relatively interesting and getting better consideration, but it is still not finely defined in the aspects of scope and areas to be focused into self-healing. Koopman (2003) summarizes some of the different views researchers have in terms of self-healing aspects. Exists vast amount of knowledge in terms of dependable system and specific techniques considered for self-healing. As shown in Figure 2, there exist four common classifications to the self-healing system problem space: fault-model, system response, system completeness, and design context. These are the aspects that have been proposed by various researchers and have been combined into one single model. Figure 3 shows the context in which self-healing aspects falls in sensor network research.

Fault Model

Self-healing systems are actually considered as dependable systems. And all dependable systems have to be defined for some kind of fault tolerant scheme. Following fault hypothesis section will explain what kind of fault the system is about to tolerate. Therefore, self-healing system should have a fault tolerant model in order to specify what kind of fault the system is to self-heal from. Without such particular model it is difficult to assess how the system is going to heal from such behavior. Fault model is categorized into fault duration which is to detect whether the fault is permanent, intermittent or transient where it is essential to state the length of fault to self-heal, fault manifestation is to identify whether severe, fail-fast, fail-silent or others, fault source is the source where the fault comes from, granularity is the size of the fault and fault profile expectations are the profile of fault occurrences.

Figure 2. Self-Healing Architecture

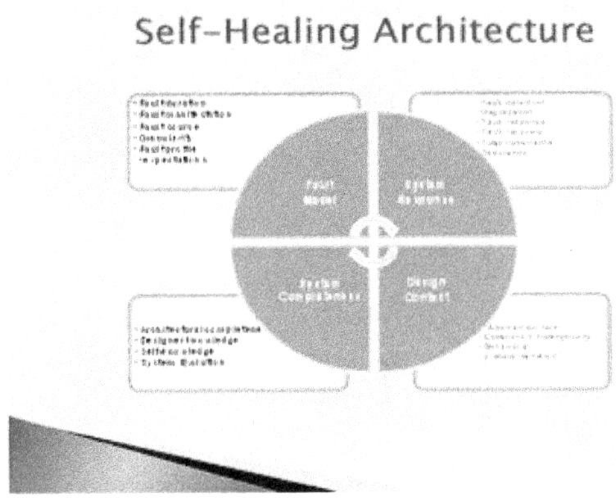

System Response

The most important thing of this model is to first detect the fault and way of responding to the types of faults. Reliability and safety or security is to be considered in order to decide on potential responses. System response is categorized into fault detection, either by internal component, peer-to-peer checking or supervisory checking, degradation is the degree of degraded operation, fault response is response mechanism, recovery is recovery operation to self-heal, constant time is to determine what kind of self-healing required and assurance is assurance for correctness.

System Completeness

Self-healing system should deal with limits in terms of knowledge, incomplete specifications, and in-complete design. System completeness is categorized into architectural completeness, designed with first implementation, designer understanding is how sound the fault model is categorized, system self-knowledge is the knowledge about themselves and environment in order to heal and system evolution should deal with changes over time.

Design Context

Several other factors need to be identified to formulate the system design. Design context is categorized into abstraction level which is self-healing to application, software or hardware, component homogeneity whether homogeneous or heterogeneous, behavioral predetermination is predetermined or deterministic behavior, user participation in healing, system linearity and system space is how big the system is. There are various research projects going on that trying to tackle the self-healing problem in the context of software systems. The model proposed here can be used as guideline to perform self- healing system in terms of software, hardware, systems or network. In conclusion, relevant aspects of self-healing systems are fault model, system response, system completeness and design context.

Figure 3. Sensor Network Overall View

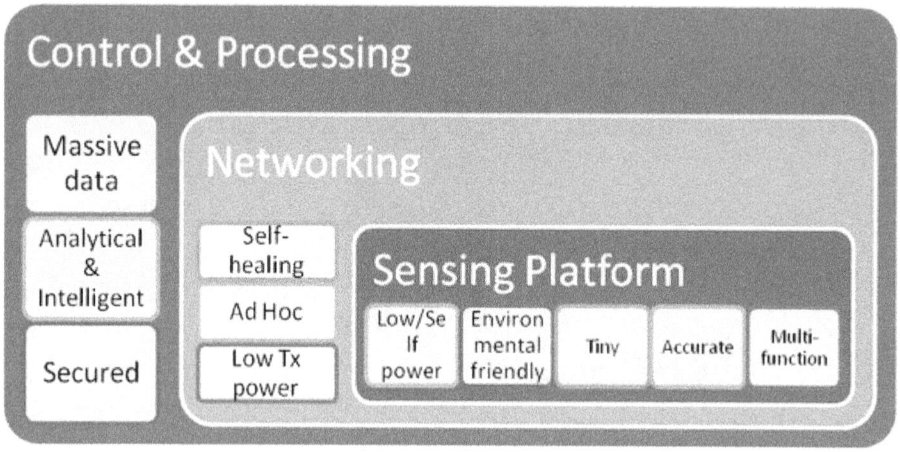

SURVEY ON SELF-HEALING WIRELESS SENSOR NETWORKS

This section comprises some of the related works in the area of self-healing, energy efficient routing and some of the bio-inspired mechanisms used for self-healing as well as energy efficient routing. The problems, suggestions and future works discussed in this paper helps in further research and enhancement of energy efficient self-healing and particularly using bio-inspired mechanisms. This survey is particularly useful to researchers looking into new areas of self-healing systems in networks. Following section starts by looking into various types of research papers followed by its protocol approach and results obtained. The results are useful in terms of making comparison with other protocols and future enhancements.

A SELF-HEALING ALGORITHM USING BEE INSPIRED ANALOGY

This section provides a brief overview of BIMAS (Ponnusamy et al, 2014). A new architecture (BIMAS) is proposed for energy constrained wireless sensor network. BIMAS is a three-tier hierarchical based clustering architecture comprised of three different layers; layer 1 being the static sensor nodes, layer 2 the mobile agents and layer 3 the base station. The proposed architecture can be a potential solution to handle routing holes as well as efficient handling of energy consumption in sensor networks. The static sensor nodes form a cluster of network, which reduces signaling and increases scalability of the network. The static nodes within the same cluster communicate with each other and forward sensed data to the cluster head. The cluster head then forwards the data to the mobile agent that is within the communication range. Mobile agent can be in any form such as mobile phone, laptop, human, vehicle or an animal as long as it is always in motion, roaming around the sensor network. There are mechanisms in which the cluster head detects a mobile agent within the proximity and initiates communication.

The proposed architecture covers the mobility of sensor nodes in which a new layer of mobile nodes is introduced. It is envisaged by D. Estrin et. al that in the near future, very large scale networks consisting of both mobile and static nodes will be deployed for applications ranging from environmental monitoring to military applications. For larger networks, this hierarchical based clustering network could save

Figure 4.

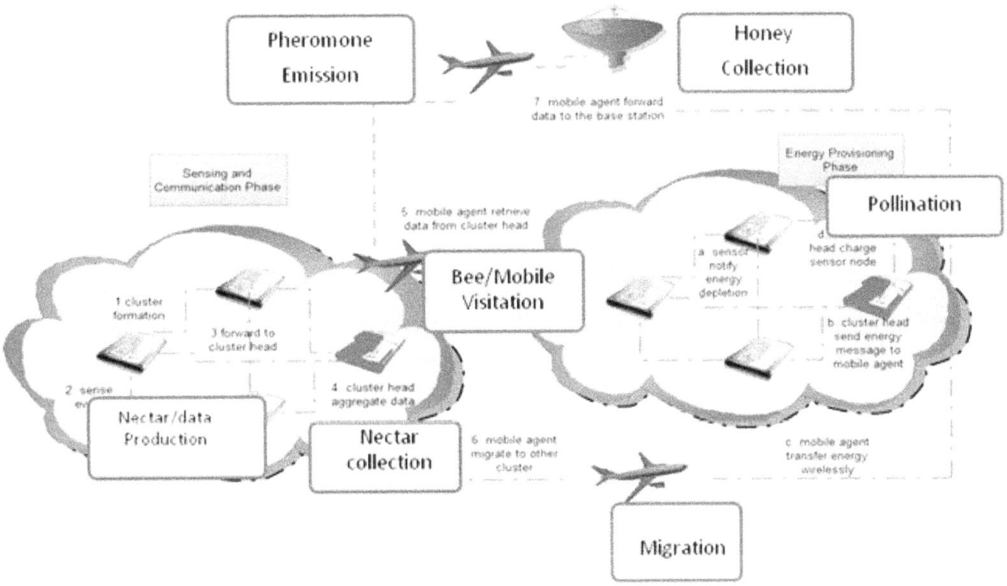

a tremendous amount of energy. The work adapts a biological system in the design of a sensor communication process, with a specific focus on self-healing characteristics. Initially, biological systems are natural systems that are able to handle many challenges of system intelligence. Most biological systems have been refined for life over many centuries, generating living organisms that are able to autonomously repair themselves when damaged, produce emergent behaviour, and survive even during drastic change in the environment.

The mechanisms of the proposed system, in this context, include balance on the internal equilibrium (homeostasis) and on the self organizing mechanism to support environmental change. During the lifespan of a communication system, various changes occur such as energy depletion, shortening of lifespan, external threat or route diversion. Based on these changes, the system should be able to self-learn, self-organize and self-heal from such events. These self-management activities are similar to living organisms such as humans, animals and plants whereby they learn and reform (self-heal). The architecture proposed in this work is motivated by the bee algorithm and animal foraging behaviour that can be applied to communication system. The autonomous and decentralized behaviour of animals has inspired the idea of using this mechanism in self-healing wireless sensor network. As can be seen in Table 1, the bee analogy can be mapped to the sensor network, in which the flower of a plant system is analogous to sensor nodes, and the bee is analogous to the mobile agent. A detailed view of our protocol and algorithm is presented in Figure 4.

The hierarchical based clustering is comprised of three different layers:

i) Layer 1 consists of plants with flowers (sensor) for intra-cluster communication, and their root mechanism and coordination with the rhizosphere for inter-cluster neighbour discovery.
ii) Layer 2 consists of communications between the flowers with the bee colony (mobile agent) for sensor data relay.

Table 1. Mapping Bee Biology to a Mobile Agent Based Sensor Network (BIMAS)

Bee	Sensor Network	Symbols
Bee	Mobile Agent	MAi
Flower	Active Sensor Node/Cluster head	CH(Si)
Flower buds	Sleeping nodes	Si or
Plant	Cluster of sensor nodes	Tid CH(Si)
Nectar	Sensor Data	Mdata
Pheromone Trail	Path visited list	Plist
Pollen	Energy	E(Sn)
Pheromone	Mobile Agent Messages	Mhello, Menergy
Honey	Aggregated data	Maggr
Bee Hive	Base Station	BS

iii) Layer 3 consists of communication of the bee colonies with the bee hive (base station) to relay the sensed data. Figure 5 depicts the proposed architecture.

Figure 5. The analogy of bee network and sensor network

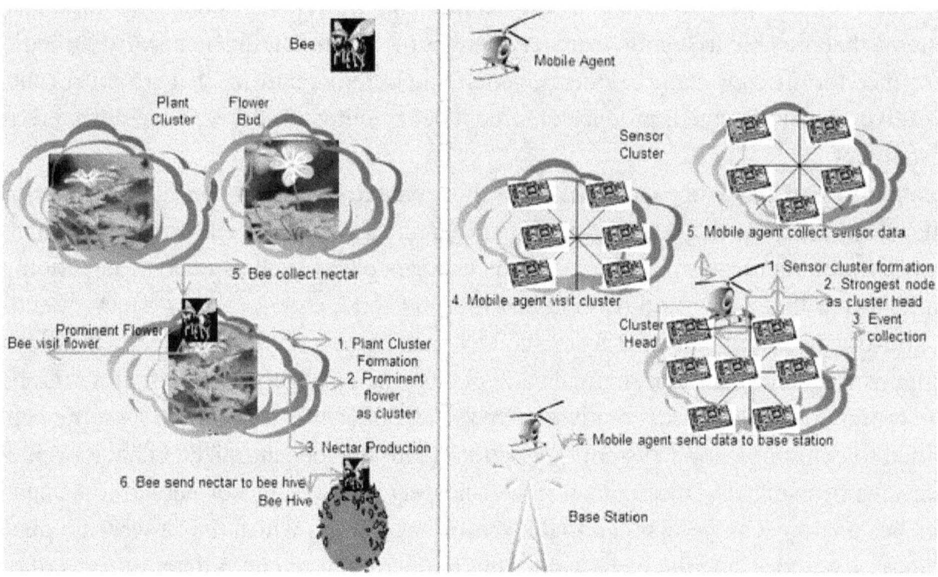

Some of the biologically inspired functions of mobile agent in BIMAS are discussed below (Figure 5):

Migration

- Energy Harvesting/Scavenging (pollination)
- Sensor Data Collection (nectar collection)

MAi migrate to other cluster

status_MAi = 'ROAMING'
Pollination (self-healing)

- MAs collect energy (pollen) from one cluster and drop it at other clusters. This behavior allows sensor nodes to equally share their energy level.
- When energy is needed, MAs migrate to other cluster for pollination. This is possible by exhibiting certain behavior such as color or scent to attract the bees. Sensor nodes send urgent message to MA for energy provisioning.

Else if (CE(CH(Si)) < Etd
MAi send MENERGY to CH(Si)
CH(Si) send MFINISH to MAi

Nectar Collection

Perform migration to collect sensor data (nectar) from one cluster to another. The collected data is aggregated into data fusion (honey) to be sent to the base station.

If (status_MAi = 'ROAMING')
agent detects sensor nodes by random walk path
MAi broadcast "MHELLO" to CH(Si)
CH(Si) inspect B_CH(Si)
if (B_CH(Si) != empty)
Bag_MAi = MAGGR
MAi send MACK to CH(Si)
CH(Si) flush buffer
CH(Si) send MFINISH to MAi

Pheromone Emission

- MAs use a path visited list (pheromone) to trace its route back to visited clusters.
- MAs also send urgent message.
  Visited_list_MAi = Tid CH(Sj, Sxyz)

Mobility Models

- Use chemotaxis walk analogous to bee chemotaxis walk together with other mobility models.

The flow of BIMAS simulation is as follows:

i. Sensor nodes are randomly dropped along the simulation area, form a cluster and elect cluster head on a rotational basis.
ii. Sensor nodes sense event, in this case, continuous trace of TCP and Ping data.
iii. Sensor triggers cluster head and forwards collected data to the cluster head.
iv. Cluster head collects data and stores into the buffer.

v. Mobile agent moves along the network by a controlled mobility since our mobile agent is assumed to be embedded on a moving entity that is deployed for other purposes. Other mobility models are also tested to study the feasibility of our bee analogy.
vi. Communication between the sensor nodes (cluster head) and mobile agent is performed and data is forwarded to the mobile agent.
vii. Energy provisioning is performed for situations in which the sensor node energy goes below the threshold value.
viii. Mobile agent moves to the nearest base station/sink.
ix. Base station processes (buffers) the data.
x. Simulation is repeated from step i to ix for all the parameter space by changing the distance, speed, number of nodes, mobility models, network size, number of clusters, with pheromone and self-healing.

Green Energy in Wireless Sensor Network

New Energy Efficient Routing Protocol for Wireless Sensor Network (Mr.Chan Hiu Fai & Dr.Rudolph Heiko, 2015)

In order to solve the energy inefficiency of WSN, both the researches had proposed a new routing algorithm for WSN by reducing the number and total transmissions distance which eventually able to ensure energy efficiency and save energy which can eventually maximize the lifetime of WSN. The WSN is a network that built up by different types of sensors and the sensors are densely deployed, thus it is very likely that the nodes are damaged in certain environment. The nature of WSN has the constraints of limited energy source, memory as well as the computational power of WSN nodes. Also, WSN does not have a fixed infrastructure nor a fixed topology, thus maintaining WSN requires mature monitoring and control strategy which is a challenging task. Since every sensor are powered by battery and are required to run remotely without human interaction, it becomes a difficult task to replace and recharge the batteries. Every node has limited energy, low storage size and low bandwidth for communication. Hence, it is very hard to ensure energy efficiency.

An energy-efficient and mobility support routing protocol for Internet of Mobile Things (Maha Bouaziz, Abderrezak Rachedi, Abdelfettah Belghith, 2017)

The Internet of Mobile Things is referring something which is inherently mobile. For instance, the unmanned aerial vehicles, autonomous truck and etc. Basically, IoMT referred to devices that applied the IoT concept which supports mobility. According to Oliveria.L et. al. (2011), the idea of making the mobility devices to be an IoT based context allows many advantages in terms of flexibility, emergence of new services, and enlargement of IoT application domains. Bouaziz.M. et. al. (2017) claimed that although the mobility nature of IoMT enables the innovative applications and allows new services, but a challenging issue still remains there as it gives a negative impact to the network performance due to the limited devices resources, unpredictable devices' movement and dynamic network topology. Problems such as the intermittent connectivity of mobile nodes and disconnection of nodes. To solve such issues, an existed routing protocol called Co-RPL proposed by Gaddour.O. et. al. (2014). It is an extension to Routing Protocol for Low Power and Lossy Network (RPL) based on the Corona mechanism to support mobility. Initially, the RPL is designed for static networks, with no support for mobility. But to address this mobility issue, the RPL based on Corona mechanism is proposed. The extensive simulation study is conducted using the Contiki simulator. The performance was used to compare with the standard RPL in terms of the node speed, packet transmission rate and number of Directed Acyclic Graphs (DAG) roots on network performance. The CoRPL routing protocol is worked by maintaining a continuous tracking of the mobile nodes' position during their dynamic movement with the aim of improving the network performance. The Co-RPL routing protocol relied on the corona mechanism which allows the localization of the standard RPL routers in motion. It is done by dividing the network into a circular area around the DAG roots and allows the RPL routers to compute its coordinates based on its distance from the DAG roots.

A Parametric Oriented Research on Routing Algorithms in Mobile Adhoc Networks (MANET). (Shailaja.P, GuruRao.CV, Nagaraju.A, 2019)

The Mobile Adhoc Networks (MANET) is a type of ad-hoc network in which composed of the wireless mobile nodes that actively form a network lacking any support of pre-existing infrastructure such as the routers or access points. Instead, each mobile node participates in routing by doing the job of data forwarding for other mobile nodes. Also, the MANET in nowadays world, it plays an important role in various applications such as the Remote Education, Visual Surveillance and etc. Since the characteristics in most of the IoMT devices are autonomous and dynamic communication paradigm, MANET would be the best solution for IoMT devices in which the mobile devices are integrated and formulated into an infrastructure-less network as well as communicated in a large distance network. Routing in MANET is very daunting because of its routing expenses, asymmetric link, restricted bandwidth, battery constraints, speed, scalability in multi-hop routing, quality of services (QoS) and packet loss. Also, the selection of an optimal path to transmit data is a critical issue in MANET as the mobile nodes are resource-constrained, constitute in a multi-hop communication and have mobility. Hence, the routing protocol to be selected must be able to cover all states of a specified network especially in a large distance communication, needs to achieve an effective QoS in terms of packet delivery, delay, and available bandwidth of link and lastly it is not allowed to consume too much network resources by the protocol overhead traffic. Because of the MANET nature, it is suitable to be applied for the network formation of IoMT. In MANET, Shailaja.P.et.al. (2019) claimed that MANET can be implemented in three routing approaches which are single metric-based routing approach, multiple metric based routing approach and learning based routing approach. Besides, a suitable routing protocol for MANET must be able to respond rapidly to the topological changes of network and few routing algorithms existed. Before starting the topic of the suitable routing algorithm for MANET, the metric for MANET route selection is the most important topic that is needed to be discussed first. According to Cao.L, Dahlberg.T and Wang.T (2007), the MANET follows the energy-related routing metric as the energy is the only factor which linked with all the objectives such as the lifetime of network and QoS related objectives like packet delivery, end-to-end delay and the optimal utilization of bandwidths. Therefore, number of earlier developed metric approaches focused on the energy efficiency routing selected were founded. First of all, the existed metric was the variance in the residual energy in which the energy left at the nodes after forwarding the data to the respective nodes can be defined as residual energy. Also, since the mobile nodes could be playing the role of multi-hop forwarder, more than one node may use the residual energy for data forwarding. Less residual energy means the mobile nodes has less capability of further data forwarding. Hence, this metric cannot result in an efficient route. Secondly, the existing metric would be the node mobility metric. It is a pivotal factor needed to be considered during the route selection. The mobile node with high mobility may result in a possibility of larger route breakups which could lead to extensive packet loss. Extra energy is also needed when a new route between the mobile nodes is needed to be established (Appiah.M, 2017). With these effects mentioned above, network may also get partitioned if one of the mobile nodes moves out of the communication range of all nodes. But the node mobility factor cannot guarantee maintaining the lowest energy path. Based on the existing metrics, it can be concluded that using a single metric is not a recommended choice for the selection of an optimal route to the destination nodes. Instead, using a composite route metric which involves multiple metrics associated with both QoS and energy would contribute to the evaluation of routes and achieving an extended network lifetime followed by a better QoS (Shailaja.P.et.al., 2019). In short, in single metric, the node or path is selected based on one metric which is the energy, while in multiple metric, the node or path is selected based on several metrics which are energy, mobility, delay and stability.

Energy-Efficient Distance Routing Algorithm

Chan.HF and Rudolph.H (2015) had proposed their routing algorithm which is called the Energy-Efficient Distance Routing Algorithm. Since the WSN is applied rapidly on a large scale of network nowadays, so the researches claimed that will focus more on the minimization of energy consumption on the long-distance transmission when designing the algorithm. Energy efficient distance routing algorithm is able to solve the problem found in the hierarchical routing algorithms in multi-hop transmission which is the energy wastage on unnecessary transmissions. As proposed by Chan.HF and Rudolph.H (2015), the network is divided into multiple cluster whereby every cluster head node of each cluster is responsible to collects and aggregates information from its respective cluster member with the aim to save energy. Then, a summary of the collected information is delivered by finding the next hop which is closest to the base station and also within the energy efficient transmission range. By doing this, it can avoid redundant transmissions and save the communication costs. Apart from that, in the case that the sensor node is very close to the base station, data is directly transmitted to the base station. This method is able to save the energy consumption as the data is delivered by finding the next hop which is closest to the base station rather than finding the shortest next hop distance method or deliver the data one by one to the next cluster head until it reaches base station which is being used in the hierarchical routing algorithms during multi hop transmission. In a conclusion, the energy efficient distance routing algorithm is using the method of finding the next hop closest to the base station which is better than the method of directly transmit data one by one to the next cluster head until it reaches base station and the method of finding the shortest next hop distance method. The first step on how it could find the next hop closest to the base station is by the respective cluster heads will wait for a fixed period before starting the transmission or wait for a certain amount of message size to be accumulated before starting the transmission. The distances to base station are to be compared. Data is sent to the base station is the distance to the base station is shorter than other cluster head. Otherwise, the cluster heads information is to be extracted from the database. Then, the next targeted cluster is to be selected within its energy efficient transmission range and also closest to the base station. The energy to be used for the data transmission of both cluster head is to be estimated. Consequently, the data is sent by the cluster head to another and update the remaining energy of the cluster heads if the remaining energy is enough for both cluster heads.

Energy Efficient and Mobility Aware Routing Protocol (EC-MRPL)

Bouaziz.M. et. al. (2017) had proposed a routing protocol named EC-mRPL which is able to overcome the problems made by mobility. It is a n energy efficient and mobility aware routing protocol. The EC-MRPL enables to better conserve the energy and sustain the connectivity of mobile nodes. EC-MRPL integrates an enhanced mobility detection method and a novel point of attachment prediction and replacement strategy aware of the resources constraints. First of all, the EC-MRPL is created with the aim of providing a continuous connectivity to ensure the reachability of mobile nodes during their movement. The proactive feature of EC-MRPL will predict the new attachment of mobile nodes before the disconnection of mobile nodes occur. This whole process is mainly relayed on the RSSI cross-layer information and the icmpv6 messages of standard RPL based on some predefined flag. In the development of EC-MRPL, in order to ensure the efficiency of energy consumption with some constraints deployed, the EC-MRPL will take the constraints into consideration and provide measures such as reduce the mobile nodes involvement to keep its energy, and share resource dissipation between the associated nodes. Besides, to ensure the transferability of higher data packet and overload, the EC-mRPL is designed by using the less exchanged signaling messages. After the EC-mRPL protocol is integrated into the Contiki simulation, Bouaziz.M. et. al. (2017) proved that the simulation result of EC-MRPL is obtained using the Contiki simulator shown that EC-mRPL outperform the Routing Protocol for Low Power and Lossy Networks based (Co-RPL) and Smart-HOP Integrated In RPL (mRPL) in terms of handover delay, data loss rate, signaling cost and energy consumption. The EC-mRPL is able to provide a seamless connectivity and reduce the signaling cost which had brought a good impact on the energy consumption of mobile nodes and the data packets transmission for the IoMT devices.

Virtual Grid Architecture (VGA) Approach Routing Algorithm

Kamal.A and Al-Karaki.J (n.d.) had proposed the efficient virtual-backbone routing for the MANET which is the Virtual Grid Architecture (VGA) approach routing algorithm. In this routing algorithm, it emphasizes on the construction of a stable and robust virtual topology and also provide mobility support. The VGA can be applied to both homogeneous and heterogeneous MANET and can be used to perform routing in MANET efficiently due to its stable structure. The methods used to route packets over VGA are the on-demand routing approach and transitive closure routing approach from discrete mathematics. In the on-demand routing approach, Kamal.A and Al-Karaki.J (n.d.) had constrained the packet broadcasting to at most four neighbors so that fewer resources is used and to match with the MANET resource constraint nature. It is expected to utilize the notion of a virtual topology to efficiently direct the route querying control traffic in a specific order. While for the transitive closure routing approach, it is used to discover paths over the VGA. It automates the calculation and uses the matrix multiplication effectively. The cluster head maintains an up-to-date image of the adjacency matrix then the adjacency matrix reflects the link status between the set of neighboring. The digit 1 and 0 are used to indicate the present and absent of ongoing link between cluster heads.

The two routing approaches over VGA which are on-demand routing approach and transitive closure routing approach were stimulated using the NS2.26 simulator by Kamal.A and Al-Karaki.J (n.d.). Results shown that both the routing approaches which have been modified to adapt the VGA was able to achieve high packet delivery ratios as well as maintaining an acceptable levels even higher loads were offered in network. In short, the stability of VGA enabled both the approaches to achieve high packet delivery. On the other hand, result also shown that the two routing approaches running over VGA generated lower normalized control overhead. It means that the two routing approaches running over VGA have better functionality in a congested or low-bandwidth environment and consumed the battery power efficiently. In terms of scalability which refers to the adaptability to larger network, the stability of VGA can enhance the scalability of the node density in the network and decrease the overhead although the number of nodes increased gradually. Hence, it can be concluded that the VGA can performed well in terms of cluster head cardinality, path length and communication overhead as well as providing mobility support.

Immune-Inspired Algorithm

Atakan et. al (2007) argues distributed node and rate selection (DNRS) systems that is nature inspired immune system. The system forms a B-cell centered designated node (DN) to sense anticipated incident and to encounter the incident signal at the sink node. This theory is very similar to immune system whereby when the system spots a pathogen, B-cells are aggravated to create some antibodies at dissimilar densities to lessen the result of the pathogen. The most important phenomenon in this protocol is the selection of DNs to ensure that they are positioned to the event source as closer as possible and they are positioned farther from each other as much as possible. There are three factors used for the selection of DNs: i) the similarity between sensor node and event source, ii) affinity between sensor node and uncorrelated neighbor nodes and iii) the affinity between sensor node and correlated neighbor nodes. Upon selection of DNs, when an event occurs, the DNs sense the event and send sensed information with sampling frequency to sink node to construct the event signal. Natural immune system based mathematical model and equations are used to tackle the sampling frequency parameter and the congestion.Besides the determination of the DNs, the algorithm also selects the most suitable sampling frequencies of sensor nodes according to the congestion. Besides the selection of the minimum number of designated nodes, the algorithm offers substantial saving of energy to sensor network.

Bokareva et. al (2005) proposes a new mechanism of immunology inspired method of self-healing in wireless sensor network. This method is based on some inherent characteristics of human immune system with five main goals in mind. The first is where there are any failures on individual component there should not be any adverse effects on the whole sensor network. In human-inspired immune system, any cells are adept to replacing their basic cells thus do not affect the entire human system. The second implies that in a large deployed sensor network, there should be some degree of independence where sensor nodes would be capable of managing themselves without the support of outside object as in human immune system. In the third approach, several layers of functionality are combined into one single layer for protection and detection of possible faults. Fourth, similar to immune system where just a small number of detector cells are needed to detect pathogens, in a resource constrained sensor network, only limited sensor nodes with limited resource are needed in order to detect failure and threat. Diversity and adaptability should be the fifth mechanism where sensor nodes should have the capability to detect possible threat in advance as similar to human system where it can detect large number of pathogens and detecting unknown pathogens and remembering them in the future. SASHA is an architecture (Figure 6) proposed for self-healing sensor network and to support the resource constrained sensor networks. Several different network objects manage to deliver resourceful and effective fault-tolerance, automatic fault recognition, network monitoring and response.

As in the natural immune system, its main role is to identify the "self" and the removal of non-self proteins. Using the same concept in sensor network, it is important to understand what is meant by self and non-self set. One of the main focuses of sensor network is to monitor unknown environments and report the correct sensor readings to the base station. One main challenge in this case is what constitutes correct sensor readings. The paper focuses on detecting defective sensor readings using some pattern recognition techniques. The architecture in SASHA comprises numerous managing components, a great number of sensor nodes, numerous monitoring nodes, base station, Thymus and Lymph (database) machines.

Figure 6. SASHA architecture

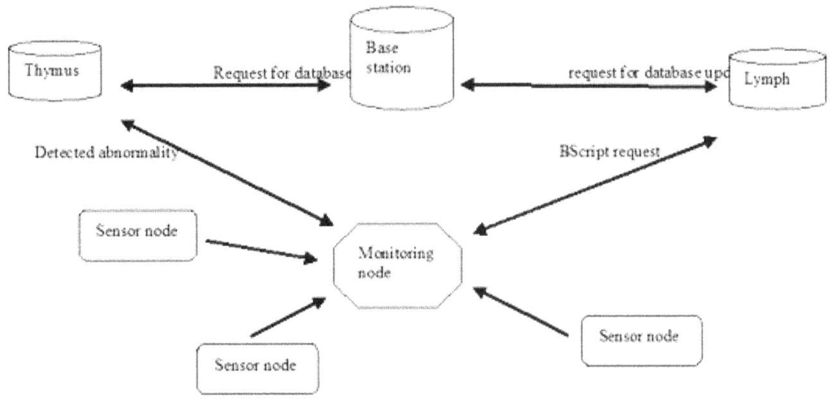

Sensor nodes are resource constrained tiny nodes to monitor and report events to the base station. Monitoring nodes have additional capability of sensing, processing and communication capability. Each monitoring node covers a certain portion of the network in the form of a tree network rooted with the monitoring node. Lymph consists of B cells in human system is responsible for detecting anomalies (infected pathogens) and to destroy them. As in sensor network, the anomalies can be detected by running mobile scripts known as B-Script. These scripts are dynamically generated to act as filter for the abnormal behaviors of the forest. Lymph machine stores databases of previous attacks, severity of damage and the need for response and the solution. Thymus as similar to human immune system is a representation of self. There are three specific aims that SASHA is looking into: i) automatic fault detection in order to achieve robustness in sensor network, it is important to identify faulty sensor readings. The fault detection mechanism comprises trivial disseminated algorithm to detect faulty readings. ii) adaptive monitoring; this process is achieved by distributing mobile scripts onto the sensor network. This process consists of generation, maturing and migration of mobile monitoring scripts. These monitoring scripts are created automatically without intervention of human. and iii) corresponding reaction by returning to various types of faults. Coordinated response to malicious behavior is handled between monitoring nodes, Lymph and Thymus.

Faulty sensor readings can be achieved by firstly identifying what constitutes correct readings. This is performed by using Self-Organizing Neural Network with Competitive Learning (SONN). The main advantage is that SONN does not need earlier knowledge of the incident being observed. In order to classify sensor readings as faulty or correct, all monitoring nodes would run SONN that is acquired from the Lymph machine. An immunology inspired self-healing architecture is proposed for wireless sensor networks. It is a holistic approach to network design for automatic fault recognition and response over possible faults. It is adaptive in nature that can evolve over observing and inference abilities.

Ant-Inspired Algorithm

Camilo et. al (1996) proposes an energy optimized protocol called the EEABR protocol based on Ant Colony Optimization heuristics by Di Caro and Dorigo (1998) for wireless sensor networks. The general idea of this algorithm is based on basic ant-inspired routing algorithm. In ant-inspired routing algorithm (Dorigo et. al, 1996), at every interval, a forward ant is initiated with the intention of discovering a path from every network until reaching the final node. The identifier of node crossed is kept in memory of the ant. At every node, a forward ant selects the advancing node using some probabilistic values. The pheromone trail of node and routes visited are kept in a routing table. This task is performed by the forward ant which is taking its route towards the destination. The same concept is applied when backward ant returns back to its origin where it was created. Upon returning, the pheromone trail is again updated in the routing table and this allows each node to identify its best neighbor to forward the data to.

This section discusses about the improved version of the ant based routing algorithm. There are two improvements made in the algorithm to reduce the quantity of energy used by sensor nodes and decide on the energy quality path identified by the ants. In the general algorithm, forward ants do not have a specific location where it is sent and due to that the amount of information kept relating to identification of neighboring nodes at the routing table is huge. But this is not suitable for large networks since a huge amount of memory is needed to keep the neighborhood information. Improvement is made to save memory in which if the forward node is sent directly to the sink node, then the routing table is only going to store information about neighboring nodes that are in the same direction with the sink node. This actually reduces the size of routing table as well as the memory needed by the nodes. In this protocol the quality of a given path is determined not only by the distance but also based on the energy level of the path. This is done by proposing a new function to determine the pheromone trail that the backward ant would take during its journey back to origin.

There are more improvements made to this protocol by trying to reduce the communication load within nodes and the energy spent for communications. It is preferred that most of the data processing are done at the ant node itself instead of sending it to the sink to be processed. In order to perform this, the memory of the ant is kept at a low rate by only keeping information of only the last two nodes visited. So in this case the path followed by the ant is no more in the memory and therefore nodes should create memory to keep records of the nodes visited. When the node receives a forward ant, the ant's identification, previous node, next node and timer value are kept in the memory. The ant is then forwarded to the next node and this continues with other nodes as well. If the record of the ant is found in the memory, the ant is then eliminated to avoid possible looping. When the backward ant reaches the node, the node inspects its memory to know whether the ant's identification is found in the memory together with timer value and finds the next node where the ant should be forwarded to. The timer value is used to ensure that the records are deleted if it is found that the ant could not find the way back to the source.

The results are collected based on the three different versions of ant based routing algorithms discussed: the basic ant-based routing algorithm (BABR), the improved ant-based routing algorithm (IABR) and energy-efficient ant-based routing algorithm (EEABR). There are four metrics, average energy, minimum energy, standard deviation, the variance between energy level on all nodes and energy efficiency are used to compare the performance of the algorithm. Based on the results obtained, EEABR and IABR perform better results as compared to BABR as both algorithms are energy conscious. EEABR lessens the energy spent in relations to communication and drop in overhead of message exchanged.

Autonomic Computing: Self-Healing

This work (Natalija and Nelson, 2007) generally looks at one specific problem which is combating routing holes in sensor networks by deploying a single mobile node. As discussed in other research papers and literature reviews, self-healing by means of mobile nodes in WSN is still not widely explored. So this paper looks at combating the routing hole through deployment of a mobile node using a simple distributed algorithm for determining the Optimal Placement of a Mobile Node (OPlaMoN) proposed. According to the author, the easiest way to achieve energy-efficient mechanism is by minimizing the number of transmissions and by utilizing optimized routing path for transmission. However achieving optimized routing path is a great challenge as this tiny-battery powered sensor nodes are scattered randomly in an irregular geographic area. Due to these natures, the network topology would eventually lead into a routing hole. The routing hole becomes a challenging area to achieve optimized routing path. And the techniques to handle these issues are considered as self-healing in WSN. This paper is generally to be the first attempt in investigating the energy aspect of self-healing by using mobile robots to handle routing holes. There are two main contributions made by this authors as to: i) use 'energy' as the main criteria for placing a mobile node within a routing hole and ii) propose an algorithm for energy-efficient optimal location of mobile node within a routing hole.

In this protocol, the optimal path between the source and the sink may come across an anomaly known as routing hole. The node at which the routing hole is identified is known as node (1) and the node that exits the hole boundary is known as node (n). Only one mobile node is used to bridge the boundary hole and the node is placed at distance r from the hole in the direction of the exit node. There are two specific types of holes considered in this protocol: circular and rectangular shaped holes.

For the circular shaped scenario, the sensor nodes on the boundary of the hole would actually form a convex polygon with radius r. As the size of the hole increases, more sensor nodes would fit into the boundary of the hole, increasing the radius r and the degree of polygon. The experiment was first done with no mobile node around the hole and the energy required to transmit a bit from the source to the sink is calculated. Then by placing a mobile node within the hole, the energy required to transmit a bit from the source to the sink through the mobile node is calculated. And based on the results, the energy required to transmit through the mobile node is not really justifiable. The same situation is applied to a rectangular shaped scenario, and produces the same result as above. If energy is used as the main criterion for this case, routing around the hole would produce better results as compared to bridging the hole through a mobile node. So rather than using energy as the main criteria, an algorithm to find the optimal position of mobile node is proposed (OPlaMoN). This algorithm uses distributed decision making by ensuring minimum exchange of information among effected nodes. Each effected boundary node would calculate a bidding-value, assumed to be a value gained by the effect if a mobile node were to be placed next to it. Each node would locally exchange its bidding-value with other nodes, and the highest node with bidding-value to be elected as the bidder. Placing a mobile node within the vicinity of the node with the highest value would assume to be energy gained. The algorithm has proven that placing a mobile node within the routing hole could possibly be a solution to combat the hole problem in WSN. However energy may not be the sole of the criterion in finding optimal solution for self-healing. Looking into other parameters such as overall transmission delay and static node failure should also be taken into consideration. So through simulation results, a distributed energy-efficient placement of mobile node within a routing hole is proposed.

The main focus of the work by (Branch et. al, 2007) is to provide a fault-tolerant and efficient routing protocol in WSN where nodes can automatically find the shortest route without deciding which neighbor to forward the data to. A transfer node announces the message to all the nodes within i transmission reach. The receiver then autonomously decides where to forward by using the knowledge of hop distance and a prioritized transmission back-off delay that ensures that the data is sent by only a single node. A time slotting mechanism is used to ensure that the available shortest route is always used and reduces the probability of packet collision. One of the main advantages of SHR is where it dynamically determines the shortest path even when topology changes occur.

This protocol is an extension of self-selective routing (SSR) where each node knows in advance its distance (hops) to the destination. In SSR, the sensor node would automatically broadcast its data to all the available neighbors within the range. Upon receiving data, the next node to send the data would be decided by using a back-off delay scheme. In this mechanism, once a node receives data it sets a random delay where the average is relative to distance from the endpoint. But based on the calculations, the algorithm ensures that only nodes nearest to the endpoint have the greatest chance of re-broadcasting the data. If a node hears the broadcast of the same packet that it is waiting to broadcast, it would automatically abandon the operation. All other nodes would be notified via an explicit acknowledgement of the broadcast so that they do not need to resend the same packet.

Though SSR has some benefits of low overhead and fault-tolerance, it has several weaknesses as well. First the back-off delay mechanism creates a probability of packet collision and due to this reason packets might need to travel along longer route while shorter routes are available in realistic cases. Moreover, SSR does not provide route healing mechanism for sending packets around damaged routes. So SHR introduces some new contributions to SSR in order to address the weaknesses discussed above.

The first improvement SHR is looking into is to introduce a time-slotted back-off delay scheme. This back-off delay is measured in transmission slots rather than using continuous time. This ensures that nodes closer to the destination would first send the packet than any other available nodes as in SSR. The second improvement would be route repair mechanism for propagating packets around severed routes. Severed route occurs when a sending node does not have any neighbor node that is closer to the destination. So SHR ensures that rerouting is performed on the shortest available routes by adjusting the hop distance of the sending node. SHR also ensures that node's hop distances are adjusted when departing node comes back to the network.

SHR protocol is operated by using a data structure of a cost table which consists of ID of the source and destination, the sequence number of the last packet and the hop distance from the destination to the current node. There are three phases uses in this protocol: i) destination request, ii) destination reply and iii) data transmission. When a node sends data to a destination in which no distance information is available, it floods a destination request packet (DREQ) to the network. The hop count value is set to 1 initially and when an intermediate node receives the DREQ packet, it creates a cost table entry if no table has been created previously. Otherwise the table is updated with packet ID, hop count field and sequence number. The intermediate node then would broadcast the packet with some random back-off delay to avoid collisions.

When the destination receives the DREQ packet, it will not rebroadcast it thus resulting in all nodes knowing their distance from the source. Destination node would broadcast a destination reply (DREP) packet. The same data field as in above is used and at this point of time all nodes would have already known the hop distance from the source node. When an intermediate node receives DREP packet, it updates its cost table with the actual hot count distance to the destination. The same process continues with any intermediate node that receives the DREP packet and the process continues until it reaches the source node where the source knows in advance that the destination node exists. Upon receiving DREP packet, the source starts sending DATA packet to the destination along the route previously used as each node knows in advance its distance to the destination.

The effect of node density is evaluated by changing the mean number of neighbors per node. As the mean value increases, more collisions occur as more neighboring nodes compete to forward the packet.

But at the same time end-to-end delay decreases as network density increases and the performance of route repair mechanism also increases in SHR. Delivery ratio also improves as density increases. Moreover, with the route repair mechanism the networks rate of failure can be handled effectively by adjusting the number of packets required to invoke route repair.

Autonomic Computing: Self-Optimizing

Mobile Ad Hoc Network (MANET) is formed by nodes communicating through a shared multihop communication network. And each node acts as a host as well as a router. Mobile Ac Hoc network is becoming a popular area since nodes in this network can be easily deployed and do not need any infrastructure to set-up the network. Recent advancement in mobile wireless devices increases the potential for mobile ad hoc networks. But some of the major challenges faced by routing in MANET are frequent changing network topology, low transmission power and low available bandwidth. Routing protocols should be able to cope with network topology changes without incurring too much of overheads of control messages.

Normal Internet routing in wired network is not suitable for ad hoc networks, thus increases the need for further research in the routing of ad hoc networks. Since there have been many routing algorithms proposed for MANET, the work done in this paper does not propose a new algorithm. But this work, by any routing algorithm provides path between source and destination, so the path can be optimized dynamically to enhance performance and consume less energy. This work (Gui & Mohapatra 2003) proposes a Self-Healing and Optimizing Routing Technique (SHORT) for ad hoc networks. There are two categories being looked at in this technique: a) Path Aware (PA) Short for optimizing the healing path and b) Energy Aware (EA) Short for conserving power usage in MANET.

The shortest path created by on-demand ad hoc routing protocols may diverge from the shortest path and due to the movement of the nodes in ad hoc networks; the shape of routing route may vary while the connectivity still remains intact. Most of the earlier proposed protocols do not initiate a new mechanism for path discovery till link failure occurs. The variations in form can be corrected by developing improved routing paths by escaping significant overheads which will be discussed in the later part of the discussion.

For extending the network life time, there is a need to look into energy efficiency of routing protocols. Many studies have been conducted at energy efficient in routing protocols such as handling minimum cost path along multihop links as well as using dissimilar power aware metrics for determining routes. The goal of the framework in this paper is to enhance power consumption through route optimization. Energy aware routing looks at extending network and nodes lifetime by routing packets to node that have sufficient energy as well as avoiding routing packets to nodes that are low on battery. Routing protocols should be able to self-heal itself by allowing nodes to come out from situations of being over-used while other nodes are less utilized. The main goal in ad hoc networking is to increase performance as well as to monitor power conservation. This paper looks at improving both measures without incurring additional overhead on the network.

Based on Figure 57, for a routing path from node A to node J, the initial routing path is discovered through some path discovery processes with the shortest path between two nodes. In this case it would take ten hops to reach from node A to node J in which, ten hops are considered the shortest path between nodes. During the execution of the process the mobility nature of the nodes may cause changes in routing path as shown in Figure 4. In this case, node I is within the reach of node A and node J is within the reach of node F. With the new path available through node I and J, there should be a possibility of identifying the shortest path which is less than ten hops in this case. But since there is no such mecha-

Figure 7. SHORT node connections

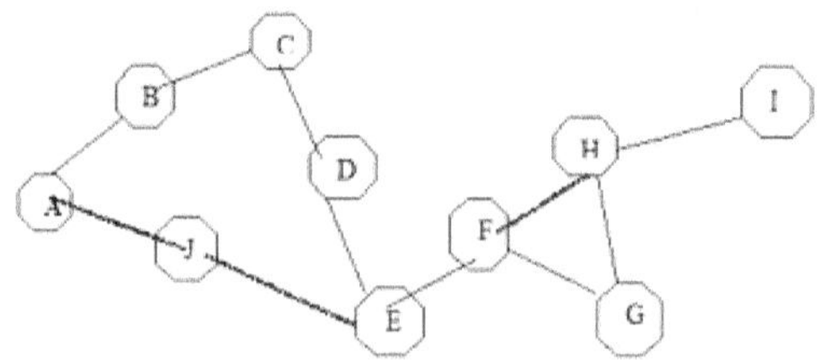

nism to update the routing table, it may not be possible for nodes to self-heal to identify such paths. So the main focus of this paper is to look into identifying such possibilities and allowing nodes to self-heal and optimize the paths available by modifying routing tables dynamically. The basic approach taken in this new protocol is known as (n,2) reduction where n hops along the path can be reduced to 2 hops only.

Autonomic Computing: Energy Aware

The main goal of energy aware load balancing is to balance the energy at each node by performing load balancing of traffic among all nodes. For example, for a connection between A and D, if the traffic continuously follows the route thus A to D ignore other available paths, then the load is just focused on this route. Nodes in other routes are left alone without incurring any energy. For example, nodes B and D can also participate in the routing process, but they are being eliminated due to no proper load balancing performed in order to distribute the traffic. This will cause nodes in the earlier route eventually be drained out of battery power. Whereas if it is to be an energy aware routing, then traffic is evenly loaded in order to consume equal amount of energy at each node. EA-Short basically looks at such mechanism to divert the traffic to other nodes.

There are two possible ways how this energy aware mechanism can be performed. In the first approach, when a node (G for example) which is closer to the current route of traffic between A, B and C overhears the message for consecutive times, it may take over the role of the intermediate node (node B in this case). Node G can perform this by comparing the energy level at node B and node G, and if the remaining energy at G is more than B, then G might take over in order to forward the messages to C. In the second approach, node G detects that node B needs help which could be due to low energy level at node B and G informs B that it may take over the role to divert traffic from A-B-C to A-G-C. This can be further explained by introducing other additional nodes along the path which may become helpers to help other nodes which are in trouble of low energy.

In this algorithm, there are two more additional fields to be included in the IP header. They are "Hop Counter" and "Residual Energy Level" together with other IP fields such as source, destination, sender and sequence number. Each node maintains an overhear table consists of source-destination pair, sequence number and "overhear list". When a node overhears a packet, it checks its source-destination field to know if the packet belongs to this new flow. If it belongs to this flow then a new entry is added into the overhear table. The residual energy field keeps the residual energy level at each transmitting node of the packet.

PA-Short algorithm is developed on both AODV and DSR with the enhanced name called AODV-SHORT and DSR-SHORT. The goal of this approach is to see how Short algorithm performs over

Table 2. A Summary of Protocols

No	Protocol	Advantages and Disadvantages
1	Yu et. al (2001)	Uniform and non-uniform Geographic energy aware routing Delivers more packets than GPSR (Intanagonwiwat et. al, 2000) For non-uniform traffic - longer network lifetime than GPSR Hole problems Picks neighbor closer to sink
2	Intanagonwiwat et. al (2000)	Geographic non-energy aware routing Avoids communication holes by delivering planar graph out of original network graph Hole problem – by using the shortest path, it enlarges hole Tends to focus traffic on the edge when it routes around holes therefore exhausting nodes on the edge
3	Wendi et. al (2000)	Uses a clustering mechanism whereby every node rotates the role as cluster head to well balance the energy distribution and to extend the network lifespan Cluster head communicates with member sensor nodes and forwards data to the base station Not appropriate for extended networks since direct communication between the cluster head and the base station incurs additional energy
4	Sha et al. (2005)	Uses right-hand rule (locating and bypassing routing holes) Guaranteed delivery and more energy efficient Avoids routing to the holes Two important notions – weight and hole
5	Youssef et al. (2002)	Uses distance to measure transmission energy and propagation delay Cluster-based approach Looks at maintaining reasonable end-to-end interval and throughput output
6	Vlajic & Moniz (2007)	Not only looking at energy but overall transmission delay and static node failure
7	Akkaya & Younis (2004)	Mobility of the gateway-dynamic maintenance of multihop Continuous adjusted network topology and update Gateway picks the best node based on its proximity to the next position it moves Notifies the last node on the current position to extend path to the new location
8	Shah & Rabaey (2002).	Not only single path but having many alternative paths for communication Energy at single node not depleted easily
9	Bokareva et. al (2005)	A naturally inspired system for hybrid sensor network by providing self-healing capabilities Automatic fault recognition and response Adaptive architecture to learn and evolve with unknown faults – self-healing Looking at faulty sensor readings which need self-healing Co-ordination between monitoring nodes
10	Gui & Mohapatra (2003)	Self-healing and optimizing techniques for mobile ad hoc networks Network performance is enhanced in relations to bandwidth and latency by not incurring extra cost Path determination can be done without using much energy Metrics such as path length, energy consumption and energy aware load balancing
11	Dressler et. al (2005)	Bio-inspired strategy in terms of self-organization and adapt to environmental changes Looking into energy efficient operation, quality of service, optimized task allocation and communication Looking into cell biology methods and mimics them into networking aspects mainly for self-organization context. Feedback loop adapted from cell biology Mobile robot systems(communication relay) and stationary sensor network
12	Branch et. al (2007)	Energy effective self-healing routing for WSN The shortest possible path is used, route repair and collision avoidance Future research-more energy efficient since random selection of neighbors which may cause more energy to be consumed.
13	Ganesan et. al (2001)	Propose two types of multipath to provide energy efficient recovery from failure Failure of primary path can be recovered without using flooding- flooding reduces network lifetime
14	Al-Fares et. al (2009)	Proposes a hierarchical routing protocol for WSN by looking into network survivability and redundancy features. Looks into node or link reliability problem and fault tolerance by introducing redundant cluster head Shows positive and better results compared to low energy adaptive clustering hierarchy (LEACH Wendi et. al (2000)).
15	Camilo et. al (2006)	Bio-inspired mechanism Uses ant pheromone trail Energy conscious. Energy consumption due to communication is reduced Lessening the overhead in terms of message interchange.
16	Tong et. al (2003)	A new concept of introducing mobile agent Sensor nodes are less burdened by reducing the load Hierarchical architecture More energy effective
17	Atakan & Akan (2007)	Biologically inspired mechanism Maps B-cell to designated nodes
18	Ponnusamy et al. (2014)	Biologically-inspired Mobile agent based system Provides self-healing in terms of energy provisioning

other on demand routing protocols. Based on the experiment, DSR performs better than AODV and AODV-SHORT performs better than AODV in terms of path optimality. In terms of delivery rate, both AODV-SHORT and DSR-SHORT achieve a stable performance for all mobility rates. As for energy aware, Short is implemented on AODV called AODV-EA. Based on the experiment, it is evident that AODV-EA performs better load balancing thus performs better energy usage at all participating nodes rather than concentrating on certain nodes only. Short algorithm achieves routing optimality by continuously observing routing paths and redirecting paths to a more optimal one. Based on the simulation results, higher delivery rate and longer network lifetime are achieved by adopting SHORT algorithm.

Summary

Table 2 and Table 3 outline a summary of protocols in the literature by considering its advantages and disadvantages. This section is discussed in brief to give an outline of self-healing protocols and their weaknesses to look for possible improvement in future research

CONCLUSION

The chapter is concluded by presenting a review of self-healing routing protocol in wireless sensor network. The protocols discussed in this chapter are categorized as energy efficient and self-healing, so that a reasonable review can be conducted. The review was conducted by analyzing the strengths and weaknesses of the protocols. The protocol is also presented in two different contexts as in bio-inspired and non bio-inspired mechanisms. Also the comparison in section 3 and 4 is done based on various metrics useful in sensor network performance. The comparative study done in terms of energy efficient, self-healing and bio-inspired methods of the protocols in the literature provides as insight for future better self-healing protocols. Moreover, the static and single layer of hierarchy should be given further thought to look into other alternative approaches such as n-tier and mobile nodes Ponnusamy et al. (2012). The given bio-inspired mechanism provides a self-healing mechanism for energy efficient protocol to increase the lifetime of wireless sensor networks. Up to now, many self-healing routing protocols have been proposed but not many have focused on biologically-inspired mechanisms for autonomic behavior.

REFERENCES

Akkaya, K., & Younis, M. (2004). *Energy-Aware Routing to a Mobile Gateway in Wireless Sensor Networks*. Paper presented at the IEEE Globecom Wireless Ad Hoc and Sensor Networks Workshop, Dallas, TX.

Akkaya, K., & Younis, M. (2005). A Survey of Routing Protocols in Wireless Sensor Networks. *Elsevier Ad Hoc Network Journal*, *3*(3), 325–349.

Al-Fares, Z., Sun, H. & Cruickshank. (2009). *A Hierarchical Routing Protocol for Survivability in Wireless Sensor Network (WSN)*. Paper presented at the International MultiConference of Engineers and Computer Scientists, IMECS 2009.

Table 3. Comparison of Protocols

	Biologically Inspired	Stationary Node	Energy Efficient	Hole Problem	Shortest Path	Fault Tolerance	Hierarchical
Wendi et. al (2000)		√	√				√
(Gui & Mohapatra 2003)		√	√		√		
Branch et. al (2007)		√	√		√	√	
Bokareva et. al,(2005)	√	√				√	
Camilo et. al (2006)	√	√	√				
Sha et al. (2005)		√	√	√	√	√	
Yu et. al (2001)		√	√	√	√		
Intanagonwiwat et. al (2000)							
Akkaya and M. Younis (2004)		Mobile Base station	√				
Al-Fares et. al (2009)		√	√			√	√
Shah and Rabaey, (2002)		√	√				
Dressler et. al (2005)	√	√	√				
Youssef et al. (2002)		√	√		√		
Ganesan et. al (2001)		√	√		√		
Vlajic and N. Moniz (2007)		Partial mobile	√	√			
Atakan and Akan (2007)	√	√	√			√	
Tong et. al (2003)		Partial mobile	√			√	√
Ponnusamy et al. (2014)	√	Partial mobile	√				√

Atakan, B., & Akan, Ö.B. (2007). Immune system-based energy efðcient andreliable communication in wireless sensor networks. Springer.

Bokareva, T., Bulusu, N., & Jha, S. (2005). SASHA: towards a self-healing hybrid sensor network architecture. *Proceedings of the Second IEEE International Workshop on Embedded Networked Sensors*, 71–78.

Bouaziz, M., Rachedi, A., & Belghith, A. (2017). *EC-MRPL: An energy-efficient and mobility support routing protocol for Internet of Mobile Things. In 14th IEEE Annual Consumer Communications & Networking Conference*. CCNC.

Branch, J. W., Lisee, M., & Szymanski, B.K. (2007). SHR: self-healing routing for wireless ad hoc sensor networks. *Proc. Int. Symposium on Performance Evaluation of Computer and Telecommunication Systems SPECTS'07*, 5–14.

Camilo, T., Carreto, C., & Silva, J. S. (2006). An Energy-Efficient Ant-Based Routing Algorithm for Wireless Sensor Network. *Proceedings of the 5th International Workshop on Ant Colony Optimization and Swarm Intelligence*, 4105, 49-59.

Chan, H.F., & Rudolph, H. (2015). *New Energy Efficient Routing Algorithm for Wireless Sensor Network.* Available at https://www.researchgate.net/publication/283507649_New_Energy_Efficient_Routing_Algorithm_for_Wireless_Sensor_Network

Di Caro, G., & Dorigo, M. (1998). AntNet: Distributed stigmergetic control forcommunication networks. *Journal of Artificial Intelligence Research, 9*, 317–365.

Dorigo, M., Maniezzo, V., & Colomi, A. (1996). Ant system: Optimization by a colony of cooperating agents. *IEEE Transactions on Systems, Man, and Cybernetics. Part B, Cybernetics, 26*(1), 29–41. PMID:18263004

Dressler, F., Kuger, B., Fuchs, G., & German, R. (2005). *Self-Organization in Sensor Networks using Bio-Inspired Mechanisms.* Paper presented at the 18th ACM/GI/ITG International Conference on Architecture of Computing Systems-System Aspects in Organic and Pervasive Computing (ARCS'05): Workshop Self-Organization and Emergence, Innsbruck, Austria.

Dressler, F. (2009). Self-Organization in Ad Hoc Networks: Overview and Classification. Univ. of Erlangen, Dept. of Computer Science 7, Technical Report 02/06.

Dressler, F., Kuger, B., Fuchs, G., & German, R. (2005). *Self-Organization in Sensor Networks using Bio-Inspired Mechanisms.* Paper presented at the 18th ACM/GI/ITG International Conference on Architecture of Computing Systems-System Aspects in Organic and Pervasive Computing (ARCS'05): Workshop Self-Organization and Emergence, Innsbruck, Austria.

Du Sha, K. J., & Shi. W. (2005). *Wear: A balanced, fault-tolerant, energy-aware routing protocol for wireless sensor networks.* Technical Report MIST-TR-2005-001, Wayne State University.

Ganesan, D., Govindan, R., Shenker, S., & Estrin, D. (2001). Highly-resilient, energy-efficient multipath routing in wireless sensor networks. *Mobile Computing and Communications Review, 5*(4), 1125.

Gui, C., & Mohapatra, P. (2003). *SHORT: Self-healing and optimizing routing techniques for mobile adhoc networks.* Paper presented at the MobiHoc. 10.1145/778415.778448

Helen, P. (2006). *A* Service Management Approach for Self-healing Wireless Sensor Networks. *LNCS, 4195*, 215–228.

Intanagonwiwat, C., Govindan, R., & Estrin, D. (2000). *Directed diffusion: a scalable and robust communication paradigm for sensor networks.* Paper presented at the ACM MobiCom '00, Boston, MA.

Jamal, Al-Karaki, & Kamal. (2000). *Routing Techniques in Wireless Sensor Networks: A Survey.* B. Karp and H. Kung. Greedy Perimeter Stateless Routing. In *Proceedings of ACM Conf. On Mobile Computing and Networking (MOBICOM).* ACM

Kamal & Al-Karaki. (n.d.). *Efficient Virtual-Backbone Routing in Mobile Ad Hoc Networks*. Dept. of Electrical and Computer Engineering lowa State University.

Koopman, P. (2003). Elements of the self-healing problem space. In *ICSE Workshop on Architecting Dependable Systems*. Available at: http://74.125.155.132/scholar?q=cache:qoVzFSeebusJ:scholar.google.com/+self-healing+wireless+sensor+netowrk+lanthaler&hl=en&as_sdt=2000

Ponnusamy, Low, & Anang. (2014). Biologically inspired mobile agent–based sensor network (BIMAS). *International Journal of Mobile Communications, 12*(5), 449-476.

Ponnusamy, V., Abdullah, A., & Downe, A. G. (2012). Energy efficient routing protocols in wireless sensor networks: a survey. *Wireless Sensor Networks and Energy Efficiency: Protocols, Routing, and Management*, 237-261.

Shah, R. C., & Rabaey, J. (2002). *Energy Aware Routing for Low Energy Ad Hoc Sensor Networks*. Paper presented at the IEEE Wireless Communications and Networking Conference (WCNC), Orlando, FL. 10.1109/WCNC.2002.993520

Shailaja, P., Guru Rao, C. V., & Nagaraju, A. (2019). A Parametric Oriented Research on Routing Algorithms in Mobile Adhoc Networks. *International Journal of Innovative Technology and Exploring Engineering*, *9*(1), 4126–4125.

Sohrabi, K., Gao, J., Ailawadhi, V., & Pottie, G. J. (2000, October). Protocols for Self-Organization of a Wireless Sensor Network. *IEEE Personal Communications*, *7*(5), 16–27. doi:10.1109/98.878532

Tong, L., Zhao, Q., & Adireddy, S. (2003). *Sensor Networks with Mobile Agent*. Paper presented at the Military Communications Intl Symp., Boston, MA.

Vlajic & Moniz. (2007). Self-Healing Wireless Sensor Networks: Results That May Surprise. In Globecom Workshops, 2007 IEEE, Washington, DC.

Wendi, R. H., Chandrakasan, A., & Balakrishnan, H. (2000). *Energy-efficient communication protocols for wireless microsensor networks*. Paper presented at the Hawaii International Conference on Systems Sciences.

Youssef, M. A., Younis, M. F., & Arisha, K. A. (2000). *A constrained shortest- path energy-aware routing algorithm for wireless sensor networks*. Paper presented at the IEEE Wireless Communications and Networking Conference (WCNC2002).

Yu, Estrin, & Govindan. (2001). *Geographical and Energy-Aware Routing: A Recursive Data Dissemination Protocol for Wireless Sensor Networks*. UCLA Computer Science Department Technical Report, UCLA-CSD TR-01-0023.

Zaman, Low, & Alghamdi. (2014). Energy efficient routing protocol for wireless sensor network. *16th International Conference on Advanced Communication Technology (ICACT),* 808-814.

Zaman, N., & Abdullah, A. (2011). Position Responsive Routing Protocol (PRRP). *IEEE The 13th International Conference on Advance Communication Technology,* 644-648.

Zaman, N., & Abdullah, A. (2012). Energy Optimization through Position Responsive Routing Protocol (PRRP) in Wireless Sensor Network (WSN). International Journal of Information and Electronics Engineering, 2(5), 748-751.

Zaman, N., Abdullah, A., & Jung, L. T. (2011). Optimization of energy usage in wireless network using Position responsive routing Protocol. *IEEE Symposium on Computers & Informatics ISCI*, 51-55.

Zaman, N., Jung, L. T., Alsaade, F., & Alghamdi, T. (2012). Wireless Sensor Network (WSN): Routing Security, Reliability and Energy Efficiency. Journal of Applied Science, 12(16), 593-597. doi:10.3923/jas.2012.593.597

Zaman & Abdullah. (2010). Energy Efficient Routing in Wireless Sensor Network: Research Issues and Challenges. *IEEE International Conference on Intelligence and Information Technology*, 239-243.

Zaman & Abdullah. (2011). Different Techniques Towards Enhancing Wireless Sensor Network (WSN) Routing Energy Efficiency and Quality of Service (QoS). *World Applied Science Journal, 13*(4), 798-805.

Zaman, Low, & Alghamdi. (2015). Enhancing Routing Energy Efficiency of Wireless Sensor Networks. *ICACT Journal*, 587-595.

Chapter 14
Globally Renewable Energy Consumption Sources

Soobia Saeed
Department of Software Engineering, UniversitiTeknologi Malaysia, Malaysia

N. Z. Jhanjhi
https://orcid.org/0000-0001-8116-4733
Taylor's University, Malaysia

Memood Naqvi
Department of Engineering and Technology, Mohwak College, Canada

ABSTRACT

Electricity consumption will be part of a large conversation about connected with international electricity demand in the next two decades. Renewable energy sources and climate change modify minimization and offer a good estimation for the technological, scientific, environmentally friendly, financial, and also societal aspects. This research is the extended version of the research presented in IGI global book Role of IoT in Green Energy Systems, 2015. A full assessment related to any profile minimization options will likely involve an evaluation of respective potential alongside minimization with the bargain with sustainable development as well as all associated risks and costs.

1. INTRODUCTION

Electrical energy usage can comprise an increasing reveal involving international energy demand above the subsequent two full decades. Recently, soaring energy price ranges and concerns regarding the environment effects involving emissions involving greenhouse gas fossils include renewed involvement in the particular development involving alternate energy sources. Renewable energy is already regarded a more appealing energy resource with regard to nuclear energy due to the absence of hazards and problems. The primary part of greenhouse gas is actually carbon dioxide; there may be international issue in relation to cutting down carbon dioxide emissions. With this impression, the different plans may be put on the

DOI: 10.4018/978-1-7998-6709-8.ch014

particular decrease involving carbon dioxide emissions, including increasing the particular rendering involving environmentally friendly energy and the advertising involving technologies. A couple of key solutions may be carried out to cut back laser emissions and conquer the situation involving climate adjust: exchanging fossil heats up having environmentally friendly energy sources wherever possible and strengthen energy productivity. This chapter discuss with regards to innovative distinct alternatives for enhance the effective use of renewable energy and also productivity in energy use. Moderately a couple of nations have begun to introduce offices by which utilize renewable good quality hotspots for energy era. The significance associated with option energy sources all meets upward with environmental change difficulties of the exorbitant make use of fossil powers. You'll locate a few essential sparks by which invigorate the off shoot of renewable good quality advancements: energy solidity, monetary effects and also co2 partial laser outflow decrease (Rask, M., 2009). The expression "elective vitality" portrays any kind of vitality other than the customary wellsprings associated with quality, including hydropower. Of late the center will keep on being in renewable quality sources. As indicated by simply Global Energy Firm portrays two critical worldwide patterns that may describe the design of renewable innovations in the medium term. Firstly, as renewable drive advancements scale upwards, from an aggregate worldwide cause of 1, 454 giga-watts (GW) inside 2011 to simply just two, 167 GW inside 2017, they additionally should spread out topographically. Second, the more the past number of years of high fossil quality asset vitality utilization has driven renewable advancements in the event that you are progressively aggressive more than an expense premise for their options in a few countries around the globe and circumstance. Agreeing with a specific end goal to Worldwide Energy Agency reckonings, wind is a standout amongst the most aggressive sort related together with renewable vitality innovation one of numerous different choices, if nearby conditions like financing, CO outflow amounts and fossil quality asset costs demonstrate viable (Organization for Financial Co-operation notwithstanding Development, 2010). Indeed, even though since you that have two essential ideas inside of your advances: vitality gives building, which consider option sources connected with renewable vitality like wind and sun oriented force and vitality capability designing. It ought to be noticed that changing the capacity division and redesigning customary vitality having renewable vitality is transformative connected to mechanical change along together with shaping markets. The changing procedure for a few types of earth inviting vitality, for example, the wind and sun oriented may happen after 2020, paying little heed to in the event that they development cost of utilization is typically firmly expanding on your one decade from now. Similarly, renewable vitality markets are not effectively shaped in view of expense inconveniences furthermore the financing of fossil gets more sultry(Muneer, T., (2007).

2. RENEWABLE ENERGY AND CLIMATE CHANGE

Aspiration for energy, as well as associated products and services, to meet besides the monetary societal development and also improved human being welfare, besides health, is actually increasing. All organizations need power services to satisfy basic human being needs like lighting, cooking, space convenience, mobility and also communication) plus serve productive treatments. Considering somewhere around 1850 world-wide fossil fuel use and also characteristics on the fossil powers, oil and also gas increased to enhance the power, resulting in a rapid expansion of beam of light fractional beam of light (CO2). [Figure 1] Types of greenhouse gas (GHG) due to the provision of a person get contributed significantly on the historical increase in every greenhouse gasoline concentrations in the atmosphere MIT. According

Assessment Report on the IPCC Final (AR4) accomplished that “predominantly inside observed raise in international conditions because the mid-20th century is probably because of the observed increase above the concentrations of greenhouse fumes anthropogenic.” New data state that the usage of fossil fuels accounts for the majority of international anthropogenic GHG emissions. Emissions generally grow and CO2 levels of MIT received risen above 390 ppm, or possibly 39% above pre-industrial levels, for the final outcome of this participation of 2, 010. You will observe multiple options the reduction of GHG emissions inside energy program although still meet international demand particularly in regards to energy providers. Most of most of these possible options, including power conservation with regards to efficiency, fossil power resources move, RE, nuclear and H2O as well as the capture and storage (CCS) is it being evaluated within just IE4. A comprehensive assessment of almost any single shareholding of mitigation solutions involve an assessment of individual mitigation potential together with its info to sustainable development together with all this risks and associated prices(Bain, R.L., 2010) . Relating Assessment Report within the IPCC Remaining (AR4) done that “predominantly on the inside observed raise in overseas conditions since the mid-20th century is most likely because of the observed enhance above your concentrations connected with greenhouse toxins anthropogenic.” New data state that the effective use of fossil energy sources accounts for most of overseas anthropogenic GHG emissions. Emissions normally grow and CO2 amounts of MIT obtained risen earlier mentioned 390 ppm, or possibly 39% earlier mentioned pre-industrial ranges, for the last outcome in this participation connected with 2, 010. You will observe multiple selections the reduction of GHG emissions on the inside energy system although still meet overseas demand particularly in terms of energy suppliers. Most of most of these possible selections, including power conservation with regards to efficiency, fossil power resources precede, RE, nuclear and H2O as well as the capture and storage (CCS) is it being evaluated in a matter of IE4. An extensive assessment of almost any single shareholding connected with mitigation remedies involve an assessment connected with individual mitigation potential as well as its details to environmentally friendly development as well as all this specific risks and associated price ranges (Bain, R.L., 2010).

3. RENEWABLE ENERGY RESOURCE TECHNOLOGIES

Ones supply of renewable energy that is continuously intensifying. A high level of investment has recently done in recent years, plus the advancement of engineering offers to eligible countries to provide additional cost effective renewable energy. It forecasted that the quality of the producing countries mentioned above 100 megawatts (MW) with respect to renewable energy increases significantly by 2017. Caused by other negative externalities irreversible besides returning to conventional production of vitality, is crucial to promote the development of renewable energy in addition to providing technologies. These technologies are in all probability, are not comparable using conventional fuels in relation to production costs, but often are comparable if many of us consider their associates, such as its environmental and social externalities. In addition, it should be noted that regular scale enterprises could play an essential role in reducing the production cost of the system. Warning expenses and distribution, as well as technologies do not differ much in the list of conventional and organic energies. Here the key points about the development around key technologies of renewable energy supply that often occur are discussed below (Van Kooten, G C, 2008):

1. Bioenergy

Figure 1. Renewable Energy Resource Technologies

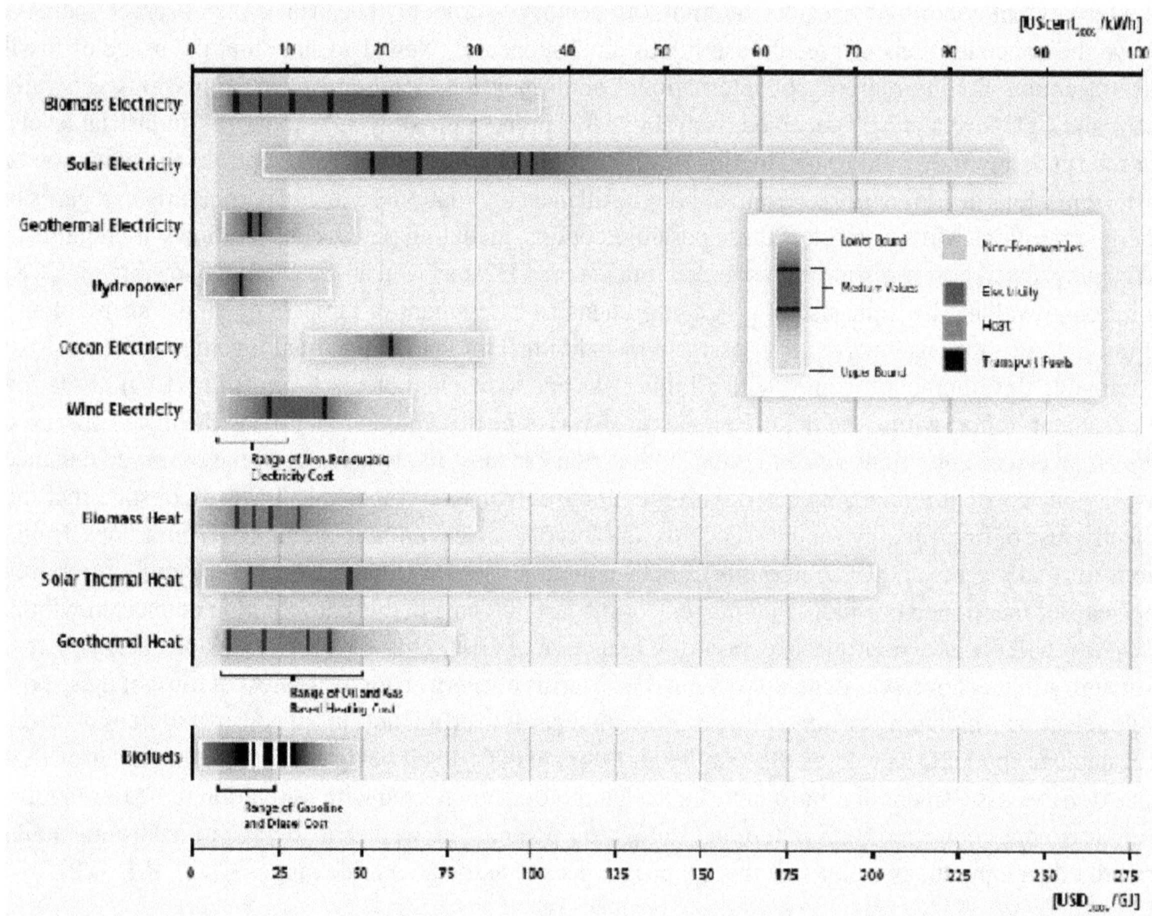

2. solar energy
3. Geothermal energy
4. Hydropower

3.1 Bioenergy

Bioenergy can be created from an assortment of biomass feedstock's, as well as materials of high rural and animals waste; backwoods estates short cycle; force edits; the natural division of metropolitan strong waste; and other natural waste streams. During a mixed bag of procedures, these materials can be utilized specifically to create vitality or warmth, or used to make vaporous, fluid, or strong energizes might. The decision of bioenergy advances is wide and specialized development fluctuates significantly. Examples of technology you can find include small boilers and also large scale heating systems dependent on domestic pellet, and output of ethanol from sugars and starch. The combined cycle power plants advanced biomass gasify incorporated action and transport fuels dependent on lignocellulose are examples of technologies which have been in a pre-commercial stage, while the production associated with liquid biofuels from algae and a few other methods of natural conversion they are investigation and development (R & D) phase. Bioenergy engineering have programs in centralized in addition to decentralized environments with the traditional use of biomass within developing countries will be the current almost all widespread iPhone app. Bioenergy usually features a constant or controllable efficiency. Bioenergy assignments usually depend on the option of method of getting local in addition to regional gas, but recent developments obtained show reliable biomass in addition to liquid biofuels are usually significantly traded internationally (Bain, Third. L., 2010).Bioenergy can be embedded in complex approaches biomass systems global foods, feed and fiber creation and forest products and also waste management and spend. Perhaps most important, bioenergy comes with an intimate and critical role from the media everyday activities of immeasurable people in creating nations throughout the world. Figure 3 shows the kinds of biomass useful for bioenergy inside developing places and produced nations. Expanding bioenergy manufacturing may not need sophisticated supervision cant land and in addition water work with; increased productivity regarding international foods commodity, supply, linens, forest products and energy; change of substantial improvements, plus a far more refined understanding of the societal, energy in addition to environmental interactions in the production and using bioenergy complex. In '08, biomass offers about 10% (3 50 EJ and year) when using the global supply primary vitality. Main uses of biomass directly enter two extensive categories: (1) small efficiency traditional biomass including wood, straw, manure fertilizers are used together with belonging to the cooking, lighting and heating in the room, usually by poorest countries in construction. This biomass is burned to produce most serious negative impacts throughout the lives and health. Increasingly, coal is every second assistance energy in rural areas to be able to create productive chains. For indicator in the extent of traditional utilization associated with biomass, Figure. 2 illustrate which cause global primary energy output from traditional industrial wood biomass will involve parallel world. (2) Greater productivity in using modern bioenergy quick solids, liquids and gases to realize that secondary energy carriers that is included in the heat, electricity, combined temperature and strength (CHP) in addition to transport heats up from unique substantial. Liquid biofuels contain ethanol and moreover biodiesel likewise for transportation by road with this particular planet and a few industrial operates by using (J. P oker. Galdo, 1997). Biomass extracted gases, largely methane, anaerobic digestion of yard waste and also treatment affecting municipal sound waste (MSW) are usually widespread to build electricity, temperature in addition to both. An important contribution of one's services depends on solids, for example motherboards, pellets, previously applied reclaimed wood and many others. It consists of heating programs district warming and trouble heating, and also. The modern-day bioenergy predicted total principal way of obtaining biomass will be 11. 3 EJ and 12 weeks and supply to extra energy end-use. Consumers can be estimated at 6. EJ and year. In addition, an industry, such as pulp industry foodstuff and paper, forestry and eats about 7. 7 EJ of biomass each year, mainly as a method to produce steam for industrial processes (EIA, 2009).

3.2 Solar Energy

Technologies to harness solar irradiation facility for you to present electricity via photovoltaic along with concentrating solar panel (csp) to be able to develop energy winter (uniform heating and cooling), to be able to meet your right illumination wishes and, potentially, to help provide fuels the idea may be taken intended for transportation along with purposes. the maturity of your technology connected with photovoltaic applications vary via R & D (e.g, Gary, fuels designed throughout solar energy), an relatively mature, with regard to example, csp, to help mature equally solar home heating and in line with silicon wafer (pv). many, but not the many technologies are usually modular throughout nature, allowing effortless use both national inside decentralized energy tactics. solar energy will be inconsistent and, for you to varying degrees, unknown, while ones temporal profile of the production connected with solar energy inside numerous ailments sensibly correlate nicely needs tenacity. Energy storage cold weather gives the button in order to improve one's output control intended for a lot of regarding you csp as well as pv technologies similar to straight home heating. Right solar energy technologies usually are regarding various other natures. filling up as well as adjusting the various methods of which a person employ energy as heating, electricity and also fuel electric power, that happen to be children related technologies. the actual team aims at a few most significant types: (a) thermal dom, containing both additionally encouraged non-aggressive heating regarding buildings, every day in addition to business driven mineral water solar heating, method heating hottest action intended for market; (b) photovoltaic (pv) power generation coming from direct change from daylight in order to electric power through photovoltaic fabric; (C) concentrated solar power (CSP) bringing in electric power through the optical focus of solar energy pertaining to high temperature water or perhaps delivers heat in order to drive a good car applications, and in addition electromechanical generators; as well as (d) pv power output methods, your current intention involving employing solar energy in order to supply helpful fuels. Solar energy at the term refers on the base involving power to its an individual RE technologies This draw towards the power of your sun directly. Specific renewable technologies, like wind and also ocean thermal, solar panel considered soon after being absorbed regarding the earth along with became otherwise.(In one's rest regarding the actual part, ones "direct" adjective applied in order to solar energy is often suppressed Just like proving for you to become understood). Merely outside your current atmosphere connected with Earth, this radiation, known Equally solar radiation, carries a good magnitude This average 1 367 W / m2 intended for single individual perpendicular to the rays of any sun surface. in ground level (usually erection Conditions Demands As sea level with the sun directly overhead), your current irradiance will be attenuated in the atmosphere related to 1, 000 interior clear sky Conditions W / m2 throughout a great short night out associated with noon from an song called 'full sun'. Internal electromagnetic waves away by the atmosphere, your durability of a sun can be done within wavelengths covering anything from about 0.20 for you to two microns. Probably your own solar irradiance is sent from the sunlight (K.-M. Chung, 2011). This reaches about the sun directly without remaining scattered of approximately inside. Irradiance, this particular "beam" and also is capable of becoming concentrated by, in addition to decorative mirrors as well as lenses, this is the lowest you may take cloud cover inside pieces. Irradiation is called diffuse irradiance. The sum of a diffuse irradiance beams along with refers to the global solar radiation. the theoretical potential of solar energy, that will indicates The amount of radiation on the surface of the Earth (land as well as sea) is usually theoretically shown intended for energy production; It is usually estimated at 3.9.106 EJ / year. This type involving number, clearly intended for illustrative purposes easily demand all use almost all available land in addition to sea location That has 100% conversion efficiency. a great further convenient metric would be the technical potential; this requires evaluating your own small group of land That is regarding hassle-free employ regarding conversion machines The item have more realistic financial transaction efficiency. Estimates of a technical potential involving energy solar farm incorporate 1 575 for you to 49, 837 EJ / year, ie exactly about two to 100 times your own consumption connected with initial electricity in the world with (Y. Lu, 2010).

3.3 Solar Fuel Manufacture

Fuel technologies convert solar PV technology throughout chemical fuels like hydrogen, synthetic gas and liquid just like methanol and diesel engine. your three easy fuels your current sun, that could possibly help work alone or maybe in any kind of combination, routes are: (1) electrochemistry; (2) photochemical / photo-biological; and also (3) thermochemical. for the first route, hydrogen is generated from electrolysis driven practice derived solar power This has been generated by just about any technique PV or perhaps CSP. Electrolysis associated with water is actually an old and also properly understood technological know-how, usually reaching 70% efficiency electricity alteration coming from hydrogen. with the second route, solar photons are designed to drive photo biological photochemical reactions as well as such time, these kinds of items in which are fuels: It is generally mimic, what plants in addition to creatures do. Alternatively, one's semiconductor material can be employed being an anode for absorbing sunlight decrease your own photo electrochemical cells, that will usually additionally quickly create hydrogen through decomposing water (Harvey, L.D.D., 2006). at the third route, your current solar heat via high temperature (as considered with any kind of ground station receiver CSP-receiver) for driving a endothermic chemical good effect of your gasoline can be used. Next, your own reagents will probably include combinations connected with water, CO2, coal, biomass and also gas. Products, that will generally constitute your own photovoltaic fuel, is actually almost any (or combinations) of an following: hydrogen, synthesis gas, along with methanol and also dimethyl ether synthesis oil. When the fossil fuel is consumed to be a reagent, your calorific c values exceed anyone solutions within reagents, so The item less fossil fuel must be burned pertaining to energy release similar. Solar fuel additionally synthesized via hydrogen and also photovoltaic CO2 to write hydrocarbons compatible in existing energy infrastructure. Currently, solar fuel production is actually for the pilot plant stage. Pilot plants within the power variety over 300 in order to 500 kW end up being produced regarding reducing thermal carbon particularly in relation to zinc oxide, sauna and steam methane reforming involving pet coke respect gasification's. The good 250-kW reactor steam reforming can be operating with Budgets (Harvey, L.D.D., 2006).

3.3.1 Photovoltaic Electricity Generation

In simpler terms, the published silicon semiconductor material, relative to the example, is positioned towards the sun. This paper, known as the cell, comprises two constituent layers by inserting impurities into the resulting silicon into any layer of type n that has a layer containing type p to form an impressive bonding interface. The cell's wonderful solar photons develop pairs of electrons that are separated across an internal electric field at its intersection. This creates negative charges using one side of the individual interface, plus optimistic charges are generally in the opposite direction. This separation from the resulting result creates tension. Provided that both sides of the personal illuminating cell are generally connected to the current filling flow on the same side of the unit by charging to the other side by generating electrical energy from the battery. Many optical technologies can be designed in parallel. For the meter, the available photovoltaic knowledge will include chip-based crystalline silicon technology and the Indian Copper Film / Indian Gallium Delphi Gallium Delphi / CIGS Silicon ion (CdTe). In addition to microcrystalline unlimited silicon), along with sharp color-based solar panels, with a significant, tangible expansion of the PV concepts of market concentration, through which very highly efficient tissue (such as materials based on equal gallium arsenide (GaAs)) corresponds to the current concentration of their window focus, as well as to additional complications in relation to, for example,

Fresnel lenses. In addition, multi-crystalline PV chips (sometimes called "polycrystalline" silicon), including tape developments, will be the dominant innovation in photovoltaic businesses, which contain a portion of the 2009 cake of about 80% (IEA, 2007).

The main deficiencies of EFI made with different types of mobile or portable devices include a large amount connected to more than 40% dedicated to cells depending on GaAs concentration, 25% of your knowledge about monocrystalline, 20% compared to polycrystalline and also cigs, 17% while It relates to cdte, in addition to belonging to 10% involving amorphous silicon. In general, clusters with respect to cells are mounted side by side on a good (usually glass) transparent sheet and also connected along the sequence to help the application form a "module" within the dimensions, all advice on how 1m by 1m (E. Martinot, 2010). When looking at deficiencies in efficiency, these are signs of identification between mobile science and perhaps portable science (mentioned above) in addition to efficiency unit science; usually the last age is 50 to be 80% of the first. Manufacturers are still able to improve performance, reduce service charges during automation, and process cells faster and at lower cost, alongside high-performance producers. Usually any module performance is guaranteed by companies for you up to 20 years for 30 years. Take advantage of photovoltaic energy so that the beneficial energy consists of more than your own cells and units; the designated photovoltaic system is often, for example, the author of its inverter to convert its correct current from a cell so that the capacitance associated with alternating current is to be compatible with recognized networks In addition to devices. For programs outside of your current network, the current software will definitely include storage models like batteries (Humayun.et.al, 2020). The deployment of this type of device is recorded more reliably, reducing costs and extending the useful life to be comparable within units such as agricultural solar power systems and is usually classified as some of the most important types: a network along with a connected network. Grid methods are grouped in two ways: distributed and also centralized. The distributed system consists of these types of small local resistance stations, a series of electricity supply mainly for the on-site good consumer, in addition to the remaining electricity for the grid. In addition, the central seed technology has certain functions as an expansion force. Frames outside of the tire are usually intended to aid an incredible individual or small meeting in terms of customers and, in most cases, an electrical storage element or even the same way requires backup power. This kind of system has great potential inside the non-electrified parts. Today's Current Solar Energy for Power Generation: Concentrated solar technologies designed electrical energy with sunlight to be able to heat the environment (Gevorkian, P., 2012). After that, the article will be moved directly or indirectly into a large thermal engine without problems running good electrical turbines. CSP effortlessly uses the portion of its ray attached to the sun's rays, so its maximum benefit tends to help provide a good small geographic range. Your current focus center provides sunlight to reach a good point (focal point) at any time consumed in all systems within the central receiver or it may be directed to the line (line focus) whenever considered within programs or perhaps a linear minimum of Fresnel. The same technology will be used in an additional case to drive the processes of refractory materials to produce energy, as defined below. The extended lines of the condenser valley in relation to your parabolic reflectors that track your current movement of sunlight focus solar radiation within about 75 to help 100 cases of the heat collector component (HCE) more beneath the focal point reflector. HCE can be black inner tube like any selective surface. In addition to an external glass tube that has a distance between only two evacuees. In current commercial designs, heat change in oil is circulated through steel tubes as its temperature rises (at approximately 400 ° C), but systems using heat transfer from other materials such as present or direct sodium vapor washing (Gevorkian, P., are shown). 2012). the second type includes the approach line system, the Fresnel linear mirror method, in addition to extended parallel strips obtained using the cube, again, using a fixed linear device. A true central receiver (also known as the "Power Tower" method), which is the focus of two systems, uses a large crowd connected by mirrors (heliostats) in the grass, and each sun tracks two axes to help focus the current sunlight on a point, as well as in a large high tower, From the central level to the large receptor, which includes a fixed inverted cavity and / or tubes, the fluid circulates through the heads of the heat transfer teeth The high temperatures (to at least 000 ° C only) that are associated with the current line approach can allow, allowing The motors for you convert heat "at least inside theory" into additional heat in the combined energy of the plate, a single parabolic reflector (as opposed to a large number of reflectors), can be used to keep track of solar energy within two axes of focus. The plate is usually created in a solar receiver. It is really unstable, but it moves from the plate, not greater than the length of the plate. Temperatures within the actual receiver motor may potentially reach speeds of up to 900 ° C in the popular notion of individual perception, the Sterling engine that drives the generator physically installed for the utility approach. Sterling plate devices are relatively small, usually building 10-20 kilowatts, but they may be overly configured to recognize a higher throughput terminal such as the main station (Gipe, P., 1995). Many of the four involving many CSP plants have advantages and disadvantages. The four are continuing to be developed as shown. The mouse click feature of CSP technologies (except for dish belonging) may be able to purchase thermal energy after being combined with this receiver and also before reaching high engine temperature (K. Hussain.et.al, 2019). The storage media taken includes molten salts, compressed air, along with steam accumulators (for exclusive short-term storage), elevated temperature for solid porcelain, phase change resources, graphite, and the original high temperature. Commercial solar thermal power plants are generally manufactured in thermal storage functions to achieve 15 hours, allowing manageable CSP to manage energy (Gordon, J., 1987).

3.1.2 Estimations For Photovoltaic Units

A photoelectric mobile is a semiconductor to alternate the sparkling pressure into electric power. This calls for electrons inside the valence band to transport to the conduction band. The exchange of electrons may be imagined if the power acquired as light is extra than incredible or equivalent to the width of the forbidden band that differentiates the valence and conduction companies such as the hollow in the strip. The main stage of photoelectric change is the absorption of the resulting light in a comfortable frightening span. The reflected light cannot be used for photoelectric switch. There are different drugs to expand absorbency. The texture, i.e. matter, mechanics, lasers, etc., produces pyramidal distances on the surface of the phone. This method increases the touch floor of the axis and reduces reflection. The reflected reflective columns can be reabsorbed via protecting the spaces. Reflection calamities are minimized by covering the surface with an anti-reflective coating. If an aluminum cover is included on the back of the phone, the photons sent in the range can be reflected as they can be accommodated. The suitability of the distinctive drugs can be evaluated by measuring reflection, permeability and absorption of the photovoltaic cells. The image in Figure 3 and 4 gives an idea of the various stages involved in the planning of optical silicon cells. The items shown are an untreated silicon wafer, a silicon wafer, a silicon wafer after steroids / treated with an anti-reflective sheath and a full optical cell with its conductive frame printed on the silk screen (H. J. Queisser and J. H. Werner, 1995).

Figure 2. Wafer of untreated silicon, wafer of textured silicon, wafer of silicon after doping/treatment with an antireflective covering, completed photovoltaic unit (H. J. Queisser and J. H. Werner, 1995).

Estimations for Silicon Units

We measure the permeability and reflection of a sheet of untreated silicon and escaping the tight silicon after steroids and medications with an anti-reflection cap, see Figures 3 and 4.

The crack quality can be determined from the transmittance range. The relationship between the wavelength of light and its associated power has been reported as an accompaniment:

$$E = \frac{(h^* c)}{\lambda}$$

Figure 3. Transmission of silicon at untreated and after medicine (T. Maruyama, Y. Shinyashiki and S. Osako, 2011)

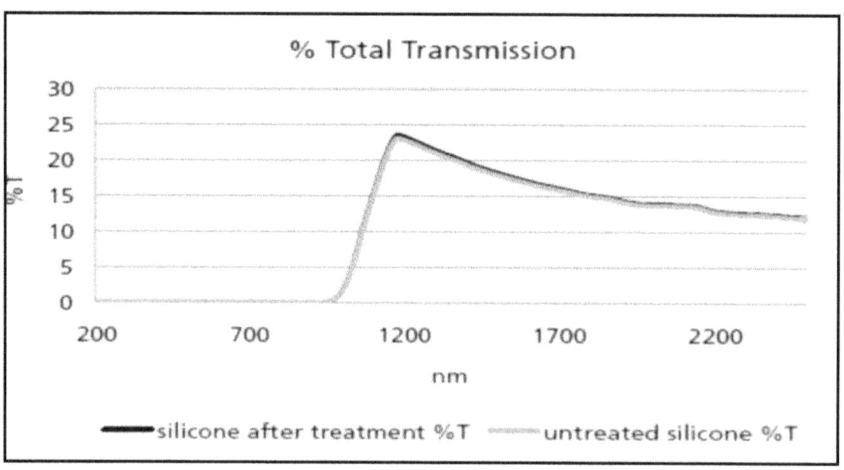

Figure 4. Reflex particle of silicon at untreated and after medication (T. Maruyama, Y. Shinyashiki and S. Osako, 2011)

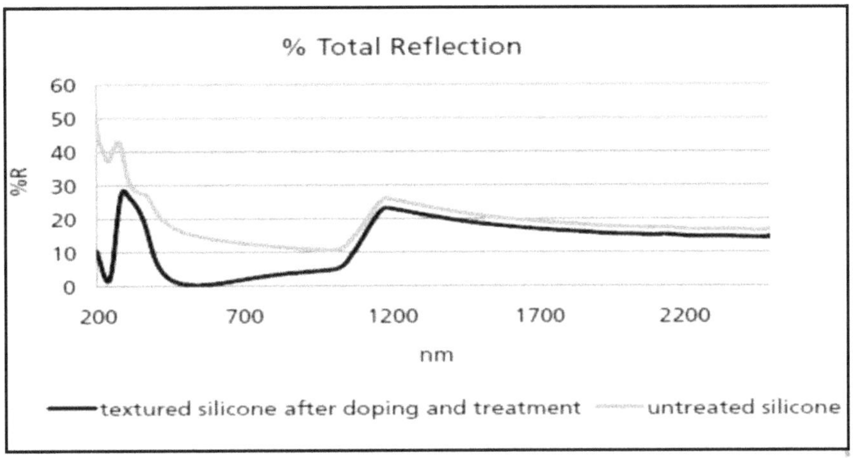

Where,

E= energy in joules
h= Planck constant for the value is 6.63 10^{-34} joules/sec
л= wavelength
c= speed of light in a vacuum for the value is »3×10^8 m/s (T. Maruyama, Y. Shinyashiki and S. Osako, 2011).

3.1.3 Other Renewable Sources

The Load separation is a critical process for achieving high efficiency in organic photovoltaic cells. The primary pair of secreted electrons must be closely separated at a sufficient speed to avoid the generation of the photoelectric current, thus losing energy conversion efficiency by suggestive recombination (Soobia Saeed, 2016). This process is done with the help of countries in transition that are between the exciting initial case and the free shipping status. Due to the rotational preservation rules, these intermediate load transfer cases usually have an individual character. Here we suggest a model for inorganic photovoltaic cells in which the separation process is formed by a flat convex lens that adjusts energy levels (Soobia Saeed.et.al, 2016). The effect of a flat convex lens is to intensify the generation of charge transfer cases with cross between systems. Since the basic condition of the system contains a whimsical monkey, triple states are protected by recombination, which increases the probability of a successful pregnancy separation. Using the modalities of open quantum systems (Soobia Saeed.et.al, 2016), we show that the population of triple-charge states grows in the presence of a convex lens and discuss the effect on the population of the vectors, and consequently on the photovoltaic stream, while highlighting its potential as a tool for researching the kinetics of the load transfer in these complex systems. The inorganic solar panel uses two different techniques, the first is the prism in a dark room with the help of a yellow bulb and the other is a flat convex lens. There are two layers on top of PVC are the prism lens and flat convex lenses. The prism strengthens the light signal (we use five different prism bands and each band has

Table 1. Reading about the Self-recuperating solar cell for day light without convex lens

S.NO	CURRENT	VOLTAGE
12.00PM	0.071	18.22
1.00PM	0.068	18.13
2.00PM	0.052	17.73
3.00PM	0.015	16.60
4.00PM	0.014	16.38

a different wavelength) while the lens picks up light from the coverage area. In other words, it is the concentrated energy source, which improves the photon emission rate. The mathematical relationship of the open circuit voltage is as follows (soobia saeed.et.al, 2015).

3.1.4 Reading About The SELF-Recuperating Solar Cell For Day Light Without Convex Lens

Now we apply the self-recovery device (blower at 1100 ° C) in a photocell in the presence of sunlight without a flat convex lens and take the readings with the help of the multimeter of the current and voltage listed below:

Reading about the Self-recuperating solar cell for day light with convex lens

We apply the self-recovery device (blower at 1100 ° C) to a photocell in the presence of a solar light with a flat convex lens and we take readings with the help of the multimeter of the current and voltage detailed below:

Reading about the Self-recuperating solar cell for day light with convex lens for current:

```
Time                 mille amp
12.00PM                   0.080
                      -0.009
```

Figure 5. Reading about the Self-recuperating solar cell for day light without convex lens

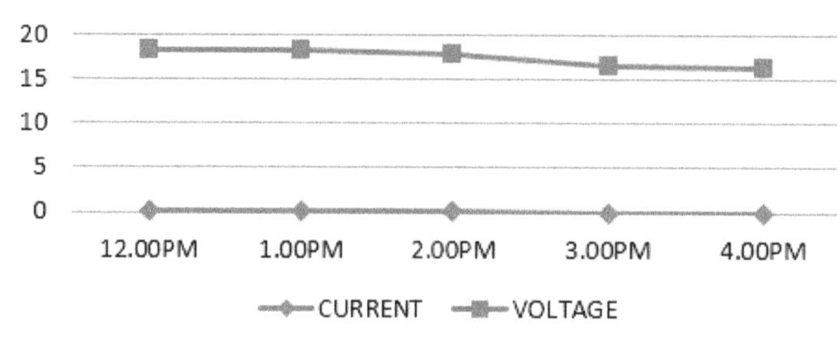

Table 2. Reading about the Self-recuperating solar cell for day light with convex lens

S.NO	CURRENT	VOLTAGE
12.00PM	0.080	18.25
1.00PM	0.071	18.14
2.00PM	0.054	17.78
3.00PM	0.016	16.64
4.00PM	0.015	16.40

```
1.00PM                 0.071                 -0.008
                         -0.017                 -0.013
2.00PM                 0.054                 -0.021
0.071
                         -0.038                  0.058
3.00PM                 0.016                 0.037
                         -0.001
4.00PM                 0.015
```

Reading about the Self-recuperating solar cell for day light with convex lens for voltage:

```
Time                          mille amp
12.00PM          18.25
                            -0.11
1.00PM                    18.14                   -0.25
                            -0.36                    -0.53
2.00PM          17.78               -0.78
2.21
                            -1.14                    1.68
3.00PM                    16.64                   0.9
                            -0.24
4.00PM                    16.4
```

Figure 6. Reading about the Self-recuperating solar cell for day light with convex lens

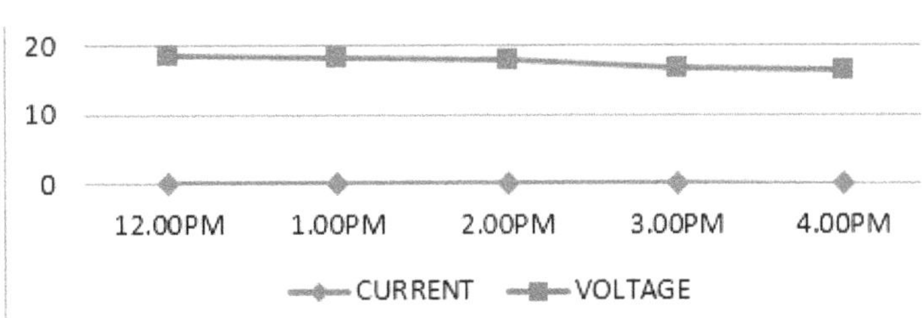

Table 3. Reading about the nail paint remover on solar cell for day light without Plano convex lens

S.NO	CURRENT	VOLTAGE
12.00PM	0.092	18.25
1.00PM	0.080	18.18
2.00PM	0.052	17.69
3.00PM	0.015	17.77
4.00PM	0.013	16.14

Photon rate of emission: The interpolation difference of these systems is [current: 0.006] and [voltage: 0.15]. The mathematical relationship is as follows. Conventional PVC minus proposed PVC for current is: Ñ I= 0.071 - 0.065 = 0.006 and voltage is Ñ I= 2.21- 2.06= 0.15.

Reading About The Nail Paint Remover on Solar Cell For Day Light Without Convex Lens

We apply nail polish remover in the photoelectric cell in the presence of sunlight without a flat convex lens and take the readings with the help of the multimeter for the current and voltage listed below:

Reading about the nail paint remover on solar cell for day light without convex lens:

```
Time                    mille amp
12.00PM        0.092
                         -0.012
1.00PM                 0.080                  -0.016
                         -0.028                    0.007
2.00PM                 0.052                  -0.009                    -0.033
                         -0.037                    -0.026
3.00PM                 0.015                  0.035
                         -0.002
4.00PM                 0.013
```

Figure 7. Reading about the nail paint remover on solar cell for day light without Plano convex lens

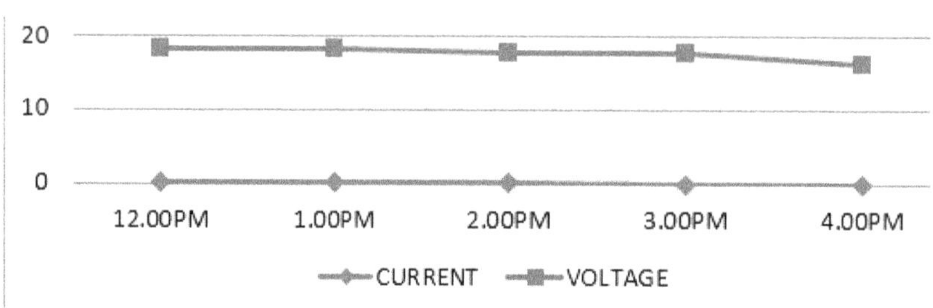

Table 4. Reading about the nail paint remover on solar cell for day light with Plano convex lens

S.NO	CURRENT	VOLTAGE
12.00PM	0.096	18.32
1.00PM	0.088	18.19
2.00PM	0.053	17.70
3.00PM	0.015	17.80
4.00PM	0.014	16.15

Reading about the nail paint remover on solar cell for day light without convex lens for voltage:

Time mille amp
12.00PM 18.25
-0.07
1.00PM 18.18 -0.42
-0.49 1
2.00PM 17.69 0.58
-3.28
0.09 -2.28
3.00PM 17.77 -1.7
-1.61
4.00PM 16.14

Reading about the nail paint remover on solar cell for day light with convex lens

We apply nail polish remover in the photoelectric cell in the presence of sunlight with flat convex lenses and we take readings with the help of the multimeter of the current and voltage shown below:

Reading about the nail paint remover on solar cell for day light with convex lens for current:

Figure 8. Reading about the nail paint remover on solar cell for day light with Plano convex lens

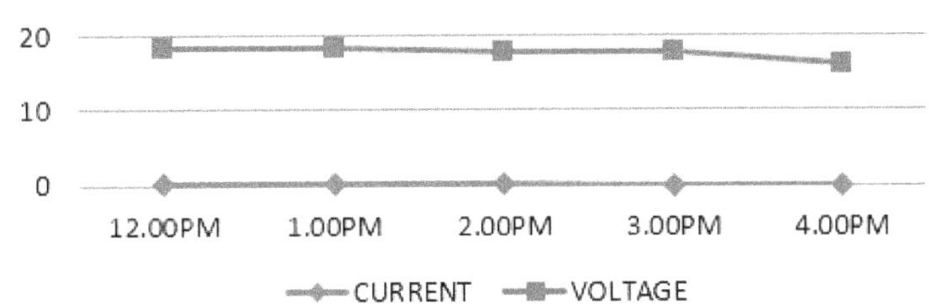

Time mille amp

12.00PM 0.096

-0.008

1.00PM 0.088 -0.027

-0.035 0.024

2.00PM 0.053 -0.003

0.016

-0.038 0.04

3.00PM 0.015 -0.037

-0.001

4.00PM 0.014

Reading about the nail paint remover on solar cell for day light with convex lens for voltage:

Time mille amp

12.00PM 18.32

-0.13

1.00PM 18.19 -0.036

-0.49 -0.95

2.00PM 17.70 -0.59 -3.29

0.1 -2.34

3.00PM 17.80 -1.75

-1.65

4.00PM 16.15

Photon rate of emission: The interpolation difference of these systems is [current: 0.049] and [voltage: -0.01].The mathematical relationship is as follows. Conventional PVC minus proposed PVC for current is: Ñ I=0.016 – (-0.033) = 0.049 and voltage is Ñ I= -3.29- (-3.28) = -0.01.

Reading about the nail paint remover with Self-recuperating (blower at 1100°C) solar cell for day light without convex lens

We apply nail polish remover with automatic recovery in the photoelectric cell in the presence of sunlight without a flat convex lens, and we take the readings with the help of the multimeter of the current and voltage detailed below:

Reading about the nail paint remover with Self-recuperating solar cell for day light without convex lens for current:

Time mille amp

12.00PM 0.064

Table 5.

S.NO	CURRENT	VOLTAGE
12.00PM	0.064	18.28
1.00PM	0.069	18.02
2.00PM	0.051	17.66
3.00PM	0.045	17.89
4.00PM	0.012	15.85

0.005
1.00PM 0.069 -0.023
-0.0018 0.035
2.00PM 0.051 0.012 -0.074
-0.006 -0.039
3.00PM 0.045 -0.027
-0.033
4.00PM 0.012

Reading about the nail paint remover with Self-recuperating solar cell for day light without convex lens for voltage:

Time mille amp
12.00PM 18.28
0.26
1.00PM 18.02 -0.1
-0.36 0.69
2.00PM 17.66 0.59 -3.55
0.23 -2.86

Figure 9. Reading about the nail paint remover with Self-recuperating solar cell for day light without Plano convex lens

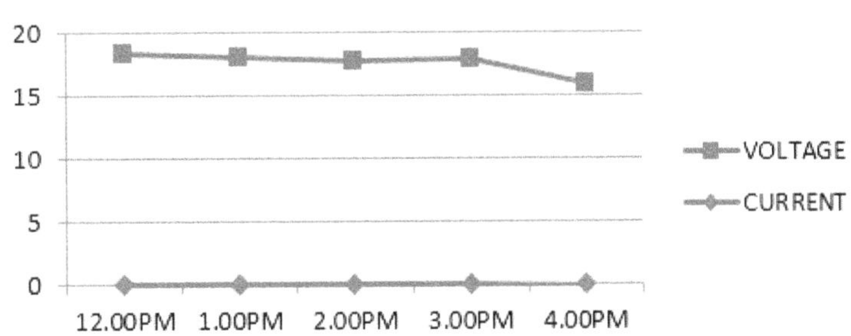

Table 6.

S.NO	CURRENT	VOLTAGE
12.00PM	0.075	18.32
1.00PM	0.078	18.06
2.00PM	0.054	17.71
3.00PM	0.048	17.92
4.00PM	0.013	15.87

3.00PM 17.89 -2.27

-2.04

4.00PM 15.85

Reading about the nail paint remover with Self-recuperating (blower at 1100°C) solar cell for day light wi convex lens

We apply nail polish remover with automatic recovery in the photoelectric cell in the presence of sunlight with a flat convex lens, and we take the readings with the help of the multimeter of the current and voltage detailed below:

Reading about the nail paint remover with Self-recuperating solar cell for day light with convex lens for current:

Time mille amp

12.00PM 0.075

0.003

1.00PM 0.078 -0.027

-0.024 0.045

2.00PM 0.054 0.018 0.002

Figure 10. Reading about the nail paint remover with Self-recuperating solar cell for day light with convex lens

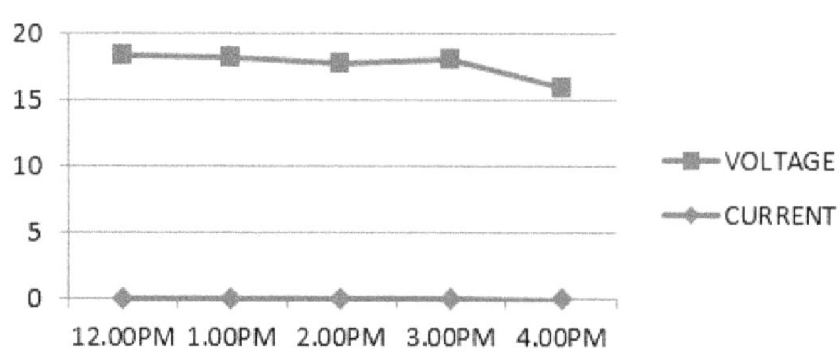

Table 7.

S.NO	CURRENT	VOLTAGE
12.00PM	0.068	18.03
1.00PM	0.088	17.97
2.00PM	0.073	17.50
3.00PM	0.062	17.59
4.00PM	0.034	17.36

-0.006 0.047
3.00PM 0.048 0.029
-0.035
4.00PM 0.013

Reading about the nail paint remover with Self-recuperating solar cell for day light with convex lens for voltage:

Time mille amp
12.00PM 18.32
-0.26
1.00PM 18.06 -0.09
-0.35 0.65
2.00PM 17.71 0.56 1.05
0.21 1.7
3.00PM 17.92 -2.26
-2.05
4.00PM 15.87

Photon rate of emission: The interpolation difference of these systems is [current: 0.94] and [voltage: 4.6].The mathematical relationship is as follows. Conventional PVC minus proposed PVC for current is: Ñ I= 0.02– (-0.074) =0.094 and voltage is Ñ I= 1.05- (-3.55) = 4.6.

Reading about the prism on solar cells for day light without convex lens

We apply the prism in a photovoltaic cell in the presence of sunlight without a flat convex lens and we take readings with the help of a multimeter of current and voltage given below:

Reading about the prism on solar cells for day light without convex lens for current:

Figure 11. Reading about the prism on powered cells for day light without convex lens

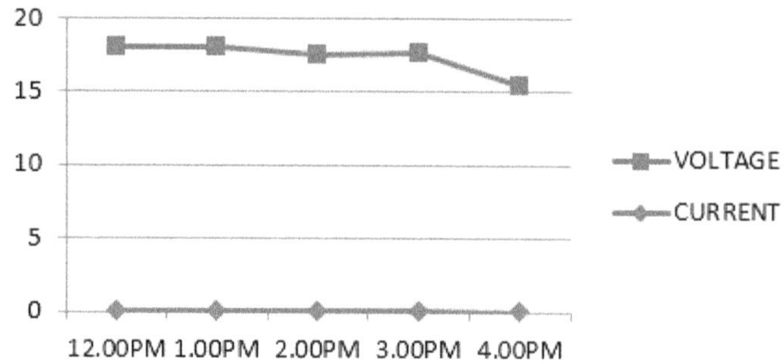

Time mille amp
12.00PM 0.068
0.02
1.00PM 0.088 -0.035
-0.015 -0.039
2.00PM 0.073 0.004
-0.059
-0.011 -0.02
3.00PM 0.062 -0.016
-0.027
4.00PM 0.034

Reading about the prism on solar cells for day light without convex lens for voltage:

Time mille amp
12.00PM 18.03
-0.06
1.00PM 17.97 -0.41
-0.47 0.97
2.00PM 17.50 0.56 -1.85
0.09 -0.88
3.00PM 17.59 -0.32
-0.23
4.00PM 17.36

Reading about the prism on solar cell for day light with convex lens

We apply the prism in the photoelectric cell in the presence of sunlight with a flat convex lens and we take readings with the help of a multimeter of current and voltage listed below:

Table 8.

S.NO	CURRENT	VOLTAGE
12.00PM	0.070	18.06
1.00PM	0.092	17.99
2.00PM	0.075	17.51
3.00PM	0.064	17.60
4.00PM	0.037	17.37

Reading about the prism on solar cell for day light with convex lens for current:

```
Time                     mille amp
12.00PM                 0.070
                          0.022
1.00PM                 0.092                  -0.039
                          -0.017                      0.045
2.00PM                 0.07
0.006                 -0.067
                          -0.011                  -0.022
3.00PM                 0.064                  -0.016
                          -0.027
4.00PM                 0.037
```

Reading about the prism on solar cell for day light with convex lens for voltage:

```
Time                     mille amp
12.00PM                 18.06
                          -0.07
```

Figure 12. Reading about the prism on solar cell for day light with convex lens

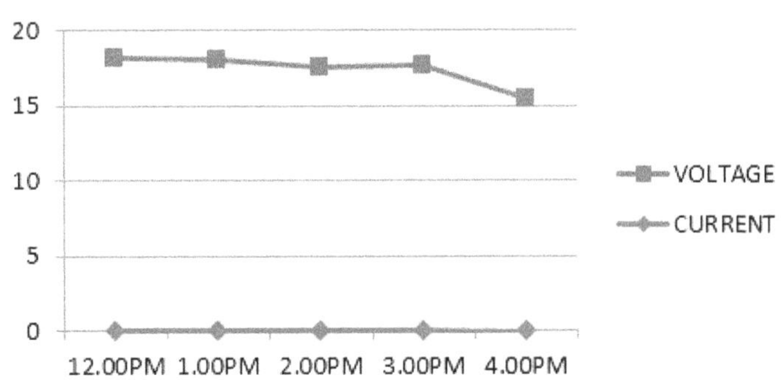

Table 9.

S.NO	CURRENT	VOLTAGE
1	0.068	17.49
2	0.065	17.45
3	0.064	17.43
4	0.062	17.40
5	0.059	17.38

```
1.00PM                  17.99                -0.41
                          -0.48                0.95
2.00PM                  17.51                0.057
-1.87
                          0.09                 -0.89
3.00PM                  17.60                -0.32
                          -0.23
4.00PM                  17.37
```

Photon rate of emission: The interpolation difference of these systems is [current: -0.008] and [voltage:-0.02].The mathematical relationship is as follows. Conventional PVC minus proposed PVC for current is: Ñ I= -0.067-(-0.059) =- 0.008 and voltage is Ñ I= -1.87- (-1.85) = -0.02.

Reading About The Coal on Solar Cell for Day Light Without Convex Lens

We apply the carbon in the photovoltaic cell in the presence of sunlight without a flat convex lens and take readings with the help of the multimeter of current and voltage listed below:

Reading about the coal on solar cell for day light without convex lens for current:

```
Time                       mille amp
1              0.068
                            -0.003
```

Figure 13. Reading about the coal on solar cell for day light without convex lens

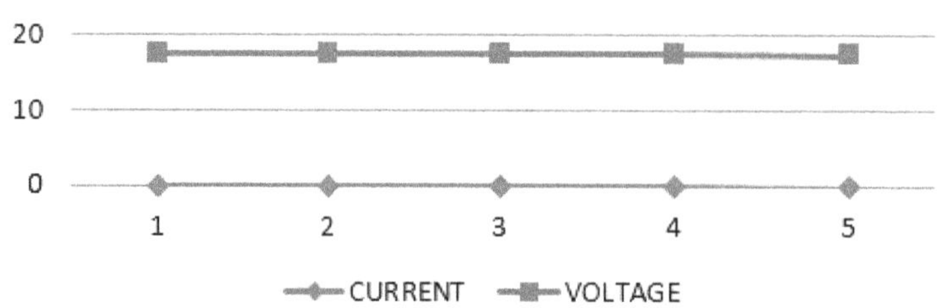

Table 10.

S.NO	CURRENT	VOLTAGE
1	0.070	17.56
2	0.069	17.51
3	0.068	17.50
4	0.065	17.46
5	0.063	17.43

2 0.065 0.002

-0.001 -0.003

3 0.064 -0.001

0.003

-0.002 0

4 0.062 -0.001

-0.003

5 0.059

Reading about the coal on solar cell for day light without Plano convex lens for voltage:

Time mille amp

1 17.48

-0.04

2 17.45 0.02

-0.02 -0.03

3 17.43 -0.01

0.05

-0.03 0.02

4 17.40 0.01

-0.02

5 17.38

Reading about the coal on sun cellular for day light with convex lens

We apply the carbon in the photovoltaic cell in the presence of sunlight with flat convex lenses and we take readings with the help of a multimeter of current and voltage shown below:

Reading approximately the coal on solar cell for day with convex lens for current:

Figure 14. Reading about the coal on solar cell for day light with convex lens

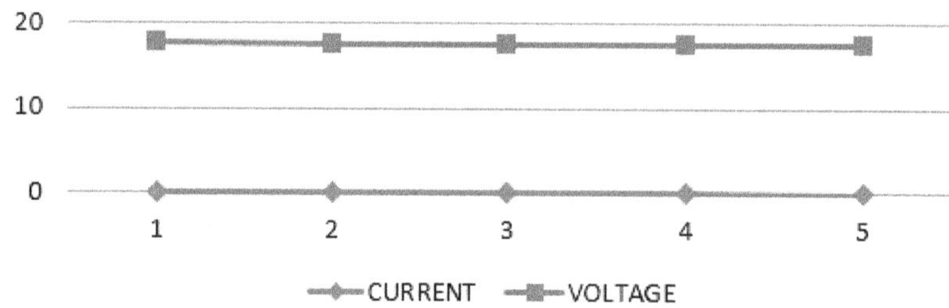

Time		mille amp			
1	0.070				
		-0.001			
2	0.069		0		
		-0.001		-0.002	
3	0.068		-0.002		0.005
		-0.003		0.003	
4	0.065		0.001		
		-0.002			
5	0.063				

Reading about the coal on solar cell for day light with convex lens for voltage:

Time		mille amp			
1	17.56				
		-0.05			
2	17.51		0.04		
		-0.01		-0.07	
3	17.50		-0.03		0.11
		-0.04		0.04	
4	17.46		0.01		
		-0.03			
5	17.43				

Photon rate of emission: The interpolation difference of these systems is [current: 0.002] and [voltage: 0.06].The mathematical relationship is as follows. Conventional PVC minus proposed PVC for current is: Ñ I= 0.005 - 0.003 = 0.002 and voltage is Ñ I= 0.11- 0.05 = 0.06.

Economic Impacts of Photovoltaic Cell Overall The World

Although capacitance additions remained solid in 2018, solar photovoltaic power generation improved via 31% in 2018 and represents the very best absolute technology growth (+136 terawatt hours) of all renewable technology, barely in advance of wind and hydropower. Despite recent coverage modifications and uncertainties in China, India, and the US, the competitiveness of sun photovoltaic has stepped forward. Solar PV continues to be heading in the right direction to attain anticipated levels within the SDS, in order to require an average annual boom price of sixteen% among 2018 and 2030. Despite an extensive drop within the marketplace in China, because of the stop of strength fees in June 2018, the new established capacity of sun PV the world over expanded by way of 7% to almost 107 GW, regardless of an annual funding lower of twenty-two%. For the ninth consecutive 12 months, sun strength attracted most of the new investments in renewable power. Investments amounting to 140 billion greenbacks (122 billion euros) in solar energy amounted to forty two. Government and commercial spending on studies and development increased by way of nearly 19% from $ eleven.8 billion (10.Three billion euros) to $ 14 billion (12.2 billion euros. Despite the urgency of accelerating the electricity transition, this equates to handiest 11.5% of Total power studies and improvement costs of US $ 121 billion (one hundred and five billion euros). The tendency of growing economies to invest more in renewable electricity is more The improvement persisted for the fourth year of investments of 140 billion greenbacks (122 billion euros) in sun energy, fifty four% or seventy five billion dollars (sixty five billion euros) have been invested in growing economies The photovoltaic enterprise has grown considerably in current years, China has end up a frontrunner in cells and sun units, observed by Taiwan and Malaysia: if Europe wants to cut its greenhouse gasoline emissions by way of half through 2030, the carbon removal technique is rapid with the aid of the energy zone via eighty%. Additional PV electricity may be among 75 and 350 GW, main to a complete restriction of 1,000 in capacities to 825 GW within the business scenario as typical, important in the European Union via 2030. This increase in sun photovoltaic power will require increase within the European Union market via approximately 14 GW in 2019 to 70 GW or greater in 2030. Currently, the manufacturing ability of sun cells and solar units inside the European Union is only 1 GW and 3 GW Respectively. From the point of view of deliver protection, the big increase in centers needs to be observed through practical local production within the union to avoid interruption of elements. The fast cost reduction in PV panel's production and the extra abilities required deserve a sparkling have a look at the potential to convey PV vegetation to Europe. The particular investment expenses required for photovoltaic units and cell manufacturing units have decreased by means of almost ninety% over the past ten years. A latest take a look at by way of the Fraunhofer Institute for Solar Energy Research (Fh ISE) commissioned by means of the German Society for the Mechanical Engineering Industry (VDMA) confirmed that the European manufacturing chain may be aggressive if sun factories with a production volume can do among five and 10 GW according to year. In 1990, the implementation of the one hundred, 000-bishop software in Germany, and the Japanese long-time period method installed in 1994, with the horizon of 2010, marked the start of incredible increase in the photovoltaic marketplace. Prior to the start of the Japanese Market Implementation Program in 1997, annual boom prices were round 10% within the photovoltaic market, pushed mainly by telecom systems, business systems, and autonomy. Since 1990, PV production has improved with the aid of 3 orders through volume, from forty six MW to over 100 GW in 2017. This corresponds to TCAC for over forty%. In 2018, amassed statistically documented facilities global formed round 520 GW driven by way of the continuing low value of sun photovoltaic electricity generation, and the range and length of latest photovoltaic markets international are growing. An increasing wide variety of massive investors are constantly increasing their investments in renewable electricity and sun photovoltaic strength, including Warren Buffett, or maybe no longer making an investment in fossil power corporations and converting this investment to renewable power, as the Rockefeller Brother Fund announced via the 2014. Concerning the destiny development of the renewable power marketplace and sun photovoltaic specially, the professional release of an alliance of nations and countries devoted to shifting the world of coal burning to cleaner power sources changed into a promising signal at COP23 59 in Bonn 2017. In the founding assertion, the Coalition says from the awesome past: "To comply with the Paris Agreement, the evaluation indicates that coal elimination is necessary no later than 2030 inside the Organization for Economic Cooperation and Development and the European Union 28, and not after 2050 within the rest of the arena". In October 2019, the alliance had ninety one contributors representing 32 national governments, 25 subnational governments, and 34 groups or organizations. Along with the general boom in power costs and the need for weather balance, this may keep maintaining demand for excessive sun structures. In the longer term, PV boom fees will retain to upward thrust, even if financial conditions fluctuate regionally and lead to a brief-time period slowdown in some markets. This view is shared by means of increasingly economic institutions turning to renewable strength as a sustainable and solid investment over the long time. Even if oil costs pulled returned from their historic highs in 2008, when they were close to $ a hundred and fifty a barrel, their capacity volatility because of deliver disruptions because of political uncertainty or natural failures creates a monetary risk. In addition, the fact that the Paris Agreement entered into force on November four, 2016 offers an additional improve to decarbonizing our power supply and draws extra investments in renewable PV, mainly. Several studies on fuel, gas and electricity subsidies have been posted considering that 2010. A running report of the International Monetary Fund for 2019 confirmed that these publish-tax subsidies extended from $ three.6 to $ five.2 billion, or 5.Four to 6.5% of GDP, among 2010 and 2017. Over 95% of those subsidies went to coal, oil and herbal fuel. These numbers are extra than 10 times the amount supplied through the International Energy Agency in World Energy Outlook 2019 as direct support for fossil fuels ($ 319 billion in 2017). In 2010, the Financial Times mentioned Fatih Birol, chief economist on the International Energy Agency in Paris, pronouncing that eliminating subsidies became a coverage that would alternate the strength sport "quick and dramatically". "I see guide for fossil fuels like appendicitis in the global electricity machine that ought to be removed 60 after making big fee discounts for the technical additives of PV structures, which include devices, inverters, BOS, and many others. Reducing the concessional fees of PV system centers, which includes permits and financing fees. Although the components of the PV gadget are global products, the actual prices for installed PV structures range substantially inside the 1/3 region of 2019, the average gadget charge for residential systems changed into round A $ 1.03 / Wp (zero.65 € / Wp), at Australia, but about $ 3.30 a Reiki (three.00 € / kW within the US). Competition and increasingly more skilled stabilizers reduce prices even extra. Some international locations, such as Germany or Italy, already set up PV ability exceeds 30% and 20% of the mounted capacity of the thermal power plant, respectively. Along with the respective wind abilties, wind and sun power collectively will exceed 60% and 30%, on Straight. To address those excessive fees of renewable energy, new technical and regulatory answers have to be applied to keep away from facing the problem of decreasing massive portions of this power. In addition to traditional pump storage options, electric powered batteries are gaining expanded interest, particularly for small extent garage answers in a low-voltage distribution community. According to BNEF, battery charges reduced 85% among 2010 and 2018. By 2025, battery prices may additionally drop beneath one hundred USD / kWh. The common range of cycles of a lithium ion battery is 5,000 cycles, and with preceding cost estimates, this will allow the level of electricity storage costs to be much less than US $ 0.05 / kWh (0.1/2 € / kWh) via 2025. Already in February in 2012, BYD (Build Your Dreams) and China State Grid Corporation (SGCC) have finished the construction of a large-scale public services mission placed in Zhangbei, Hebei Province, which combines 100MW of wind energy,

40MW of sun photovoltaic machine and 36MWh Of lithium-ion strength storage. It is interesting to note that the batteries used in this gadget are Li-ion batteries for automobiles, which were previously used inside the BYD 36 for round 4000 cycles. The largest photovoltaic flowers with masses extra than 50 hours with a partial storage space of as much as four hours are surely pretty not unusual. According to investment analysts and enterprise forecasts, solar strength will continue to grow at excessive fees in the coming years. Various associations within the photovoltaic industry, as well as Greenpeace, the European Renewable Energy Council (EREC), the Energy Monitoring Group with the Lappeenranta University of Technology (LUT), Bloomberg New Energy Financing (BNEF), the International Energy Agency and evolved the International Renewable Energy Agency (IRENA) IRENA), Scenarios for Future Growth of PV Systems . The special scenarios for the Greenpeace / EREC have a look at, Energy Watch Group / LUT research, BNEF New Energy Outlook (NEO) 2019 and IEA World Energy Outlook 2016 and 2018 scenarios. It is interesting to notice the expected PV amplitude in the IEA eventualities, the greater the increase Great from 2016 to 2018, however it is nevertheless at the bottom quit. Older scenarios may be located in preceding photoelectric reputation. With new facilities deliberate around the arena between 230 and 295 GW in 2019 and 2020, even a one hundred% energy area situation for 2020 is within reach. These projections show that there are great opportunities for PV within the destiny if the right coverage measures are taken, but we must undergo in thoughts that this improvement will not show up on its personal. The ongoing efforts and assist of all parties concerned could be important to put into effect the planned exchange to a sustainable electricity supply the usage of PV by using imparting an essential component. The fundamental boundaries to these traits are the belief, regulatory frameworks and regulations imposed on modern strength transmission and distribution structures. In light of the urgent need to cast off carbon from our energy and power an awful lot quicker than previously concept (IRENA, 2019).

In the closing ten years, we've now not best noticed a continuous boom in energy fees, but also greater volatility. This highlights the vulnerability caused by our contemporary dependence on fossil electricity assets, and will increase the burden that growing international locations face of their battle for future development. On the other hand, we are witnessing a continuous lower in the charges of manufacturing the resulting renewable energy and LCOE technologies, due to the industry's gaining knowledge of curves. It is vital to understand that simplest LCOE elements of photovoltaic come from in a single day funding expenses. Since external energy expenses are commonly not taken into consideration subsidies for classic strength risks and price volatility, renewables and photovoltaic are still much less mature within the marketplace than strength sources. Conventional risk charges ought to be paid for additional financing. Meanwhile, monetary costs, lets in, and administrative charges are extra related to the very last costs of photovoltaic. If access to financing is at the equal stage, the charges of the LCOE may also drop considerably. However, energy production from solar photovoltaic structures has already shown that it could be less expensive than residential consumer fees in a wide variety of nations. In addition, in evaluation to standard electricity sources, renewable energies continue to be the handiest ones that provide the opportunity of decreasing prices within the future in place of growing them(Kroposki, B., 2017).

The deployment of sun photovoltaic systems at the floor has extended considerably in recent years, thanks in big component to assist regulations, along with internet metering and tax incentives that make PV in a few markets greater attractive than the monetary outlook that buys power from the grid. Small hybrid optical networks, digital energy plants, and PPA service for the general public service. The competitiveness of dispensed sun energy is sincerely evident amid the growing proliferation in big markets, which includes Brazil, China, Germany and Mexico; but, there are important differences between international locations, highlighting the opportunity of in addition development. The sun photovoltaic enterprise will appoint greater than 18 million people by way of 2050, five times extra than the entire of three.6 million jobs in 2018. However, to maximize electricity transmission effects, a comprehensive policy framework is wanted. Implementation regulations must be coordinated and aligned with integration and qualification rules. Under the umbrella of permitting rules, a unique cognizance is needed on industrial, financial, academic policies and abilities to maximize the advantages of transition. Education and abilities policies can assist provide the body of workers with the ideal capabilities and boom neighborhood employment possibilities. Likewise, robust industrial regulations based totally on country wide deliver chains can enable earnings increase and employment possibilities by means of leveraging current financial sports to aid the development of the solar photovoltaic enterprise. This can only be done by way of easing current boundaries at various tiers (political, political, economic, technical, organizational, political, and social). Network integration, grid resilience, economies of scale, get right of entry to finance, lack of first-class standards and measures, and consumer focus are some of the primary boundaries that can preclude the deployment of solar photovoltaic talents inside the next 3 a long time. Immediately lessen existing limitations, thru a fixed of assist policies and implementation measures that encompass innovative commercial enterprise fashions and monetary equipment are necessary to drive the future deployment of sun photovoltaic skills to permit transition to the destiny low carbon and sustainable energy(Shravan, K., Chunduri, K. ,2019).

Solar PV Deployment Barriers and Their Solutions:

This record virtually shows that solar PV is one of the renewable strategic technologies had to implement the global energy transformation consistent with the weather goals of Paris. Technology is now to be had, and may be quick and broadly implemented as price-effective. However, notwithstanding the strong momentum, sun PV initiatives retain to stand serious constraints which could restrict the extended growth wished in the coming a long time. In general, renewables are to various ranges stricken by troubles arising from extraordinary undertaking specifications, geographical contexts, and maturity levels. For solar PV, limitations at one of a kind tiers (technological, financial, regulatory, normative, and socio-political) could avert the unfold of sun photovoltaic electricity within the subsequent three decades. Immediately lowering those barriers, through a hard and fast of help rules and implementation measures, is crucial to force implementation within the future (Soobia Saeed, 2015).

While power transformation can yield comprehensive socio-financial benefits at the global level, a deeper perception at the local degree highlights how the advantages (and costs) of transformation will have an effect on otherwise in unique components of the sector. These international locations with specific starting points for electricity transmission depths, strengths and diversity of supply chains, the diploma to which economies rely on fossil fuels) exceptional degrees of countrywide ambition and approach of implementation. The agreement of nearby and countrywide effects of energy conversion depends on a -the front coverage framework. The framework must encourage the deployment of renewable power to reduce power consumption and growth energy get charge to. At the identical time, diffusion of renewable energies must be incorporated into broader guidelines that make electricity a catalyst for inclusive, sustainable and economic growth. The "honest transition" coverage framework is based totally on three sets of transformational rules: implementation guidelines, integrative guidelines and allowing guidelines. It is crucial that these policy agencies paintings together to ensure a fair and inclusive energy transition, mainly to triumph over all the existing barriers (political, marketplace, monetary, technological, regulatory, political, and social). The device should be chosen and designed in keeping with the United States of America specific conditions and desires. Solutions and measures any more are broader and require particular examine that specialize in addressing handiest limitations inside the industry. This document provides an outline of a few key aspects that must be considered to accelerate the deployment of sun PV power alongside some success examples primarily based on IRENA's complete work in renewable energy to date, alongside an overview of extra literature (IRENA, 2019).

3.4 Geothermal Energy

Geothermal heat is usually an application form connected with energy created in addition to store at the Earth. That is still taken throughout history for the bathroom, heating and kitchen. Geothermal energy is made from radioactive corrosion, within achievement 4 temperatures, 000C with the core of any planet Earth. while geothermal energy is usually exhibited worldwide, This really is called carefully assessed the geothermal gradient The item indicates whether or not a good region is really a privileged place with regard to enactment. Rate at steps your own temperature increases to its reason how the depth of your planet earth increases. regarding example, the average geothermal gradient throughout France can be 4 ° C / 100 m within number 10 / 100m in the Alsace region inside Personal on a couple of ° C / 100 m inside the mountains of any Pyrenees. within Iceland plus the volcanic regions, your own gradient will probably reach Just as high As 30 ° C / 100 m (Lund, J.W., 1995). your own geothermal gradient is usually not lone your sole tool used to measure your availability involving geothermal energy. concerning permeability rocks, of which determines your rate regarding heat flowing on the surface; This can be acknowledged As another mouse clicks step about the availability connected with geothermal energy. Geothermal energy features an big advantage than wind in addition to solar energy This really is available from Whenever through the season. approximated CO2 emissions which are generated because of the geothermal resources tend to be 55 g / kWh make use of information from just about any statement regarding 73% associated with geothermal power plants. There is plausibility that the worth could be really decreased to zero if geothermal fluid is re-injected into the ground. A sum of 24 nations is at present being utilized as geothermal force plants. The majority of the world's geothermal limit (GW) can be introduced in eight global territories: United Claims (V. Lukosevicius, 2000). The Philippines (right around 2 0) furthermore Indonesia (February 1), South America (0 9), Italy (0 9), totally New Zealand (0 8), Iceland (0 7) and in addition Japan (0. 5) (BP, 2012). It genuinely is foreseen that geothermal ability to increment with advances in the co-generation of geothermal vitality, utilizing water through coal and oil wells all through 2015. Fridleifsson and Freeston (1994) researched the advancement with the geothermal vitality as per encounters from around the world, and are required to full speculation costs from 15 for you to 20 billion dollars in geothermal vitality worldwide over the next decade. Notwithstanding, this assessment enhanced five times in the IEA report, which appraises that the aggregate speculation expenses to get 104 billion dollars worldwide between 2010-2020 so you can meet the destinations perfect (IEA, 2007). Fridleifsson practice and Freeston, who view self as adequacy geothermal conditions diverse time furthermore capacity limit, geothermal vitality, could well be utilized by every base furthermore crest burden force plants. They guaranteed to which natural issues made by the freedom of steam furthermore gas, and additionally challenges from the stream, it could be affected by cutting edge building. To make utilizing innovation of geothermal vitality accessible in the nations that procure, troubles of deficiency of subsidizing and comprehension of the framework should be determined. Murphy and Niitsuma talk about techniques to balance the higher expenses of geothermal power as a result of attractive natural advantages. That they proposed actualizing financial approach measures to fortify the feasibility, including the usage of a carbon as opposed to weight and adaptation of some incredible advantages of geothermal vitality (Murphy, H., & Niitsuma, H., 1999). The Prompting Okazaki, japan, Indonesia and the Philippines as nations around the globe with high rates of geothermal increment, halfway because of bolster arrangements founded by its overseeing figures Utilizes a factual and information used in Iceland to help assess the speculation expenses taught to manufacture an innovation geothermal quality plant. The aggregate expense of procurement in geothermal vitality establishments inside of the scope of 20-60 MW, evaluated for 1 267 USD/ kW in an endlessly more known field and 1 440 USD/ kW in an obscure field(Reif, T., 2008). Assessed the general programming of geothermal power with information from 72 nations and their outcomes demonstrated that the utilization of cement 273, 372 TJ (terajoule)/ year regarding geothermal vitality, and expecting uncover 6. 06 × 109 J of one's for each clasp or barrel of oil could spare 128.9 trillion barrels of oil or 19, 2 million tons for every year. They additionally contended that extra staff may be a sparing of 2 million barrels (41. 6. 2 million tons) contrasted and fuel oil and 7, 000, 000 heaps of carbon outflows if the utilization of geothermal warmth pumps is alluded to as amid the cooling mode. Frick connected a specialized examination in the life cycle for enhanced geothermal quality era (EGS) having low temperature stores that will evaluate the impact of the land conditions inside of nature framework (Frick, & Schröder, G., 2010). The outcomes demonstrate the way that life cycle of a double geothermal force plant is subordinate upon the materials and vitality inputs. Consequently, the effective access in your base store boring is overweight to have minimal ecological effect. There is confirmation that maybe less positive geothermal warmth and power can bring about a procedure of vitality alongside enhance solidness. In a comparable study and Saner examined the essentialness utilization and outflows of nursery gasses after which you can connect an existence cycle appraisal to take a gander at the ecological effects as a result of area source savings introduced contrasted all together with traditional warming strategies. furthermore, explored the German dialect geothermal segment, with expand exposure from the significance associated with political assist, in addition to the general conditions from the electricity market (Purkus and also Barth, 2011) .They argued that despite the disadvantages associated with high investment costs almost no risk of heat from geothermal engineering, the advantage associated with flash could not allow him to become reliable base-load power. Kaya et is the reason. (2011) investigated the reinjection in geothermal fields with data via 91 plants (worldwide) found which geothermal and reinjection plan should reduce the possibility of contamination of groundwater. Results showed the response of different strategies for geothermal reservoirs reinjection determined through geothermal technology. In the most recent study, now the status from the (global) geothermal energy and found the high temperature technologies (Flash, plus dry steam) generation technologies incorporate most developed geothermal energy. These folks grow define four models, as well as 1FMP (single flash) 2FMP (double flash), 3FMP (triple flash) plus DSMP (dry steam), to manage various geothermal techniques. The results showed the DSMP model receives the best amount NPV and the difficulty TIR (22. 8%). In addition, it is estimated that the expenses associated with electricity to turn into a model (Nawaz, I., & Tiwari, G., 2006).

3.5 Hydropower

Hydropower makes use of energy which involves far more water to reduce elevations, especially to reach electricity. Pursuits include hydroelectric dams together with reservoirs river management, in addition to current jobs, as well as protecting a procession in the dimensions of the project. This assortment gives hydropower to meet up with urban desires large centralized in addition to decentralized non-urban desires. Hydro technologies are generally mature. Hydroelectric things to do ranging manipulate a company temporarily. By comparison, the controllable output given by hydroelectric facilities that may have reservoirs allows you to meet maximum power demands and turn into a help to balance the particular electrical systems that may have large portions relative age group RE variable. The function on hydroelectric reservoirs commonly reflects its many works by using, for illustration, water, irrigation in addition to Drought flood control and navigation, as well as thoroughly clean energy present(IEA (2010).

3.6 Ocean Energy

water power leads to the potential, kinetic, chemical energy along with power involving sea water, along with this really is transformed to provide the electricity, heat, or perhaps drinking bottled water(Charlier, R.H., 2003). a great wide variety of technologies will be achieved, like dams respect tidal range, sub turbines for you to ocean currents as well as tides in addition heat exchangers for its conversion of thermal energy associated with water, and also a great menu of models to help harness ones vitality involving waves as well as salinity gradients. water technology, from the exception of the genuine tidal dams is actually near your own phases of an pilot in addition to demonstration project, and also requires quite a few additional research in addition to development. a number of any technologies have changed profit energy production ensemble des within some other degrees of predictability (e.g, waves, tides and also the current selection), whilst others can give almost constant operation or maybe possibly controllable for such as ocean thermal as well as salinity gradient(ETSAP, 2010).

3.7 Wind Energy

Wind power uses the specific kinetic energy in the stirring air. The main request of significance to help climate change mitigation ought to be to create electricity from large wind generators located on the property (land) or perhaps fresh seasonal or perhaps (offshore) water. Onshore wind technology was being made and deployed a huge scale. Offshore wind power technologies will probably continue technical progress. Wind power will be variable along both equally, to some quantity, unpredictable, but experience with comprehensive studies as of many provinces has demonstrated that the incorporation connected with wind energy normally involves not pose insurmountable barriers approach Soobia Saeed, 2015).

3.8 Mutual Heat and Power

Consolidated warmth or maybe joined with force (CHP) is the system of warmth and power together. It is required to get an extensive pick up in effectiveness for every source independently. All force dispersion organizations novel power easily to manage or sauna. For right around 30-40% of the aggregate vitality heap of any nation pays for warming, cogeneration is a proficient utilization of fuel every time a piece of the imperativeness is removed, while the residual heat. Capture some or all residual energy as a byproduct involving heating. In Reykjavik end users can easily invest in both branches, together with the thermal energy into energy (Tester, 2005). Undoubtedly one of cogeneration can be particularly cogeneration unit Avedore, Denmark multifuel grow (Humayun, M., Niazi, M., Jhanjhi, N. et al, 2020). Shipley ET age (2008) estimated the increase in cogeneration capacity America and 20% in 2030 simply cause a reduction 5.3 Quads (quadrillion British temperate units) and the use of only one of 848

MMT fractional laser light. Depending on the results, America more than 1.9 Quads of energy resource consumption and emissions of 248 million metric numerous CO2 by using Generation Company saves. Based around the WEO (2012) Registration, the average efficiency of the vegetation of energy will be a 41% worldwide, with practically 60% of the primary energy is converted to heat loss (IEA, 2012E). CHP could extensively alter the residual heat standard in a positive financial significance for professional technical or business waste and heating constructions. It is anticipated that the new elements of cogeneration could help the energy efficiency of its highest and not 85%. (Madiment and Tozer, 2002) investigated the effective use of the combined temperature and cooling power (CCHP) for supermarkets in England and directly compared with the energy content saving price label / capital while using conventional engineering. The results suggest that CCHP provides a significant level of primary energy despite the reduction of CO2 released compared to conservative schemes, but also intended to compete with the most useful technologies regarding long help. They argued that innovative technologies such as fuel cells can deliver updates over the efficiency of the CCHP force belonging to refrigeration. Hawkes and Leach looked profitable trading strategies associated with three alternative micro-cogeneration engineering such as Sterling engine, gas engine with the fuel system of solar panels dependent solid oxide (SOFC) for residential use in the United together. They assessed the financial and ecological attributes of the technologies mentioned above led by the heat, electricity brought with techniques minimum load operation (Leach, M., 2007). The effects showed the greatest reduction in CO2 emissions following approach cheap operating costs were good high SOFC cost computer, the same. An individual considered this export volume in relation to electricity aggregates reveal the micro-cogeneration unit as a virtual power plant (VPP) through participation in the energy of low-cost sector. They found that the export capacity connected with micro-cogeneration systems depends largely around the technical parameters, energy prices along with affiliate demand profile. Based on the real model applied, the marginal price was conjectured with just a small cogeneration system is above the placement price of high season. Kiviluoma and Meibom (2010) studied the impact associated with variable power generation through wind turbines and electricity stored in electric vehicles the opportunity to improve the elasticity of its influence grating. Using the results, cogeneration units could actually be viable to produce more elastic in terms of invention techniques and using heat energy options (Noor Zaman.et.al, 2017). Is investigated this contribution Christidiset high temperature storage to boost cogeneration units in liberalized areas of energy, the implementation of particular design which measures the ability of custody of probable heat and optimal tax. They concluded that the separation of electricity production and heat stipulate could make available a stretch of cost recovery time for storage devices within the proposed program of energy (Soobia Saeed, 2015, Christidis, A. et.al, 2012).

4. CONCLUSION

The actual improvement regarding climate change will be simply just regarding the main challenges of the last century. Transition to help the single end user exactly who supports the higher power green global energy program incase be a great integral section of a reply involving humanity to live. Your chapter offers a great tick's basis for these types of chapters generally transition. the particular a complete in addition to reliable contribution to the debate at whether or not renewable energy can easily solve your climate problem with a financially attractive method. that is a model intended for further development of the renewable energy sector, as well as absolutely establishes the role within minimizing climate change everything. Ongoing queries about climate change have intended renewable energy a switch towards the overall collection of world energy consumption component. Technology regarding renewable energy might reduce CO_2 emissions from replacing fossil fuels fractional power generation industry and the transport sector. because of its negative in addition to irreparable externalities at the production associated with conventional energy, which is to be vital regarding development, along with promote renewable energy technologies provide. power generation while in friendly energy sources on the environment must end up being updated to reduce the unit cost associated with energy and also make sure these are generally compatible using a competitive substitute regarding conventional energy sources. A couple main solutions might be implemented to lessen CO_2 emissions and to overcome the particular dilemma involving climate alter: the substitution of fossil heating system by alternative energy sources regularly and strengthen energy performance, no matter what kind. In this specific chapter, we should instead be, wind flow, geothermal along with hydroelectric sun places, due to important contribution in the energy generated through the production involving renewable energy and gives sources. Renewable is increasing on the global level. Following extreme increase in oil prices as well as effects about the prices involving coal along with propane, the lots of investment it has recently been made in the recent past on alternative energy. These advancements in technologies have permitted countries to generate renewable power in larger plus more profitable levels.

Due to the negative along with irretrievable externalities linked to conventional energy taking out along with consumption, you will need to promote along with develop option vitality produce and usage. The IEA expects developing a positive growth involving alternative power means. They become substitutes related to fossil stimulates and decrease emissions. At any time, some option technologies might not be comparable that you can conventional stimulates, the area of manufacturing costs along with transmission, but may very well be comparable in the case we look at the many beneficial externalities, for example the environment along with social effects. Also, understand that should turn out to be financial regions of degree can play an essential role in reducing the cost of manufacturing program. Costs along with transmission technology and presentation will not differ considerably from typical and alternative energy options. We have at the moment presented detailed home elevators key technology of option development energy supply, such as hydro, the wind, solar, geothermal as well as length. The emphasis may be on latest production potential, in addition on the estimated, along with development expenses sink potential. Hydropower may be the largest source of renewable energy for energy generation throughout the world. Despite its important factor of energy generation, its growth is difficult on account of high expense costs as well as relocation in the initial set cost in the environment and the population. Hydro power is attractive as a result of combined way to obtain water with regard to agriculture, people, and recreation along with industrial employ. Moreover, you will keep water along with energy that can be used for inserts electrical power generation basic and high two. Having access to financial, politics and market risks, the focus with the allocation connected with resources along with ecological neighborhood concerns are viewed as obstacles to the development connected with hydropower potential. The capacity with the installed blowing wind power in addition has been growing, especially in countries just like China, the USA all, Germany and Denmark. Advantages connected with wind electrical power plants including contracts along with turnkey installing within a short period of period, lower when compared with nuclear along with hydropower investment decision, economies connected with production in lean muscle, lack connected with fuel prices and low operation along with maintenance prices. The issues connected with the aid of intermittent blowing wind energy contain wind electricity and added cost for that transmission of energy to buyers. Generation cost is determined by the location, viability about the minimum speed required for wind generators. China is promoting a volume of only solar panel technology, decreased era load as a result of availability connected with cheap labor and the general public subsidies. Geothermal energy has been used to be able to record more than bathroom, home heating and cooking food. The geothermal gradient, as well as the rock permeability app determines their particular financial viability. Unlike wind along with solar, geothermal solar panel is consistently available over the year, despite the technological know-how has some adverse environment effects. Improving electricity efficiency is surely an important method to reduce electricity use and as a consequence CO2 emissions, so that you can overcome neighborhood problem rescheduling. We outlined the state of the art methods connected with technical along with financial viability within the usage of green electricity, plus to be able to use along with combined replacement inside the first part of this evaluation. In the particular latter part of energy efficiency technologies are discussed. The efficiency connected with electricity networks is seen at distinct levels, including power creation, transmission, demonstration and consumption. To do this, Energy Efficiency, as well as solar cells with non-organic materials, electric vehicles, combined temperature and electrical power, virtual electrical control plants is different technologies. Each of the technologies is discussed in depth and when comparing their activities especially in the emission rate and method of perception of photons. The reasoning behind this execution is intensified daylight; in reality, by fake materials, further photon assimilation results in more electrons or power being released. This study examines the economic growth of photovoltaic cells (PVC) and the role of PVC in energy source renewal. It will be one of the components of our examination's victory. Our group will soon be producing more inorganic photovoltaic cells; it will be superior to our proposed PVC. This technological innovation was critical of green energy and efficiency provided complete and useful information which they can use in doing the determination from the different actors in the present rapid current market development. Each technological innovation has its advantages and disadvantages vary through location, availability, ease connected with technology suppliers, financial restrictions and environmental items to consider. Each metropolis, region or even nation has different first conditions in which conclude the mixture of action that may produce less costly costs, while minimizing the damage a result of the atmosphere. Therefore, there is absolutely no single solution for all energy must have and the situation but a remedy of ideal location between some achievable renewable options.

REFERENCES

Almulhim, Islam, & Zaman. (n.d.). A Lightweight and Secure Authentication Scheme for IoT Based E-Health Applications. *International Journal of Computer Science and Network Security, 19*(1), 107-120.

Almulhim, M., & Zaman, N. (2018). Proposing secure and lightweight authentication scheme for IoT based E-health applications. *20th International Conference on Advanced Communication Technology (ICACT)*, 481-487. 10.23919/ICACT.2018.8323802

Alshammari, M. O., Almulhem, A. A., & Zaman, N. (2017). Internet of Things (IoT): Charity Automation. *International Journal of Advanced Computer Science and Applications*, *8*(2).

Andersen, P. H., Mathews, J. A., & Rask, M. (2009). Integrating private transport into renewable energy policy: The strategy of creating intelligent recharging grids for electric vehicles. *Energy Policy*, *37*(7), 2481–2486. doi:10.1016/j.enpol.2009.03.032

Asif, M., & Muneer, T. (2007). Energy supply, its demand and security issues for developed and emerging economies. *Renewable & Sustainable Energy Reviews*, *11*(7), 1388–1413. doi:10.1016/j.rser.2005.12.004

Bain, R. L. (2007). World Biofuels Assessment, Worldwide Biomass Potential: Technology Characterizations" NREL/MP-510-42467. *National Renewable EnergyLaboratory, Golden, CO, USA*, *140*, 144–150.

Benitez, L. E., Benitez, P. C., & Van Kooten, G. C. (2008). The financials of wind power with energy storage. *Energy Financials*, *30*(4), 1973–1989.

Blanco, M. I. (2009). the financials of wind energy. *Renewable & Sustainable Energy Reviews*, *13*(6-7), 1372–1382. doi:10.1016/j.rser.2008.09.004

Chang, K.-C., Lin, W.-M., Lee, T.-S., & Chung, K.-M. (2011). Subsidy programs on diffusion of solar water heaters: Taiwan's exper283 ience. *Energy Policy*, *39*(2), 563–567. doi:10.1016/j.enpol.2010.10.021

Charlier, R. H. (2003). Sustainable co-generation from the tides: A review. *Renewable & Sustainable Energy Reviews*, *7*(3), 187–213. doi:10.1016/S1364-0321(02)00011-4

Christidis, A., Koch, C., Pottel, L., & Tsatsaronis, G. (2012). The contribution of heat storage to the profitable operation of combined heat and power plants in liberalized electricity Markets. *Energy*, *41*(1), 75–82. doi:10.1016/j.energy.2011.06.048

DeMeo, E. A., & Galdo, J. F. (1997). *Renewable Energy Technology Characterizations. TR-109496*. U.S. Department of Energy and Electric Power Research Institute.

EIA. (2009). *Energy Consumption by Manufacturers—Data Tables*. EIA.

ETSAP. (2010). *Marine Energy Technology Brief E13 - November, 2010, Energy Technology Systems Analysis Programme*. International Energy Agency.

Frick, S., Kaltschmitt, M., & Schröder, G. (2010). Life cycle assessment of geothermal binary power plants using enhanced low-temperature reservoirs. *Energy*, *35*(5), 2281–2294. doi:10.1016/j.energy.2010.02.016

Fridleifsson, I. B., & Freeston, D. H. (1994). geothermal energy research and development. *Geothermics*, *23*(2), 175–214. doi:10.1016/0375-6505(94)90037-X

Gevorkian, P. (2012). *Large Scale Solar Power Systems: Construction and Financials*. Cambridge University Press.

Gipe, P. (1995). Wind energy comes of age (Vol. 4). Wiley. doi:10.1017/CBO9781139226608

Gordon, J. (1987). Optimal sizing of stand-alone photovoltaic solar power systems. *Solar Cells*, *20*(4), 295–313. doi:10.1016/0379-6787(87)90005-6

Han, J., Mol, A. P. J., & Lu, Y. (2010). Solar water heaters in China: A new day dawning. *Energy Policy*, *38*(1), 383–391. doi:10.1016/j.enpol.2009.09.029

Harvey, L. D. D. (2006). *A Handbook on Low-Energy Buildings and District-Energy Systems: Fundamentals, Techniques and Examples. Earthscan*. Sterling.

Hawkes, A., & Leach, M. (2007). Cost-effective operating strategy for residential microcombined heat and power. *Energy*, *32*(5), 711–723. doi:10.1016/j.energy.2006.06.001

Humayun, Jhanjhi, Alamri, & Khan. (2020). Smart Cities and Digital Governance. In *Employing Recent Technologies for Improved Digital Governance*. IGI Global. . doi:10.4018/978-1-7998-1851-9.ch005

Humayun & Jhanjhi. (2019). Exploring The Relationship Between GSD, Knowledge Management, Trust And Collaboration. *Journal of Engineering Science and Technology, 14*(2), 820-843.

Humayun, M., Niazi, M., Jhanjhi, N., Alshayeb, M., & Mahmood, S. (2020). Cyber Security Threats and Vulnerabilities: A Systematic Mapping Study. *Arabian Journal for Science and Engineering*, *45*(4), 3171–3189. Advance online publication. doi:10.100713369-019-04319-2

Hussain, K., Hussain, S. J., Jhanjhi, N. Z., & Humayun, M. (2019). SYN Flood Attack Detection based on Bayes Estimator (SFADBE) For MANET. *International Conference on Computer and Information Sciences (ICCIS)*, 1-4. 10.1109/ICCISci.2019.8716416

IEA. (2007). *Renewables for Heating and Cooling – Untapped Potential*. International Energy Agency.

IRENA. (2019). *Future of Solar Photovoltaic: Deployment, investment, technology, grid integration and socio-economic aspects (A Global Energy Transformation: paper)*. International Renewable Energy Agency.

Kerr, D. (2007), Marine energy. *Philosophical Transactions of the Royal Society London, Series A (Mathematical, Physical and Engineering Sciences), 365*(1853), 971-92.

Kroposki, B. (2017). Integrating high levels of variable renewable energy into electric power systems. *Journal of Modern Power Systems and Clean Energy*, *5*(6), 831–837. doi:10.100740565-017-0339-3

Lund, J. W. (1995). Onion dehydration. *Transactions - Geothermal Resources Council*, *19*, 69–74.

Maidment, G., & Tozer, R. (2002). Combined cooling heat and power in supermarkets. *Applied Thermal Engineering*, *22*(6), 653–665. doi:10.1016/S1359-4311(01)00117-X

McGowin, C. (2008). *Renewable Energy Technical Assessment Guide. TAG-RE: 2007, Electric Power Research Institute*. EPRI.

Murphy, H., & Niitsuma, H. (1999). Strategies for compensating for higher costs of geothermal electricity with environmental benefits. *Geothermics*, *28*(6), 693–711. doi:10.1016/S0375-6505(99)00018-8

Nawaz, I., & Tiwari, G. (2006). Embodied energy analysis of photovoltaic (PV) system based on macro- and micro-level. *Energy Policy*, *34*(17), 3144–3152. doi:10.1016/j.enpol.2005.06.018

Ngô, C., & Natowitz, J. (2009). *Our energy future: resources, alternatives and the environment* (Vol. 11). Wiley. doi:10.1002/9780470473795

Ponnusamy, Ramachandran, Jung, & Zaman. (2017). Bio-Inspired Energy Scavenging in Wireless Ad Hoc Network. *IEEE Innovations in Electrical Engineering and Computational Technologies (ICIEET)*.

Purkus, A., & Barth, V. (2011). Geothermal power production in future electricity markets— A scenario analysis for Germany. *Energy Policy*, *39*(1), 349–357. doi:10.1016/j.enpol.2010.10.003

Radeckas, B., & Lukosevicius, V. (2000), Klaipeda Geothermal demonstration project. *Proceedings World Geothermal Congress 2000*, 3547-3550.

Reif, T. (2008). Profitability analysis and risk management of geothermal projects. *Geo-Heat Center Quarterly Bulletin*, *28*(4), 1–4.

Saeed. (2016). Performance Analysis of Single Mono-Crystalline Photovoltaic Cell Using Ultra- Violet Glass. *Journal of Information & Communication Technology, 10*(2), 19-25.

Saeed, S. (2015). Different Resources Consumption of Renewable Energy. Biologically-Inspired Energy Harvesting through Wireless Sensor Technologies, 85-102.

Saeed, S., Syed, M. R. N., & Khan, H. A. (2016). Estimation of Multi-Crystalline Photovoltaic Cell using Solar Spectrum of Five Different Bands. *Life Science Journal, Newyork*, *13*(8), 56–6.

Saeed & Jaffri. (2015). Estimation of freeze solar panel with Comparison of two different type's photo-voltaic cell. *International Journal of Technology & Research, 3*(3), 59-64.

Sara, Bushra, Jhanjhi, Mamoona, & Malik. (n.d.). Improving Knowledge Sharing in Distributed Software Development. *International Journal of Advanced Computer Science and Applications, 10*(6), 434-443.

Shravan, K., & Chunduri, K. (2019). *Advanced solar module technology – The time for a new generation of solar modules has come*. TaiyangNews.

Zaman & Almusalli. (2017). Review: Smartphone Power Consumption & Energy Saving Techniques. *IEEE Innovations in Electrical Engineering and Computational Technologies (ICIEET).*

Zaman, N., & Ahmed, M. (2017). Towards the Evaluation of Authentication Protocols for Mobile Command and Control Unit in Healthcare. *Journal of Medical Imaging and Health Informatics*, *7*(3), 739–742. doi:10.1166/jmihi.2017.2070

Zhang, X., Ruoshui, W., Molin, H., & Martinot, E. (2010). A study of the role played by renewable energies in China's sustainable energy supply. *Energy*, *35*(11), 4392–4399. doi:10.1016/j.energy.2009.05.030

Compilation of References

Aarts, E., & Wichert, R. (2009). *Ambient intelligence*. Springer Berlin Heidelberg.

Abbas, A. (2004). *Grid Computing: A Practical Guide to Tech. and App*. Firewall Media.

Abdelkareem, M. A. A., Xu, L., Ahmed Ali, M. K., Elagouz, A., Mi, J., Guo, S., Liu, Y., & Zuo, L. (2018a). Vibration energy harvesting in automotive suspension system: A detailed review. *Applied Energy*, *229*, 672–699. doi:10.1016/j.apenergy.2018.08.030

Abdelkareem, M. A. A., Xu, L., Guo, X., Ahmed Ali, M. K., Elagouz, A., Hassan, M. A., Essa, F. A., & Zou, J. (2018b). Energy harvesting sensitivity analysis and assessment of the potential power and full car dynamics for different road modes. *Mechanical Systems and Signal Processing*, *110*, 307–332. doi:10.1016/j.ymssp.2018.03.009

Abrol, A., & Jha, R. K. (2016). Power optimization in 5G networks: A step towards GrEEn communication. *IEEE Access: Practical Innovations, Open Solutions*, *4*, 1355–1374. doi:10.1109/ACCESS.2016.2549641

Abu Hassan, M., Kudus, N., Sidek, S., Mohamed, S., & Aripin, M. A. (2018). Factors That Influence The Sustainability Of IoT Usage For Smart Living. *International Journal of Pure and Applied Mathematics*, *119*, 573–587.

Adhikari, R., Pipattanasomporn, M., & Rahman, S. (2018). An algorithm for optimal management of aggregated HVAC power demand using smart thermostats. *Applied Energy*, *217*, 166–177. doi:10.1016/j.apenergy.2018.02.085

Adimalla, N., Dhakate, R., Kasarla, A., & Taloor, A. K. (2020). Appraisal of groundwater quality for drinking and irrigation purposes in Central Telangana, India. *Groundwater for Sustainable Development, 10*, 100334. https://state.bihar.gov.in/krishi/CitizenHome.html

Afghan, S. A., Husam, A., & Husi, G. (2018). Towards the self-powered Internet of Things (IoT) by energy harvesting: Trends and technologies for green IoT. *Proc. 2nd Int. Symp. Small-Scale Intell. Manuf. Syst. (SIMS),* 1–5.

Agarwal, V., Banerjee, N., Chakraborty, D., & Mittal, S. (2013). USense--A Smartphone Middleware for Community Sensing. In *2013 IEEE 14th International Conference on Mobile Data Management* (Vol. 1, pp. 56–65). IEEE.

Aggarwal, A., Du, P., & Robert, D. K. (2010). Grid Scheduling Optimization Based on Resource Characteristics. *Journal of Computer Information Systems*, *6*(14), 4609–4616.

Aguilera, J. J., Kazanci, O. B., & Toftum, J. (2019). Thermal adaptation in occupant-driven HVAC control. *Journal of Building Engineering*, *25*, 100846. doi:10.1016/j.jobe.2019.100846

Agyemang, J. O., Kponyo, J. J., Klogo, G. S., & Boateng, J. O. (2020). Lightweight Rogue Access Point Detection Algorithm for WiFi-Enabled Internet of Things (IoT) Devices. Internet of Things, 100200.

Ahmed, K. Z., Amir, M. F., Ko, J. H., & Mukhopadhyay, S. (2016). Reconfigurable 96x128 Active Pixel Sensor with 2.1 μW/mm2 Power Generation and Regulated MultiDomain Power Delivery for Self-Powered Imaging. *42nd European Solid-State Circuits Conference*, 507 - 510.

Ahmed, S. (2016). Exploring Current Trends of Energy Harvesting. In *Biologically-Inspired Energy Harvesting through Wireless Sensor Technologies* (pp. 268–278). IGI Global. doi:10.4018/978-1-4666-9792-8.ch014

Ainslie, M. A., & McColm, J. G. (1998). A simplified formula for viscous and chemical absorption in sea water. *The Journal of the Acoustical Society of America*, *103*(3), 1671–1672. doi:10.1121/1.421258

Akkaya, K., & Younis, M. (2004). *Energy-Aware Routing to a Mobile Gateway in Wireless Sensor Networks*. Paper presented at the IEEE Globecom Wireless Ad Hoc and Sensor Networks Workshop, Dallas, TX.

Akkaya, K., & Younis, M. (2005). A Survey of Routing Protocols in Wireless Sensor Networks. *Elsevier Ad Hoc Network Journal*, *3*(3), 325–349.

Akyildiz, I. F., Pompili, D., & Melodia, T. (2005). Underwater acoustic sensor networks: Research challenges. *Ad Hoc Networks*, *3*(3), 257–279. doi:10.1016/j.adhoc.2005.01.004

Akyildiz, I. F., Pompili, D., & Melodia, T. (2006). State-of-the-art in protocol research for underwater acoustic sensor networks. *Proceedings of the 1st ACM international workshop on Underwater networks*. 10.1145/1161039.1161043

Al Suleimani, Z., & Rao, N. R. (2000). Wind-powered electric water-pumping system installed in a remote location. *Applied Energy*, *65*(1-4), 339–347. doi:10.1016/S0306-2619(99)00099-9

Al-Ali, A.R., Al Nabulsi, A., Mukhopadhyay, S., Awal, M.S., Fernandes, S. & Ailabouni, K. (2020). IoT-solar energy powered smart farm irrigation system. *Journal of Electronic Science and Technology*, 100017.

Alamri, M. (2019). Blockchain for Internet of Things (IoT) Research Issues Challenges & Future Directions: A Review. *Int. J. Comput. Sci. Netw. Secur*, *19*, 244–258.

Albreem, M. A., El-Saleh, A. A., Isa, M., Salah, W., Jusoh, M., Azizan, M. M., & Ali, A. (2017, November). Green internet of things (IoT): An overview. In *2017 IEEE 4th International Conference on Smart Instrumentation, Measurement and Application (ICSIMA)* (pp. 1-6). IEEE.

Albreem, M. A. (2015, April). 5G wireless communication systems: Vision and challenges. In *2015 International Conference on Computer, Communications, and Control Technology (I4CT)* (pp. 493-497). IEEE. 10.1109/I4CT.2015.7219627

Al-Fares, Z., Sun, H. & Cruickshank. (2009). *A Hierarchical Routing Protocol for Survivability in Wireless Sensor Network (WSN)*. Paper presented at the International MultiConference of Engineers and Computer Scientists, IMECS 2009.

Alippi, C., Anastasi, G., Di Francesco, M., & Roveri, M. (2009). Energy Management in Wireless Sensor Networks with Energy-hugry Sensors. *IEEE Instrumentation & Measurement Magazine*, *12*(2), 16–23. doi:10.1109/MIM.2009.4811133

Alippi, C., & Galperti, C. (2008). *An Adaptive System for Optimal Solar Energy Harvesting in Wireless Sensor Network Nodes. IEEE Transactions on Circuits and Systems*, *55*(6), 1742–1750. doi:10.1109/TCSI.2008.922023

All about circuits. (2020). https://www.allaboutcircuits.com/industry-articles/utilizing-the-different-types-of-common-iot-connection-methods/

Almeida, A., Mulero, R., Rametta, P., Urošević, V., Andrić, M., & Patrono, L. (2019). A critical analysis of an IoT—Aware AAL system for elderly monitoring. *Future Generation Computer Systems*, *97*, 598–619. doi:10.1016/j.future.2019.03.019

Almulhim, Islam, & Zaman. (n.d.). A Lightweight and Secure Authentication Scheme for IoT Based E-Health Applications. *International Journal of Computer Science and Network Security*, *19*(1), 107-120.

Almulhim, M., Islam, N., & Jhanjhi, N. Z. (2019). A Lightweight and Secure Authentication Scheme for IoT Based E-Health Applications. *International Journal of Computer Science and Network Security*, *19*(1), 107–120.

Almulhim, M., & Zaman, N. (2018). Proposing secure and lightweight authentication scheme for IoT based E-health applications. *20th International Conference on Advanced Communication Technology (ICACT)*, 481-487. 10.23919/ICACT.2018.8323802

Almusaylim, Z. A., & Jhanjhi, N. Z. (2019). A review on smart home present state and challenges: Linked to context-awareness internet of things (IoT). *Wireless Networks*, *25*(6), 3193–3204. doi:10.100711276-018-1712-5

Alsamhi, S., Ma, O., Ansari, M. S., & Meng, Q. (2018). *Greening internet of things for smart everythings with a green-environment life: A survey and future prospects.* arXiv preprint arXiv:1805.00844.

Alsamhi, S. H., Ma, O., Ansari, M. S., & Meng, Q. (2019). Greening internet of things for greener and smarter cities: A survey and future prospects. *Telecommunication Systems*, *72*(4), 609–632. doi:10.100711235-019-00597-1

Alshammari, M. O., Almulhem, A. A., & Zaman, N. (2017). Internet of Things (IoT): Charity Automation. *International Journal of Advanced Computer Science and Applications*, *8*(2).

Al-Turjman, F., Altrjman, C., Din, S., & Paul, A. (2019). Energy monitoring in IoT-based ad hoc networks: An overview. *Computers & Electrical Engineering*, *76*, 133–142. doi:10.1016/j.compeleceng.2019.03.013

Amazon Elastic Compute Cloud (EC2). (n.d.). Retrieved July, 2008, from https://www.amazon.com/cc2/

Aminov, P., & Agrawal, J. P. (2014, May). RF energy harvesting. In *Electronic Components and Technology Conference (ECTC), 2014 IEEE 64th* (pp. 1838-1841). IEEE. 10.1109/ECTC.2014.6897549

An E.-A.-U. (2011). *A Discussion of Smart Meters And RF Exposure Issues.* Academic Press.

Anastasi, G., Conti, M., Di Francesco, M., Passarella, A., Denko, M., & Yang, L. (2006). How to prolong the lifetime of wireless sensor networks. *Mobile Ad Hoc and Pervasive Communications*, 1-26.

Anastasi, G., Conti, M., Di Francesco, M., & Passarella, A. (2009). Energy conservation in wireless sensor networks: A survey. *Ad Hoc Networks*, *7*(3), 537–568. doi:10.1016/j.adhoc.2008.06.003

Andersen, P. H., Mathews, J. A., & Rask, M. (2009). Integrating private transport into renewable energy policy: The strategy of creating intelligent recharging grids for electric vehicles. *Energy Policy*, *37*(7), 2481–2486. doi:10.1016/j.enpol.2009.03.032

Andolfatto, D. (2018). Blockchain: What it is, what it does, and why you probably don't need one. *Review - Federal Reserve Bank of St. Louis*, *100*(2), 87–95. doi:10.20955/r.2018.87-95

Andoulssi, R., Draou, A., Jerbi, H., Alghonamy, A., & Khiari, B. (2013). Non linear control of a photovoltaic water pumping System. *Energy Procedia*, *42*, 328–336. doi:10.1016/j.egypro.2013.11.033

Androulaki, E., Barger, A., Bortnikov, V., Cachin, C., Christidis, K., De Caro, A., Enyeart, D., Ferris, C., Laventman, G., Manevich, Y., & Muralidharan, S. (2018, April). Hyperledger fabric: a distributed operating system for permissioned blockchains. In *Proceedings of the Thirteenth EuroSys Conference* (pp. 1-15). 10.1145/3190508.3190538

Angelopoulos, C. M., Evangelatos, O., Nikoletseas, S., Raptis, T. P., Rolim, J. D. P., & Veroutis, K. (2015). A user-enabled testbed architecture with mobile crowdsensing support for smart, green buildings. In *2015 IEEE International Conference on Communications (ICC)* (pp. 573–578). IEEE. 10.1109/ICC.2015.7248383

Anitha, R., & Rajkumar, N. (2006). *Compensation Based Grid Scheduling.* Paper presented at the 2006 Annual IEEE India Conference. 10.1109/INDCON.2006.302816

Ansari, M. H., & Karami, M. A. (2017). Experimental investigation of fan-folded piezoelectric energy harvesters for powering pacemakers. *Smart Materials and Structures*, *26*(6), 065001. doi:10.1088/1361-665X/aa6cfd PMID:29674807

Aristeidis, K., Joannopoulos, J. D., & Marin, S. (2008). *Efficient wireless non-radiative mid-range energy transfer*. Elsevier.

Armbrust, M., Fox, A., Griffith, R., Joseph, A. D., Katz, R., Konwinski, A., ... Stoica, I. (2010). A view of cloud computing. *Communications of the ACM*, *53*(4), 50–58. doi:10.1145/1721654.1721672

Arshad, R., Zahoor, S., Shah, M. A., Wahid, A., & Yu, H. (2017). Green IoT: An investigation on energy saving practices for 2020 and beyond. *IEEE Access: Practical Innovations, Open Solutions*, *5*, 15667–15681. doi:10.1109/ACCESS.2017.2686092

Ashraf, A., & Porres, I. (2018). Multi-objective dynamic virtual machine consolidation in the cloud using ant colony system. *International Journal of Parallel. Emergent and Distributed Systems*, *33*(1), 103–120. doi:10.1080/17445760.2017.1278601

Ashton, K. (2009). *That "Internet of Things" thing in the real world, things matter more than ideas*. RFID Journal.

Asif, M., & Muneer, T. (2007). Energy supply, its demand and security issues for developed and emerging economies. *Renewable & Sustainable Energy Reviews*, *11*(7), 1388–1413. doi:10.1016/j.rser.2005.12.004

Atakan, B., & Akan, Ö.B. (2007). Immune system-based energy efðcient andreliable communication in wireless sensor networks. Springer.

Atallah, R., Khabbaz, M., & Assi, C. (2016). Energy harvesting in vehicular networks: A contemporary survey. *IEEE Wireless Communications*, *23*(2), 70–77. doi:10.1109/MWC.2016.7462487

Avgerinou, M., Bertoldi, P., & Castellazzi, L. (2017). Trends in data centre energy consumption under the european code of conduct for data centre energy efficiency. *Energies*, *10*(10), 1470. doi:10.3390/en10101470

Awan, Shah, & Iqbal, Gillani, Ahmad, & Nam. (2019). Underwater Wireless Sensor Networks: A Review of Recent Issues and Challenges. *Wireless Communications and Mobile Computing*.

Azevedo, J. A. R., & Santos, F. E. S. (2012). Energy harvesting from wind and water for autonomous wireless sensor nodes. *IET Circuits, Devices & Systems*, *6*(6), 413–420. doi:10.1049/iet-cds.2011.0287

Bahinipati, C. S., & Viswanathan, P. K. (2019). Incentivizing resource efficient technologies in India: Evidence from diffusion of micro-irrigation in the dark zone regions of Gujarat. *Land Use Policy*, *86*, 253–260. doi:10.1016/j.landusepol.2019.04.017

Bahmanyar, A., Jamali, S., Estebsari, A., Pons, E., Bompard, E., Patti, E., & Acquaviva, A. (2016). Emerging smart meters in electrical distribution systems: Opportunities and challenges. In *2016 24th Iranian Conference on Electrical Engineering (ICEE)* (pp. 1082–1087). IEEE.

Bain, R. L. (2007). World Biofuels Assessment, Worldwide Biomass Potential: Technology Characterizations" NREL/MP-510-42467. *National Renewable EnergyLaboratory, Golden, CO, USA*, *140*, 144–150.

Baker, M., & Carpenter, B. (2000). *Mpj: A proposed java message passing api and environment for high performance computing*. Paper presented at the International Parallel and Distributed Processing Symposium. Bal, D. H. DAS-2 team. Retrieved from http://gwa.ewi.tudelft.nl/pmwiki/reports/gwa-t-1/trace_analysis_report.html

Balasubramaniam, R. (2013). *Wireless energy transfer using resonant inductive coupling*. In *Proceedings of Sixth IRAJ International Conference* (pp. 44-47). Pune, India: Power Publishers.

Baliga, A., Subhod, I., Kamat, P., & Chatterjee, S. (2018). *Performance evaluation of the quorum blockchain platform.* arXiv preprint arXiv:1809.03421.

Bandara, S., Yashiro, T., Koshizuka, N., & Sakamura, K. (2016, August). Access control framework for api-enabled devices in smart buildings. In *2016 22nd Asia-Pacific Conference on Communications (APCC)* (pp. 210-217). IEEE. 10.1109/APCC.2016.7581479

Bansal, K. (2011). A critical review on concept of green database. *Global Journal of Business Management and Information Technology*, *1*, 113–118.

Barroso, L. A. (2005). The price of performance. *ACM Queue, 7*, 48-53.

Basset, P., Galayko, D., Paracha, A. M., Marty, F., Dudka, A., & Bourouina, T. (2009). A batch-fabricated and electret-free silicon electrostatic vibration energy harvester. *Journal of Micromechanics and Microengineering*, *19*(11), 115025. doi:10.1088/0960-1317/19/11/115025

Bawaneh, Nezami, Rasheduzzaman, & Deken. (2019). Energy Consumption Analysis and Characterization of Healthcare Facilities in the United States. *Energies, 12*(9), 3775.

Beeby, S. P., Torah, R. N., Tudor, M. J., Glynne-Jones, P., O'Donnell, T., Saha, C. R., & Roy, S. (2007). A micro-electromagnetic generator for vibration energy harvesting. *Journal of Micromechanics and Microengineering*, *17*(7), 1257–1265. doi:10.1088/0960-1317/17/7/007

Beeby, S. P., Tudor, M. J., & White, N. M. (2006). Energy harvesting vibration sources for microsystems applications. *Measurement Science & Technology*, *17*(12), R175–R195. doi:10.1088/0957-0233/17/12/R01

Behera, H., Khuntia, S., & Nayak, S. (2012). An improved least-laxity-first scheduling algorithm for real-time tasks. *International Journal of Engineering Science and Technology*, *4*(04).

Bello, A. & Mahadevan, V. (2019). A cloud based conceptual identity management model for secured Internet of Things operation. *Journal of Cyber Security and Mobility*, 53-74.

Ben Elghali, S. E., Benbouzid, M. E. H., & Charpentier, J. F. (2007). Marine tidal current electric power generation technology: state of the art and current status. IEEE international electric machines & drives conference, 1407–1412. doi:10.1109/IEMDC.2007.383635

Benbelkacem, S., Belhocine, M., Bellarbi, A., Zenati-Henda, N., & Tadjine, M. (2013). Augmented reality for photovoltaic pumping systems maintenance tasks. *Renewable Energy*, *55*, 428–437. doi:10.1016/j.renene.2012.12.043

Benitez, L. E., Benitez, P. C., & Van Kooten, G. C. (2008). The financials of wind power with energy storage. *Energy Financials*, *30*(4), 1973–1989.

Berlian, M. H., Sahputra, T. E. R., Ardi, B. J. W., Dzatmika, L. W., Besari, A. R. A., Sudibyo, R. W., & Sukaridhoto, S. (2017). Design and implementation of smart environment monitoring and analytics in real-time system framework based on internet of underwater things and big data. *Proceedings - 2016 International Electronics Symposium, IES 2016*, 403–408. 10.1109/ELECSYM.2016.7861040

Berman, F., Fox, G., & Hey, T. J. M. G. I. R. (2003). *Overview of the book: grid computing–making the global infrastructure a reality*. Retrieved from http://www.grid5000.org/

Bernabé, J. B., Ramos, J. L. H., & Gómez-Skarmeta, A. F. (2017). Holistic Privacy-Preserving Identity Management System for the Internet of Things. *Mobile Information Systems*, *2017*, 6384186–1. doi:10.1155/2017/6384186

Bhardwaj, M., Garnett, T., & Chandrakasan, A. P. (2001). *Upper bounds on the lifetime of sensor networks.* Paper presented at the ICC 2001. IEEE International Conference on Communications. Conference Record (Cat. No. 01CH37240). 10.1109/ICC.2001.937346

Blanco, M. I. (2009). the financials of wind energy. *Renewable & Sustainable Energy Reviews*, *13*(6-7), 1372–1382. doi:10.1016/j.rser.2008.09.004

Boehm, A. B., Grant, S., Kim, J., Mowbray, S., McGee, C., Clark, C., Foley, D. M., & Wellman, D. (2002). Decadal and shorter period variability of surf zone water quality at Huntington Beach, California. *Environmental Science & Technology*, *36*(18), 3885–3892. doi:10.1021/es020524u PMID:12269739

Boisseau, S., Despesse, G., Monfray, S., Puscasu, O., & Skotnicki, T. (2013). Semi-flexible bimetal-based thermal energy harvesters. *Smart Materials and Structures*, *22*(2), 025021. doi:10.1088/0964-1726/22/2/025021

Boisseau, S., Despesse, G., Ricart, T., Defay, E., & Sylvestre, A. (2011). Cantilever-based electret energy harvesters. *Smart Materials and Structures*, *20*(10), 105013. doi:10.1088/0964-1726/20/10/105013

Bokareva, T., Bulusu, N., & Jha, S. (2005). SASHA: towards a self-healing hybrid sensor network architecture. *Proceedings of the Second IEEE International Workshop on Embedded Networked Sensors*, 71–78.

Bol, Boufouss, Flandre, & De Vos. (2015). A 0.48mm2 - 5μW-10mW Indoor/Outdoor PV Energy-Harvesting Management Unit in a 65nm SoC based on a Single Bidirectional Multi-Gain/Multi-Mode Switched-Cap Converter with Supercap Storage. *41st European Solid-State Circuits Conference*, 241 - 244.

Boonma, P., & Suzuki, J. (2007). BiSNET: A biologically-inspired middleware architecture for self-managing wireless sensor networks. *Computer Networks*, *51*(16), 4599–4616. doi:10.1016/j.comnet.2007.06.006

Bottaccioli, L., Aliberti, A., Ugliotti, F., Patti, E., Osello, A., Macii, E., & Acquaviva, A. (2017). Building energy modelling and monitoring by integration of IoT devices and building information models. In *2017 IEEE 41st Annual Computer Software and Applications Conference (COMPSAC)* (Vol. 1, pp. 914–922). IEEE. 10.1109/COMPSAC.2017.75

Bouaziz, M., Rachedi, A., & Belghith, A. (2017). *EC-MRPL: An energy-efficient and mobility support routing protocol for Internet of Mobile Things. In 14th IEEE Annual Consumer Communications & Networking Conference.* CCNC.

Bourdakis, E., Olesen, B. W., & Grossule, F. (2015). *Night time cooling by ventilation or night sky radiation combined with in-room radiant cooling panels including phase change materials.* Paper presented at 36th AIVC Conference, Madrid, Spain.

Bowen, C. R., Taylor, J., LeBoulbar, E., Zabek, D., Chauhan, A., & Vaish, R. (2014). Pyroelectric materials and devices for energy harvesting applications. *Energy & Environmental Science*, *7*(12), 3836–3856. doi:10.1039/C4EE01759E

Bradai, S., Naifar, S., Viehweger, C., & Kanoun, O. (2018). Electromagnetic Vibration Energy Harvesting for Railway Applications. *MATEC Web of Conference, ICoEV 2017*, *148*(8). 10.1051/matecconf/201814812004

Branch, J. W., Lisee, M., & Szymanski, B.K. (2007). SHR: self-healing routing for wireless ad hoc sensor networks. *Proc. Int. Symposium on Performance Evaluation of Computer and Telecommunication Systems SPECTS'07*, 5–14.

Brock, D. (2001). *The electronic product code (epc) a naming scheme for physical objects.* Auto-ID Center, White Paper.

Bruneo, D., Lhoas, A., Longo, F., & Puliafito, A. (2014). Modeling and evaluation of energy policies in green clouds. *IEEE Transactions on Parallel and Distributed Systems*, *26*(11), 3052–3065. doi:10.1109/TPDS.2014.2364194

Bu, L., Wu, X. M., Wang, X. H., & Liu, L. T. (2011). Liquid encapsulated energy harvester for low frequency vibrations. *Proceedings of the 16th International Solid-State Sensors,Actuators and Microsystems Conference (TRANSDUCERS '11)*, 1673–1676. 10.1109/TRANSDUCERS.2011.5969863

Burrus. (2014). *The Internet of Things Is Far Bigger Than Anyone Realizes*. Retrieved from Wired web site: https://www.wired.com/2014/11/the-internet-of-things-bigger/

Camilo, T., Carreto, C., & Silva, J. S. (2006). An Energy-Efficient Ant-Based Routing Algorithm for Wireless Sensor Network. *Proceedings of the 5th International Workshop on Ant Colony Optimization and Swarm Intelligence*, 4105, 49-59.

Cao, L., & Li, Z. Yao, & Lian. (2019). Smartfloat: A Multimodal Underwater Vehicle Combining Float and Glider Capabilities. *IEEE Access, 7*, 77825–38. doi:10.1109/ACCESS.2019.2922171

Cao, S., & Li, J. (2017). A survey on ambient energy sources and harvesting methods for structural health monitoring applications. *Advances in Mechanical Engineering*, *9*(4), 1–14. doi:10.1177/1687814017696210

Carreira, P., Costa, A. A., Mansur, V., & Arsénio, A. (2018). Can HVAC really learn from users? A simulation-based study on the effectiveness of voting for comfort and energy use optimization. *Sustainable Cities and Society*, *41*, 275–285. doi:10.1016/j.scs.2018.05.043

Carrera, E. V. (2003). *Conserving disk energy in network servers*. ICS.

Catarinucci, L., De Donno, D., Mainetti, L., Palano, L., Patrono, L., Stefanizzi, M. L., & Tarricone, L. (2015). An IoT-aware architecture for smart healthcare systems. *IEEE Internet of Things Journal*, *2*(6), 515–526. doi:10.1109/JIOT.2015.2417684

Cetin, K. S., & Kallus, C. (2016). Data-Driven Methodology for Energy and Peak Load Reduction of Residential HVAC Systems. *Procedia Engineering*, *145*, 852–859. doi:10.1016/j.proeng.2016.04.205

Chalasani, S., & Conrad, J. M. (2008). A survey of energy harvesting sources for embedded systems. *Proceedings of IEEE Southeastcon*, 442–447.

Chamanian, S., Baghaee, S., Ulusan, H., Zorlu, O., Kulah, H., & Biyikoglu, E. U. (2014). Powering-up Wireless Sensor Nodes Utilizing Rechargeable Batteries and an Electromagnetic Vibration Energy Harvesting System. *Energies*, *7*(10), 6323–6339. doi:10.3390/en7106323

Chan, H.F., & Rudolph, H. (2015). *New Energy Efficient Routing Algorithm for Wireless Sensor Network*. Available at https://www.researchgate.net/publication/283507649_New_Energy_Efficient_Routing_Algorithm_for_Wireless_Sensor_Network

Chan, C. A., Gygax, A. F., Wong, E., Leckie, C. A., Nirmalathas, A., & Kilper, D. C. (2013). Methodologies for assessing the use-phase power consumption and greenhouse gas emissions of telecommunications network services. *Environmental Science & Technology*, *47*(1), 485–492. doi:10.1021/es303384y PMID:23211093

Chang, K.-C., Lin, W.-M., Lee, T.-S., & Chung, K.-M. (2011). Subsidy programs on diffusion of solar water heaters: Taiwan's exper283 ience. *Energy Policy*, *39*(2), 563–567. doi:10.1016/j.enpol.2010.10.021

Chao, C. M., Jiang, C. H., & Li, W. C. (2017). DRP: An energy-efficient routing protocol for underwater sensor networks. *International Journal of Communication Systems*, *30*(15), 1–10. doi:10.1002/dac.3303

Chapter 7: Agriculture and Food Management. (n.d.). *Economic Survey 2019-20*. https://www.indiabudget.gov.in/economicsurvey/eco_sur_vol(iiallchapter).php

Charlier, R. H. (2003). Sustainable co-generation from the tides: A review. *Renewable & Sustainable Energy Reviews*, *7*(3), 187–213. doi:10.1016/S1364-0321(02)00011-4

Chean, L. T., Ponnusamy, V., & Fati, S. M. (2018, April). Authentication scheme using unique identification method with homomorphic encryption in Mobile Cloud Computing. In *2018 IEEE Symposium on Computer Applications & Industrial Electronics (ISCAIE)* (pp. 195-200). IEEE. 10.1109/ISCAIE.2018.8405469

Chen, C. (2012). *Green databases through integration of renewable energy*. Nanyang Technological University & Sun Yat-sen University.

Chen, C. L., Chen, H. C., & Lee, J. Y. (2016). Application of a generic superstructure-based formulation to the design of wind-pumped-storage hybrid systems on remote islands. *Energy Conversion and Management*, *111*, 339–351.

Cheng, J. C. P., Chen, W., Chen, K., & Wang, Q. (2020). Data-driven predictive maintenance planning framework for MEP components based on BIM and IoT using machine learning algorithms. *Automation in Construction*, *112*, 103087. doi:10.1016/j.autcon.2020.103087

Chen, J., & Wang, Z. L. (2017). Reviving Vibration Energy Harvesting and Self-Powered Sensing by a Triboelectric Nanogenerator. *Joule*, *1*(3), 480–521. doi:10.1016/j.joule.2017.09.004

Chen, J., Zhu, G., Yang, W., Jing, Q., Bai, P., Yang, Y., Hou, T. C., & Wang, Z. L. (2013). Harmonic-Resonator-Based Triboelectric Nanogenerator as a Sustainable Power Source and a Self-Powered Active Vibration Sensor. *Advanced Materials*, *25*(42), 6094–6099. doi:10.1002/adma.201302397 PMID:23999798

Chen, M., Mao, S., & Liu, Y. (2014). Big data: A survey. *Mobile Networks and Applications*, *19*(2), 171–209. doi:10.100711036-013-0489-0

Chen, Y. J., Wang, L. C., & Wang, S. (2018). Stochastic Blockchain for IoT Data Integrity. *IEEE Transactions on Network Science and Engineering*.

Chhabra, A., Vashishth, V., & Sharma, D. K. (2017). Seir: A stackelberg game based approach for energy-aware and incentivized routing in selfish opportunistic networks. In *2017 51st Annual Conference on Information Sciences and Systems (CISS)* (pp. 1–6). IEEE. 10.1109/CISS.2017.7926113

Chhabra, A., Vashishth, V., & Sharma, D. K. (2018). A fuzzy logic and game theory-based adaptive approach for securing opportunistic networks against blackhole attacks. *International Journal of Communication Systems*, *31*(4), e3487. doi:10.1002/dac.3487

Chih-Lin, I., Rowell, C., Han, S., Xu, Z., Li, G., & Pan, Z. (2014). Toward green and soft: A 5G perspective. *IEEE Communications Magazine*, *52*(2), 66–73. doi:10.1109/MCOM.2014.6736745

Choi, J., Cho, J. K., & Seo, C. (2011). Analysis on Transmission efficiency of wireless energy transmission resonator based on magnetic resonance. In Microwave Workshop Series on Innovative Wireless Power Transmission: Technologies, Systems, and Applications (IMWS), 2011 IEEE MTT-S International (pp. 199-202). IEEE. doi:10.2528/PIERM11050903

Chong, F. T., Heck, M. J., Ranganathan, P., Saleh, A. A., & Wassel, H. M. (2013). Data center energy efficiency: Improving energy efficiency in data centers beyond technology scaling. *IEEE Design & Test*, *31*(1), 93–104. doi:10.1109/MDAT.2013.2294466

Chowdhury, B. H., Ula, S., & Stokes, K. (1993). Photovoltaic-powered water pumping-design, and implementation: Case studies in Wyoming. *IEEE Transactions on Energy Conversion*, *8*(4), 646–652.

Christidis, A., Koch, C., Pottel, L., & Tsatsaronis, G. (2012). The contribution of heat storage to the profitable operation of combined heat and power plants in liberalized electricity Markets. *Energy*, *41*(1), 75–82. doi:10.1016/j.energy.2011.06.048

Closas, A., & Rap, E. (2017). Solar-based groundwater pumping for irrigation: Sustainability, policies, and limitations. *Energy Policy*, *104*, 33–37.

Colarelli, D., & Grunwald, D. (2002). *Massive arrays of idle disks for storage archives*. Academic Press.

Cook & Zhang. (2019). Dynamics Modeling of AUVs. *Mobile Robots*, 269–89. doi:10.1002/9781119534839.ch11

Cook-Chennault, K. A., Thambi, N., & Sastry, A. M. (2008). Powering MEMS portable devices—A review of non-regenerative and regenerative power supply systems with special emphasis on piezoelectric energy harvesting systems. *Smart Materials and Structures*, *17*(4), 043001. doi:10.1088/0964-1726/17/4/043001

Cook, G. (2011). *How dirty is your data: A look at the energy choices that power cloud computing*. Greenpeace.

Corke, P., Valencia, P., Sikka, P., Wark, T., & Overs, L. (2007, June). Long-duration solar-powered wireless sensor networks. In *Proceedings of the 4th workshop on Embedded networked sensors* (pp. 33-37). ACM. 10.1145/1278972.1278980

Corrêa, T. P., Seleme, S. I. Jr, & Silva, S. R. (2012). Efficiency optimization in stand-alone photovoltaic pumping system. *Renewable Energy*, *41*, 220–226.

Cottone, F., Vocca, H., & Gammaitoni, L. (2009). Nonlinear energy harvesting. *Physical Review Letters*, *102*(8), 080601. doi:10.1103/PhysRevLett.102.080601 PMID:19257728

Cuadras, A., Gasulla, M., & Ferarri, V. (2010). Thermal energy harvesting through pyroelectricity. *Sensors and Actuators. A, Physical*, *159*(1), 132–139. doi:10.1016/j.sna.2009.12.018

Cui, J.-H., Kong, J., Gerla, M., & Zhou, S. (2006). The challenges of building mobile underwater wireless networks for aquatic applications. *IEEE Network*, *20*(3), 12–18. doi:10.1109/MNET.2006.1637927

Cui, Y., Zhang, Q., Yao, M., Dong, W., & Gao, S. (2015). Vibration piezoelectric energy harvester with multi-beam. *AIP Advances*, *5*(4), 041332. doi:10.1063/1.4919049

Curry, E., Guyon, B., Sheridan, C., & Donnellan, B. (2012). Developing a sustainable it capability: Lessons from intel's journey. *MIS Quarterly Executive*, *11*(2), 61–74.

Dahl, P. H., Miller, J. H., Cato, D. H., & Andrew, R. K. (2007). Underwater ambient noise. *Acoustics Today*, *3*(1), 23–33. doi:10.1121/1.2961145

Daniels, A., Giuliano, A., Zhu, M., & Tiwari, A. (2013). Modeling, Validation and Design Analyses of a Piezoelectric Cymbal Transducer for Non-Resonant Energy Harvesting. *IEEE International Conference on Green Computing and Communications and IEEE Internet of Things and IEEE Cyber, Physical and Social Computing*, 1665–1667. 10.1109/GreenCom-iThings-CPSCom.2013.303

Dash, S., Swain, A. R., & Ajay, A. (2012). Reliable energy aware multi token based MAC protocol for WSN. *Proceedings of the26th IEEE International Conference on Advanced Information Networking and Applications (AINA '12)*, 144–151. 10.1109/AINA.2012.61

Data Flair training. (2020). https://data-flair.training/blogs/how-iot-works/

Dave, B., Buda, A., Nurminen, A., & Främling, K. (2018). A framework for integrating BIM and IoT through open standards. *Automation in Construction*, *95*, 35–45. doi:10.1016/j.autcon.2018.07.022

Dayarathna, M., Wen, Y., & Fan, R. (2015). Data center energy consumption modeling: A survey. *IEEE Communications Surveys and Tutorials*, *18*(1), 732–794. doi:10.1109/COMST.2015.2481183

Dechant, E., Fedulov, F., Chashin, D. V., Fetisov, L. Y., Fetisov, Y. K., & Shamonin, M. (2017). Low-frequency, broadband vibration energy harvester using coupled oscillators and frequency up-conversion by mechanical stoppers. *Smart Materials and Structures*, *26*(6), 065021. doi:10.1088/1361-665X/aa6e92

DeMeo, E. A., & Galdo, J. F. (1997). *Renewable Energy Technology Characterizations. TR-109496*. U.S. Department of Energy and Electric Power Research Institute.

Deterre, M. (2013). *Towards an energy harvester for leadless pacemakers* (PhD Thesis). Paris: Universite Paris-Sud.

Di Caro, G., & Dorigo, M. (1998). AntNet: Distributed stigmergetic control forcommunication networks. *Journal of Artificial Intelligence Research*, *9*, 317–365.

Diène, B., Rodrigues, J. J., Diallo, O., Ndoye, E. H. M., & Korotaev, V. V. (2020). Data management techniques for Internet of Things. *Mechanical Systems and Signal Processing*, *138*, 106564.

Dietrich, J., Schmidt, R., Hogan, M., & Allen, G. (2007). The Green Data Center. IBM Global Services, 1-20.

Ding, G., Zhao, X., Wang, J., & Xu, C. (2020). Vibration energy harvesting from roads under traffic loads. *Road Materials and Pavement Design*, *21*(3), 780–799. doi:10.1080/14680629.2018.1527719

Diro, A., Reda, H., Chilamkurti, N., Mahmood, A., Jhanjhi, N. Z., & Nam, Y. (2020). Lightweight Authenticated-Encryption Scheme for Internet of Things Based on Publish-Subscribe Communication. *IEEE Access: Practical Innovations, Open Solutions*, *8*, 60539–60551. doi:10.1109/ACCESS.2020.2983117

Dittmann, G., & Jelitto, J. (2019, June). A Blockchain Proxy for Lightweight IoT Devices. In *2019 Crypto Valley Conference on Blockchain Technology (CVCBT)* (pp. 82-85). IEEE. 10.1109/CVCBT.2019.00015

Dorigo, M., Maniezzo, V., & Colomi, A. (1996). Ant system: Optimization by a colony of cooperating agents. *IEEE Transactions on Systems, Man, and Cybernetics. Part B, Cybernetics*, *26*(1), 29–41. doi:10.1109/3477.484436 PMID:18263004

Dorzhiev, V., Karami, A., Basset, P., Marty, F., Dragunov, V., & Galayko, D. (2015). Electret-Free Micromachined Silicon Electrostatic Vibration Energy Harvester With the Bennet's Doubler as Conditioning Circuit. *IEEE Electron Device Letters*, *36*(2), 183–185. doi:10.1109/LED.2014.2387213

dos Santos, M. P. S., Ferreira, J. A. F., Simões, J. A. O., Pascoal, R., Torrão, J., Xue, X., & Furlani, E. P. (2016). Magnetic levitation-based electromagnetic energy harvesting: A semi-analytical non-linear model for energy transduction. *Scientific Reports*, *6*(1), 18579. doi:10.1038rep18579 PMID:26725842

Dotti, F. E., & Sosa, M. D. (2019). Pendulum systems for harvesting vibration energy from railroad tracks and sleepers during the passage of a high-speed train: A feasibility evaluation. *Theoretical & Applied Mechanics Letters*, *9*(4), 229–235. doi:10.1016/j.taml.2019.03.005

Dressler, F. (2009). Self-Organization in Ad Hoc Networks: Overview and Classification. Univ. of Erlangen, Dept. of Computer Science 7, Technical Report 02/06.

Dressler, F., Kuger, B., Fuchs, G., & German, R. (2005). *Self-Organization in Sensor Networks using Bio-Inspired Mechanisms*. Paper presented at the 18th ACM/GI/ITG International Conference on Architecture of Computing Systems-System Aspects in Organic and Pervasive Computing (ARCS'05): Workshop Self-Organization and Emergence, Innsbruck, Austria.

Du Sha, K. J., & Shi. W. (2005). *Wear: A balanced, fault-tolerant, energy-aware routing protocol for wireless sensor networks*. Technical Report MIST-TR-2005-001, Wayne State University.

Du Toit, N. E. (2005). *Modeling and Design of a MEMS Piezoelectric Vibration Energy Harvester* (MSc. thesis). Massachusetts Institute of Technology.

Duan, B., Xin, K., & Zhong, Y. (2019). Optimal Dispatching of Electric Vehicles Based on Smart Contract and Internet of Things. *IEEE Access: Practical Innovations, Open Solutions*, *8*, 9630–9639.

Dutta, J., & Roy, S. (2017). IoT-fog-cloud based architecture for smart city: Prototype of a smart building. In *2017 7th International Conference on Cloud Computing, Data Science & Engineering-Confluence* (pp. 237–242). IEEE.

Dzone. (2020). https://dzone.com/articles/12-iot-platforms-for-building-iot-projects

Ebaid, M. S., Qandil, H., & Hammad, M. (2013). A unified approach for designing a photovoltaic solar system for the underground water pumping well-34 at Disi aquifer. *Energy Conversion and Management*, *75*, 780–795.

EIA. (2009). *Energy Consumption by Manufacturers—Data Tables*. EIA.

Elasaad, H., Bilton, A., Kelley, L., Duayhe, O., & Dubowsky, S. (2015). Field evaluation of a community scale solar powered water purification technology: A case study of a remote Mexican community application. *Desalination*, *375*, 71–80.

Electronics and ICT. (2020). https://electronicsandict.com/iot-in-hvac-systems/

Erturk, A., & Inman, D. J. (2009). An experimentally validated bimorph cantilever model for piezoelectric energy harvesting from base excitations. Iopscience Smart Materials and Structures, 18(2). doi:10.1088/0964-1726/18/2/025009

Erturk, A., Vieira, W. G. R., DeMarqui, C. Jr, & Inman, D. J. (2010). On the energy harvesting potential of piezoaeroelastic systems. *Applied Physics Letters*, *96*(18), 184103. doi:10.1063/1.3427405

Eshghi, A. T., Lee, S., Sadoughi, M. K. M., Hu, C., Kim, Y. C., & Seo, J. H. (2017). Design optimization under uncertainty and speed variability for a piezoelectric energy harvester powering a tire pressure monitoring sensor. *Smart Materials and Structures*, *26*(10), 105037. doi:10.1088/1361-665X/aa8886

Ethier, A., & Bedard, J. (2007). *Development of a real-time water quality buoy for the fraser river estuary.* Paper presented at the OCEANS 2007. 10.1109/OCEANS.2007.4449424

ETSAP. (2010). *Marine Energy Technology Brief E13 - November, 2010, Energy Technology Systems Analysis Programme*. International Energy Agency.

European Editors. (2013). *Digi-Key Corporation*. https://www.digikey.com/en/articles/techzone/2013/sep/smaller-smarter-ultralowpower-sensors-raise-potential-for-energy-harvesting-in-medical-implants

Eu, Z. A., Tan, H. P., & Seah, W. K. (2011). Design and performance analysis of MAC schemes for wireless sensor networks powered by ambient energy harvesting. *Ad Hoc Networks*, *9*(3), 300–323. doi:10.1016/j.adhoc.2010.07.014

Eu, Z. A., Tan, H. P., & Seah, W. K. G. (2009). *Routing and relay nodeplacement in wireless sensor networks powered by ambient energy har-vesting. In Proc. of the IEEE Wireless Communications & Networking Conference*. WCNC.

Evangelatos, O., Samarasinghe, K., & Rolim, J. (2013). Syndesi: A framework for creating personalized smart environments using wireless sensor networks. In 2013 IEEE international conference on distributed computing in sensor systems (pp. 325–330). IEEE. doi:10.1109/DCOSS.2013.35

Fadel, E., Gungor, V. C., Nassef, L., Akkari, N., Malik, M. G. A., Almasri, S., & Akyildiz, I. F. (2015). A survey on wireless sensor networks for the smart grid. *Computer Communications*, *71*, 22–33. doi:10.1016/j.comcom.2015.09.006

Faheem, M., Tuna, G., & Gungor, V. C. (2017). LRP: Link quality-aware queue-based spectral clustering routing protocol for underwater acoustic sensor networks. *International Journal of Communication Systems*, *30*(12), 1–17. doi:10.1002/dac.3257

Fan, F. R., Tian, Z. Q., & Wang, Z. L. (2012). Flexible triboelectric generator. *Nano Energy*, *1*(2), 328–334. doi:10.1016/j.nanoen.2012.01.004

Farahnakian, F., Ashraf, A., Pahikkala, T., Liljeberg, P., Plosila, J., Porres, I., & Tenhunen, H. (2014). Using ant colony system to consolidate VMs for green cloud computing. *IEEE Transactions on Services Computing*, *8*(2), 187–198. doi:10.1109/TSC.2014.2382555

Fathabadi, H. (2019). Solar energy harvesting in buildings using a proposed novel electrochemical device as an alternative to PV modules. *Renewable Energy*, *133*, 118–125. doi:10.1016/j.renene.2018.10.010

Fei F., Mai J. D., & Li W. J. (2012). A wind-utter energy converter for powering wireless sensors. *Sensors and Actuators A: Physical, 173*(1), 163-171.

Feng, W., Alshaer, H., & Elmirghani, J. M. (2010). Green information and communication technology: Energy efficiency in a motorway model. *IET Communications*, *4*(7), 850–860. doi:10.1049/iet-com.2009.0153

Feng, X., & Hu, J. (2020). Research on the identification and management of vehicle behaviour based on Internet of things technology. *Computer Communications.*

Fernández-Ramos, J., Narvarte-Fernández, L. & Poza-Saura, F. (2010). Improvement of photovoltaic pumping systems based on standard frequency converters by means of programmable logic controllers. *Solar Energy, 84*(1), 101-109.

Ferrara, F., Gimelli, A., & Luongo, A. (2014). Small-scale Concentrated Solar Power (CSP) Plant: ORCs Comparison for Different Organic Fluids. *Energy Procedia, 45*, 217–226.

Filip, N., & Simu, I. (2011). About the noise energy conversion from agriculture tractor engines. *Proceedings of the 39thInternational Symposium on Agricultural Engineering*, 29–37.

Fisher, F., & Simmons, V. (1977). Sound absorption in sea water. *The Journal of the Acoustical Society of America*, *62*(3), 558–564. doi:10.1121/1.381574

Foong, F. M., Thein, C. K., Ooi, B. L., & Yurchenko, D. (2019). Increased power output of an electromagnetic vibration energy harvester through anti-phase resonance. *Mechanical Systems and Signal Processing*, *116*, 129–145. doi:10.1016/j.ymssp.2018.06.012

Foster, I., & Kesselman, C. (1999). *The globus toolkit*. Academic Press.

Frick, S., Kaltschmitt, M., & Schröder, G. (2010). Life cycle assessment of geothermal binary power plants using enhanced low-temperature reservoirs. *Energy*, *35*(5), 2281–2294. doi:10.1016/j.energy.2010.02.016

Fridleifsson, I. B., & Freeston, D. H. (1994). geothermal energy research and development. *Geothermics*, *23*(2), 175–214. doi:10.1016/0375-6505(94)90037-X

Friswell, M. I., Ali, S. F., Bilgen, O., Adhikari, S., Lees, A. W., & Litak, G. (2013). Non-linear piezoelectric vibration energy harvesting from a vertical cantilever beam with tip mass. *Journal of Intelligent Material Systems and Structures*, *23*(13), 1505–1521. doi:10.1177/1045389X12455722

Fruhlinger, J. (2018). *The Mirai botnet explained: How teen scammers and CCTV cameras almost brought down the internet.* CSO Online. Disponible en: https://www. csoonline. com/article/3258748/security/the-mirai-botnet-explainedhow-teen-scammers-and-cctv-cameras-almost-brought-down-the-internet. html

Fulk, C., Hobar, G., Olsen, K., El-Tawab, S., Rahman, F., & Ghazizadeh, P. (2019). Cloud-based Low-cost Energy Monitoring System through the Internet of Things. In *2019 IEEE International Conference on Pervasive Computing and Communications Workshops (PerCom Workshops)* (pp. 322–327). IEEE. 10.1109/PERCOMW.2019.8730823

Fu, Y., Ouyang, H., & Davis, R. B. (2019). Triboelectric energy harvesting from the vibro-impact of three cantilevered beams. *Mechanical Systems and Signal Processing*, *121*, 509–531. doi:10.1016/j.ymssp.2018.11.043

Gaikwad, P. P., Gabhane, J. P., & Golait, S. S. (2015). A survey-based on the Smart Homes system using Internet-of-Things. In *2015 International Conference on Computation of Power, Energy, Information, and Communication (ICCPEIC)* (pp. 330–335). IEEE. 10.1109/ICCPEIC.2015.7259486

Galinina, O., Mikhaylov, K., Andreev, S., Turlikov, A., & Koucheryavy, Y. (2015). Smart home gateway system over Bluetooth low energy with wireless energy transfer capability. *EURASIP Journal on Wireless Communications and Networking*, *2015*(1), 178. doi:10.118613638-015-0393-3

Galstyan, A., Czajkowski, K., & Lerman, K. (2005). Resource allocation in the grid with learning agents. *Journal of Grid Computing*, *3*(1-2), 91–100. doi:10.100710723-005-9003-7

Ganesan, D., Govindan, R., Shenker, S., & Estrin, D. (2001). Highly-resilient, energy-efficient multipath routing in wireless sensor networks. *Mobile Computing and Communications Review*, *5*(4), 1125.

Gao, X., & Pishdad-Bozorgi, P. (2019). BIM-enabled facilities operation and maintenance: A review. *Advanced Engineering Informatics*, *39*, 227–247. doi:10.1016/j.aei.2019.01.005

Gapchup, A., Wani, A., Wadghule, A., & Jadhav, S. (2017). Emerging trends of green IoT for smart world. *International Journal of Innovative Research in Computer and Communication Engineering*, *5*(2), 2139–2148.

Garcia, G. T., Sanchez, V. M., Marin, C. N. L., Cortez, J. I., Acevedo, C. A. R., Gonzalez, G. S. A., Ameca, J. L. H., & Garcia, M. D. C. M. (2018). Wireless sensor network for monitoring physical variables applied to green technology (IoT green technology). *European Journal of Electrical Engineering and Computer Science*, *2*(2). Advance online publication. doi:10.24018/ejece.2018.2.2.15

Garrido-Rubio, J., González-Piqueras, J., Campos, I., Osann, A., González-Gómez, L., & Calera, A. (2020). Remote sensing–based soil water balance for irrigation water accounting at plot and water user association management scale. *Agricultural Water Management*, *238*, 106236.

Geraldi, R. T., Reinehr, S., & Malucelli, A. (2020). Software product line applied to the internet of things: A systematic literature review. *Information and Software Technology*, 106293.

Gevorkian, P. (2012). *Large Scale Solar Power Systems: Construction and Financials*. Cambridge University Press.

Ghayvat, H., Liu, J., Babu, A., Alahi, E. E., Gui, X., & Mukhopadhyay, S. C. (2015). Internet of Things for smart homes and buildings: Opportunities and Challenges. *Journal of Telecommunications and the Digital Economy*, *3*(4), 33–47. doi:10.18080/jtde.v3n4.23

Ghayvat, H., Mukhopadhyay, S. C., & Gui, X. (2016). Issues and mitigation of interference, attenuation, and direction of arrival in IEEE 802.15. 4/ZigBee to wireless sensors and networks based smart building. *Measurement*, *86*, 209–226. doi:10.1016/j.measurement.2016.01.045

Gholamhosseinian, A., & Khalifeh, A. (2012). *Cloud computing and sustainability: Energy efficiency aspects*. Academic Press.

Ghosh, P., & Ghosh, S. (2020). IoT and Machine Learning in Green Smart Home Automation and Green Building Management. *Journal of Alternate Energy Sources and Technologies*, *10*(3), 8–36.

Ghosh, S., Kolady, D. E., Das, U., Gorain, S., Srivastava, S. K., & Mondal, B. (2019). Spatio-temporal variations in effects of participatory irrigation management (PIM) reform in India: A panel data analysis. *Agricultural Water Management*, *222*, 48–61.

Gilbert, J. M., & Balouchi, F. (2008). Comparison of energy harvesting systems for wireless sensor networks. *International Journal of Automation and Computing*, *5*(4), 334–347. doi:10.100711633-008-0334-2

Gipe, P. (1995). Wind energy comes of age (Vol. 4). Wiley. doi:10.1017/CBO9781139226608

Glasgo, B., Hendrickson, C., & Azevedo, I. L. (2017). Assessing the value of information in residential building simulation: Comparing simulated and actual building loads at the circuit level. *Applied Energy*, *203*, 348–363. doi:10.1016/j.apenergy.2017.05.164

Goiri, I. n. (2011). Greenslot: scheduling energy consumption in green datacenters. *International Conference for High Performance Computing, Networking, Storage and Analysis, SC '11*, 20:1–20:11. 10.1145/2063384.2063411

Goiri, I. n. (2012). Greenhadoop: leveraging green energy in data-processing frameworks. *7th ACM european conference on Computer Systems, EuroSys '12*, 57–70. 10.1145/2168836.2168843

Goldstein, E. A., Nasuta, D., Li, S., Martin, C., & Raman, A. (2018). Free Subcooling with the Sky: Improving the efficiency of air conditioning systems. *International Refrigeration and Air Conditioning Conference*. Paper 1913, https://docs.lib.purdue.edu/iracc/1913

Gondchawar, N., & Kawitkar, R. S. (2016). IoT based smart agriculture. *International Journal of Advanced Research in Computer and Communication Engineering*, *5*(6), 838–842.

Gordon, J. (1987). Optimal sizing of stand-alone photovoltaic solar power systems. *Solar Cells*, *20*(4), 295–313. doi:10.1016/0379-6787(87)90005-6

Gray, A. R. (2020, February 4). *Automated monitoring for changes in energy consumption patterns*. Google Patents.

Greenberg, A., Hamilton, J., Maltz, D. A., & Patel, P. (2008). The cost of a cloud: Research problems in data center networks. *Computer Communication Review*, *39*(1), 68–73. doi:10.1145/1496091.1496103

Green, P. L., Papatheou, E., & Sims, N. D. (2013). Energy harvesting from human motion and bridge vibrations: An evaluation of current nonlinear energy harvesting solutions. *Journal of Intelligent Material Systems and Structures*, *24*(12), 1494–1505. doi:10.1177/1045389X12473379

Greenspan, G. (2015). *Multichain private blockchain-white paper.*

Grid Workloads Archive. (n.d.). Retrieved from http://gwa.ewi.tudelft.nl/pmwiki/

GSMA-Green Power for Mobile. (2015). *The Global Telecom Tower ESCO Market.* Technical Report. Retrieved June 5, 2020 from https://www.gsma.com/mobilefordevelopment/wp-content/uploads/2015/01/140617-GSMA-report-draft-vF-KR-v7.pdf

Gui, C., & Mohapatra, P. (2003). *SHORT: Self-healing and optimizing routing techniques for mobile adhoc networks.* Paper presented at the MobiHoc. 10.1145/778415.778448

Guin, U., Cui, P., & Skjellum, A. (2018, July). Ensuring proof-of-authenticity of IoT edge devices using blockchain technology. In *2018 IEEE International Conference on Internet of Things (iThings) and IEEE Green Computing and Communications (GreenCom) and IEEE Cyber, Physical and Social Computing (CPSCom) and IEEE Smart Data (SmartData)* (pp. 1042-1049). IEEE. 10.1109/Cybermatics_2018.2018.00193

Gungor, V. C. & Hancke, G. P. (2009). Industrial wireless sensor networks: Challenges, design principles, and technical approaches. *Industrial Electronics, IEEE Transactions on, 56*(10), 4258-4265.

Guo, H., & Heidemann, J. (2018, August). IP-based IoT device detection. In *Proceedings of the 2018 Workshop on IoT Security and Privacy* (pp. 36-42). 10.1145/3229565.3229572

Guoqiang, S., Yanming, C., Chao, Z., & Yanxu, Z. (2013). Design and implementation of a smart IoT gateway. In *2013 IEEE International Conference on Green Computing and Communications and IEEE Internet of Things and IEEE Cyber, Physical and Social Computing* (pp. 720–723). IEEE. 10.1109/GreenCom-iThings-CPSCom.2013.130

Gupta, E. (2019). The impact of solar water pumps on energy-water-food nexus: Evidence from Rajasthan, India. *Energy Policy*, *129*, 598–609.

Gurumurthi, S. (2003a). Interplay of energy and performance for disk arrays running transaction processing workloads. *ISPASS*, 123–132.

Gurumurthi, S. (2003b). DRPM: Dynamic speed control for power management in server class disks. *ISCA*, 169–179.

Hajiheidari, S., Wakil, K., Badri, M., & Navimipour, N. J. (2019). Intrusion detection systems in the Internet of things: A comprehensive investigation. *Computer Networks*, *160*, 165–191.

Hakimi, S. M. (2017). A novel intelligent control of HVAC system in smart microgrid. *Journal of Electrical Systems and Information Technology*, *4*(2), 299–309. doi:10.1016/j.jesit.2017.01.005

Hale, P. (1986). Acceleration and time to fail. *Quality and Reliability Engineering International*, *2*(4), 259–262. doi:10.1002/qre.4680020409

Halim, M. A., & Park, J. Y. (2014). A non-resonant, frequency up-converted electromagnetic energy harvester from human-body-induced vibration for hand-held smart system applications. *Journal of Applied Physics*, *115*(9), 094901. doi:10.1063/1.4867216

Hamid, H. M. A., & Butler, Z. C. (2019). Design and Optimization of a MEMS Triboelectric Energy Harvester for Nanosensor Applications. *IEEE Sensors Applications Symposium. SAS 2018*, 1–6. 10.1109/SAS.2019.8706072

Han, L., & Wanhui, H. A. N. (2019, May 2). *Electric automobile energy monitoring and swapping network in remote monitoring of cloud computing network*. Google Patents.

Han, X., & Zhao, C. (2014). Distributing monitor system based on WIFI and GSM supporting SCPI. In *2014 13th International Symposium on Distributed Computing and Applications to Business, Engineering, and Science* (pp. 272–274). IEEE. 10.1109/DCABES.2014.59

Hande, A., Polk, T., Walker, W., & Bhatia, D. (2007). Indoor solar energyharvesting for sensor network router nodes. *Microprocessors and Microsystems*, *31*(6), 420–432. doi:10.1016/j.micpro.2007.02.006

Han, J., Mol, A. P. J., & Lu, Y. (2010). Solar water heaters in China: A new day dawning. *Energy Policy*, *38*(1), 383–391. doi:10.1016/j.enpol.2009.09.029

Han, N. D., Chung, Y., & Jo, M. (2015). Green data centers for cloud-assisted mobile ad hoc networks in 5G. *IEEE Network*, *29*(2), 70–76. doi:10.1109/MNET.2015.7064906

Harizopoulos, S. (2009). *Energy efficiency: the new holy grail of data management systems research*. HP Labs & UCLA.

Harmon, R. R., & Auseklis, N. (2009). *Sustainable IT services: Assessing the impact of green computing practices*. Paper presented at the PICMET'09-2009 Portland International Conference on Management of Engineering & Technology. 10.1109/PICMET.2009.5261969

Haruna, A. A., Zakaria, N. B., Jung, L. T., Pal, A. J., Naono, K., & Okitsu, J. (2014). *Performance comparison of least slack time based heuristics for job scheduling on computational grid*. Paper presented at the 2014 International Conference on Computer and Information Sciences (ICCOINS). 10.1109/ICCOINS.2014.6868841

Harvey, L. D. D. (2006). *A Handbook on Low-Energy Buildings and District-Energy Systems: Fundamentals, Techniques and Examples. Earthscan.* Sterling.

Hawkes, A., & Leach, M. (2007). Cost-effective operating strategy for residential microcombined heat and power. *Energy*, *32*(5), 711–723. doi:10.1016/j.energy.2006.06.001

Heer, R., Wissenwasser, J., Milnera, M., & Farmer, L. (2010). Wireless powered electronic sensors for biological applications. Proceedings of IEEE EMBC 2010, 700-703.

Helen, P. (2006). *A* Service Management Approach for Self-healing Wireless Sensor Networks. *LNCS*, *4195*, 215–228.

Hendijanizadeh, M., Sharkh, S. M., Elliott, S. J., & Torbati, M. M. (2013). Output power and efficiency of electromagnetic energy harvesting systems with constrained range of motion. *Smart Materials and Structures*, *22*(12), 125009. doi:10.1088/0964-1726/22/12/125009

Hernández-Ramos, J. L., Moreno, M. V., Bernabé, J. B., Carrillo, D. G., & Skarmeta, A. F. (2015). SAFIR: Secure access framework for IoT-enabled services on smart buildings. *Journal of Computer and System Sciences*, *81*(8), 1452–1463. doi:10.1016/j.jcss.2014.12.021

Hersent, O., Boswarthick, D., & Elloumi, O. (2011). The internet of things: Key applications and protocols. John Wiley & Sons.

Hi-globe. (n.d.). Retrieved September 15, 2015, from http://hi-globe.com/?page_id=85

Hill, J. L., & Culler, D. E. (2002). Mica: A wireless platform for deeply embedded networks. *IEEE Micro*, *22*(6), 12–24. doi:10.1109/MM.2002.1134340

Hobbs, W. B., & Hu, D. L. (2012). Tree-inspired piezoelectric energy harvesting. [tree inspired]. *Journal of Fluids and Structures*, *28*, 103–114. doi:10.1016/j.jfluidstructs.2011.08.005

Holman, R., Stanley, J., & Ozkan-Haller, T. (2003). Applying video sensor networks to nearshore environment monitoring. *IEEE Pervasive Computing*, *2*(4), 14–21. doi:10.1109/MPRV.2003.1251165

Huang, X. (2020). Intelligent remote monitoring and manufacturing system of production line based on industrial Internet of Things. *Computer Communications*, *150*, 421–428.

Hua, W. S., Li-Dong, D., De-Yi, K., Hao-Yue, P., Zhen, F., & Zhan, Z. (2014). Hybrid device for acoustic noise reduction and energy harvesting based on a silicon micro-perforated panel structure. *Chinese Physics B*, *23*(4), 044302. doi:10.1088/1674-1056/23/4/044302

Hubert, T., & Grijalva, S. (2011). Realizing smart grid benefits requires energy optimization algorithms at residential level. *Innovative Smart Grid Technologies (ISGT)*, 1–8.

Humayun & Jhanjhi. (2019). Exploring The Relationship Between GSD, Knowledge Management, Trust And Collaboration. *Journal of Engineering Science and Technology, 14*(2), 820-843.

Humayun, Jhanjhi, Alsayat, & Ponnusamy. (2020). Internet of things and ransomware: Evolution, mitigation and prevention. *Egyptian Informatics Journal.*

Humayun, M., Jhanjhi, N. Z., Malak, Z., & Khan, A. (2020). Smart Cities and Digital Governance. In *Employing Recent Technologies for Improved Digital Governance*. IGI Global. doi:10.4018/978-1-7998-1851-9.ch005

Humayun, M., Jhanjhi, N. Z., Hamid, B., & Ahmed, G. (2020). Emerging Smart Logistics and Transportation Using IoT and Blockchain. *IEEE Internet of Things Magazine*, *3*(2), 58–62.

Humayun, M., Niazi, M., Jhanjhi, N., Alshayeb, M., & Mahmood, S. (2020). Cyber Security Threats and Vulnerabilities: A Systematic Mapping Study. *Arabian Journal for Science and Engineering*, *45*(4), 3171–3189. Advance online publication. doi:10.100713369-019-04319-2

Hussain, K., Hussain, S. J., Jhanjhi, N. Z., & Humayun, M. (2019). SYN Flood Attack Detection based on Bayes Estimator (SFADBE) For MANET. *International Conference on Computer and Information Sciences (ICCIS)*, 1-4. 10.1109/ICCISci.2019.8716416

Hu, W., Liu, J., Zhu, Y., Dong, J., Ma, X., Li, S., Zhang, J., Zhu, X., & Chen, W. (2019). Analysis of energy monitoring for a double-pulsed CO 2 integrated path differential absorption lidar at 1.57 μm. *Applied Optics*, *58*(3), 616–625. doi:10.1364/AO.58.000616 PMID:30694245

Hu, Y., Yang, J., Jing, Q., Niu, S., Wu, W., & Wang, Z. L. (2013). Triboelectric Nanogenerator Built on Suspended 3D Spiral Structure as Vibration and Positioning Sensor and Wave Energy Harvester. *ACS Nano*, *7*(11), 10424–10432. doi:10.1021/nn405209u PMID:24168315

Hwang, M., Choi, D., & Kim, P. (2010). *Least slack time rate first: New scheduling algorithm for multi-processor environment.* Paper presented at the 2010 International Conference on Complex, Intelligent and Software Intensive Systems. 10.1109/CISIS.2010.20

Hwang, Y., Radermacher, R., Al Alili, A., & Kubo, I. (2008). Review of Solar Cooling Technologies. *HVAC & R Research*, *14*(3), 507–528. doi:10.1080/10789669.2008.10391022

IEA. (2007). *Renewables for Heating and Cooling – Untapped Potential.* International Energy Agency.

IEEE Computer Society. (2017). Retrieved June 5, 2020 from https://www.computer.org/publications/tech-news/research/internet-of-military-battlefield-things-iomt-iobt

Imran, M., Khan, L. U., Yaqoob, I., Qureshi, M. A., & Ahmed, A. (2019). Energy harvesting in 5G networks: Taxonomy, requirements, challenges and future directions. *IEEE Wirel. Commun.* https://arxiv.org/pdf/1910.00785.pdf

Intanagonwiwat, C., Govindan, R., & Estrin, D. (2000). *Directed diffusion: a scalable and robust communication paradigm for sensor networks.* Paper presented at the ACM MobiCom '00, Boston, MA.

International Energy Agency. (2020). https://www.iea.org/reports/the-future-of-cooling

International Telecommunication Union. (2005). *ITU internet report 2005: the internet of things.* International Telecommunication Union, *Workshop Report.*

Internet of Thing (IoT). (n.d.). Retrieved from Cisco web site: https://www.cisco.com/web/solutions/trends/iot/overview.html

Internet of Thing. (1994-2014). Retrieved from Tech Target web site: https://whatis.techtarget.com/definition/Internet-of-Things

Internet of Things. (2014). Retrieved from Microsoft web site: http://www.microsoft.com/windowsembedded/en-us/internet-of-things.aspx

IRENA. (2019). *Future of Solar Photovoltaic: Deployment, investment, technology, grid integration and socio-economic aspects (A Global Energy Transformation: paper).* International Renewable Energy Agency.

J., V., G., V. H., D., S., & L., H. (2017). Model selection for continuous commissioning of HVAC-systems in office buildings: A review. *Renewable and Sustainable Energy Reviews, 76*, 673–686.

Jain, R. K., Banerjee, P., Baksi, D., & Samanta, S. K. (2019, July). IoT Based Interface Device for Automatic Molding Machine towards SMART FOUNDRY-2020. In *2019 10th International Conference on Computing, Communication and Networking Technologies (ICCCNT)* (pp. 1-6). IEEE.

Jain, R., Goel, V., Rekhi, J. K., & Alzubi, J. A. (2020). IoT-Based Green Building: Towards an Energy-Efficient Future. In Green Building Management and Smart Automation (pp. 184–207). IGI Global.

Jamal, Al-Karaki, & Kamal. (2000). *Routing Techniques in Wireless Sensor Networks: A Survey.* B. Karp and H. Kung. Greedy Perimeter Stateless Routing. In *Proceedings of ACM Conf. On Mobile Computing and Networking (MOBICOM)*. ACM

Jan, M.A., Nanda, P., He, X., & Liu, R.P. (2015, August). A sybil attack detection scheme for a centralized clustering-based hierarchical network. In *2015 IEEE Trustcom/BigDataSE/ISPA* (Vol. 1, pp. 318-325). IEEE.

Jan, M. A., Nanda, P., He, X., & Liu, R. P. (2018). A Sybil attack detection scheme for a forest wildfire monitoring application. *Future Generation Computer Systems*, *80*, 613–626. doi:10.1016/j.future.2016.05.034

Jayakody, T. Chatzinotas, & Durrani. (2017). Wireless Information and Power Transfer: A New Paradigm for Green Communications. Springer.

Jaziri, N., Boughamoura, A., Müller, J., Mezghani, B., Tounsi, F., & Ismail, M. (2019). A comprehensive review of thermoelectric generators: Technologies and common applications. *Energy Reports*. Advance online publication. doi:10.1016/j.egyr.2019.12.011

Jeyasheeli, P. G., & Selva, J. V. J. (2017). An IOT design for smart lighting in green buildings based on environmental factors. In *2017 4th International Conference on Advanced Computing and Communication Systems (ICACCS)* (pp. 1–5). IEEE. 10.1109/ICACCS.2017.8014559

Jhanjhi, N. Z., Almusali, F. A., Brohi, S. N., & Abdullah, A. (2018). Middleware Power Saving Scheme for Mobile Applications. *IEEE Fourth International Conference on Advances in Computing, Communication & Automation (ICACCA)*, 1-6.

Jhanjhi, N. Z., & Almusalli, F. A. (2017). Review: Smartphones power consumption & energy saving techniques. *2017 International Conference on Innovations in Electrical Engineering and Computational Technologies (ICIEECT)*, 1-7. 10.1109/ICIEECT.2017.7916593

Jia, D., & Liu, J. (2009). Human power-based energy harvesting strategies for mobile electronic devices. *Frontiers of Energy and Power Engineering in China*, *3*(1), 27–46. doi:10.100711708-009-0002-4

Jiang, X., Polastre, J., & Culler, D. (2005). Perpetual Environmentally Powered Sensor Networks. *Fourth International Symposium on Information Processing in Sensor Networks*, *36*(5), 463–468.

Jiang, X., Polastre, J., & Culler, D. (2005, April). Perpetual environmentally powered sensor networks. In *Proceedings of the 4th international symposium on Information processing in sensor networks* (p. 65). IEEE Press.

Jiao, P., Borchani, W., Hasni, H., & Lajnef, N. (2017). Enhancement of quasi-static strain energy harvesters using non-uniform cross-section post-buckled beams. *Smart Materials and Structures*, *26*(8), 085045. doi:10.1088/1361-665X/aa746e

Jin, X., Zhang, F., Vasilakos, A. V., & Liu, Z. (2016). *Green data centers: A survey, perspectives, and future directions.* arXiv preprint arXiv:1608.00687.

Joseph, A. D. (2005). Works in Progress-Energy Harvesting Projects. *IEEE Pervasive Computing*, *4*(1), 69–71. doi:10.1109/MPRV.2005.8

Jouhari, M., Ibrahimi, K., Tembine, H., & Ben-Othman, J. (2019). Underwater Wireless Sensor Networks: A Survey on Enabling Technologies, Localization Protocols, and Internet of Underwater Things. *IEEE Access: Practical Innovations, Open Solutions*, *7*, 96879–96899. doi:10.1109/ACCESS.2019.2928876

Juang, S. S.-Y., Lin, P.-Y., Lin, Y.-C., Chen, Y.-S., Shen, P.-S., Guo, Y.-L., Wu, Y.-C., & Chen, P. (2019). Energy Harvesting Under Dim-Light Condition With Dye-Sensitized and Perovskite Solar Cells. *Frontiers in Chemistry*, *7*, 209. doi:10.3389/fchem.2019.00209 PMID:31024895

Jurdak, R., Lopes, C. V., & Baldi, P. (2004). Battery lifetime estimation and optimization for underwater sensor networks. *IEEE Sensor Network Operations*, *2006*, 397–420.

Kalpakis, K., Dasgupta, K., & Namjoshi, P. (2002). Maximum lifetime data gathering and aggregation in wireless sensor networks. In *Networks* (pp. 685–696). World Scientific. doi:10.1142/9789812776730_0055

Kamal & Al-Karaki. (n.d.). *Efficient Virtual-Backbone Routing in Mobile Ad Hoc Networks*. Dept. of Electrical and Computer Engineering lowa State University.

Kamalam, G., & Bhaskaran, V. M. (2011). An efficient hybrid job scheduling algorithm for computational grids. *International Journal of Computers and Applications*.

Kamalinejad, P., Mahapatra, C., Sheng, Z., Mirabbasi, S., Leung, V. C., & Guan, Y. L. (2015). Wireless energy harvesting for the internet of things. *IEEE Communications Magazine*, *53*(6), 102–108. doi:10.1109/MCOM.2015.7120024

Kanna, P. R., Ahmad, S. S., & Khan, M. A. U. (n.d.). *A Study of Green Building-Renewable Energy Harvesting Technique*. Academic Press.

Kansal, A., & Srivastava, M. B. (2003). An environmental energy harvesting framework for sensor networks. In *Low Power Electronics and Design, 2003. ISLPED'03, Proceedings of the 2003 International Symposium on* (pp. 481-486). 10.1145/871506.871624

Kansal, A., Hsu, J., Zahedi, S., & Srivastava, M. B. (2007). Power management in energy harvesting sensor networks. *ACM Transactions on Embedded Computing Systems*, *6*(4), 32. doi:10.1145/1274858.1274870

Kao, C. C., Lin, Y. S., De Wu, G., & Huang, C. J. (2017). A comprehensive study on the internet of underwater things: Applications, challenges, and channel models. *Sensors (Switzerland)*, *17*(7), 1375–1378. doi:10.339017071477 PMID:28640220

Karami, M. A., Bilgen, O., Inman, D. J., & Friswell, M. I. (2011). Experimental and analytical parametric study of single-crystal unimorph beams for vibration energy harvesting. *IEEE Transactions on Ultrasonics, Ferroelectrics, and Frequency Control*, *58*(7), 1508–1520. doi:10.1109/TUFFC.2011.1969 PMID:21768034

Kareem, A., Kijewski, T., & Tamura, Y. (1999). Mitigation of motions of tall buildings with specific examples of recent applications. *Wind and Structures*, *2*(3), 201–251. doi:10.12989/was.1999.2.3.201

Kerr, D. (2007), Marine energy. *Philosophical Transactions of the Royal Society London, Series A (Mathematical, Physical and Engineering Sciences)*, *365*(1853), 971-92.

Khalilnejad, A., French, R. H., & Abramson, A. R. (2020). Data-driven evaluation of HVAC operation and savings in commercial buildings. *Applied Energy*, *278*, 115505. doi:10.1016/j.apenergy.2020.115505

Khan, A., Jhanjhi, N. Z., Humayun, M., & Ahmad, M. (2020). The Role of IoT in Digital Governance. In *Employing Recent Technologies for Improved Digital Governance*. IGI Global. . doi:10.4018/978-1-7998-1851-9.ch007

Khan, Z. A., & Abbasi, U. (2016). Evolution of wireless sensor networks toward internet of things. Emerging Communication Technologies Based on Wireless Sensor Networks: Current Research and Future Applications, 179–200. doi:10.1201/b13092

Khan, A. (2020). N. Z. Jhanjhi, Mamoona Humayun, and Muneer Ahmad. "The Role of IoT in Digital Governance. In *Employing Recent Technologies for Improved Digital Governance* (pp. 128–150). IGI Global.

Khan, A., Jhanjhi, N. Z., & Humayun, M. (2020). *Secure Smart and Remote Multipurpose Attendance Monitoring System, EW*. EAI. doi:10.4108/eai.13-7-2018.164583

Khan, F. U., & Iqbal, M. (2018). Electromagnetic Bridge Energy Harvester Utilizing Bridge's Vibrations and Ambient Wind for Wireless Sensor Node Application. *Journal of Sensors*, *2018*, 3849683. doi:10.1155/2018/3849683

Khanna, A., Arora, S., Chhabra, A., Bhardwaj, K. K., & Sharma, D. K. (2019). IoT architecture for preventive energy conservation of smart buildings. In *Energy Conservation for IoT Devices* (pp. 179–208). Springer. doi:10.1007/978-981-13-7399-2_8

Khan, W. Z., Rehman, M. H., Zangoti, H. M., Afzal, M. K., Armi, N., & Salah, K. (2020). Industrial internet of things: Recent advances, enabling technologies and open challenges. *Computers & Electrical Engineering*, *81*, 106522.

Kim, J. Y., Lee, H.-J., Son, J.-Y., & Park, J.-H. (2015). Smart home web of objects-based IoT management model and methods for home data mining. In *2015 17th Asia-Pacific Network Operations and Management Symposium (APNOMS)* (pp. 327–331). IEEE. 10.1109/APNOMS.2015.7275448

Kim, S. G., Priya, S., & Kanno, I. (2012). Piezoelectric MEMS for energy harvesting. *MRS Bulletin*, *37*(11), 1039–1050. doi:10.1557/mrs.2012.275

King, R. R., Law, D. C., Edmondson, K. M., Fetzer, C. M., Kinsey, G. S., Yoon, H., Sherif, R. A., & Karam, N. H. (2007). 40% efficient metamorphic GaInP/GaInAs/Ge multijunction solar cells. *Applied Physics Letters*, *90*(18), 3. doi:10.1063/1.2734507

Kinsler, L. E., Frey, A. R., Coppens, A. B., & Sanders, J. V. (1999). *Fundamentals of acoustics*. Academic Press.

Klein, J., & Zuo, L. (2017). A velocity-amplified electromagnetic energy harvester for small amplitude vibration. *Smart Materials and Structures*, *26*(9), 095057. doi:10.1088/1361-665X/aa7e54

Kolias, C., Kambourakis, G., Stavrou, A., & Voas, J. (2017). DDoS in the IoT: Mirai and other botnets. *Computer*, *50*(7), 80–84. doi:10.1109/MC.2017.201

Kong, J., Cui, J.-h., Wu, D., & Gerla, M. (2005). *Building underwater ad-hoc networks and sensor networks for large scale real-time aquatic applications*. Paper presented at the MILCOM 2005-2005 IEEE Military Communications Conference.

Kong, J., Chung, S. W., & Skadron, K. (2012). Recent thermal management techniques for microprocessors. *ACM Computing Surveys*, *44*(3), 1–42. doi:10.1145/2187671.2187675

Konya, D. B. (n.d.). *The NorduGrid team*. Retrieved from http://www.nordugrid.org/

Koomey, J. (2011). *Growth in data center electricity use 2005 to 2010*. Analytical Press.

Koomey, J. (2011). *Growth in data center electricity use 2005 to 2010*. Academic Press.

Koopman, P. (2003). Elements of the self-healing problem space. In *ICSE Workshop on Architecting Dependable Systems*. Available at: http://74.125.155.132/scholar?q=cache:qoVzFSeebusJ:scholar.google.com/+self-healing+wireless+senso r+netowrk+lanthaler&hl=en&as_sdt=2000

Koutitas, G. (2010). Green network planning of single frequency networks. *IEEE Transactions on Broadcasting*, *56*(4), 541–550. doi:10.1109/TBC.2010.2056252

Kowalska-Pyzalska, A., & Byrka, K. (2019). Determinants of the willingness to energy monitoring by residential consumers: A case study in the city of Wroclaw in Poland. *Energies*, *12*(5), 907. doi:10.3390/en12050907

Krioukov, A. (2011). Integrating renewable energy using data analytics systems: Challenges and opportunities. Bulletin of the IEEE Computer Society Technical Committee, 1–9.

Kroposki, B. (2017). Integrating high levels of variable renewable energy into electric power systems. *Journal of Modern Power Systems and Clean Energy*, *5*(6), 831–837. doi:10.100740565-017-0339-3

Kubba, A. E., & Jiang, K. (2014). A Comprehensive Study on Technologies of Tyre Monitoring Systems and Possible Energy Solutions. *Sensors (Basel)*, *14*(6), 10306–10345. doi:10.3390140610306 PMID:24922457

Kumar, A., & Hancke, G. P. (2014). Energy efficient environment monitoring system based on the IEEE 802.15. 4 standard for low cost requirements. *IEEE Sensors Journal*, *14*(8), 2557–2566. doi:10.1109/JSEN.2014.2313348

Kumar, A., Sharma, A., Kumar, R., Vaish, R., & Chauhan, V. S. (2014). Finite element analysis of vibration energy harvesting using lead-free piezoelectric materials: A comparative study. *Journal of Asian Ceramic Societies*, *2*(2), 138–143. doi:10.1016/j.jascer.2014.02.001

Lachhab, F., Bakhouya, M., Ouladsine, R., & Essaaidi, M. (2018). Towards an Intelligent Approach for Ventilation Systems Control using IoT and Big Data Technologies. *Procedia Computer Science*, *130*, 926–931. doi:10.1016/j.procs.2018.04.091

Lai, S.-K., Wang, C., & Zhang, L. H. (2019). A nonlinear multi-stable piezomagnetoelastic harvester array for low-intensity, low-frequency, and broadband vibrations. *Mechanical Systems and Signal Processing*, *122*, 87–102. doi:10.1016/j.ymssp.2018.12.020

Lam, K. Y., & Chi, C. H. (2016, November). Identity in the Internet-of-Things (IoT): New challenges and opportunities. In *International Conference on Information and Communications Security* (pp. 18-26). Springer. 10.1007/978-3-319-50011-9_2

Lee, R.-G., Chen, K.-C., Chiang, S.-S., Lai, C.-C., Liu, H.-S., & Wei, M.-S. (2006). A Backup Routing with Wireless Sensor Network for Bridge Monitoring System. In *Annual Conference on Communication Networks and Services Research*. IEEE Computer Society.

Lee, V. C. (2012). *Energy Harvesting for Wireless Sensor Network* (Doctoral dissertation). University of California, Berkeley, CA.

Lee, I. (2019). The Internet of Things for enterprises: An ecosystem, architecture, and IoT service business model. *Internet of Things*, *7*, 100078. doi:10.1016/j.iot.2019.100078

Lee, Y. J., Zhou, G., & Lua, K. B. (2019). Vortex-induced vibration wind energy harvesting by piezoelectric MEMS device in formation. *Nature*, *9*, 20404. PMID:31892701

Leung, V. (2015). Green Internet of Things for smart cities. *International workshop on smart cities and urban informatics*.

Li, B., & Shen, B. (2010). *Slack-based advance reservation for Grid jobs*. Paper presented at the 2010 3rd International Conference on Advanced Computer Theory and Engineering (ICACTE). Retrieved from http://lcg.web.cern.ch/LCG/

Li, L., Zhang, Y., San, H., Guo, Y. I., & Chen, X. (2008). MEMS for vibration energy harvesting. *MEMS/MOEMS Technologies and Applications III*, vol. 6836 *of Proceedings of the SPIE*.

Li, K., He, Q., Wang, J., Zhou, Z., & Li, X. (2018). Wearable energy harvesters generating electricity from low-frequency human limb movement. *Microsystems & Nanoengineering*, *4*(24), 24. doi:10.103841378-018-0024-3 PMID:31057912

Li, L., Zheng, Y., Zheng, S., & Ke, H. (2020). The new smart city programme: Evaluating the effect of the internet of energy on air quality in China. *The Science of the Total Environment*, *714*, 136380.

Lin, K., Yu, J., Hsu, J., Zahedi, S., Lee, D., Friedman, J., ... Srivastava, M. (2005, November). Heliomote: enabling long-lived sensor networks through solar energy harvesting. In *Proceedings of the 3rd international conference on Embedded networked sensor systems* (pp. 309-309). ACM. 10.1145/1098918.1098974

LinkedIn. (2020). *IoT characteristics*. https://www.linkedin.com/pulse/internet-things-iot-characteristics-kavyashree-g-c

Lin, L., Guan, X., Hu, B., Li, J., Wang, N., & Sun, D. (2020). Deep reinforcement learning and LSTM for optimal renewable energy accommodation in 5G internet of energy with bad data tolerant. *Computer Communications*.

Lin, T., Wang, J. J., & Zuo, L. (2018). Efficient electromagnetic energy harvester for railroad transportation. *Mechatronics*, *53*, 277–286. doi:10.1016/j.mechatronics.2018.06.019

Lin, Y. D., Yeh, J. H., Yang, T. H., Ku, C. Y., Tsao, S. L., & Lai, Y. C. (2009). Efficient dynamic frame aggregation in IEEE 802.11 s mesh networks. *International Journal of Communication Systems*, *22*(10), 1319–1338. doi:10.1002/dac.1028

Lin, Y. H., Chou, Z. T., Yu, C. W., & Jan, R. H. (2015). Optimal and maximized configurable power saving protocols for corona-based wireless sensor networks. *IEEE Transactions on Mobile Computing*, *14*(12), 2544–2559. doi:10.1109/TMC.2015.2404796

Liou, E. C., Kao, C. C., Chang, C. H., Lin, Y. S., & Huang, C. J. (2018). Internet of underwater things: Challenges and routing protocols. *Proceedings of 4th IEEE International Conference on Applied System Innovation 2018, ICASI 2018*, 1171–1174. 10.1109/ICASI.2018.8394494

Liu, Z. (2011). Greening geographical load balancing. ACM SIGMETRICS joint international conference on Measurement and modeling of computer systems, SIGMETRICS'11, 233–244. doi:10.1145/1993744.1993767

Liu, K., Bi, Y., & Liu, D. (2020). Internet of Things based acquisition system of industrial intelligent bar code for smart city applications. *Computer Communications*, *150*, 325–333.

Liu, L., Wang, H., Liu, X., Jin, X., He, W. B., Wang, Q. B., & Chen, Y. (2009, June). GreenCloud: a new architecture for green data center. In *Proceedings of the 6th international conference industry session on Autonomic computing and communications industry session* (pp. 29-38). 10.1145/1555312.1555319

Liu, Q., Ma, Y., Alhussein, M., Zhang, Y., & Peng, L. (2016). Green data center with IoT sensing and cloud-assisted smart temperature control system. *Computer Networks*, *101*, 104–112. doi:10.1016/j.comnet.2015.11.024

Liu, X., & Ansari, N. (2018). Dual-Battery Enabled Green Proximal M2M Communications in LPWA for IoT. *Proc. IEEE Int. Conf. Commun. (ICC)*, 1–6. 10.1109/ICC.2018.8422203

Liu, X., & Ansari, N. (2019). Toward green IoT: Energy solutions and key challenges. *IEEE Communications Magazine*, *57*(3), 104–110. doi:10.1109/MCOM.2019.1800175

Liu, Y.-P., & Vasic, D. (2012). Self-Powered Electronics for Piezoelectric Energy Harvesting Devices. In M. Lallart (Ed.), *Small Scale Energy Harvesting* (p. 327). InTech. doi:10.5772/51211

Liu, Y., & Zhang, S. (2020). Information security and storage of Internet of Things based on block chains. *Future Generation Computer Systems*, *106*, 296–303.

Li, W., Koo, C., Hong, T., Oh, J., Cha, S. H., & Wang, S. (2020). A novel operation approach for the energy efficiency improvement of the HVAC system in office spaces through real-time big data analytics. *Renewable & Sustainable Energy Reviews*, 127.

Li, X., Guo, M., & Dong, S. (2011). A flex-compressive-mode piezoelectric transducer for mechanical vibration/strain energy harvesting. *IEEE Transactions on Ultrasonics, Ferroelectrics, and Frequency Control*, *58*(4), 698–703. doi:10.1109/TUFFC.2011.1862 PMID:21507747

Li, X., Hehn, T., Thewes, M., Kuehne, I., Frey, A., Scholl, G., & Manoli, Y. (2013a). Non-resonant electromagnetic energy harvester for car-key applications. *Journal of Physics: Conference Series*, *476*, 012096. doi:10.1088/1742-6596/476/1/012096

Li, Y., & Shi, R. (2015). An intelligent solar energy-harvesting system for wireless sensor networks. *EURASIP Journal on Wireless Communications and Networking*, *2015*(1), 179. doi:10.118613638-015-0414-2

Li, Y., Yu, H., Su, B., & Shang, Y. (2008). Hybrid micro power source for wireless sensor network. *IEEE Sensors Journal*, *8*(6), 678–681. doi:10.1109/JSEN.2008.922692

Li, Z., Zuo, L., Kuang, J., & Luhrs, G. (2013b). Energy-harvesting shock absorber with a mechanical motion rectifier. *Smart Materials and Structures*, *22*(2), 025008. doi:10.1088/0964-1726/22/2/025008

Lloret, J., Macías, E., Suárez, A., & Lacuesta, R. (2012). Ubiquitous monitoring of electrical household appliances. *Sensors (Basel)*, *12*(11), 15159–15191. doi:10.3390121115159 PMID:23202205

Lohndorf, M., Kvisteroy, T., Westby, E., & Halvorsen, E. (2007). Evaluation of energy harvesting concepts for tire pressure monitoring systems. Proceedings PowerMEMS 2007, 331 – 334.

Lu, C., Park, S. P., Raghunathan, V., & Roy, K. (2010). Analysis and design of ultra low power thermoelectric energy harvesting systems. *Proceedings of the ACM/IEEE International Symposium on Low-Power Electronics and Design (ISLPED '10)*, 183–188. 10.1145/1840845.1840882

Lu, C., Raghunathan, V., & Roy, K. (2010). Micro-scale energy harvesting: A system design perspective. *Design and Automation Conference (ASPDAC) IEEE*, 89 – 94. 10.1109/ASPDAC.2010.5419913

Lucy, E. A. (2010). *Innowattech: Harvesting Energy and Data A stand-alone technology*. Retrieved March 2015, 2, from https://www.academia.edu/29338723/Innowattech_Harvesting_Energy_and_Data_A_stand_alone_technology

Lu, F., Lee, H. P., & Lim, S. P. (2004). Modeling and analysis of micro piezoelectric power generators for micro-electromechanical systems applications. *Smart Materials and Structures*, *13*(1), 57–63. doi:10.1088/0964-1726/13/1/007

Lund, J. W. (1995). Onion dehydration. *Transactions - Geothermal Resources Council*, *19*, 69–74.

Maeguchi, T., Masuda, A., Katsumura, H., Kagata, H., & Okumura, H. (2015). Band widening of piezoelectric vibration energy harvesters by utilizing mechanical stoppers and magnets. *Journal of Physics: Conference Series*, *660*, 012099. doi:10.1088/1742-6596/660/1/012099

Mahalle, P., Babar, S., Prasad, N. R., & Prasad, R. (2010, July). Identity management framework towards internet of things (IoT): Roadmap and key challenges. In *International Conference on Network Security and Applications* (pp. 430-439). Springer. 10.1007/978-3-642-14478-3_43

Mahmud, K., Khan, B., Ravishankar, J., Ahmadi, A., & Siano, P. (2020). An internet of energy framework with distributed energy resources, prosumers and small-scale virtual power plants: An overview. *Renewable & Sustainable Energy Reviews*, *127*, 109840.

Maidment, G., & Tozer, R. (2002). Combined cooling heat and power in supermarkets. *Applied Thermal Engineering, 22*(6), 653–665. doi:10.1016/S1359-4311(01)00117-X

Mainwaring, A., Culler, D., Polastre, J., Szewczyk, R., & Anderson, J. (2002). Wireless Sensor Networks for Habitat Monitoring. In *Proc. 1st ACM International Workshop on Wireless Sensor Networks and Applications*. ACM.

Maksimovic, M. (2017). Green Internet of Things (G-IoT) at engineering education institution: the classroom of tomorrow. *Green Internet of Things*, 270-273.

Maksimovic, M. (2017). Greening the future: Green Internet of Things (G-IoT) as a key technological enabler of sustainable development. In Internet of Things and Big Data Analytics toward Next Generation Intelligence. Springer.

Maksimović, M., & Omanović-Mikličanin, E. (2017). Green internet of things and green nanotechnology role in realizing smart and sustainable agriculture. In *VIII International Scientific Agriculture Symposium "AGROSYM 2017* (pp. 2290-2295). Academic Press.

Maksimovic, M. (2017). The role of green internet of things (G-IoT) and big data in making cities smarter, safer and more sustainable. *International Journal of Computing and Digital Systems, 6*(04), 175–184. doi:10.12785/IJCDS/060403

Mancelos, N., Correia, J., Pires, L. M., Oliveira, L. B., & Oliveira, J. P. (2014). Piezoelectric Energy Harvester for a CMOS Wireless Sensor. In *Technological Innovation for Collective Awareness Systems* (pp. 470–477). Springer Berlin Heidelberg. doi:10.1007/978-3-642-54734-8_52

Mandal, S., Turicchia, L., & Sarpeshkar, R. (2010). A low-power, battery-free tag for body sensor networks. *IEEE Pervasive Computing, 9*(1), 71-77.

Mann, B. P., & Owens, B. A. (2010). Investigations of a nonlinear energy harvester with a bistable potential well. *Journal of Sound and Vibration, 329*(9), 1215–1226. doi:10.1016/j.jsv.2009.11.034

Manual, X. U. (n.d.). *Citrix: Zen Server*. Retrieved June, 3, 2013, from http://bits.xensource.com/Xen/docs/user.pdf

Manyika, J. (2013). *Disruptive technologies: Advance that will transform life, business and the global economy*. Retrieved from McKinsey & Company web site: https://www.mckinsey.com/insights/business_technology/disruptive_technologies

Mao, G. (2017). 5G green mobile communication networks. *China Communications, 14*(2), 183–184. doi:10.1109/CC.2017.7868166

Mardamutu, K., Ponnusamy, V., & Zaman, N. (2018). Green energy in data centers. In Sustainable Development: Concepts, Methodologies, Tools, and Applications (pp. 386-401). IGI Global. doi:10.4018/978-1-5225-3817-2.ch019

Markit, I. H. S. (2017). *Number of Connected IoT Devices Will Surge to 125 Billion by 2030*. Retrieved June 5, 2020 from https://news.ihsmarkit.com/prviewer/release_only/slug/number-connected-iot-devices-will-surge-125-billion-2030-ihs-markit-says

Maruyama, M., Takahashi, A., Yajima, H., Takeuchi, A., & Yamashita, N. (2013). Reducing electric power consumption for air conditioning by improving temperature distribution in telecom equipment rooms. *NTT Technical Review, 11*, 10.

Mascarenas, D. L. (2006). *Development of an impedance-based wireless sensor node for monitoring of bolted joint preload* (M.S. Thesis). Dept. of Structural Engineering, University of California, San Diego, CA.

Matova, S. P., Elfrink, R., Vullers, R. J. M., &Schaijk, R. V. (2011). Harvesting energy from airow with a michromachined piezoelectric harvester inside a Helmholtzresonator. *Journal of Micromechanics and Microengineering, 21*(10), 1-6.

Matsui, K. (2017). Proposal and implementation of real-time certification system for smart home using IoT technology. *Energy Procedia, 142*, 2027–2034. doi:10.1016/j.egypro.2017.12.406

Matsui, K. (2018). An information provision system to promote energy conservation and maintain indoor comfort in smart homes using sensed data by IoT sensors. *Future Generation Computer Systems*, *82*, 388–394. doi:10.1016/j.future.2017.10.043

McGowin, C. (2008). *Renewable Energy Technical Assessment Guide. TAG-RE: 2007, Electric Power Research Institute*. EPRI.

Mehrban, S., Nadeem, M. W., Hussain, M., Ahmed, M. M., Hakeem, O., Saqib, S., ... Khan, M. A. (2020). Towards Secure FinTech: A Survey, Taxonomy, and Open Research Challenges. *IEEE Access: Practical Innovations, Open Solutions*, *8*, 23391–23406. doi:10.1109/ACCESS.2020.2970430

Mell, P., & Grance, T. (2011). *The NIST definition of cloud computing*. Academic Press.

Melodia, T., Kulhandjian, H., Kuo, L.-C., & Demirors, E. (2013). *Advances in Underwater Acoustic Networking*. Wiley-IEEE Press. doi:10.1002/9781118511305.ch23

Mica, Mica2, Mica2Dot, MicaZ, Telos. (n.d.). http://www.xbow.com/products

Middei, R., Bianchi, S., Petrucci, P. O., Ursini, F., Cappi, M., De Marco, B., ... Matt, G. (2019). High-energy monitoring of NGC 4593 II. Broad-band spectral analysis: Testing the two-corona model. *Monthly Notices of the Royal Astronomical Society*, *483*(4), 4695–4705. doi:10.1093/mnrasty3379

Miglani, A., Kumar, N., Chamola, V., & Zeadally, S. (2020). Blockchain for Internet of Energy management: Review, solutions, and challenges. *Computer Communications*, *151*, 395–418.

Mikulic, M. (2019). *Global number of pacemakers in 2016 and 2023*. https://www.statista.com/statistics/800794/pacemakers-market-volume-in-units-worldwide/

Miragliotta, G., & Shrouf, F. (2012, September). Using Internet of Things to improve eco-efficiency in manufacturing: a review on available knowledge and a framework for IoT adoption. In *IFIP International Conference on Advances in Production Management Systems* (pp. 96-102). Springer.

Miranda, M. (2004). *Electrostatic vibration-to-electric energy conversion* (PhD thesis). Massachusetts Institute of Technology.

MIT Technology Review. (2003). *10 Emerging Technologies That Will Change the World*. Retrieved September 15, 2015 from http://www2.technologyreview.com/featured-story/401775/10-emerging-technologies-that-will-change-the/2/

Mitcheson, P. D., & Green, T. C. (2012). Maximum Effectiveness of Electrostatic Energy Harvesters When Coupled to Interface Circuits. *IEEE Transactions on Circuits and Systems*, *59*(12), 3098–3111. doi:10.1109/TCSI.2012.2206432

Mitcheson, P. D., Yeatman, E. M., Rao, G. K., Holmes, A. S., & Green, T. C. (2008). Energy harvesting from human and machine motion for wireless electronic devices. *Proceedings of the IEEE*, *96*(9), 1457–1486. doi:10.1109/JPROC.2008.927494

Mittal, S. (2014). *Power management techniques for data centers: A survey*. Retrieved from https://www.sharcnet.ca/my/front/

Mohamed, M. I., Wu, W. Y., & Moniri, M. (2011). Power harvesting for smart sensor networks in monitoring water distribution system. *Proceedings of the International Conference onNetworking, Sensing and Control (ICNSC '11)*, 393–398. 10.1109/ICNSC.2011.5874897

Mohammadi, E., Karimi, M., & Heikalabad, S. R. (2011). A novel virtual machine placement in cloud computing. *Australian Journal of Basic and Applied Sciences*, *5*(10), 1549–1555.

Mohammedi, A., Mezzai, N., Rekioua, D., & Rekioua, T. (2014). Impact of shadow on the performances of a domestic photovoltaic pumping system incorporating an MPPT control: A case study in Bejaia, North Algeria. *Energy Conversion and Management*, *84*, 20–29.

Mohanty, P., Bhuvaneswari, G., Balasubramanian, R., & Dhaliwal, N. K. (2014). MATLAB based modeling to study the performance of different MPPT techniques used for solar PV system under various operating conditions. *Renewable & Sustainable Energy Reviews*, *38*, 581–593.

Mokeddem, A., Midoun, A., Kadri, D., Hiadsi, S., & Raja, I. A. (2011). Performance of a directly-coupled PV water pumping system. *Energy Conversion and Management*, *52*(10), 3089–3095.

Monfray, S., Puscasu, O., Savelli, G., & Sapremanian, U. (2012). Innovative thermal energy harvesting for zero power electronics. In *Silicon Nano electronics Workshop (SNW)* (pp. 1–4). IEEE. doi:10.1109/SNW.2012.6243313

Morabito, A., & Hendrick, P. (2019). Pump as turbine applied to micro energy storage and smart water grids: A case study. *Applied Energy*, *241*, 567–579.

Morais, R., Matos, S. G., Fernandes, M. A., Valente, A. L. G., Soares, S. F. S. P., Ferreirac, P. J. S. G., & Reis, M. J. C. S. (2008). Sun, wind and water flow as energy supply for small stationary data acquisition platforms. *Computers and Electronics in Agriculture*, *64*(2), 120–132. doi:10.1016/j.compag.2008.04.005

Murphy, H., & Niitsuma, H. (1999). Strategies for compensating for higher costs of geothermal electricity with environmental benefits. *Geothermics*, *28*(6), 693–711. doi:10.1016/S0375-6505(99)00018-8

Mylona, Z., Kolokotroni, M., & Tassou, S. A. (2018). Coupling night ventilative and active cooling to reduce energy use in supermarkets with high refrigeration loads. *Energy and Building*, *171*, 26–39. doi:10.1016/j.enbuild.2018.04.021

Nacef, A. B., Senouci, S.-M., Doudane, Y. G., & Beylot, A.-L. (2011). A cooperative low power Mac protocol for wireless sensor networks. *Proceedings of the IEEE InternationalConference on Communications (ICC '11)*, 1–6. 10.1109/icc.2011.5962416

Naeem, M., Pareek, U., Lee, D. C., & Anpalagan, A. (2013). Estimation of distribution algorithm for resource allocation in green cooperative cognitive radio sensor networks. *Sensors (Basel)*, *13*(4), 4884–4905. doi:10.3390130404884 PMID:23584119

Naito, Y., & Uenishi, K. (2019). Electrostatic MEMS Vibration Energy Harvesters inside of Tire Treads. *Sensors (Basel)*, *19*(4), 890. doi:10.339019040890 PMID:30795502

Nandugudi, A., Maiti, A., Ki, T., Bulut, F., Demirbas, M., Kosar, T., . . . Challen, G. (2013). Phonelab: A large programmable smartphone testbed. In *Proceedings of First International Workshop on Sensing and Big Data Mining* (pp. 1–6). Academic Press.

Nandyala, C. S., & Kim, H. K. (2016). Green IoT agriculture and healthcare application (GAHA). *International Journal of Smart Home*, *10*(4), 289–300. doi:10.14257/ijsh.2016.10.4.26

Nasir, A. A., Zhou, X., Durrani, S., & Kennedy, R. A. (2013). Relaying protocols for wireless energy harvesting and information processing. *IEEE Transactions on Wireless Communications*, *12*(7), 3622–3636. doi:10.1109/TWC.2013.062413.122042

Nat. Phy. Lab. (NPL), UK. (n.d.). Retrieved from http://resource.npl.co.uk/acoustics/techguides/concepts/sonarequ.html

Nati, M., Gluhak, A., Abangar, H., & Headley, W. (2013). Smartcampus: A user-centric testbed for the internet of things experimentation. In *2013 16th International Symposium on Wireless Personal Multimedia Communications (WPMC)* (pp. 1–6). IEEE.

Nawaz, I., & Tiwari, G. (2006). Embodied energy analysis of photovoltaic (PV) system based on macro-and micro-level. *Energy Policy, 34*(17), 3144–3152. doi:10.1016/j.enpol.2005.06.018

Nechibvute, A., Chawanda, A., & Luhanga, P. (2012). Piezoelectric Energy Harvesting Devices: An Alternative Energy Source for Wireless Sensors. *Smart Materials Research*, 1–13.

New Hampshire Department of Environmental Services. (2017). *Universal Waste, Batteries: Management Requirements for Handlers and Transporters.* Environmental Fact Sheet, WMD-HW-18.

Ngô, C., & Natowitz, J. (2009). *Our energy future: resources, alternatives and the environment* (Vol. 11). Wiley. doi:10.1002/9780470473795

Ning, H., Li, Y., Shi, F., & Yang, L. T. (2020). Heterogeneous edge computing open platforms and tools for internet of things. *Future Generation Computer Systems, 106*, 67–76.

Ni, T., Zuo, L., & Kareem, A. (2011). Assessment of Energy Potential and Vibration Mitigation of Regenerative Tuned Mass Dampers on Wind Excited Tall Buildings. *Proceedings of the ASME 2011 International Design Engineering Technical Conferences & Computers and Information in Engineering Conference, IDETC/CIE 2011*, 333–342. 10.1115/DETC2011-48728

Nwaigwe, K. N., Anthony, O. C., Ogueke, N., & Anyanwu, E. E. (2010). Review of nocturnal cooling systems. *International Journal of Energy for a Clean Environment, 11*(1-4), 117–143. doi:10.1615/InterJEnerCleanEnv.2011003225

Okitsu, J., Naono, K., Khamis, M. F. I., Haruna, A. A., & Zakaria, N. (2013). *Towards an architecture for integrated Gas District Cooling with Data Center control to reduce CO2 emission.* Paper presented at the 2013 International Green Computing Conference Proceedings. 10.1109/IGCC.2013.6604468

Olivo, J., Carrara, S., & De Micheli, G. (2011). Energy harvesting and remote powering for implantable biosensors. *IEEE Sensors Journal, 11*, 1573–1586.

Omar, A. S., & Basir, O. (2018, July). Identity management in IoT networks using Blockchain and Smart Contracts. In *2018 IEEE International Conference on Internet of Things (iThings) and IEEE Green Computing and Communications (GreenCom) and IEEE Cyber, Physical and Social Computing (CPSCom) and IEEE Smart Data (SmartData)* (pp. 994-1000). IEEE. 10.1109/Cybermatics_2018.2018.00187

Ooi, B. L. (2010). *Optimisation and Frequency Tuning Concepts for a Vibration Energy Harvester* (PhD Thesis). University of Hull.

Ooi, B. A., Aziz, R., & Gilbert, J. (2014). Analytical and numerical modelling for dual-resonator vibration-based electromagnetic generator. In *5th International Conference on Intelligent and Advanced Systems (ICIAS)* (pp. 1-5). Kuala Lumpur: IEEE. 10.1109/ICIAS.2014.6869443

Ooi, B., & Gilbert, J. (2014). Design of wideband vibration-based electromagnetic generator by means of dual-resonator. *Sensors and Actuators. A, Physical, 213*(0), 9–18. doi:10.1016/j.sna.2014.03.037

Orekan, T., & Zhang, P. (2019). *Underwater Wireless Power Transfer* (1st ed.). Springer. doi:10.1007/978-3-030-02562-5

Orgerie, A.-C., Lefevre, L., & Gelas, J.-P. (2010). *Demystifying energy consumption in grids and clouds.* Paper presented at the International Conference on Green Computing. 10.1109/GREENCOMP.2010.5598295

Ouyang, H., Liu, Z., Li, N., Shi, B., Zou, Y., Xie, F., Ma, Y., Li, Z., Li, H., Zheng, Q., Qu, X., Fan, Y., Wang, Z. L., Zhang, H., & Li, Z. (2019). Symbiotic cardiac pacemaker. *Nature Communications, 10*(1), 1821. doi:10.103841467-019-09851-1 PMID:31015519

Pacis, D. M. M., Subido, E. D., Jr., & Bugtai, N. T. (2017). Research on the Application of Internet of Things (IoT) Technology Towards A Green Manufacturing Industry: A Literature Review. In *DLSU Research Congress 2017* (pp. 1-11). Academic Press.

Padhy, P., Martinez, K., Riddoch, A., Hart, J., & Ong, H. (2005). Glacial environment monitoring using sensor networks. *Proceedings of the Conference on Real-World Wireless Sensor Networks.*

Palosaari, J., Leinonen, M., Hannu, J., Juuti, J., & Jantunen, H. (2012). Energy harvesting with a cymbal type piezoelectric transducer from low frequency compression. *Journal of Electroceramics*, *28*(4), 214–219. doi:10.100710832-012-9713-8

Pan, J., Jain, R., Paul, S., Vu, T., Saifullah, A., & Sha, M. (2015). An internet of things framework for smart energy in buildings: Designs, prototype, and experiments. *IEEE Internet of Things Journal*, *2*(6), 527–537. doi:10.1109/JIOT.2015.2413397

Paradiso, J. A., & Starner, T. (2005). Energy scavenging for mobile and wireless electronics. *Pervasive Computing, IEEE*, *4*(1), 18–27.

Park, C., & Chou, P.H. (2007). AmbiMax: Autonomous Energy Harvesting Platform for Multi-Supply Wireless Sensor Nodes. *3rd Annual IEEE Communications Society on Sensor and Ad Hoc Communications and Networks, 1*, 168–177.

Park, S., Koh, B., Wang, Y., Jaemin, K., Kim, Y., Pedram, M., & Chang, N. (2013). Maximum power transfer tracking in a solar USB charger for smartphones. *Symposium on low power electronics and design, IEEE 2013*, 88 – 93. 10.1109/ISLPED.2013.6629272

Park, G., Rosing, T., Todd, M. D., Farrar, C. R., & Hodgkiss, W. (2008). Energy harvesting for structural health monitoring sensor networks. *Journal of Infrastructure Systems*, *14*(1), 64–79. doi:10.1061/(ASCE)1076-0342(2008)14:1(64)

Parra, G. D. L. T., Rad, P., & Choo, K. K. R. (2019). Implementation of deep packet inspection in smart grids and industrial Internet of Things: Challenges and opportunities. *Journal of Network and Computer Applications*, *135*, 32–46.

Patnaik, S., & Biswal, B. (2020). Importance of Nutrient Loading and Irrigation in Gross Primary Productivity Trends in India. *Journal of Hydrology (Amsterdam)*, 125047.

Patti, E., Acquaviva, A., & Macii, E. (2013). Enable sensor networks interoperability in smart public spaces through a service-oriented approach. In *5th IEEE International Workshop on Advances in Sensors and Interfaces IWASI* (pp. 2–7). IEEE. 10.1109/IWASI.2013.6576081

Pawlish, M., Varde, A. S., Robila, S. A., & Ranganathan, A. (2014). A call for energy efficiency in data centers. *SIGMOD Record*, *43*(1), 45–51. doi:10.1145/2627692.2627703

Pazowski, P. (2015). Green computing: latest practices and technologies for ICT sustainability. In *Managing Intellectual Capital and Innovation for Sustainable and Inclusive Society, Joint International Conference, Bari, Italy* (pp. 1853-1860). Academic Press.

Peigney, M., & Siegert, D. (2013). Piezoelectric energy harvesting from traffic-induced bridge vibrations. *Smart Materials and Structures*, *22*(9), 095019. doi:10.1088/0964-1726/22/9/095019

Penteado, D., Costa, L., & Pedroza, A. (2010). *Deep-ocean data acquisition using underwater sensor networks.* Paper presented at the The Twentieth International Offshore and Polar Engineering Conference.

Peoples, C., Parr, G., McClean, S., Scotney, B., & Morrow, P. (2013). Performance evaluation of green data centre management supporting sustainable growth of the internet of things. *Simulation Modelling Practice and Theory*, *34*, 221–242. doi:10.1016/j.simpat.2012.12.008

Playán, E., Salvador, R., Bonet, L., Camacho, E., Intrigliolo, D. S., Moreno, M. A., Rodríguez-Díaz, J. A., Tarjuelo, J. M., Madurga, C., Zazo, T., & Sánchez-de-Ribera, A. (2018). Assessing telemetry and remote control systems for water users associations in Spain. *Agricultural Water Management, 202*, 89–98.

Png, E., Srinivasan, S., Bekiroglu, K., Chaoyang, J., Su, R., & Poolla, K. (2019). An internet of things upgrades for smart and scalable heating, ventilation and air-conditioning control in commercial buildings. *Applied Energy, 239*, 408–424. doi:10.1016/j.apenergy.2019.01.229

Pobering, S., & Schwesinger, N. (2004). A novel hydropower harvesting device. *Proceedings of the International Conference onMEMS, NANO and Smart Systems (ICMENS '04)*, 480–485.

Pobering, S., & Schwesinger, N. (2009). Power supply for wireless sensor systems. *Proceedings of the IEEE Sensors Conference(SENSORS '08)*, 685–688.

Polastre, J., Szewczyk, R., & Culler, D. (2005, April). Telos: enabling ultra-low power wireless research. In *Proceedings of the 4th international symposium on Information processing in sensor networks* (p. 48). IEEE Press.

Ponnusamy, Low, & Anang. (2014). Biologically inspired mobile agent–based sensor network (BIMAS). *International Journal of Mobile Communications, 12*(5), 449-476.

Ponnusamy, Ramachandran, Jung, & Zaman. (2017). Bio-Inspired Energy Scavenging in Wireless Ad Hoc Network. *IEEE Innovations in Electrical Engineering and Computational Technologies (ICIEET).*

Ponnusamy, V., Abdullah, A., & Downe, A. G. (2012). Energy efficient routing protocols in wireless sensor networks: A survey. In Wireless Sensor Networks and Energy Efficiency: Protocols, Routing and Management (pp. 237-261). IGI Global. doi:10.4018/978-1-4666-0101-7.ch010

Ponnusamy, V., Abdullah, A., & Downe, A. G. (2012). Energy efficient routing protocols in wireless sensor networks: a survey. *Wireless Sensor Networks and Energy Efficiency: Protocols, Routing, and Management*, 237-261.

Ponnusamy, V., Tay, Y. P., Lee, L. H., Low, T. J., & Zhao, C. W. (2020). Energy harvesting methods for Internet of Things. In Securing the Internet of Things: Concepts, Methodologies, Tools, and Applications (pp. 956-976). IGI Global. doi:10.4018/978-1-5225-9866-4.ch042

Ponnusamy, V. (2014). Energy analysis in Wireless Sensor Network: A comparison. *International Journal of Computer Networks and Communications Security, 2*(9), 328–338.

Ponnusamy, V. (Ed.). (2016). *Biologically-inspired Energy Harvesting Through Wireless Sensor Technologies.* IGI Global. doi:10.4018/978-1-4666-9792-8

Ponnusamy, V., Low, T. J., & Amin, A. H. M. (2014). Biologically inspired mobile agent–based sensor network (BIMAS). *International Journal of Mobile Communications, 12*(5), 449–476. doi:10.1504/IJMC.2014.064593

Ponnusamy, V., Ramachandran, T., Jung, L. T., & Jhanjhi, N. Z. (2017). Bio-Inspired Energy Scavenging in Wireless Ad Hoc Network. In *IEEE Innovations in Electrical Engineering and Computational Technologies.* ICIEET. doi:10.1109/ICIEECT.2017.7916600

Popa, D., Popa, D. D., & Codescu, M. M. (2017). Reliability for a green internet of things. *Buletinul AGIR*, (1).

Powers, C. (2011). *Leadership in Green IT. U.S. Department of ENERGY, The National Renewable Energy Laboratory.* NREL.

Pozo, B., Garate, J. I., Araujo, J. Á., & Ferreiro, S. (2019). Energy Harvesting Technologies and Equivalent Electronic Structural Models - Review. *Electronics (Basel), 8*(5), 486. doi:10.3390/electronics8050486

Prauzek, M., Konecny, J., Borova, M., Janosova, K., Hlavica, J., & Musilek, P. (2018). Energy Harvesting Sources, Storage Devices and System Topologies for Environmental Wireless Sensor Networks: A Review. *Sensors (Basel), 18*(8), 2446. doi:10.339018082446 PMID:30060513

ProductsP. C. M. (2020). https://www.pcmproducts.net/files/cool_thermal_energy_storage.pdf

PSM Market Research. (2020). https://www.psmarketresearch.com/market-analysis/hvac-market

Purkus, A., & Barth, V. (2011). Geothermal power production in future electricity markets— A scenario analysis for Germany. *Energy Policy, 39*(1), 349–357. doi:10.1016/j.enpol.2010.10.003

Qiu, T., Zhao, Z., Zhang, T., Chen, C., & Chen, C. L. P. (2019). Underwater Internet of Things in Smart Ocean: System Architecture and Open Issues. *IEEE Transactions on Industrial Informatics, PP*, (c), 1–1. doi:10.1109/TII.2019.2946618

Qolomany, B., Al-Fuqaha, A., Gupta, A., Benhaddou, D., Alwajidi, S., Qadir, J., & Fong, A. C. (2019). Leveraging machine learning and big data for smart buildings: A comprehensive survey. *IEEE Access: Practical Innovations, Open Solutions, 7*, 90316–90356. doi:10.1109/ACCESS.2019.2926642

Rabaey, J. M., Ammer, M. J., da Silva, J. L. Jr, Patel, D., & Roundy, S. (2000). PicoRadio supports ad hoc ultra-low power wireless networking. *Computer, 33*(7), 42–48. doi:10.1109/2.869369

Radeckas, B., & Lukosevicius, V. (2000), Klaipeda Geothermal demonstration project. *Proceedings World Geothermal Congress 2000*, 3547-3550.

Raghunathan, V., Kansal, A., Hsu, J., Friedman, J., & Srivastava, M. (2005). Design considerations for solar energy harvesting wireless embedded systems. *Proceedings of the 4th InternationalSymposium on Information Processing in Sensor Networks(IPSN '05)*, 457–462. 10.1109/IPSN.2005.1440973

Rahman, R. A., & Shah, B. (2016). Security analysis of IoT protocols: A focus in CoAP. In *2016 3rd MEC international conference on big data and smart city (ICBDSC)* (pp. 1–7). IEEE. 10.1109/ICBDSC.2016.7460363

Ramachandran, T., Jung, L. T., Ponnusamy, V., & Hudaya, A. (2014, June). ZigBee wireless communication for monitoring renewable street light system. In *Computer and Information Sciences (ICCOINS), 2014 International Conference on* (pp. 1-6). IEEE.

Ramasur, D., & Hancke, G. P. (2012). A wind energy harvester for low power wireless sensor networks. *Proceedings of the IEEE International Instrumentation and Measurement Technology Conference (I2MTC '12)*, 2623–2627. 10.1109/I2MTC.2012.6229698

Ramesh, M., & Saini, R. P. (2020). Dispatch strategies based performance analysis of a hybrid renewable energy system for a remote rural area in India. *Journal of Cleaner Production*, 120697.

Ranganathan, P., Rivoire, S., & Moore, J. (2009). Models and metrics for energy-efficient computing. *Advances in Computers, 75*, 159–233. doi:10.1016/S0065-2458(08)00803-6

Rathore, P.K.S., Das, S.S. & Chauhan, D.S. (2018). Perspectives of solar photovoltaic water pumping for irrigation in India. *Energy Strategy Reviews, 22*, 385-395.

Rault, T., Bouabdallah, A., & Challal, Y. (2014). Energy efficiency in wireless sensor networks: A top-down survey. *Computer Networks, 67*, 104–122. doi:10.1016/j.comnet.2014.03.027

Rawat, P., Singh, K. D., Chaouchi, H., & Bonnin, J. M. (2014). Wireless sensor networks: A survey on recent developments and potential synergies. *The Journal of Supercomputing, 68*(1), 1–48. doi:10.100711227-013-1021-9

Reason. (n.d.). *Support > Self-help > Essential Acoustic Formulae.* Retrieved from http://www.reson.com/support/self-help/essential-acoustic-formulae/

Reddy, V. R., Rout, S. K., Shalsi, S., Pavelic, P., & Ross, A. (2020). Managing underground transfer of floods for irrigation: A case study from the Ramganga basin, India. *Journal of Hydrology (Amsterdam), 583*, 124518.

Regan, F., Lawler, A., & McCarthy, A. (2009). *SmartCoast Project- Smart Water Quality Monitoring System.* Environmental Protection Agency.

Reif, T. (2008). Profitability analysis and risk management of geothermal projects. *Geo-Heat Center Quarterly Bulletin, 28*(4), 1–4.

Reinisch, H., Gruber, S., Unterassinger, H., Wiessecker, M., Hofer, G., Pribyl, W., & Holweg, G. (2011). An electromagnetic energy harvesting system with 190 nW idle mode power consumption for a BAW based wireless sensor node. *IEEE Journal of Solid-State Circuits, 46*(7), 1728-1741.

Rekha, R. V., & Sekar, J. R. (2016). An unified deployment framework for realization of green Internet of Things (GIoT). *Middle East Journal of Scientific Research, 24*(2), 187–196.

ReVibe Energy. (2020). *The modelC.* https://revibeenergy.com/modelc/

Review of solar assisted HVAC systems; Its performance analysis using CO_2 as a refrigerant. (2019). *Energy Sources, Part A: Recovery, Utilization and Environmental Effects, 41*(24), 2957-2974.

Rhimi, M., & Lainef, R. (2012). Tunable Energy Harvesting from Ambient Vibrations in Civil Structures. *Journal of Energy Engineering, 134*(4), 185–193. doi:10.1061/(ASCE)EY.1943-7897.0000077

Riahi, M., Papaioannou, T. G., Trummer, I., & Aberer, K. (2013). Utility-driven data acquisition in participatory sensing. In *Proceedings of the 16th International Conference on Extending Database Technology* (pp. 251–262). 10.1145/2452376.2452407

Rice, J. A., & Diego, S. (2001). *ADAPTIVE Shallow Water Acoustic Networks.* Academic Press.

Rocha, J. G. Goncalves, L. M. Rocha, P. F. Silva, M. P. & Lanceros-Mendez, S. (2008). Energy harvesting from piezoelectric materials fully integrated in footwear. *Industrial Electronics, IEEE Transactions on, 57*(3), 813-819.

Rohr, Lipton, & Kong, Stephen, Taylor, & Ro. (2020). Efficiency Limits of Underwater Solar Cells. *Joule, 4*, 1–10.

Rose, T., Kifayat, K., Abbas, S., & Asim, M. (2020). A hybrid anomaly-based intrusion detection system to improve time complexity in the Internet of Energy environment. *Journal of Parallel and Distributed Computing.*

Roundy, S. J. (2003). *Energy scavenging for wireless sensor nodes with a focus on vibration to electricity conversion* (Doctoral dissertation). University of California, Berkeley, CA.

Roundy, S. J. (2003). *Energy Scavenging for Wireless Sensor Nodes with a Focus on Vibration to Electricity Conversion* (PhD. thesis). Berkeley, CA: University of California.

Roundy, S. J. (2003). *Energy scavenging for wireless sensor nodes with a focus on vibration to electricity conversion* [Ph.D. dissertation]. Department of Mechanical Engineering, University of California, Berkeley, CA.

Roy, A., Datta, A., Siddiquee, J., Poddar, B., Biswas, B., Saha, S., & Sarkar, P. (2016, October). Energy-efficient Data Centers and smart temperature control system with IoT sensing. In *2016 IEEE 7th Annual Information Technology, Electronics and Mobile Communication Conference (IEMCON)* (pp. 1-4). IEEE. 10.1109/IEMCON.2016.7746251

Rubinstein, R. Y., & Kroese, D. P. (2016). *Simulation and the Monte Carlo method* (Vol. 10). John Wiley & Sons. doi:10.1002/9781118631980

Saadon, S., & Sidek, O. (2011). A review of vibration-based MEMS piezoelectric energy harvesters. *Energy Conversion and Management, 52*(1), 500–504. doi:10.1016/j.enconman.2010.07.024

Saadon, S., & Sidek, O. (2011). Vibration-based MEMS piezoelectric energy harvester (VMPEH) modeling and analysis for green energy source. *Proceedings of the 4th International Conference on Developments in eSystems Engineering (DeSE '11)*, 527–531. 10.1109/DeSE.2011.85

Saeed & Jaffri. (2015). Estimation of freeze solar panel with Comparison of two different type's photovoltaic cell. *International Journal of Technology & Research, 3*(3), 59-64.

Saeed, S. (2015). Different Resources Consumption of Renewable Energy. Biologically-Inspired Energy Harvesting through Wireless Sensor Technologies, 85-102.

Saeed. (2016). Performance Analysis of Single Mono-Crystalline Photovoltaic Cell Using Ultra- Violet Glass. *Journal of Information & Communication Technology, 10*(2), 19-25.

Saeed, S., Syed, M. R. N., & Khan, H. A. (2016). Estimation of Multi-Crystalline Photovoltaic Cell using Solar Spectrum of Five Different Bands. *Life Science Journal, Newyork, 13*(8), 56–6.

Safaei, M., Sodano, H. A., & Anton, S. R. (2019). A review of energy harvesting using piezoelectric materials: State-of-the-art a decade later (2008–2018). *Smart Materials and Structures, 28*(11), 113001. doi:10.1088/1361-665X/ab36e4

Sahana, S., Singh, K., Das, S., & Kumar, R. (2017). Energy efficient shortest path routing protocol in underwater sensor networks. *Proceeding - IEEE International Conference on Computing, Communication and Automation, ICCCA 2016*, 546–550. 10.1109/CCAA.2016.7813780

Saleem, M., Di Caro, G. A., & Farooq, M. (2011). Swarm intelligence based routing protocol for wireless sensor networks: Survey and future directions. *Information Sciences, 181*(20), 4597–4624. doi:10.1016/j.ins.2010.07.005

Samuel, S. S. I. (2016). A review of connectivity challenges in IoT-smart home. In *2016 3rd MEC International conference on big data and smart city (ICBDSC)* (pp. 1–4). IEEE. 10.1109/ICBDSC.2016.7460395

Sander & Saulny. (2007). *Bridge collapse in Minneapolis Kills at least 7*. Retrieved from New York Times web site: https://www.nytimes.com/2007/08/02/us/02bridge.html?_r=0

Sara, Bushra, Jhanjhi, Mamoona, & Malik. (n.d.). Improving Knowledge Sharing in Distributed Software Development. *International Journal of Advanced Computer Science and Applications, 10*(6), 434-443.

Sardini, E., &Serpelloni, M. (2011). Self-powered wireless sensor for air temperature and velocity measurements with energy harvesting capability. *IEEE Transactions on Instrumentation and Measurement, 60*(5), 1838-1844.

Sarkar, S., & Misra, S. (2016). Theoretical modelling of fog computing: A green computing paradigm to support IoT applications. *Iet Networks, 5*(2), 23–29. doi:10.1049/iet-net.2015.0034

Savaglio, C., Ganzha, M., Paprzycki, M., Bădică, C., Ivanović, M., & Fortino, G. (2020). Agent-based Internet of Things: State-of-the-art and research challenges. *Future Generation Computer Systems, 102*, 1038–1053.

Schmidt, R. R., & Shaukatullah, H. (2002). *Computer and telecommunications equipment room cooling: a review of literature.* Paper presented at the ITherm 2002. Eighth Intersociety Conference on Thermal and Thermomechanical Phenomena in Electronic Systems (Cat. No. 02CH37258). 10.1109/ITHERM.2002.1012531

Seah, W. K. G., Zhi, A. E., & Tan, H.-P. (2009). Wireless Sensor Networks Powered by Ambient Energy Harvesting (WSNHEAP)—survey and challenges. *Proceedings of the 1stInternational Conference on Wireless Communication, Vehicular Technology, Information Theory and Aerospace and Electronic Systems Technology, Wireless (VITAE '09)*, 1–5.

Sethi, P., & Sarangi, S. R. (2017). Internet of things: Architectures, protocols, and applications. *Journal of Electrical and Computer Engineering*, *2017*, 2017. doi:10.1155/2017/9324035

Seyedi, A., & Sikdar, B. (2008). Energy Efficient Transmission Strategies for Body Sensor Networks with Energy Harvesting. *Proc. of the42nd Annual Conference on Information Sciences and Systems (CISS)*, 704–709. 10.1109/CISS.2008.4558613

Shah, R. C., & Rabaey, J. (2002). *Energy Aware Routing for Low Energy Ad Hoc Sensor Networks*. Paper presented at the IEEE Wireless Communications and Networking Conference (WCNC), Orlando, FL. 10.1109/WCNC.2002.993520

Shah, S. T., Choi, K. W., Hasan, S. F., & Chung, M. Y. (2016). Throughput analysis of two-way relay networks with wireless energy harvesting capabilities. *Ad Hoc Networks*, *53*, 123–131. doi:10.1016/j.adhoc.2016.09.024

Shaikh, F. K., & Zeadally, S. (2016). Energy harvesting in wireless sensor networks: A comprehensive review. *Renewable & Sustainable Energy Reviews*, *55*, 1041–1054. doi:10.1016/j.rser.2015.11.010

Shaikh, F. K., Zeadally, S., & Exposito, E. (2015). Enabling technologies for green internet of things. *IEEE Systems Journal*, *11*(2), 983–994. doi:10.1109/JSYST.2015.2415194

Shailaja, P., Guru Rao, C. V., & Nagaraju, A. (2019). A Parametric Oriented Research on Routing Algorithms in Mobile Adhoc Networks. *International Journal of Innovative Technology and Exploring Engineering*, *9*(1), 4126–4125.

Shang, Y., Li, Y., Yu, H., Sun, H., & Su, B. (2007). Experimental investigation on thermoelectric generator of micro hybrid power source. *Proceedings of the 3rd International Symposium on Advanced Optical Manufacturing and Testing Technologies: Design, Manufacturing, and Testing of Micro- and Nano-Optical Devices and Systems*, 6724. 10.1117/12.782827

Sharad, S., Sivakumar, P. B., & Anantha Narayanan, V. (2015, July). A novel IoT-based energy management system for large scale data centers. In *Proceedings of the 2015 ACM Sixth International Conference on Future Energy Systems* (pp. 313-318). ACM.

Sharma, P., Pegus, I. I. P. II, Irwin, D., Shenoy, P., Goodhue, J., & Culbert, J. (2017). Design and operational analysis of a green data center. *IEEE Internet Computing*, *21*(4), 16–24. doi:10.1109/MIC.2017.2911421

Shin, & Inwhee. (2016). Energy management algorithm for solar-powered energy harvesting wireless sensor node for Internet of things. *IET Journals the Institution of Engineering and Technology*, *10*(122), 1508–1521.

Shinohara, N. (2014). *Wireless Power Transfer via Radio waves*. Wiley Press.

Shirai, H., Mitamura, H., Arai, N., & Moriya, K. (2020). Study of Energy Harvesting from Low-Frequency Vibration with Ferromagnetic Powder and Non-magnetic Fluid. *Plasmonics*, *15*(2), 559–571. doi:10.100711468-019-01067-9

Shravan, K., & Chunduri, K. (2019). *Advanced solar module technology – The time for a new generation of solar modules has come*. TaiyangNews.

Shuja, J., Bilal, K., Madani, S. A., Othman, M., Ranjan, R., Balaji, P., & Khan, S. U. (2014). Survey of techniques and architectures for designing energy-efficient data centers. *IEEE Systems Journal*, *10*(2), 507–519. doi:10.1109/JSYST.2014.2315823

Sidhu, B. S., Kandlikar, M., & Ramankutty, N. (2020). Power tariffs for groundwater irrigation in India: A comparative analysis of the environmental, equity, and economic tradeoffs. *World Development*, *128*, 104836.

Silva, R., Farinholt, K., & Park, G. (2011). Hybrid energy sources for embedded sensor nodes. *Proceedings of the Industrial and Commercial Applications of Smart Structures Technologies*. 10.1117/12.882010

Silva, S., & Ruano, A. (2018). The IMBPC HVAC system: Wireless Sensors and IoT Platform. *IFAC-PapersOnLine*, *51*(10), 1–8. doi:10.1016/j.ifacol.2018.06.227

Simjee, F. I., & Chou, P. H. (2008). Efficient Charging of Supercapacitors forExtended Lifetime of Wireless Sensor Nodes. *IEEE Transactions on Power Electronics*, *23*(3), 1526–1536. doi:10.1109/TPEL.2008.921078

Simjee, F., Pai, H., & Chou, P. H. (2006). Everlast: Long-life, Supercapacitor operated Wireless Sensor Node. In *Proc 2006 International Symposium on Low Power Electronics and Design*. ACM. 10.1145/1165573.1165619

Singh, A., Sinha, U., & Sharma, D. K. (2020). Cloud-Based IoT Architecture in Green Buildings. In *Green Building Management and Smart Automation* (pp. 164–183). IGI Global. doi:10.4018/978-1-5225-9754-4.ch008

Singh, M., Muljadi, E., & Jonkman, J. (2013). Hybrid electro-mechanical simulation tool for wind turbine generators. *Proc. of IEEE Green Technologies Conf.*, 266–270. 10.1109/GreenTech.2013.48

Singi, K., Kaulgud, V., Bose, R. J. C., & Podder, S. (2019, May). ShIFt-Software Identity Framework for Global Software Delivery. In *2019 ACM/IEEE 14th International Conference on Global Software Engineering (ICGSE)* (pp. 122-128). IEEE. 10.1109/ICGSE.2019.00032

Smoker, J., Nouh, M., Aldraihem, O., & Baz, A. (2012). Energy harvesting from a standing wave thermo acoustic-piezoelectric resonator. *Journal of Applied Physics*, *111*(10), 104901. doi:10.1063/1.4712630

Sodano, H. A., Inman, D. J., & Park, G. (2005). Generation and storage of electricity from power harvesting devices. *Journal of Intelligent Material Systems and Structures*, *16*(1), 67–75. doi:10.1177/1045389X05047210

Sohrabi, K., Gao, J., Ailawadhi, V., & Pottie, G. J. (2000, October). Protocols for Self-Organization of a Wireless Sensor Network. *IEEE Personal Communications*, *7*(5), 16–27. doi:10.1109/98.878532

Sojan, S., & Kulkarni, R.K. (2016). A comprehensive review of Energy Harvesting techniques and its potential application. *International Journal of Computer Applications, 139*(3), 14– 19.

Song, Y., Han, B., Zhang, X., & Yang, D. (2012). Modeling and simulation of smart home scenarios based on the Internet of Things. In *2012 3rd IEEE international conference on network infrastructure and digital content* (pp. 596–600). IEEE. 10.1109/ICNIDC.2012.6418824

Song, W., Xiao, Z., Chen, Q., & Luo, H. (2013). Adaptive resource provisioning for the cloud using online bin packing. *IEEE Transactions on Computers*, *63*(11), 2647–2660. doi:10.1109/TC.2013.148

Sozer, E. M., Stojanovic, M., & Proakis, J. G. (2000). Underwater acoustic networks. *IEEE Journal of Oceanic Engineering*, *25*(1), 72–83. doi:10.1109/48.820738

Srujana, Sai, & Neha, Mathews, & Harigovindan. (2015). Multi-Source Energy Harvesting System for Underwater Wireless Sensor Networks. *Procedia Computer Science*, *46*, 1041–1048.

Statista. (2020). *IoT Number of connected devices worldwide*. dhttps://www.statista.com/statistics/471264/iot-number-of-connected-devices-worldwide/

Stojcev, M. K., Kosanovic, M. R., & Golubovic, L. R. (2009). Power management and energy harvesting techniques for wireless sensor nodes. *Proceedings of the 9th International Conference on Telecommunication in Modern Satellite, Cable, and Broadcasting Services (TELSIKS '09)*, 65–72. 10.1109/TELSKS.2009.5339410

Stokes, C. D., Duff, A., Mantini, M. J., Grant, B. A., & Venkatasubramanian, R. (2010). Nanostructured thermoelectric material and device technology for energy harvesting applications. *Proceedings of the 4th IEEE Nanotechnology Materials and Devices Conference (NMDC '10)*, 154–159. 10.1109/NMDC.2010.5652061

Sudevalayam, S., & Kulkarni, P. (2011). Energy harvesting sensor nodes: Survey and implications. *IEEE Communications Surveys and Tutorials*, *13*(3), 443–461. doi:10.1109/SURV.2011.060710.00094

Sun, H., Yin, M., Wei, W., Li, J., Wang, H., & Jin, X. (2018). MEMS based energy harvesting for the Internet of Things: A survey. *Microsystem Technologies*, *24*(7), 2853–2869. doi:10.100700542-018-3763-z

Sun, K., & Ryoo, I. (2015, October). A study on medium access control scheme for energy efficiency in wireless smart sensor networks. In *2015 International Conference on Information and Communication Technology Convergence (ICTC)* (pp. 623-625). IEEE. 10.1109/ICTC.2015.7354625

Sun, Ooi, Kang, Kong, Alkhazragi, Guo, & Ouhssain, Weng, Jones, & Ng. (2020). A Review on Practical Considerations and Solutions in Underwater Wireless Optical Communication. *Journal of Lightwave Technology*, *38*(2), 421–431.

Su, Ullah, Liu, & Choi. (2020). A Review of Underwater Localization Techniques, Algorithms, and Challenges. *Journal of Sensors*.

Tabbakh, S. R. K., Maarefdoust, R., Kyun, N. C., & Ali, B. M. (2010). Environmental taxonomy of power scavenging techniques for autonomous self powered wireless sensors. *Proceedings of the Asia Pacific Conference on Circuit and System (APCCAS '10)*, 1031–1034.

Tan, H. P., Eu, Z. A., & Seah, W. K. (2009). Impact of Power Control in Wireless Sensor Networks Powered by Ambient Energy Harvesting for Railroad Health Monitoring. *Proc. of the 2nd Intl Workshop on Applications of Ad hoc and Sensor Networks (AASNET)*. 10.1109/WAINA.2009.113

Tan, Y. K., & Panda, S. K. (2011a). Energy harvesting from hybrid indoor ambient light and thermal energy sources for enhanced performance of wireless sensor nodes. *IEEE Transactions on Industrial Electronics, 58*(9), 4424-4435.

Tan, Y. K., & Panda, S. K. (2011b). Self-autonomous wireless sensor nodes with wind energy harvesting for remote sensing of wind-driven wild re spread. *IEEE Transactions on Instrumentation and Measurement, 60*(4), 1367-1377.

Taneja, J., Jeong, J., & Culler, D. (2008, April). Design, modeling, and capacity planning for micro-solar power sensor networks. In *Proceedings of the 7th international conference on Information processing in sensor networks* (pp. 407-418). IEEE Computer Society. 10.1109/IPSN.2008.67

Tang, X., Lei, T., & Zuo, L. (2014). Design and Optimization of a Tubular Linear Electromagnetic Vibration Energy Harvester. *IEEE/ASME Transactions on Mechatronics*, *19*(2), 615–622. doi:10.1109/TMECH.2013.2249666

Tang, X., & Zuo, L. (2011). Enhanced vibration energy harvesting using dual-mass systems. *Journal of Sound and Vibration*, *330*(21), 5199–5209. doi:10.1016/j.jsv.2011.05.019

Tang, X., & Zuo, L. (2012a). Vibration energy harvesting from random force and motion excitations. *Smart Materials and Structures*, *21*(7), 075025. doi:10.1088/0964-1726/21/7/075025

Tang, X., & Zuo, L. (2012b). Simultaneous energy harvesting and vibration control of structures with tuned mass dampers. *Journal of Intelligent Material Systems and Structures*, *23*(18), 2117–2127. doi:10.1177/1045389X12462644

Tang, Z. B., Deng, Y. D., Su, C. Q., Shuai, W. W., & Xie, C. J. (2015). A research on thermoelectric generator's electrical performance under temperature mismatch conditions for automotive waste heat recovery system. *Case Studies in Thermal Engineering*, *5*, 143–150. doi:10.1016/j.csite.2015.03.006

Tan, Y. K., & Panda, S. K. (2007). A novel piezoelectric based wind energy harvester for low-power autonomous wind speed sensor. *Proceedings of the 33rd Annual Conference of the IEEE Industrial Electronics Society (IECON '07)*, 2175–2180. 10.1109/IECON.2007.4460120

Taylan, O., Kaya, D., Bakhsh, A. A., & Demirbas, A. (2018). Bioenergy life cycle assessment and management in energy generation. *Energy Explor. Exploit.*, *36*(1), 166–181. doi:10.1177/0144598717725871

Technica, A. (2008). NVIDIA denies rumors of faulty chips, mass GPU failures. Academic Press.

Terroso-Saenz, F., González-Vidal, A., Ramallo-González, A. P., & Skarmeta, A. F. (2019). An open IoT platform for the management and analysis of energy data. *Future Generation Computer Systems*, *92*, 1066–1079. doi:10.1016/j.future.2017.08.046

Thein, C. K., Foong, F. M., & Shu, Y. C. (2019a). Damping ratio and power output prediction of an electromagnetic energy harvester designed through finite element analysis. *Sensors and Actuators. A, Physical*, *286*, 220–231. doi:10.1016/j.sna.2018.12.041

Thein, C. K., Foong, F. M., & Shu, Y. C. (2019b). Spring amplification and dynamic friction modelling of a 2DOF/2SDOF system in an electromagnetic vibration energy harvester – Experiment, simulation, and analytical analysis. *Mechanical Systems and Signal Processing*, *132*, 232–252. doi:10.1016/j.ymssp.2019.06.028

Tong, L., Zhao, Q., & Adireddy, S. (2003). *Sensor Networks with Mobile Agent*. Paper presented at the Military Communications Intl Symp., Boston, MA.

Torres, E. O., & Rincon-Mora, G. A. (2009). Electrostatic Energy-Harvesting and Battery-Charging CMOS System Prototype. *IEEE Transactions on Circuits and Systems. I, Regular Papers*, *56*(9), 1938–1948. doi:10.1109/TCSI.2008.2011578

Trinchero, D., Stefanelli, R., Brunazzi, D., Casalegno, A., Durando, M., & Galardini, A. (2011). Integration of smart house sensors into a fully networked (web) environment. In *SENSORS, 2011 IEEE* (pp. 1624–1627). IEEE. doi:10.1109/ICSENS.2011.6127374

Trnka, M., Cerny, T., & Stickney, N. (2018). Survey of Authentication and Authorization for the Internet of Things. *Security and Communication Networks*, *2018*. doi:10.1155/2018/4351603

Tso, M. H., Yuan, J., & Wong, W. O. (2013). Design and experimental study of a hybrid vibration absorber for global vibration control. *Engineering Structures*, *56*, 1058–1069. doi:10.1016/j.engstruct.2013.06.017

Tushar, W., Wijerathne, N., Li, W.-T., Yuen, C., Poor, H. V., Saha, T. K., & Wood, K. L. (2018). *Iot for green building management*. arXiv Preprint arXiv:1805.10635.

Tushar, W., Wijerathne, N., Li, W. T., Yuen, C., Poor, H. V., Saha, T. K., & Wood, K. L. (2018). Internet of things for green building management: Disruptive innovations through low-cost sensor technology and artificial intelligence. *IEEE Signal Processing Magazine*, *35*(5), 100–110. doi:10.1109/MSP.2018.2842096

Ucar, A., & Balo, F. (2009). Investigation of wind characteristics and assessment of wind generation potentiality in Uluda˘g-Bursa, Turkey. *Applied Energy*, *86*(3), 333–339. doi:10.1016/j.apenergy.2008.05.001

ur Rehman, M.H., Yaqoob, I., Salah, K., Imran, M., Jayaraman, P.P. & Perera, C. (2019). The role of big data analytics in industrial Internet of Things. *Future Generation Computer Systems, 99*, 247-259.

Urick, R. J. (1975). *Principles of underwater sound-2*. W.E. Johns, Notes on batteries. Retrieved from http://www.gizmology.net/batteries.htm

US Department of Energy. (2013). *Pacific Northwest National Laboratory*. Retrieved June, 28, 2013, from http://eioc.pnnl.gov/ research/sitawareness.stm

US Environment Protection Agency. (2013). *Report to Congress on Server and Data Center Efficiency: Public Law 109-431*. Energy Star. http://www.energystar.gov/ia/ partners/prod_development/downloads/EPA_Datacenter_Report_Congress_Final1.pdf

Vasseur, J.-P., & Dunkels, A. (2010). *Interconnecting Smart Objects with IP - The Next Internet*. Morgan Kaufmann.

Venkatesh, V., Morris, M. G., Gordon, B., & Davis, F. D. (2003, September). User acceptance of information technology: Toward a unified view. *Management Information Systems Quarterly*, *27*(3), 425–478. doi:10.2307/30036540

Vermesan, O., Friess, P., & Guillemin, P. (2011). Internet of things strategic research roadmap. *Global Technological and Societal Trends*, *1*, 9–52.

Viet, N. V., Xie, X. D., Liew, K. M., Banthia, N., & Wang, Q. (2016). Energy harvesting from ocean waves by a floating energy harvester. *Energy*, *112*, 1219–1226. doi:10.1016/j.energy.2016.07.019

Vijayakumari, A. (2020). A non-iterative MPPT of PV array with online measured short circuit and open circuit quantities. *Journal of King Saud University-Engineering Sciences.*

Vlajic & Moniz. (2007). Self-Healing Wireless Sensor Networks: Results That May Surprise. In Globecom Workshops, 2007 IEEE, Washington, DC.

Vullers, R. J. M., Schaijk, R. V., Doms, I., Hoof, C. V., & Mertens, R. (2009). Micropower energy harvesting. *Solid-State Electronics*, *53*(7), 684–693. doi:10.1016/j.sse.2008.12.011

Vullers, R. J. M., Schaijk, R., Visser, H. J., Penders, J., & Hoof, C. V. (2010). Energy Harvesting for autonomous wireless sensor networks. *IEEE Solid-State Circuits Magazine*, *2*(2), 29–38. doi:10.1109/MSSC.2010.936667

Wang, D. (2008). *Meeting green computing challenges*. Paper presented at the 2008 10th Electronics Packaging Technology Conference. 10.1109/EPTC.2008.4763421

Wang, D.-A., & Chang, K.-H. (2010). Electromagnetic energy harvesting from flow induced vibration. *Microelectronics Journal*, *41*(6), 356–364. doi:10.1016/j.mejo.2010.04.005

Wang, D.-A., & Ko, H.-H. (2010). Piezoelectric energy harvesting from flow-induced vibration. *Journal of Micromechanics and Microengineering*, *20*(2), ID025019. doi:10.1088/0960-1317/20/2/025019

Wang, D.-A., & Liu, N.-Z. (2011). A shear mode piezoelectric energy harvester based on a pressurized water flow. *Sensors and Actuators, A. Physical*, *167*(2), 449–458.

Wang, H. I. (2014). Constructing the green campus within the internet of things architecture. *International Journal of Distributed Sensor Networks*, *10*(3), 804627. doi:10.1155/2014/804627

Wang, J. J., Penamalli, G. P., & Zuo, L. (2012). Electromagnetic Energy Harvesting from Train Induced Railway Track Vibrations. *8th IEEE/ASME International Conference on Mechatronic and Embedded Systems and Applications*, 29–34. 10.1109/MESA.2012.6275532

Wang, L., Luo, G., Jiang, Z., Zhang, F., Zao, L., Yang, P., Lin, Q., & Maeda, R. (2019). Broadband vibration energy harvesting for wireless sensor node power supply in train container. *The Review of Scientific Instruments*, *90*(12), 125003. doi:10.1063/1.5127243 PMID:31893793

Wang, M., Qiu, S., Dong, H., & Wang, Y. (2017). Design an IoT-based building management cloud platform for green buildings. In *2017 Chinese Automation Congress (CAC)* (pp. 5663–5667). IEEE. 10.1109/CAC.2017.8243793

Wang, T., Xia, Y., Muppala, J., Hamdi, M., & Foufou, S. (2014, December). A general framework for performance guaranteed green data center networking. In *2014 IEEE Global Communications Conference* (pp. 2510-2515). IEEE. 10.1109/GLOCOM.2014.7037185

Wang, W., Cao, J., Zhang, N., Lin, J., & Liao, W. H. (2017). Magnetic-spring based energy harvesting from human motions: Design, modeling and experiments. *Energy Conversion and Management*, *132*, 189–197. doi:10.1016/j.enconman.2016.11.026

Wang, X. (2012). Piezoelectric nanogenerators-Harvesting ambient mechanical energy at the nanometer scale. *Nano Energy*, *1*(1), 13–24. doi:10.1016/j.nanoen.2011.09.001

Wang, X., Liang, X., Hao, Z., Du, H., Zhang, N., & Qian, M. (2016). Comparison of electromagnetic and piezoelectric vibration energy harvesters with different interface circuits. *Mechanical Systems and Signal Processing*, *72–73*, 906–924. doi:10.1016/j.ymssp.2015.10.016

Wang, Z. (2018). A privacy-preserving and accountable authentication protocol for IoT end-devices with weaker identity. *Future Generation Computer Systems*, *82*, 342–348. doi:10.1016/j.future.2017.09.042

Wan, Z. G., Tan, Y. K., & Yuen, C. (2011). Review on energy harvesting and energy management for sustainable wirelesssensor networks. *Proceedings of the IEEE 13th International Conference on Communication Technology (ICCT '11)*, 362–367.

Water Framework Directive, W. F. D. (n.d.). *Council Directive 2000/60/EC*. Retrieved from http://www.defra.gov.uk/environment/water/wfd/

Weddell, A. (2010). *Notes from "Energy Harvesting for Wireless Automation" Conference*. Hanson Wade.

Wendi, R. H., Chandrakasan, A., & Balakrishnan, H. (2000). *Energy-efficient communication protocols for wireless microsensor networks*. Paper presented at the Hawaii International Conference on Systems Sciences.

Werner-Allen, G., Lorincz, K., Ruiz, M., Marcillo, O., Johnson, J., Lees, J., & Welsh, M. (2006, March-April). Deploying a Wireless Sensor Network on an Active Volcano. *IEEE Internet Computing*, *10*(2), 18–25. doi:10.1109/MIC.2006.26

Wikipedia. (2020). *Internet of Things*. https://en.wikipedia.org/wiki/Internet_of_things

Williams, C. B., & Yates, R. B. (1996). Analysis of a micro-electric generator for microsystems. *Sensors and Actuators. A, Physical*, *52*(1-3), 8–11. doi:10.1016/0924-4247(96)80118-X

Wischke, M., Masur, M., Kroner, M., & Woias, P. (2011). Vibration harvesting in traffic tunnels to power wireless sensor nodes. *Smart Materials and Structures*, *20*(8), 085014. doi:10.1088/0964-1726/20/8/085014

Wu, Yu. Yuan, & Tan. (2019). Towards a Gliding Robotic Dolphin: Design, Modeling, and Experiments. *IEEE/ASME Transactions on Mechatronics, 24*(1), 260–70. doi:10.1109/TMECH.2019.2891290

Wu, J., Guo, S., Li, J., & Zeng, D. (2016). Big Data Meet Green Challenges: Greening Big Data. *IEEE Systems Journal*, *10*(3), 873–887. doi:10.1109/JSYST.2016.2550538

Wu, Q., Li, G. Y., Chen, W., Ng, D. W. K., & Schober, R. (2017). An overview of sustainable green 5G networks. *IEEE Wireless Communications*, *24*(4), 72–80. doi:10.1109/MWC.2017.1600343

Wu, Y.-W., Yang, C.-L., Yuh, P.-H., & Chang, Y.-W. (2005). Joint exploration of architectural and physical design spaces with thermal consideration. *Proceedings of the 2005 international symposium on Low power electronics and design*. 10.1145/1077603.1077636

Xia, M., Li, T., Zhang, Y., & De Silva, C. W. (2016). Closed-loop design evolution of engineering system using condition monitoring through internet of things and cloud computing. *Computer Networks*, *101*, 5–18. doi:10.1016/j.comnet.2015.12.016

Xiao, L., Qian, J. J., Lian, Z., Long, Z., Tang, X., Chen, T., Du, X., & Cao, L. (2016). RF energy powered wireless temperature sensor for monitoring electrical equipment. *Sensors and Actuators. A, Physical*, *249*, 276–28. doi:10.1016/j.sna.2016.08.022

Xiao, L., Zhu, Y., Ni, L. M., & Xu, Z. (2008). Incentive-based scheduling for market-like computational grids. *IEEE Transactions on Parallel and Distributed Systems*, *19*(7), 903–913. doi:10.1109/TPDS.2007.70791

Xiong, N., Han, W., & Vandenberg, A. (2012). Green cloud computing schemes based on networks: A survey. *IET Communications*, *6*(18), 3294–3300. doi:10.1049/iet-com.2011.0293

Xu, R., Chen, Y., Blasch, E., & Chen, G. (2018, July). Blendcac: A blockchain-enabled decentralized capability-based access control for iots. In *2018 IEEE International Conference on Internet of Things (iThings) and IEEE Green Computing and Communications (GreenCom) and IEEE Cyber, Physical and Social Computing (CPSCom) and IEEE Smart Data (SmartData)* (pp. 1027-1034). IEEE. 10.1109/Cybermatics_2018.2018.00191

Xu, H., & Yang, B. (2015). An incentive-based heuristic job scheduling algorithm for utility grids. *Future Generation Computer Systems*, *49*, 1–7. doi:10.1016/j.future.2015.02.002

Xu, J., Tao, F., Liu, Y., Hu, C., Xu, Y., Keivanimehr, F., & Nabipour, N. (2020). Data transmission method for sensor devices in internet of things based on multivariate analysis. *Measurement*, 107536.

Xu, Z. (2011). *Building a power-aware database management system*. University of South Florida.

Yang, B., & Lee, C. (2010). Non-resonant electromagnetic wideband energy harvesting mechanism for low frequency vibrations. *Microsystem Technologies*, *16*(6), 916–966. doi:10.100700542-010-1059-z

Yang, C.-T., Chen, S.-T., Liu, J.-C., Liu, R.-H., & Chang, C.-L. (2020). On construction of an energy monitoring service using big data technology for the smart campus. *Cluster Computing*, *23*(1), 265–288. doi:10.100710586-019-02921-5

Yang, J., Chen, J., Yang, Y., Zhang, H., Yang, W., Bai, P., Su, Y., & Wang, Z. L. (2014). Broadband vibrational energy harvesting on a triboelectric nanogenerator. *Advanced Energy Materials*, *4*(6), 1301322. doi:10.1002/aenm.201301322

Yang, J., Xiao, W., Jiang, C., Hossain, M. S., Muhammad, G., & Amin, S. U. (2018). AI-powered green cloud and data center. *IEEE Access: Practical Innovations, Open Solutions*, *7*, 4195–4203. doi:10.1109/ACCESS.2018.2888976

Yang, W. Q., Chen, J., Zhu, G., Wen, X., Bai, P., Su, Y., Lin, Y., & Wang, Z. L. (2013). Harvesting vibration energy by a triple-cantilever based triboelectric nanogenerator. *Nano Research*, *6*(12), 880–886. doi:10.100712274-013-0364-0

Yang, Y. W., Wang, S., Stein, P., Xu, B. X., & Yang, T. (2017). Vibration-based energy harvesting with a clamped piezoelectric circular diaphragm: Analysis and identification of optimal structural parameters. *Smart Materials and Structures*, *26*(4), 045011. doi:10.1088/1361-665X/aa5fda

Ye, G., & Soga, K. (2012). Energy harvesting from water distribution systems. *Journal of Energy Engineering*, *138*(1), 7–17. doi:10.1061/(ASCE)EY.1943-7897.0000057

Ye, W., Heidemann, J., & Estrin, D. (2002, June). An energy-efficient MAC protocol for wireless sensor networks. In *Proceedings. Twenty-First Annual Joint Conference of the IEEE Computer and Communications Societies* (Vol. 3, pp. 1567-1576). IEEE.

Yick, J., Mukherjee, B., & Ghosal, D. (2008). Wireless sensor network survey. *Computer Networks*, *52*(12), 2292–2330. doi:10.1016/j.comnet.2008.04.002

Yong, Y.-T., Chow, C.-O., Kanesan, J., & Ishii, H. (2009). A survey on design of self-powered wireless sensor network. *Proceedings of the International Conference for Technical Postgraduates (TECHPOS '09)*, 1–5. 10.1109/TECHPOS.2009.5412044

Yoshida, J. (2008). *The truth about last year's xbox 360 recall*. EE Times.

Youssef, M. A., Younis, M. F., & Arisha, K. A. (2000). *A constrained shortest- path energy-aware routing algorithm for wireless sensor networks*. Paper presented at the IEEE Wireless Communications and Networking Conference (WCNC2002).

Yu, Estrin, & Govindan. (2001). *Geographical and Energy-Aware Routing: A Recursive Data Dissemination Protocol for Wireless Sensor Networks*. UCLA Computer Science Department Technical Report, UCLA-CSD TR-01-0023.

Zafar, S., Hussain, R., Hussain, F., & Jangsher, S. (2019). Interplay between Big Spectrum Data and Mobile Internet of Things: Current solutions and future challenges. *Computer Networks*, *163*, 106879.

Zaman & Abdullah. (2010). Energy Efficient Routing in Wireless Sensor Network: Research Issues and Challenges. *IEEE International Conference on Intelligence and Information Technology*, 239-243.

Zaman & Abdullah. (2011). Different Techniques Towards Enhancing Wireless Sensor Network (WSN) Routing Energy Efficiency and Quality of Service (QoS). *World Applied Science Journal, 13*(4), 798-805.

Zaman & Almusalli. (2017). Review: Smartphone Power Consumption & Energy Saving Techniques. *IEEE Innovations in Electrical Engineering and Computational Technologies (ICIEET).*

Zaman, Low, & Alghamdi. (2014). Energy efficient routing protocol for wireless sensor network. *16th International Conference on Advanced Communication Technology (ICACT),* 808-814.

Zaman, Low, & Alghamdi. (2015). Enhancing Routing Energy Efficiency of Wireless Sensor Networks. *ICACT Journal*, 587-595.

Zaman, N., & Abdullah, A. (2012). Energy Optimization through Position Responsive Routing Protocol (PRRP) in Wireless Sensor Network (WSN). International Journal of Information and Electronics Engineering, 2(5), 748-751.

Zaman, N., Abdullah, A., & Jung, L. T. (2011). Optimization of energy usage in wireless network using Position responsive routing Protocol. *IEEE Symposium on Computers & Informatics ISCI*, 51-55.

Zaman, N., Jung, L. T., Alsaade, F., & Alghamdi, T. (2012). Wireless Sensor Network (WSN): Routing Security, Reliability and Energy Efficiency. Journal of Applied Science, 12(16), 593-597. doi:10.3923/jas.2012.593.597

Zaman, N., & Abdullah, A. (2011). Position Responsive Routing Protocol (PRRP). *IEEE The 13th International Conference on Advance Communication Technology,* 644-648.

Zaman, N., & Ahmed, M. (2017). Towards the Evaluation of Authentication Protocols for Mobile Command and Control Unit in Healthcare. *Journal of Medical Imaging and Health Informatics*, *7*(3), 739–742. doi:10.1166/jmihi.2017.2070

Ze, L., Ruichun, T., Ye, T., & Xin, S. (2017). Design of multicast protocol for underwater multimedia data transmission. *2016 3rd International Conference on Systems and Informatics, ICSAI 2016, Icsai*, 710–715. 10.1109/ICSAI.2016.7811045

Zeadally, S., Shaikh, F. K., Talpur, A., & Sheng, Q. Z. (2020). Design architectures for energy harvesting in the Internet of Things. *Renewable & Sustainable Energy Reviews*, *128*, 109901.

Zhang, C., Harne, R. L., Li, B., & Wang, K. W. (2020b). Harmonic analysis and experimental validation of bistable vibration energy harvesters interfaced with rectifying electrical circuits. *Communications in Nonlinear Science and Numerical Simulation*, *82*, 105069. doi:10.1016/j.cnsns.2019.105069

Zhang, H., & Afzalul, K. (2014). Design and Analysis of a Connected Broadband Multi-Piezoelectric-Bimorph Beam Energy Harvester. *IEEE Transactions on Ultrasonics, Ferroelectrics, and Frequency Control*, *61*(6), 1016–1023. doi:10.1109/TUFFC.2014.2997 PMID:24859665

Zhang, X., Pan, H., Qi, L., Zhang, Z., Yuan, Y., & Liu, Y. (2017). A renewable energy harvesting system using a mechanical vibration rectifier (MVR) for railroads. *Applied Energy*, *204*, 1535–1543. doi:10.1016/j.apenergy.2017.04.064

Zhang, X., Ruoshui, W., Molin, H., & Martinot, E. (2010). A study of the role played by renewable energies in China's sustainable energy supply. *Energy*, *35*(11), 4392–4399. doi:10.1016/j.energy.2009.05.030

Zhang, Y., Phuong, P. T. T., Roake, E., Khanbareh, H., Wang, Y., Dunn, S., & Bowen, C. (2020a). Thermal Energy Harvesting Using Pyroelectric-Electrochemical Coupling in Ferroelectric Materials. *Joule*, *4*(2), 301–309. doi:10.1016/j.joule.2019.12.019

Zhang, Y., Wang, T., Zhang, A., Peng, Z., Luo, D., Chen, R., & Wang, F. (2016). Electrostatic energy harvesting device with dual resonant structure for wideband random vibration sources at low frequency. *The Review of Scientific Instruments*, *87*(12), 125001. doi:10.1063/1.4968811 PMID:28040962

Zhang, Z. K., Cho, M. C. Y., & Shieh, S. (2015, April). Emerging security threats and countermeasures in IoT. In *Proceedings of the 10th ACM Symposium on Information, Computer and Communications Security* (pp. 1-6). 10.1145/2714576.2737091

Zhang, Z., Pang, H., Georgiadis, A., & Cecati, C. (2019). Wireless Power Transfer - An Overview. *IEEE Transactions on Industrial Electronics*, *66*(2), 1044–1058. doi:10.1109/TIE.2018.2835378

Zheng, J., Gao, D. W., & Lin, L. (2013). Smart meters in the smart grid: An overview. In 2013 IEEE Green Technologies Conference (GreenTech) (pp. 57–64). IEEE.

Zhihong, J., Jian, H., Wenzhou, L., Zhe, C., Ning, L., Siyuan, W., Xiao, Z., & Chang, L. (2018). Energy internet-a new driving force for sustainable urban development. *Energy Procedia*, *152*, 1206–1211.

Zhou, Z. (2019, October). GreenEdge: Greening Edge Datacenters with Energy-Harvesting IoT Devices. In *2019 IEEE 27th International Conference on Network Protocols (ICNP)* (pp. 1-6). IEEE. http://www.datacenterknowledge.com/inside-facebooks-lulea-data-center/ http://www.theguardian.com/technology/2014/oct/22/cloud-storage-data-centre-sweden-arctic-hydro66

Zhou, G., Huang, L., Li, W., & Zhu, Z. (2014). Harvesting Ambient Environmental Energy for Wireless Sensor Networks: A Survey. *Journal of Sensors*, *2014*, 815467. doi:10.1155/2014/815467

Zhou, H.-Y., Wu, F., & Hou, K.-M. (2008). An event-driven multi-threading real-time operating system dedicated to wireless sensor networks. *Proceedings of the International Conferenceon Embedded Software and Systems (ICESS '08)*, 3–12. 10.1109/ICESS.2008.58

Zhou, K., Dai, H. L., Abdelkefi, A., & Ni, Q. (2020). Theoretical modeling and nonlinear analysis of piezoelectric energy harvesters with different stoppers. *International Journal of Mechanical Sciences*, *166*, 105233. doi:10.1016/j.ijmecsci.2019.105233

Zhou, X., Zhang, R., & Ho, C. K. (2013). Wireless information and power transfer: Architecture design and rate-energy tradeoff. *IEEE Transactions on Communications*, *61*(11), 4754–4767. doi:10.1109/TCOMM.2013.13.120855

Zhou, Z., Qin, W., Du, W., Zhu, P., & Liu, Q. (2019). Improving energy harvesting from random excitation by nonlinear flexible bi-stable energy harvester with a variable potential energy function. *Mechanical Systems and Signal Processing*, *115*, 162–172. doi:10.1016/j.ymssp.2018.06.003

Zhuang, L., Chen, X., & Guan, X. (2020). A decentralized method for energy conservation of an HVAC system. *Building Simulation*, *13*(1), 155–170. doi:10.100712273-019-0569-6

Zhu, C., Leung, V. C., Shu, L., & Ngai, E. C. H. (2015). Green internet of things for smart world. *IEEE Access: Practical Innovations, Open Solutions*, *3*, 2151–2162. doi:10.1109/ACCESS.2015.2497312

Zhu, D., Beeby, S., Tudor, J., White, N., & Harris, N. (2010). A novel miniature wind generator for wireless sensing applications. *Proceedings of the 9th IEEE Sensors Conference (SENSORS '10)*, 1415–1418.

Zhu, G., Peng, B., Chen, J., Jing, Q., & Wang, Z. L. (2015). Triboelectric nanogenerators as a new energy technology: From fundamentals, devices, to applications. *Nano Energy*, *14*, 126–138. doi:10.1016/j.nanoen.2014.11.050

Zhu, Q. (2004). *Reducing energy consumption of disk storage using power-aware cache management*. Department of Computer Science, University of Illinois at Urbana Champaign.

Zhu, X., Badr, Y., Pacheco, J., & Hariri, S. (2017, September). Autonomic identity framework for the internet of things. In *2017 International Conference on Cloud and Autonomic Computing (ICCAC)* (pp. 69-79). IEEE. 10.1109/ICCAC.2017.14

Zoha, A., Gluhak, A., Imran, M. A., & Rajasegarar, S. (2012). Non-intrusive load monitoring approaches for disaggregated energy sensing: A survey. *Sensors (Basel)*, *12*(12), 16838–16866. doi:10.3390121216838 PMID:23223081

Zou, H., Zhang, Y., Guo, L., Wang, P., He, X., Dai, G., Zheng, H., Chen, C., Wang, A. C., Xu, C., & Wang, Z. L. (2019). Quantifying the triboelectric series. *Nature Communications*, *10*(1), 1427. doi:10.103841467-019-09461-x PMID:30926850

Zou, H., Zhou, Y., Yang, J., & Spanos, C. J. (2018). Device-free occupancy detection and crowd counting in smart buildings with Wi-Fi-enabled IoT. *Energy and Building*, *174*, 309–322. doi:10.1016/j.enbuild.2018.06.040

Zuo, L., & Cui, W. (2013). Dual-Functional Energy-Harvesting and Vibration Control: Electromagnetic Resonant Shunt Series Tuned Mass Dampers. *Journal of Vibration and Acoustics*, *135*(5), 510181–510189. doi:10.1115/1.4024095 PMID:23918165

Zuo, L., Scully, B., Shestani, J., & Zhou, Y. (2010). Design and characterization of an electromagnetic energy harvester for vehicle suspensions. *Smart Materials and Structures*, *19*(4), 045003. doi:10.1088/0964-1726/19/4/045003

About the Contributors

Noor Zaman is currently working as Associate Professor with Taylor's University, Malaysia, have 21 years of teaching, research, and administrative experience internationally. He has an intensive background of academic quality accreditation in higher education besides scientific research activities. He had worked for academic accreditation for more than a decade. He earned ABET accreditation twice for three programs at College of computer sciences and IT King Faisal University Saudi Arabia. He also worked for the Education Evaluation Commission Higher Education Sector (EECHES) formerly NCAAA, Saudi Arabia, for institutional accreditation. He also worked for the National Computing Education Accreditation Council (NCEAC). Dr NZ Jhanjhi is currently working as Associate Professor at School of Computer Science and Engineering, Taylor's University, Malaysia. Besides teaching, research, and administrative responsibilities, he is responsible for the Cybersecurity research Cluster, where he is supervising a great number of Postgraduate students, mainly in the cybersecurity area. Cybersecurity research cluster has extensive research collaboration globally with several professional. He is the link between industry partners and academia as a cybersecurity specialist to represent Taylor's. A good number of postgraduate students are graduated under his supervision. Dr NZ Jhanjhi is Associate Editor and Editorial Assistant Board for several reputable journals including IEEE Access Journal, PC member for several IEEE conferences worldwide, and guest editor for the reputed indexed journals. Active reviewer for a series of Q1 journals has been awarded globally as a top 1% reviewer by Publons (Web of Science). He has high indexed publications in WoS/ISI/SCI/Scopus, and his collective research Impact factor is more than 150 points. He has international Patents on his account, edited/authored more than 25 research books published by world-class publishers. He has a great experience of supervising and co-supervising postgraduate students, an ample number of PhD and Master students graduated under his supervision. He is an external PhD/Master thesis examiner/evaluator for several universities globally. He has completed more than 20 international funded research grants successfully. He has served as Keynote speaker for several international conferences, presented several Webinars worldwide, chaired international conference sessions. He has achieved consecutively higher performance award at Taylor's. He has experienced teaching advanced-era specialized courses, including Wireless Network Security, Secured Software Systems, Computer Forensic, Research Methodology Mobile Programming (Android), Mobile Computing, Cloud Computing and .Net Framework programming besides other undergraduate and postgraduate courses, graduation projects and thesis supervision. He has strong analytical, problem solving, interpersonal and communication skills. His areas of interest include Cyber Security, Wireless Network Security, Internet of Things (IoT), Mobile Application Development, Ad hoc Networks, Cloud Computing, Big Data, Mobile Computing, and Software Engineering.

Low Tang Jung obtained his bachelor's degree in Computer Technology from Teesside University (UK 1989), MSc IT from National University Malaysia (2001), and PhD IT from Universiti Teknologi PETRONAS (Malaysia 2012). Dr. Low has been in the academic line for more 25 years involving with various engineering and ICT courses. He is currently an Assoc. Prof. in UTP Computer and Information Sciences Department. His research interest includes wireless sensor network, embedded systems, and IoT. Some of his current R&D include blockchain consensus model and IoT applications.

Anang Hudaya Muhamad Amin is an assistant lecturer in the Computer and Information Science (CIS) Division at Higher Colleges of Technology, United Arab Emirates. He received a BTech (Hons.) in information technology from Universiti Teknologi PETRONAS and a masters in network computing and PhD from Monash University. His research interests include artificial intelligence with specialization in distributed pattern recognition and bio-inspired computational intelligence, wireless sensor networks, and distributed computing.

* * *

Nabih Abdelmajid obtained his PhD in Computer Science with a specialization in Information Security and Internet Technology from Bradford University in Britain in 2010. He published a set of research and scientific publications. Certified as a digital forensic investigator. He has issued many specialized courses in the field of information security and has excellent experience in the field of training, and he has a certificate as a certified trainer.

Sumana Chattaraj is pursuing M.Tech. in Sensors and Internet of Things at Indian Institute of Technology Jodhpur, Rajasthan, India. She has actively involved in IoT based research on electrical products.

Muhammad Faruq Foong obtained a degree in Bachelor of Aerospace Engineering in 2015 from Monash University. After receiving his bachelor's degree, he joined Aerospace Malaysian Innovation Centre as a researcher. Shortly afterwards, he left the position to pursue a PhD degree in Mechanical Engineering at Heriot-Watt University, focusing on electromagnetic vibration energy harvesting. His main research interest involves vibration energy harvesting, linear vibrations, anti-phase vibrations and vibration-friction interactions. Muhammad Faruq Foong received his PhD degree in 2020 and currently is a senior lecturer at Universiti Teknologi Malaysia.

Ming-Lee Gan completed his Ph.D. in Computer Science studies at Universiti Tunku Abdul Rahman, Malaysia in 2013. He received his B.Eng. in Electrical & Electronics from Universiti Tenaga Nasional, Malaysia in 2004 and M.Sc. in Systems Engineering & Management from the Malaysia University of Science and Technology in 2006. In March 2012, he joined Universiti Tunku Abdul Rahman, where he is currently an Assistant Professor. His research focus is in network path protection, network routing algorithms and network reliability analysis.

Goh Hock Guan (HG Goh) received his Bachelor of Information Technology majoring in Multimedia from University of Malaya in 2003, Master of Science in Information Technology from Multimedia University in 2006, and Doctor of Philosophy in Electronic and Electrical Engineering from University of Strathclyde in 2014. He is currently working as an Assistant Professor in Faculty of Information and

Communication Technology (FICT), Kampar Campus, Universiti of Tunku Abdul Rahman (UTAR). His research interest includes Cognitive Wireless Sensor-Actor Networks, Agriculture/Environmental Monitoring System, Internet of Things, and Data Analysis/Processing.

Saif Hamad Alkaabi is a nineteen year old CIS student at HCT majoring in security and forensics. Saif loves to play football, computer games, and volleyball. He also trains parrots in his free time. Saif has been diagnosed with diabetes type 1 when he was very young. The disease was a this was challenging. He has Participated in American University of Sharjah Physics Open Day competition, Python competition, Robotics club, tutoring program in the college, and tries to find solutions for others when they face problems. He volunteered as an organizer and led a group of marshals in the Arabian Gulf League. He gained leadership experience, and to be more focused during work.

Ahmed Abba Haruna, FHEA, FIMC, CMC, acquired his Ph.D. and M.Sc. in Information Technology from Universiti Teknologi PETRONAS (UTP), Malaysia. He obtained his Bachelor Degree in Information Technology from Infrastructure University Kuala Lumpur, Malaysia. He has recently acquired his Post Graduate Certificate in Higher Education (PGCHE), at Middlesex University. Moreover, he has achieved the status of Fellow of the Higher Education Academy, UK, Fellow of Instituite of Management and Consultant, and a Certified Management Consultant. He has about eight years of invaluable experience in lecturing in Mauritius and in Malaysia. Currently he serves as a Senior Lecturer in Computer Science Department in Skyline University Nigeria and as a Governing Council Member in the same university. His research interest includes Green & Sustainable Computing, Scheduling, Resource Allocation, Machine Learning, and Higher Education Teaching and Learning Support.

Muzammil Hussain is currently working as an Assistant Professor in the Department of Computer Science, School of Systems and Technology, University of Management and Technology, Lahore, Pakistan. He is also serving as a 'Director Graduate Studies and Research' in the School of Systems and Technology, University of Management and Technology. He completed his Ph.D. from the University of Malaya, Kuala Lumpur, Malaysia (and he got Bright Sparks Program scholarship from University of Malaya for his PhD studies). He received his Bachelor of Science in Computer Science from COMSATS University Islamabad in 2013, and awarded with Gold Medal. He has served as a Research Associate for more than three years (2013-2017) in the Department of Computer Systems and Technology, Faculty of Computer Science and Information Technology, University of Malaya, Kuala Lumpur, Malaysia. He led or member for many funded research projects and he has published more than 25 research articles in prestigious international conferences and journals. His potential research areas include, IoT, network security, FinTech Security, Blockchain, Android Security, Bioinformatics, IoT Security, SDN Security, and Internet Security.

Sultan Khalid Abdulqader Rashed Ahli is an undergraduate student at Higher Colleges of Technology, Dubai Mens Campus. He is currently studying in the Information Technology Program, under the Security and Forensics specialization.

Md. Irfan Khan is Regional Manager-Southeast Asia, Supreme & Co. Pvt. Ltd.. His area of interest includes the marketing of T&D line products for sustainable development.

Muhammad Imran Majid is Head of Electrical Engineering Department, Institute of Business Management, Senior Member IEEE, a Chartered Engineer registered with Engineering Council UK and Commonwealth Scholarship Committee Advisory Council Member. He has a PhD in Electronics from University of Surrey, UK, MSc in Space Engineering from Umea University, Sweden and BSc in Electrical Engineering from UET Lahore. After his PhD, Imran consulted and later worked at University of Toronto Canada as Optics Team Lead for two of Canada's most advanced remote sensing satellites, NEMO-AM and NEMO-HD. He is also currently involved with Ontario Centre of Excellence technical Reviewer. Imran has about 10 years' experience in industrial consultancy, teaching and mentoring young professionals in high technology sector. He has worked for Telenor, SSBV, Surrey Satellite and various consultancies during the course of his career. He is Co-founder of UN NPO Space Generation Advisory Council where he coordinated activities in the Asia Pacific region from 1999 to 2008. He represented the Space Generation Forum Declaration during the 3rd UN Global Conference on Space, chaired sessions at 1st Ground Station Network Workshop in Tokyo Japan, UN/ESA Sympoias in Austria (2000-2003). Previously he was selected as a Buraq Space Cadet (1996), actively participated in 9th International Aerospace School Uzbekistan 1997 and pioneered student exchanges at the world's oldest space port at Baikonur Kazakhstan in 1998. Imran also helped found the IET on Surrey Campus, UK for students at the University of Surrey, UK, and IET Karachi, mentored IEng candidates and is currently on the Information Communication and Technology Council IEP Advisory Panel based in Toronto Canada. His research interests are in resource optimization for future wireless systems, engineering goal driven project management and academic-industrial partnerships to aid in economic growth. He is currently teaching Electronic Communication Systems, Opto-Electronics for BE and Project Management, Thesis I,2 courses for MS (EM) at IoBM.

Sushri Mukherjee is pursuing M.Tech. in Power Systems at Indian Institute of Technology Delhi, New Delhi, India. She is actively involved in the development and standardization of indigenous products associated with T&D Sector.

Kapilan N. is working as Professor and Head of Nagarjuna College of Engineering and Technology. He has published more than 50 research papers in Journals and authored one text book and five book chapters. He has completed several sponsored projects. He has received Best Paper Presentation Award and filed a patent. He is the member of several professional societies.

Mahfooz Osama is Senior Lecturer in the Institute of Business Management, Karachi Pakistan. He is currently pursuing his PhD in Telecommunication Engineering from the University of Malaga, Spain, He holds an MBA degree in Telecommunication Management from IoBM, Karachi Pakistan and Bachelors in Electronics Engineering from PAF-KIET, Karachi Pakistan. He is also the Branch Counselor of IEEE Student branch in IoBM.

Vidhya P. is working as a Assistant Professor in Department of Computer Science of SRSMN Government First Grade College. She has published several papers in Journals and Conferences.

Dharmbir Prasad is Assistant Professor, Asansol Engineering College (Dept. of EE). His area of research interest is the Economic Operation of Power System.

Rudra Pratap Singh is Assistant Professor, Asansol Engineering College (Dept. of EE). His research interest includes the application of state estimation and optimization techniques in various fields of engineering.

Chung Ket Thein is an Assistant Professor at School of Aerospace Engineering at University of Nottingham Ningbo China. He obtained his BEng and PhD Mechanical Engineering from University of Hull, UK, focused on the structural optimisation. His research interests include (1) vibration energy harvesting design (both electromagnetic and piezoelectric), (2) multi-objective and multidisciplinary structural optimisation using finite-element analysis, (3) fibre-reinforce composite material using 3D printing technology.

Index

H

I

J

K

L

M

N

P

R

S

T

U

V

W

www.ingramcontent.com/pod-product-compliance
Lightning Source LLC
LaVergne TN
LVHW060402300926
849162LV00006B/1327
9781799867098